FIFTH EDITION

Sensory Evaluation Techniques

FIFTH EDITION

Sensory Evaluation Techniques

Morten C. Meilgaard
Gail Vance Civille • B. Thomas Carr

CRC Press is an imprint of the
Taylor & Francis Group, an **informa** business

CRC Press
Taylor & Francis Group
6000 Broken Sound Parkway NW, Suite 300
Boca Raton, FL 33487-2742

Printed on acid-free paper
Version Date: 20150513

International Standard Book Number-13: 978-1-4822-1690-5 (Hardback)

Library of Congress Cataloging-in-Publication Data

Civille, Gail Vance.
Sensory evaluation techniques / Gail Vance Civille and B. Thomas Carr. -- Fifth edition.
pages cm
Earlier editon: Sensory evaluation techniques / authors, Morten Meilgaard, Gail Vance Civille, B. Thomas Carr.
Includes bibliographical references and index.
ISBN 978-1-4822-1690-5
1. Sensory evaluation. I. Carr, B. Thomas. II. Meilgaard, Morten. Sensory evaluation techniques. III. Title.

TA418.5.M45 2016
658.5'62--dc23 2015018524

Visit the Taylor & Francis Web site at
http://www.taylorandfrancis.com

and the CRC Press Web site at
http://www.crcpress.com

Dedication

to

Manon, Frank, and Cathy

CONTENTS

PREFACE TO THE FIFTH EDITION

How does one plan, execute, complete, analyze, interpret, and report sensory tests? The practices and recommendations in this book are intended to cover all of those phases of sensory evaluation. The text is meant as a personal reference volume for food scientists, research and development scientists, cereal chemists, perfumers, and other professionals working in industry, academia, or government who need to conduct good sensory evaluation. The book should also supply useful background to marketing research, advertising, and legal professionals who need to understand the results of sensory evaluation. It could also give a sophisticated general reader the same understanding.

Because the first edition was used as a textbook at the university and professional level, partly in courses taught by the authors, the second, third, fourth, and fifth editions incorporate a growing number of ideas and improvements arising from questions from students. The objective of the book is now twofold. First, as a "how to" text for professionals, it aims for a clear and concise presentation of practical solutions, accepted methods, standard practices, and some advanced techniques. Second, as a textbook for courses at the academic level, it aims to provide just enough theoretical background to enable the student to understand which sensory methods are best suited to particular research problems and situations and how tests can best be implemented.

The authors do not intend to devote text and readers' time to resolving controversial issues, but a few had to be tackled. We take a fresh look at all statistical methods used for sensory tests, and we hope you like our straightforward approach. The second edition was the first book to provide an adequate solution to the problem of similarity testing. This was adopted and further developed by ISO TC34/SC12 on Sensory Evaluation, resulting in the current "unified" procedure (Section 7.2) in which the user's choice of α- and β-risks defines whether difference or similarity is tested for. Another "first" is the unified treatment of all ranking tests with the Friedman statistic, in preference to Kramer's tables.

Chapter 12 on the Spectrum™ method of descriptive sensory analysis, developed by Civille, has been expanded. The philosophy behind Spectrum is threefold: (1) the test should be tailored to suit the objective of the study (and not to suit a prescribed format); (2) the choice of terminology and reference standards should make use not only of the senses and imagination of the panelists but also of the accumulated experience of the sensory profession as recorded in the literature; and (3) a set of calibrated intensity scales is provided, which permits different panels at different times and locations to obtain comparable and reproducible profiles. The chapter contains full descriptive lexicons suitable for descriptive analysis of a number of products, for example, cheese, mayonnaise, spaghetti sauce, white bread, cookies, and toothpaste. There are updated intensity scales for attributes such as crispness, juiciness, and some common aromatics; and two training exercises.

The authors wish the book to be cohesive and readable; we have tried to substantiate our directions and organize each section so as to be meaningful. We do not want the book to be a turgid set of tables, lists, and figures. We hope we have provided structure to the methods, reason to the procedures, and coherence to the outcomes. Although our aim is to describe all tests in current use, we want this to be a reference book that can be read for understanding as well as a handbook that can serve to describe all major sensory evaluation practices.

The organization of the chapters and sections is also straightforward. Chapter 1 lists the steps involved in a sensory evaluation project, and Chapter 2 briefly reviews the workings of our senses. In Chapter 3, we list what is required of the equipment, the tasters, and the samples, while in Chapter 4, we have collected a list of those psychological pitfalls that invalidate many otherwise good studies. Chapter 5 discusses how sensory responses can be measured in quantitative terms. In Chapter 6, we provide guidelines for selecting the appropriate sensory approach. Chapter 7 describes all the common sensory tests for difference—the triangle, duo–trio, and so on—and Chapter 8 outlines the various attribute tests, such as ranking and numerical intensity scaling. New to the fifth edition is a discussion of tetrad testing, both specified and unspecified, which spans both Chapters 7 and 8. Thresholds and just-noticeable differences are briefly discussed in Chapter 9, followed by what we consider the main chapters: Chapter 10 on the selection and training of panelists, Chapters 11 and 12 on descriptive testing, and Chapter 13 on affective tests (consumer tests). All the descriptive references have been reviewed and revised for the Spectrum references in Chapter 12. Chapter 13 defines in detail several classic qualitative and quantitative methods for testing with consumers and includes a substantial review of Internet research. Chapter 17 offers an update to the existing texts on the application of sensory to quality control. This refined and streamlined adaptation provides more practical approaches for setting up rigorous in-plant sensory tests that match the organization's skill set and resources. Chapter 18 provides an introduction to and discussion of some of the latest methods used by sensory researchers in many consumer products companies.

The body of text on statistical procedures is found in Chapters 14 and 15, but, in addition, each method (triangle, duo–trio, etc.) in Chapters 7 and 8 is followed by a number of examples showing how statistics are used in the interpretation of each method. Basic concepts for tabular and graphical summaries, hypothesis testing, and the design of sensory panels are presented in Chapter 14. We refrain from detailed discussion of statistical theory, preferring instead to give examples. Chapter 15 discusses multifactor experiments that can be used, for example, to screen for variables that have large effects on a product, to identify variables that interact with each other in how they affect product characteristics, or to identify the combination of variables that maximize some desirable product characteristic, such as consumer acceptability. Chapter 15 also contains a discussion of multivariate techniques that can be used to efficiently summarize large sets of sensory data; to identify relationships among responses that might otherwise go unnoticed, especially those involving instrumental, sensory, and consumer data; and to group respondents or samples that exhibit similar patterns of behavior. Chapter 15 also includes a discussion of Thurstonian scaling, which can be used to study differences among products and to uncover the decision processes used by assessors during their sensory evaluations.

New in the fifth edition, Chapter 20 offers some challenges to students in sensory evaluation. The questions are real life and ask students to think through complete sensory research questions. Answers provide some feedback to students who are using this book to make decisions.

At the end of the book, the reader will find guidelines for reporting results, plus the usual glossaries, indexes, and statistical tables.

With regard to terminology, the terms "assessor," "judge," "panelist," "respondent," "subject," and "taster" are used interchangeably, as are "he," "she," and "(s)he" for the sensory analyst (the sensory professional, the panel leader) and for individual panel members.

Morten Meilgaard
Gail Vance Civille
B. Thomas Carr

AUTHORS

Gail Vance Civille is president of Sensory Spectrum, Inc., a management consulting firm involved in the field of sensory evaluation of foods, beverages, pharmaceuticals, paper, fabrics, personal care, and other consumer products. Sensory Spectrum provides guidance in the selection, implementation, and analysis of test methods for solving problems in quality control, research, development, production, and marketing. She has trained several flavor and texture descriptive profile panels in her work with industry, universities, and government.

As a course director for the Center for Professional Advancement and Sensory Spectrum, Ms. Civille has conducted several workshops and courses in basic sensory evaluation methods as well as in advanced methods and theory. In addition, she has been invited to speak to several professional organizations and universities about various facets of sensory evaluation. She is a founding member and former chair of the Society of Sensory Professionals.

Ms. Civille has published books and articles on general sensory methods, as well as sophisticated descriptive flavor and texture techniques. A graduate of the College of Mount Saint Vincent, New York, for which she serves on the Board of Trustees, Ms. Civille earned a BS degree in chemistry and began her career in product evaluation with the General Foods Corporation.

B. Thomas Carr is principal of Carr Consulting, a research consulting firm that provides project management, product evaluation, and statistical support services to the food, beverage, personal care, and home care industries. He has over 30 years of experience in applying statistical techniques to all phases of research on consumer products. Prior to founding Carr Consulting, Mr. Carr held a variety of business and technical positions in the food and food ingredient industries. As director of contract research for NSC Technologies/NutraSweet, he identified and coordinated outside research projects that leveraged the technical capabilities of all the groups within NutraSweet research and development, particularly in the areas of product development, analytical services, and sensory evaluation. Prior to that, as manager of statistical services at both NutraSweet and Best Foods, Inc., he worked closely with the sensory, analytical, and product development groups on the design and analysis of a full range of research studies in support of product development, quality assurance/quality control, and research guidance consumer tests.

Mr. Carr is a member of the US delegation to the ISO TC34/SC12. He is actively involved in the statistical training of scientists and has been an invited speaker to several professional organizations on the topics of statistical methods and statistical consulting in industry. Since 1979, Mr. Carr has supported the development of new food ingredients, consumer food products, and over-the-counter drugs by integrating the statistical and sensory evaluation functions into the mainstream of the product development effort.

This has been accomplished through the application of a wide variety of statistical techniques, including design of experiments, response surface methodology, mixture designs, sensory/instrumental correlation, and multivariate analysis.

Mr. Carr received his BA degree in mathematics from the University of Dayton and his master's degree in statistics from Colorado State University.

ACKNOWLEDGMENTS

We wish to thank our associates at work and our families at home for thoughts and ideas, for material assistance with typing and editing, and for emotional support. Many people have helped with suggestions and discussion over the years. Contributors at the concept stage were Andrew Dravnieks, Jean Eggert, Roland Harper, Derek Land, Elizabeth Larmond, Ann Noble, Rosemarie Pangborn, John J. Powers, Patricia Prell, and Elaine Skinner. Improvements in later editions were often suggested by readers and were given form with help from our colleagues from two Subcommittees on Sensory Evaluation, ASTM E-18 and ISO TC34/SC12, of whom we would like to single out Louise Aust, Edgar Chambers, Daniel Ennis, Sylvie Issanchou, Sandy MacRae, Magni Martens, Suzanne Pecore, Benoit Rousseau, Rick Schifferstein, and Pascal Schlich. We also thank our colleagues Clare Dus, Kathy Foley, Kernon Gibes, Stephen Goodfellow, Dan Grabowski, Lynne Hare, Annlyse Retiveau Krogmann, Marura Lenjo, Ruta Ona Lesniauskas, Marie Rudolph, Lee Stapleton, Joanne Seltsam, Bob Baron, Emily Guzman, and especially Lydia Lawless and Katelyn Scoular for help with editing, writing, illustrations, and ideas.

IN MEMORY OF MORTEN MEILGAARD, D.SC.

From the mid-1980s, when we started this venture, now commonly called the *SET*, we were impressed with the leadership, gentility, intelligence, and doggedness of our lead author, Morten Meilgaard. Although we dreaded the large manila envelopes with the Stroh's logo—because it meant a detailed worksheet and timeline for "next steps"—we knew Morten was taking on as much, if not more, than we were. Because he led by example and was meticulous and precise in his own contributions, we knew we needed to up our game to match his. Although we worked on this text for over 20 years until his passing in 2009, we *never* had "words" or contentious disagreements. It was hard work, but the culture and tone were always agreeable, friendly, *and* productive. We are not alone in branding Morten a gentleman and scholar. All who knew him offer a similar description. It was our pleasure to watch Morten work, challenge ideas, formulate solutions, and keep us on task and on time. Morten remains listed as the first author on this text because he is this text's primary author.

Morten C. Meilgaard, M.Sc., D.Sc., F.I. Brew, was visiting professor (emeritus) of sensory science at the Agricultural University of Denmark and vice president of research (also emeritus) at the Stroh Brewery Co., Detroit, MI. He studied biochemistry and engineering at the Technical University of Denmark, to which he returned in 1982 to receive a doctorate for a dissertation on beer flavor compounds and their interactions. After 6 years as a chemist at the Carlsberg Breweries, he worked from 1957 to 1967 and again from 1989 as a worldwide consultant on brewing and sensory testing. He served for 6 years as director of research for Cervecería Cuauhtémoc in Monterrey, Mexico, and for 25 years with Stroh. At the Agricultural University of Denmark, his task was to establish sensory science as an academic discipline for research and teaching.

Dr. Meilgaard's professional interest was the biochemical and physiological basis of flavor, and more specifically the flavor compounds of hops and beer and the methods by which they can be identified, namely, chemical analysis coupled with sensory evaluation techniques. He had published over 70 papers and developed the beer wheel lexicon that is still used in the beer industry. He was the recipient of the Schwarz Award and the Master Brewers Association Award of Merit for studies of compounds that affect beer flavor. He was founder and past president of the Hop Research Council of the United States and was past chairman of the Scientific Advisory Committee of the US Brewers Association. For 14 years, he was chairman of the Subcommittee on Sensory Analysis of the American Society of Brewing Chemists. He had chaired the US delegation to the ISO TC34/SC12 Subcommittee on Sensory Evaluation. Dr. Meilgaard died April 11, 2009 and is survived by his sons, Stephen Goodfellow and Justin Meilgaard.

1

Introduction to Sensory Techniques

1.1 INTRODUCTION

This introduction is in three parts. The first part lists some reasons why sensory tests are performed and briefly traces the history of their development. The second part introduces the basic approach of modern sensory analysis, which is to treat the panelists as measuring instruments. As such, they are highly variable and very prone to bias, but they are the only instruments that will measure what needs to be measured; therefore, the variability must be minimized and the bias must be controlled by making full use of the best existing techniques in psychology and psychophysics. In the third part, a demonstration is provided of how these techniques are applied with the aid of seven practical steps.

1.2 DEVELOPMENT OF SENSORY TESTING

Sensory tests, of course, have been conducted for as long as there have been human beings evaluating the goodness and badness of food, water, weapons, shelters, and everything else that can be used and consumed.

The rise of trading inspired slightly more formal sensory testing. A buyer, hoping that a part would represent the whole, would test a small sample of a shipload. Sellers began to set their prices on the basis of an assessment of the quality of goods. With time, ritualistic schemes of grading wine, tea, coffee, butter, fish, and meat developed, some of which survive to this day.

Grading gave rise to the professional taster and consultant to the budding industries of foods, beverages, and cosmetics in the early 1900s. Literature was developed that used the term *organoleptic testing* (Pfenninger, 1979) to denote the supposedly objective measurement of sensory attributes. In reality, tests were often subjective, tasters too few, and interpretations open to prejudice.

Pangborn (1964) traces the history of systematic "sensory" analysis that is based on wartime efforts to provide acceptable food to American forces (Dove, 1946, 1947) and on the development of the triangle test in Scandinavia (Bengtsson and Helm, 1946; Helm and Trolle, 1946). A major role in the development of sensory testing was played by the

Food Science Department at the University of California at Davis, resulting in the book by Amerine et al. (1965).

Scientists have only recently developed sensory testing as a formalized, structured, and codified methodology, and they continue to develop new methods and refine existing ones. The current state of sensory techniques is recorded in the dedicated journals *Chemical Senses, Journal of Sensory Studies, Journal of Texture Studies, Food Quality,* and *Journal of Cosmetic Studies;* in the proceedings of the Pangborn Symposia (triennial) and the International Sensometrics Group (biannual), both usually published as individual papers in the journal *Food Quality & Preference;* and the proceedings of the Weurman Symposia (triennial, but published in book form, e.g., Martens et al., 1987; Bessière and Thomas, 1990). Sensory papers presented to the Institute of Food Technologists (IFT) are usually published in the IFT's *Journal of Food Science* or *Food Technology.* Papers presented at the Society of Sensory Professionals are typically published in the *Journal of Sensory Studies.*

The methods that have been developed serve economic interests. Sensory testing can develop a level of acceptability for a commodity or help determine the value of a commodity. Sensory testing evaluates alternative courses to select the one that optimizes value for money. The principal uses of sensory techniques are in quality control, product development, and research. They find application not only in the characterization and evaluation of foods and beverages but also in other fields such as household products, environmental odors, personal hygiene products, diagnosis of illnesses, testing of pure chemicals, and so on. The primary function of sensory testing is to conduct valid and reliable tests that provide data on the basis of which sound decisions can be made.

1.3 HUMAN SUBJECTS AS INSTRUMENTS

Dependable sensory analysis is based on the skill of the sensory analyst in optimizing the four factors of such analysis, which we all recognize because they are the ones that govern any measurement (Pfenninger, 1979).

1. *Definition of the problem*: We must define precisely what it is we wish to measure; important as this is in "hard" science, it is much more so with senses and feelings.
2. *Test design*: Not only must the design leave no room for subjectivity and take into account the known sources of bias, but it also must minimize the amount of testing required to produce the desired accuracy of results. Test controls for subjects, site, samples, and sensory methods must be in place (Civille and Oftedal 2012).
3. *Instrumentation*: The test subjects must be selected and trained to give a reproducible verdict; the analyst must work with them until he/she knows their sensitivity and bias in the given situation.
4. *Interpretation of results*: Using statistics, the analyst chooses the correct null hypothesis and the correct alternative hypothesis and draws only those conclusions that are warranted by the results.

Tasters, as measuring instruments, are (1) quite variable over time, (2) very variable among themselves, and (3) highly prone to bias. To account adequately for these shortcomings requires (1) that measurements be repeated, (2) that enough subjects (often 20–50) are made available so that verdicts are representative, and (3) that the sensory analyst respects the many rules and pitfalls that govern panel attitudes (see Chapter 4). Subjects vary innately in sensitivity by a factor of 2–10 or more (Meilgaard and Reid, 1979; Pangborn, 1981) and should not be interchanged halfway through a project. Subjects must be selected for sensitivity and must be trained and retrained (see Chapter 9) until they fully understand the task at hand. The annals of sensory testing are replete with results that are unreliable because many of the panelists did not understand the questions and/or the terminology used in the test, did not recognize the tactile and fragrance parameters in the products, or did not feel comfortable with the mechanics of the test or the numerical expressions used.

For these reasons and others, it is very important for the sensory analyst to be actively involved in the development of the scales and the terminology/lexicons used to measure the panelists' responses. A good scale requires much study, must be based on a thorough understanding of the physical and chemical factors that govern the sensory variable in question, and requires several reference points and thorough training of the panel on that scale. It is unreasonable to expect that even an experienced panelist would possess the necessary knowledge and skill to develop a lexicon that is consistently accurate and precise. Only through the direct involvement of a knowledgeable sensory professional in the development of scales can one obtain descriptive analyses, for example, that will mean the same in 6 months' time as they do today.

1.3.1 Chain of Sensory Perception

When sensory analysts study the relationship between a given physical stimulus and the subject's response, the outcome is often regarded as a one-step process. In fact, there are at least three steps in the process: The stimulus hits the sense organ and is converted to a nerve signal that travels to the brain. With previous experiences in memory, the brain then interprets, organizes, and integrates the incoming sensations into perceptions. Finally, a response is formulated based on the subject's perceptions (Schiffman, 1996).

In dealing with the fact that humans often yield varied responses to the same stimulus, sensory professionals need to understand that differences between two people's verdicts can be caused either by a difference in the sensation they receive because their sense organs differ in sensitivity or by a difference in their mental treatment of the sensation, for example, because of a lack of knowledge of the particular odor, taste, and so on, or because of lack of training in expressing what they sense in words and numbers. Through training and the use of references, sensory professionals can attempt to shape the mental process so that subjects move toward showing the same response to a given stimulus.

A commendable critical review of the psychophysical measurement of human olfactory function (with 214 references) can be found in Chapter 10 of Doty and Laing (2003).

1.4 CONDUCTING A SENSORY STUDY

The best products are developed in organizations where the sensory professional is more than the provider of a specialized testing service. Only through a process of total involvement can he or she be in the position of knowing what tests are necessary and appropriate at every point during the life of a research project. The sensory professional (like the statistician) must take an active role in developing the research program, collaborating with the other involved parties on the development of the experimental designs that ultimately will be used to answer the questions posed. Erhardt (1978) divides the role of the sensory analyst into the following seven practical tasks:

1. Determine the project objective. Defining the needs of the project leader is the most important requirement for conducting the correct test. Were the samples submitted as a product improvement, to permit cost reduction or ingredient substitution, or as a match of a competitor's product? Is one sample expected to be similar or different from others, preferred or at parity, variable in one or more attributes? If this critical step is not carried out, the sensory analyst is unlikely to use the appropriate test or to interpret the data correctly.
2. Determine the test objective. Once the objective of the project can be clearly stated, the sensory analyst and the project leader can determine the test objective: overall difference, attribute difference, relative preference, acceptability, and so on. Avoid attempting to answer too many questions in a single test. A good idea is for the sensory analyst and project leader to record in writing, before the test is initiated, the project objective, the test objective, the specifics of the test, the set of samples, and a brief statement of how the test results will be used.
3. Screen the samples. During the discussion of project and test objectives, the sensory analyst should examine all of the sensory properties of the samples to be tested. This enables the sensory analyst to choose test methods that take into account any sensory biases introduced by the samples. For example, visual cues (color, thickness, sheen) may influence overall difference responses, such as those provided in a triangle test, for example, to measure differences due to sweetness of sucrose versus aspartame. In such a case, an attribute test would be more appropriate. In addition, product screening provides information on possible terms to be included in the score sheet.
4. Design the test. After defining the project and test objectives and screening the samples, the sensory analyst can proceed to design the test. This involves selection of the test technique (see Chapter 6 for general guidelines; see Chapters 7, 8, 9, 11, 12, 13, 17, and 18 for specific guidelines for different sensory techniques); selecting and training panelists (see Chapter 10); designing the accompanying score sheet (ballot, questionnaire); specifying the criteria for sample preparation and presentation (see Chapter 3); and determining how the data will be analyzed (see Chapters 13 and 14). Care must be taken, in each step, to adhere to the principles of statistical design of experiments to ensure that the most sensitive evaluation of the test objective is attained.
5. Conduct the test. Even when technicians are used to carry out the test, the sensory analyst is responsible for ensuring that all the requirements of the test design are met.
6. Analyze the data. Because the procedure for analysis of the data was determined at the test design stage, the necessary expertise and statistical programs, if used, will be ready to begin data analysis as soon as the study is completed. The data

should be analyzed for the main treatment effect (test objective) as well as other test variables, such as order of presentation, time of day, different days, and/or subject variables such as age, sex, geographic area, and so on (see Chapters 14 and 15).

7. Interpret and report results. The initial clear statement of the project and test objectives will enable the sensory analyst to review the results, express them in terms of the stated objectives, and make any recommendations for action that may be warranted. The latter should be stated clearly and concisely in a written report that also summarizes the data, identifies the samples, and states the number and qualification of subjects (see Chapter 16).

The main purpose of this book is to help the sensory analyst develop the methodology, subject pool, facilities, and test controls required to conduct analytical sensory tests with trained and/or experienced tasters. In addition, Chapters 13 and 19 discuss the organization of consumer tests, that is, the use of naïve consumers (nonanalytical) for large-scale evaluations, which are structured to represent the population of the product market. The role of sensory evaluation in the development of advertising claims is also addressed.

The role of sensory evaluation and quality is to provide valid and reliable information to research and development (R&D), production, and marketing in order for management to make sound business decisions about the perceived sensory properties of products. The ultimate goal of any sensory program should attempt to find the most cost-effective and efficient method with which to obtain the most sensory information. When possible, internal laboratory difference or descriptive techniques are used in place of more expensive and time-consuming consumer tests to develop cost-effective sensory analysis. Further cost savings may be realized by correlating as many sensory properties as possible with instrumental, physical, or chemical analyses. In some cases, it may be possible to replace a part of routine sensory testing with cheaper and quicker instrumental techniques.

REFERENCES

Amerine, M. A., R. M. Pangborn, and E. B. Roessler (1965). *Principles of Sensory Evaluation of Food.* New York: Academic Press.

Bengtsson, K., and E. Helm (1976). Principles of taste testing. *Wallerstein Lab Commun* 9: 171.

Bessière, Y., and A. F. Thomas (eds.) (1990). *Flavour Science and Technology.* Chichester: Wiley.

Civille, G. V., and K. N. Oftedal (2012). Sensory evaluation techniques: Makes "good for you" taste "good." *Physiol Behav* 107: 598–605.

Doty, R. L., and D. G. Laing. (2003). In R. L. Doty (ed.), *Handbook of Olfaction and Gustation* (2nd edn., pp. 203–28). New York: Marcel Dekker.

Dove, W. E. (1946). Developing food acceptance research. *Science* 103: 187.

Dove, W. E. (1947). Food acceptability: Its determination and evaluation. *Food Technol* 1: 39.

Erhardt, J. P. (1978). The role of the sensory analyst in product development. *Food Technol* 32: 11, 57.

Helm, E., and B. Trolle (1946). Selection of a taste panel. *Wallerstein Lab Commun* 9: 181.

Martens, M., G. A. Dalen, and H. Russwurm, Jr. (1987). *Flavour Science and Technology.* Chichester: Wiley.

Meilgaard, M. C., and D. S. Reid (1979). Determination of personal and group thresholds and the use of magnitude estimation in beer flavour chemistry. In D. G. Land and H. E. Nursten (eds.), *Progress in Flavour Research* (pp. 67–73). London: Applied Science Publishers.

Pangborn, R. M. (1964). Sensory evaluation of food: A look backward and forward. *Food Technol* 18: 1309.
Pangborn, R. M. (1981). Individuality in response to sensory stimuli. In J. Solms and R. L. Hall (eds.), *Criteria of Food Acceptance. How Man Chooses What He Eats* (p. 177). Zürich: Forster-Verlag.
Pfenninger, H. B. (1979). Methods of quality control in brewing. *Schweizer Brauerei-Rundschau* 90: 121.
Schiffman, H. R. (1996). *Sensation and Perception. An Integrated Approach* (4th edn.). New York: Wiley.

2

Sensory Attributes and the Way We Perceive Them

2.1 INTRODUCTION

This chapter reviews (1) the sensory attributes with which the book is concerned, for example, the appearance, odor, flavor, and feel of different products; and (2) the mechanisms that people use to perceive those attributes, for example, the visual, olfactory, gustatory, and tactile/kinesthetic senses and sometimes sound. This chapter is brief, not because it is less important, but because many other books cover in depth the psychological and physiological mechanisms of sensation and perception. The sensory professional is urged to study the references for this chapter (e.g., Amerine et al., 1965; ASTM, 1968; Civille and Lyon, 1996; Lawless and Heymann, 1998; and Stone and Sidel, 2004) and to build a good library of books and journals on sensory perception. Sensory testing is an interdisciplinary science comprised of information and methods adapted from psychology, physiology, statistics, linguistics, medicine, chemistry, physics, sociology, anthropology, and a host of other fields. Experimental designs need to be based on a thorough knowledge of the physical, chemical, and psychophysiological factors behind the attributes of interest. Results of sensory tests often have several possible explanations; therefore, when interpreting these results, it is important to consider new knowledge about the senses, the true nature of product attributes, probability and risk management, and consumer behavior.

2.2 SENSORY ATTRIBUTES

The attributes of a food item are typically perceived in the following order:

- Appearance
- Odor/aroma/fragrance
- Consistency and texture
- Flavor (aromatics, chemical feelings, basic tastes)

However, in the process of perception, most or all of the attributes overlap; that is, the subject receives a jumble of near-simultaneous sensory impressions, and without training, he or she will not be able to provide an independent evaluation of each. This section gives examples of the types of sensory attributes that exist in terms of the way that they are perceived and the words or phrases that may be used to describe them.

In this book, *flavor* is the combined impression perceived via the chemical senses from a product in the mouth and does not include appearance and texture. The term *aromatics* is used to indicate those volatile constituents that originate from food in the mouth and are perceived by the olfactory system via the posterior nares.

2.2.1 Appearance

As every shopper knows, the appearance of the product and/or the package often is the only attribute on which the decision to purchase or consume a product is based. Hence, people become adept at making wide and risky inferences from small clues, and test subjects will do the same in the booth. It follows that the sensory analyst must pay meticulous attention to every aspect of the appearance of test samples (Amerine et al., 1965, 399; McDougall, 1983; Lawless and Heymann, 2010) and often must attempt to obliterate or mask much of it with lighting, opaque containers, and so on. Although it is easy to "justify" letting the panel see the product because "that is the way it will be used," this justification is deceptive. That interaction may be critical if consumer understanding is desired but is misleading if it shrouds the truth about the sensory properties of the product.

General appearance characteristics are listed below, and an example of the description of appearance with the aid of scales is given in Appendix 12.1.

Color	A phenomenon that involves both physical and psychological components—the perception by the visual system of light of wavelengths 400–500 nm (blue), 500–600 nm (green and yellow), and 600–800 nm (red), commonly expressed in terms of the hue, value, and chroma of the Munsell color system. The evenness of color, as opposed to an uneven or blotchy appearance, is important. Deterioration of food is often accompanied by a color change. Good descriptions of procedures for the sensory evaluation of appearance and color are given by Clydesdale (1984), McDougall (1988), and Lawless and Heymann (1998). Because color can be impacted by the surroundings, the lighting, and the angle of the observer, these need to be standardized when testing.
Size and shape	The length, thickness, width, particle size, geometric shape (square, circular, etc.), and distribution of pieces, for example, of vegetables, pasta, specks, and so on. Size and shape often are key indicators of quality aspects of the product and also may indicate defects (Kramer and Twigg, 1973; Gatchalian, 1981).

Surface texture	The dullness or shininess of a surface; its roughness versus evenness; and its appearance of wetness, dryness, smoothness, crust, or depth may be key characteristics, either of the product, particularly for many nonfood products such as paint, personal care products (e.g., nail polish or lipstick), and so on; or for the effects of the product, such as that of shampoo or conditioner on hair.
Clarity	The haze (Siebert et al., 1981) or opacity (McDougall, 1988) of liquids, gels, or solids or the presence or absence of particles of visible size. Similarly, the ability to see multiple layers in some products such food, personal care products, polishes, and so on can be an indication of other sensory attributes.
Carbonation	For beverages with gas, the degree of effervescence observed (during pouring or after sitting for a specified period) (Descoins et al., 2006).

2.2.2 Odor/Aroma/Fragrance

The odor of a product is detected when its volatiles enter the nasal passage and are perceived by the olfactory system. The term *odor* or one of its variations is used when the volatiles are sniffed through the nose. Aroma is the odor of a food product, and fragrance is the odor of a perfume, cosmetic, or personal or household good. As mentioned earlier, aromatics are the volatiles perceived by the olfactory system from a substance in the mouth. The term smell is not used in this book because it has a negative connotation (e.g. malodor) to some people, whereas to others, it is the same as odor.

The amount of volatiles that escape from a product is affected by the temperature as well as the nature of the compound. The vapor pressure of a substance exponentially increases with temperature according to the following formula:

$$\log p = -0.05223a / T + b \tag{2.1}$$

where:

p is the vapor pressure in mmHg
T is the absolute temperature ($T = t$°C+273.1)
a and b are substance constants that can be found in handbooks (Howard, 1996)

Volatility also is influenced by the condition of a surface; at a given temperature, more volatiles escape from a soft, porous, humid surface than from a hard, smooth, dry one.

Many odors are released only when an enzymic reaction takes place at a freshly cut surface (e.g., the smell of an onion). Odorous molecules must be transmitted by a gas, for example the atmosphere, water vapor, or an industrial gas; and the intensity of the perceived odor is determined by the proportion of odorous molecules in the gas that come into contact with the observer's olfactory receptors (Laing, 1983).

Sensory professionals continue to be challenged by the sorting of fragrance/aroma sensations into identifiable terms (see Chapter 11 on descriptive analysis and Civille and Lyon (1996) for a database of descriptors for many products). There is not, at this point, any internationally standardized odor terminology, in part because of the number of odors that can be found. According to Harper (1972), some 17,000 odorous compounds are known. A trained descriptive panel often can differentiate and name several thousands of odors, and trained flavorists or perfumers can name thousands more. Many terms may be ascribed to a single compound (e.g., thymol = herb-like, green, rubber-like), and a single term may be associated with many compounds (lemon = α-pinene, β-pinene, α-limonene, β-ocimene, citral, citronellal, linalool, α-terpineol, etc.). Examples of this can be found in a number of different publications (e.g., Lota et al., 2002; Vekiari et al., 2002; Smadja et al., 2005).

2.2.3 Consistency and Texture

The third set of attributes to be considered are those perceived by sensors in the mouth other than taste and chemical feelings. Texture also is perceived by the skin and muscles of the body, other than those in the mouth, when evaluating food products (e.g., squeezing fruit or bread) personal care products (e.g., rubbing lotion onto the skin), home care products (e.g., scrubbing a hard surface during cleaning), or other products (e.g., the force needed to tear or cut a paper or textiles).

Viscosity typically refers to the rate of a Newtonian liquid's flow under some force such as gravity or stirring. It can be measured accurately (at a standard temperature) and varies from a low of approximately 1 cP (centipoise) for water or beer to 1000s of cP for thick products such as some sugar syrups, honey, or molasses.

Consistency (of fluids such as purees, sauces, juices, syrups, jellies, and cosmetics), in principle, must be measured by sensory evaluation (Kramer and Twigg, 1973). However, with practice, some standardization is possible with the aid of consistometers (Kramer and Twigg, 1973; Mitchell, 1984) to measure the flow properties.

Texture is complex, as demonstrated by the existence of an entire journal devoted to the subject, the *Journal of Texture Studies*. *Texture* can be defined as the sensory manifestation of the structure or inner makeup of products in terms of their

- Reaction to stress, measured as mechanical properties (such as hardness/firmness, adhesiveness, cohesiveness, gumminess, springiness/resilience, viscosity) by the kinesthetic sense in the muscles of the hand, fingers, tongue, jaw, or lips
- Tactile feel properties, measured as geometrical particles (grainy, gritty, crystalline, flaky) or moisture properties (wetness, oiliness, moistness, dryness) by the tactile nerves in the surface of the skin of the hand, lips, or tongue (Szczesniak, 1963)

Table 2.1 lists general mechanical, geometrical, and moisture properties of foods, skincare products, and fabrics. Note that across such a wide variety of products, the textural properties are all derived from the same general classes of texture terms measured kinesthetically or in terms of tactile qualities. Additional texture terms are listed in Appendixes 12.1, 12.2, and 12.4. A recommended review of texture perception and measurement is that of De Man et al. (1976).

Table 2.1 Components of Texture

Mechanical Properties: Reaction to Stress, Measured Kinesthetically		
Foods	**Skincare**	**Fabrics**
Hardness: Force to Attain a Given Deformation		
Firmness (compression)	Force to compress	Force to compress
Hardness (bite)	Force to spread	Force to stretch
Cohesiveness: Degree to Which Sample Deforms (Rather than Ruptures)		
Cohesive	Cohesive	Stiffness
Chewy	Short	
Fracturable (crispy/crunchy)	Viscosity	
Viscosity		
Adhesiveness: Force Required to Remove Sample from a Given Surface		
Sticky (tooth/palate)	Tacky	Fabric/fabric friction
Toothpack	Drag	Hand friction (drag)
Denseness: Compactness of Cross-Section		
Dense/heavy	Dense/heavy	Fullness/flimsy
Airy/puffy/light	Airy/light	
Springiness: Rate of Return to Original Shape after Some Deformation		
Springy/rubbery	Springy	Resilient (tensile and compression)
		Cushy (compression)

Geometrical Properties: Perception of Particles (Size, Shape, Orientation) Measured by Tactile Means	
Smoothness	Absence of all particles
Gritty	Small, hard, sharp particles
Grainy	Small, round particles
Chalky/powdery	Fine particles (film)
Fibrous	Long, stringy particles (fuzzy fabric)
Lumpy/bumpy	Large, even pieces or protrusions

Moisture Properties: Perception of Water, Oil, Fat, Measured by Tactile Means		
Foods	**Skincare**	**Fabrics**
Moistness: Amount of Wetness/ Oiliness Present, When Not Certain Whether Oil and/or Water Moisture Release: Amount of Wetness/ Oiliness Exuded		
Juicy	Wets down	Moisture release
Oily	Amount of liquid fat	
Greasy	Amount of solid fat	

2.2.4 Flavor

Flavor, as an attribute of foods, beverages, and seasonings, has been defined (Amerine et al., 1965, 549) as the sum of perceptions resulting from stimulation of the sense ends that are grouped together at the entrance of the alimentary and respiratory tracts. However, for purposes of practical sensory analysis, Caul (1957) is followed, and the term is restricted to the impressions perceived via the chemical senses from a product in the mouth. Defined in this manner, flavor includes

- The aromatics, that is, olfactory perceptions caused by volatile substances released from a product in the mouth via the posterior nares
- The tastes, that is, gustatory perceptions (salty, sweet, sour, bitter, umami) caused by soluble substances in the mouth
- The chemical feeling factors that stimulate nerve ends in the soft membranes of the buccal and nasal cavities (astringency, spice heat, cooling, bite, metallic feel, burn)

A large number of individual flavor words are listed in Chapter 11 and in Civille and Lyon (1996).

2.2.5 Noise

The noise produced during the mastication of foods or the handling of fabrics or paper products is a minor, but not negligible, sensory attribute in those products. However, in products such as automobiles, sound systems, and industrial factories, sound can be a primary sensory factor (either positive or negative). It is common to measure the pitch, loudness, and persistence of sounds produced by foods or fabrics. The pitch and loudness of the sound contribute to the overall sensory impression. Differences in pitch of some rupturing foods (crispy, crunchy, brittle) provide sensory input that is used in the assessment of freshness/staleness. Oscilloscopic measurements by Vickers and Bourne (1976) permitted a sharp differentiation between products described as crispy and those described as crunchy. Kinesthetically, these differences correspond to measurable differences in hardness, denseness, and the force of rupture (fracturability) of a product. A crackly or crisp sound on handling can cause a subject to expect stiffness in a fabric. The duration or persistence of sound from a product often suggests other properties, for example, strength (crisp fabric), freshness (crisp apples, potato chips), toughness (squeaky clams), or thickness (plopping liquid). Table 2.2 lists common noise characteristics of foods, skincare products, and fabrics.

In other products such as motorcycles, where the sound is a key consumer attribute, the sound can become almost synonymous with the brand; for example, Harley-Davidson Corp. attempted to trademark the sound of a Harley-Davidson motorcycle. Stereo equipment manufacturers must evaluate the output sound of the systems they create (Smith, 2010), and companies that manufacture so-called noise-cancelling headphones must ensure that ambient noise does not compete with the sounds the listener wants to enjoy. Acoustics is a developing area for sensory scientists.

Table 2.2 Common Noise Characteristics of Foods, Skincare Products, and Fabrics

Noise Properties[a]		
Foods	**Skincare**	**Fabrics**
Crispy	Squeak	Crisp
Crunchy		Crackle
Squeak		Squeak

Pitch: Frequency of sound.
Loudness: Intensity of sound.
Persistence: Endurance of sound over time.
[a] Perceived sounds (pitch, loudness, persistence) and auditory measurement.

2.3 HUMAN SENSES

The five senses are so well covered in textbooks (Piggott, 1988; Kling and Riggs, 1971; Sekuler and Blake, 1990; Geldard, 1972) that a description here is superfluous. Therefore, this discussion will be limited to pointing out some characteristics that are of particular importance in designing and evaluating sensory tests. A clear and brief account of the sensors and neural mechanisms that are used to perceive odor, taste, vision, and hearing, followed by a chapter on the intercorrelation of the senses, is found in *Basic Principles of Sensory Evaluation* (American Society for Testing and Materials [ASTM], 1968). Lawless and Heymann (1998) review what is known about sensory interaction within and between the sensory modalities.

2.3.1 Sense of Vision

Light entering the lens of the eye (see Figure 2.1) is focused on the retina, where the rods and cones convert it to neural impulses that travel to the brain via the optic nerve. Some aspects of color perception that must be considered in sensory testing are that

- Subjects often give consistent responses about an object's color even when filters are used to mask differences (perhaps because the filters mask hues but generally do not mask brightness and chroma). Thus, using red lights for meat may work to mask differences in perceived degree of doneness (i.e., red vs. pink vs. brown), but will not mask differences in the visual perception of dryness, the appearance of fat or muscle fibers, or the outside "char" that is on a grilled product.
- Subjects are influenced by adjoining or background color, and the relative sizes of areas of contrasting color (i.e., a blotchy appearance, as distinct from an even distribution of color) affects perception.
- The gloss and texture of a surface also affect the perception of color.
- Color vision differs among subjects: Degrees of color blindness exist, for example, the inability to distinguish between red and orange or blue and green; and exceptional color sensitivity also exists, allowing certain subjects to discern visual differences that the panel leader cannot see.

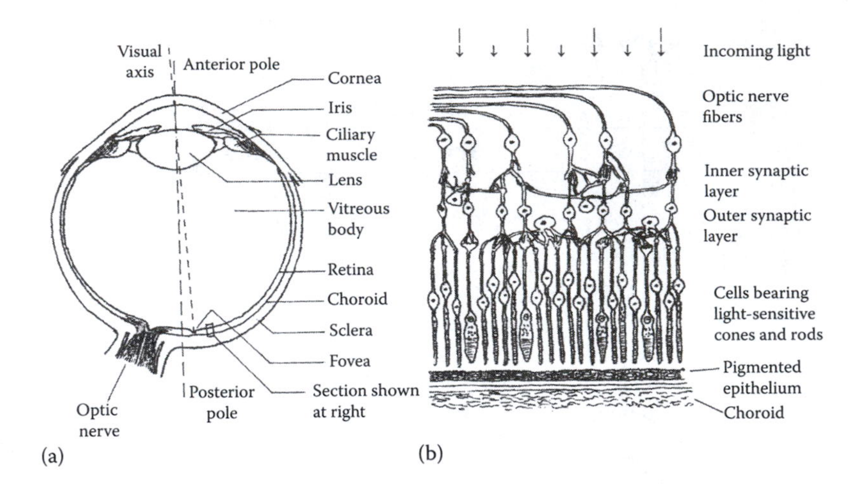

Figure 2.1 The eye, showing the lens, retina, and optic nerve. The entrance of the optic nerve is the blind spot. The fovea is a small region, central to the retina, that is highly sensitive to detail and consists entirely of cones. (Modified from Hochberger, J. E. [1964] *Perception*. Englewood Cliffs, NJ: Prentice-Hall. With permission.)

The chief lesson to be learned from this is that attempts to mask differences in color or appearance are often unsuccessful, and if undetected, they can cause the experimenter to erroneously conclude that a difference in flavor or texture exists, when the difference is actually in a visual appearance characteristic.

2.3.2 Sense of Touch

The group of perceptions generally described as the sense of touch can be divided into *somesthesis* (tactile sense, skinfeel) and *kinesthesis* (deep pressure sense or proprioception), with both sensing variations in physical pressure. Figure 2.2 shows the several types of nerve endings in the skin surface, epidermis, dermis, and subcutaneous tissue. These surface nerve ends are responsible for the somesthetic sensations called touch, pressure, heat, cold, itching, and tickling. Deep pressure, kinesthesis, is felt through nerve fibers in muscles, tendons, and joints, the main purpose of which is to sense the tension and relaxation of muscles. Figure 2.3 shows how the nerve fibers are buried within a tendon. Kinesthetic perceptions corresponding to the mechanical movement of muscles (heaviness, hardness, stickiness, etc.) result from stress exerted by muscles of the hand, jaw, or tongue and the sensation of the resulting strain (compression, shear, rupture) within the sample being handled, masticated, and so on. The surface sensitivity of the lips, tongue, face, and hands is much greater than that of other areas of the body, resulting in the ease of detection of small force differences, particle size differences, and thermal and chemical differences from hand and oral manipulation of products.

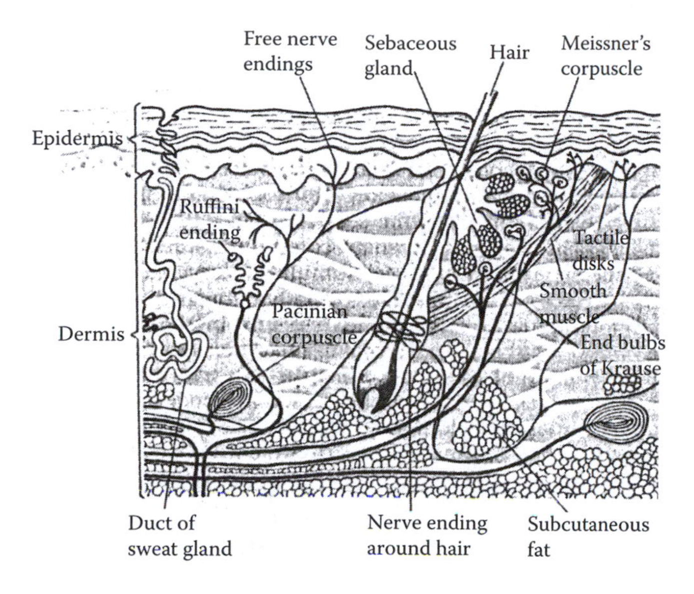

Figure 2.2 Composite diagram of the skin in cross section. Tactile sensations are transmitted from a variety of sites; for example, the free nerve endings and the tactile discs in the epidermis, and, in the dermis, the Meissner corpuscles, end bulbs of Krause, Ruffini endings, and Pacinian corpuscles. (From Gardner, E. [1968] *Fundamentals of Neurology*, 5th ed. Philadelphia: W.B. Saunders.)

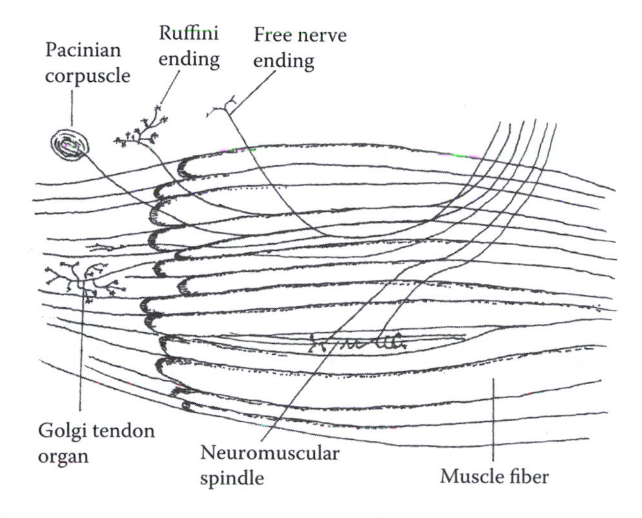

Figure 2.3 Kinesthetic sensors in a tendon and muscle joint. (Modified from Geldard, F.A. [1972]. *The Human Senses*. New York: Wiley. With permission.)

2.3.3 Olfactory Sense

2.3.3.1 General

Airborne odorants are sensed by the olfactory epithelium, which is located in the roof of the nasal cavity (see Figure 2.4). Odorant molecules are sensed by the millions of tiny, hair-like cilia that cover the epithelium and are enervated by more than 1000 different olfactory receptor types (Buck and Axel, 1991). The anatomy of the nose is such that only a small fraction of inspired air reaches the olfactory epithelium via the nasal turbinates or via the back of the mouth on swallowing (Maruniak, 1988). Optimal contact is obtained by moderate inspiration (sniffing) for 1–2 s (Laing, 1983). At the end of 2 s, the receptors have adapted to the new stimulus, and one must allow 5–20 s or longer for them to de-adapt before a new sniff can produce a full-strength sensation. A complication is that the odorant(s) can fill the location where a stimulus is to be tested, thereby reducing the subject's ability to detect a particular odorant or differences among similar odorants. Cases of total odor blindness, *anosmia*, are rare, but specific anosmia, the inability to detect specific odors, is common (Harper, 1972). For this reason, potential panelists should be screened for sensory acuity using odors similar to those to be tested.

Whereas the senses of hearing and sight can accommodate and distinguish stimuli that are 10^4- to 10^5-fold apart, the olfactory sense has trouble accommodating a 10^2-fold difference between the threshold and the concentration that produces saturation of the receptors. On the other hand, whereas the ear and the eye each can sense only one type of signal (namely, oscillations of air pressure and electromagnetic waves of 400–800 nm wavelength), the nose has enormous discriminating power, discriminating many thousands of different odors.

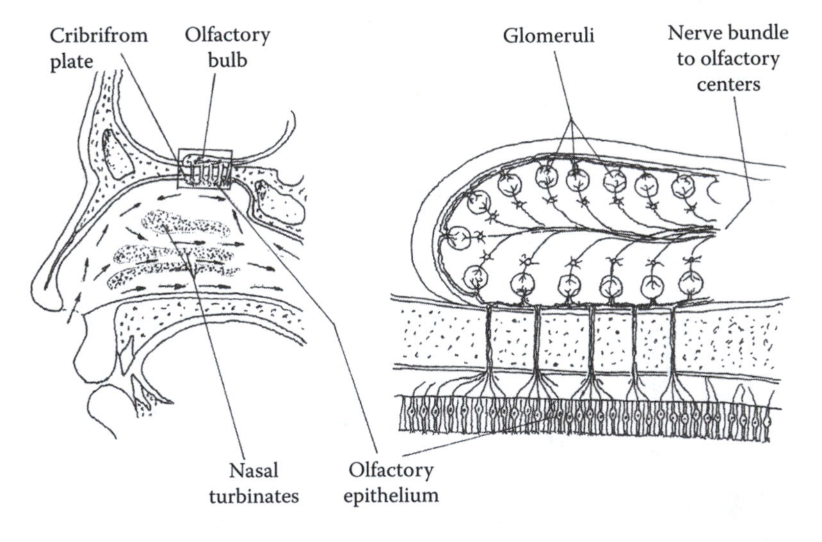

Figure 2.4 Anatomy of the olfactory system. Signals generated by the approximately 1000 types of sensory cells pass through the cribriform plate into the olfactory bulb where they are sorted through the glomeruli before passing on to the higher olfactory centers. (Modified from Axel, R. [1995]. *Sci Am* (October): 154–9.)

The receptors' sensitivity to different chemicals varies over a range of 10^{12} or more (Harper, 1972; Meilgaard, 1975). Typical thresholds (see Table 2.3) vary from 1.3×10^{19} molecules per milliliter air for ethane to 6×10^{7} molecules per milliliter for allyl mercaptan, and it is very likely that substances exist or will be discovered that are more potent. Note that pure water and pure air are not in the list because these bathe the sensors and cannot be sensed.

The table illustrates how easily a chemical standard can be misflavored by impurities. For example, an average observer presented with a concentration of 1.5×10^{17} molecules per milliliter of methanol that is 99.99999% pure but contains 0.00001% ionone would perceive a 10 × threshold of methanol but a 100 × threshold odor of ionone. Purification by distillation and charcoal treatment might reduce the level of ionone impurity tenfold, but it would still be at 10 × threshold or as strong as the odor of methanol itself.

Table 2.3 Some Typical Threshold Values in Air

Chemical Substance	Molecules/mL Air
Allyl mercaptan	6×10^{7}
Ionone	1.6×10^{8}
Vanillin	2×10^{9}
sec-Butyl mercaptan	2×10^{8}
Butyric acid	1.4×10^{11}
	6.9×10^{9}
Acetaldehyde	9.6×10^{12}
Camphor	5×10^{12}
	6.4×10^{12}
	4×10^{14}
Trimethylamine	2.2×10^{13}
Phenol	7.7×10^{12}
	2.6×10^{13}
	1×10^{13}
	1.3×10^{15}
Methanol	1.1×10^{16}
	1.9×10^{16}
Ethanol	2.4×10^{15}
	2.3×10^{15}
	1.6×10^{17}
Phenyl ethanol	1.7×10^{17}
Ethane	1.3×10^{19}

Source: From Harper, R. (1972). *Human Senses in Action*, 253. Churchill Livingstone, London. With permission.

Note: The figures quoted should be treated as orders of magnitude only because they may have been derived by different methods.

Gas chromatographic methods are constantly improving and now can detect in the parts per billion ranges. However, there are numerous odor substances, probably thousands, that occur in nature at parts per trillion levels to which the nose is more sensitive than the gas chromatograph. In addition, the complexity of odor mixtures that result in unexpected sensory perceptions is enormous. Researchers are a long way away from being able to predict an odor from gas chromatographic analysis (Chambers and Koppel, 2013).

Although Buck and Axel (1991) identified the genes and the protein receptors that comprise the olfactory epithelium, nothing definite is known about the way the brain handles the incoming information to produce in humans' minds the perception of a given odor quality and the strength of that quality. Much less is known about how the brain handles mixtures of different qualities whose signals arrive simultaneously via the olfactory nerve (Lawless, 1986). For a detailed review of the perception of odorant mixtures, see Doty and Laing (2003).

Moncrieff (1951) lists 14 conditions that any theory of olfaction must fulfill. Odorous molecular compounds on the incoming air, in their many orientations and conformations, are attracted and briefly interact with particular sites in the pattern. An attractive theory is that of Luca Turin (1996).

Buck and Axel (1991) received the 2004 Nobel Prize (Altman, 2004) for their discovery in mammalian olfactory mucosa of a family of approximately 1000 genes, coding for as many different olfactory receptor proteins. This group then found (Axel, 1995) that each olfactory neuron expresses one, and only one, receptor protein. They also found that the neurons that express a given protein all terminate in two and only two of the approximately 2000 glomeruli in the olfactory bulb. It seems to follow that the work of the brain is one of sorting and learning. For example, it may learn that if glomeruli numbers 205, 464, and 1,723 are strongly stimulated, then geraniol's odor has been identified.

Human sensitivity to various odors may be measured by dual flow olfactometry, using *n*-butanol as a standard (Moskowitz et al., 1974). Subjects show varying sensitivity to odors depending on hunger, satiety, mood, concentration, the presence or absence of respiratory infections, and, in women, menstrual cycle and pregnancy (Maruniak, 1988).

Given the complexity of the receptors, the enormous range shown by the thresholds for different compounds, and the sociocultural background of the individual, it is not surprising that different people may receive different perceptions from a given odorant or describe perceptions differently. The largest study ever in this area with consumers was The National Geographic Smell Survey; see Gibbons (1986), Gilbert and Wysocki (1987), Wysocki and Gilbert (1989), and Wysocki, et al. (1991). Recent data using two similarly trained panels from two different cultural and ethnic backgrounds to evaluate a complex food product showed that, generally, differences in the words or names associated with a perception were related to experiences associated with cultural background rather than completely different perceptions of the product (Cherdchu et al., 2013). The lesson to be learned from this is that if the job is to characterize or identify a new odor, one needs a large consumer panel or a group of highly trained experts to describe the perception. A panel of one, regardless of qualifications, can be biased and ultimately misleading.

2.3.3.2 Retronasal Odor

An important part of what is called *flavor-by-mouth* is retronasal odor. When people chew and swallow, some of the volatiles in the mouth pass via the nasopharyngeal passage into the nose where they contact the olfactory epithelium (Figure 2.4). For more detail, see Mozell et al. (1969).

Retronasal perception is often responsible for one's ability to identify an odor or a flavor. As an example, Lawless et al. (2004) showed that the so-called metallic taste of solutions of $FeSO_4$ disappears when both nares are blocked.

2.3.3.3 Odor Memory

A first encounter with an odor often is remembered over a very long time. Factors that affect its acquisition and retention are discussed by Köster et al. (2002). Short-term and long-term odor memory are highly important for an animal's survival in the wild as they are for a human subject's performance on a panel (see Parr, Heatherbell, and White, 2002). A selection of odorants useful in panel selection and training are those of ISO Standards 5496 and 22935, "Initiation and Training of Assessors in the Detection and Recognition of Odours."

A problem in odor memory is that, whereas an odor may be perfectly remembered, people tend to forget its name or to apply to it the name of a similar odor (Jönsson and Olsson, 2003). Similarly, when subjects do recall a name but apply it to a different odor, they may mentally transfer characteristics associated with the name to the new odor (see Köster, et al., 2002).

2.3.4 Chemical/Trigeminal Sense

Chemical irritants such as ammonia, ginger, horseradish, onion, chili peppers, menthol, and so on stimulate the trigeminal nerve ends (see Figure 2.5), causing perceptions of burn, heat, cold, pungency, and so on in the mucosa of the eyes, nose, and mouth. Subjects often have difficulty separating trigeminal sensations from olfactory and/or gustatory ones. Experiments that seek to determine olfactory sensitivity among subjects can be confounded by responses to trigeminal rather than olfactory sensations.

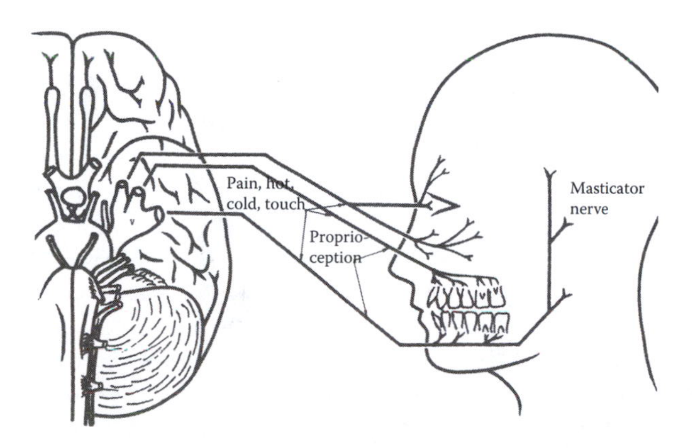

Figure 2.5 Pathway of the trigeminus (V) nerve. (Modified from Netter, F. H. [1973]. *CIBA Collection of Medical Illustrations*, vols. 1 and 3. Summit, NJ: Ciba-Geigy Corporation; readers interested in greater detail are referred to Boudreau, J. C. [1986] *J Sens Stud* 1, 185–202.)

For most compounds, the trigeminal response requires a concentration of the irritant that is higher in orders of magnitude than one that stimulates the olfactory or gustatory receptors. Trigeminal effects assume practical significance (1) when the olfactory or gustatory threshold is high, for example, for short-chain compounds such as formic acid or for persons with partial anosmia or ageusia; and (2) when the trigeminal threshold is low, for example, for capsaicin.

The trigeminal response to mild irritants (such as carbonation, mouth burn caused by high concentrations of sucrose and salt in confections and snacks, the heat of peppers and other spices) may contribute to, rather than distract from, acceptance of a product (Carstens et al., 2002; for review, see Viana, 2011).

2.3.5 Sense of Gustation/Taste

Like olfaction, gustation is a chemical sense (see review by Drewnowski, 2001). It involves the detection of stimuli dissolved in water, oil, or saliva by the taste buds that are primarily located on the surface of the tongue as well as in the mucosa of the palate and areas of the throat. Figure 2.6 shows the taste system in three different perspectives. Compared with olfaction, the contact between a solution and the taste epithelium on the tongue and walls of the mouth is more regular in that every receptor is immersed for at least some seconds. There is no risk of the contact being too brief, but there is ample opportunity for oversaturation. Molecules causing strong bitterness probably bind to the receptor proteins, and some may remain for hours or days (the cells of the olfactory and gustatory epithelium are renewed on average every 6–8 days; Beidler, 1960; Oakley and Riddle, 1992). The prudent taster should take small sips and keep each sip in the mouth for only a couple of seconds, then wait (depending on the perceived strength) for 15–60 s before tasting again. The first and second sips are the most sensitive, and one should train oneself to accomplish in those first sips all the mental comparisons and adjustments required by the task at hand. Where this is not possible, for example, in a lengthy questionnaire with more than eight or ten questions and untrained subjects, the experimenter must be prepared to accept a lower level of discrimination.

The gustatory sensors are bathed in a complex solution, the saliva (which contains water, amino acids, proteins, sugars, organic acids, salts, etc.), and they are fed and maintained by a second solution, the blood (which contains an even more complex mixture of the same substances). Hence, humans' sensitivities to levels (e.g., of salt) that are lower than those in saliva is low and ill defined. Typical thresholds for taste substances are shown in Figure 2.7.

The range between the weakest tastant, sucrose, and the strongest, Strophantin (a bitter alkaloid) is no more than 104, much smaller than the range of 1012 shown by odorants. The figure also shows the range of thresholds for 47 individuals, and it is seen that the most and least sensitive individuals generally differ by a factor of 102. In the case of phenylthiocarbamide (also phenylthiourea), a bimodal distribution is seen (Amerine et al., 1965, 109): The population consists of two groups, one with an average threshold of 0.16 g/100 mL and another with an average threshold of 0.0003 g/100 mL. Vanillin is another substance that appears to show two peaks (Meilgaard, et al., 1982), but the total number of compounds for which bimodal distributions have been reported is small, and

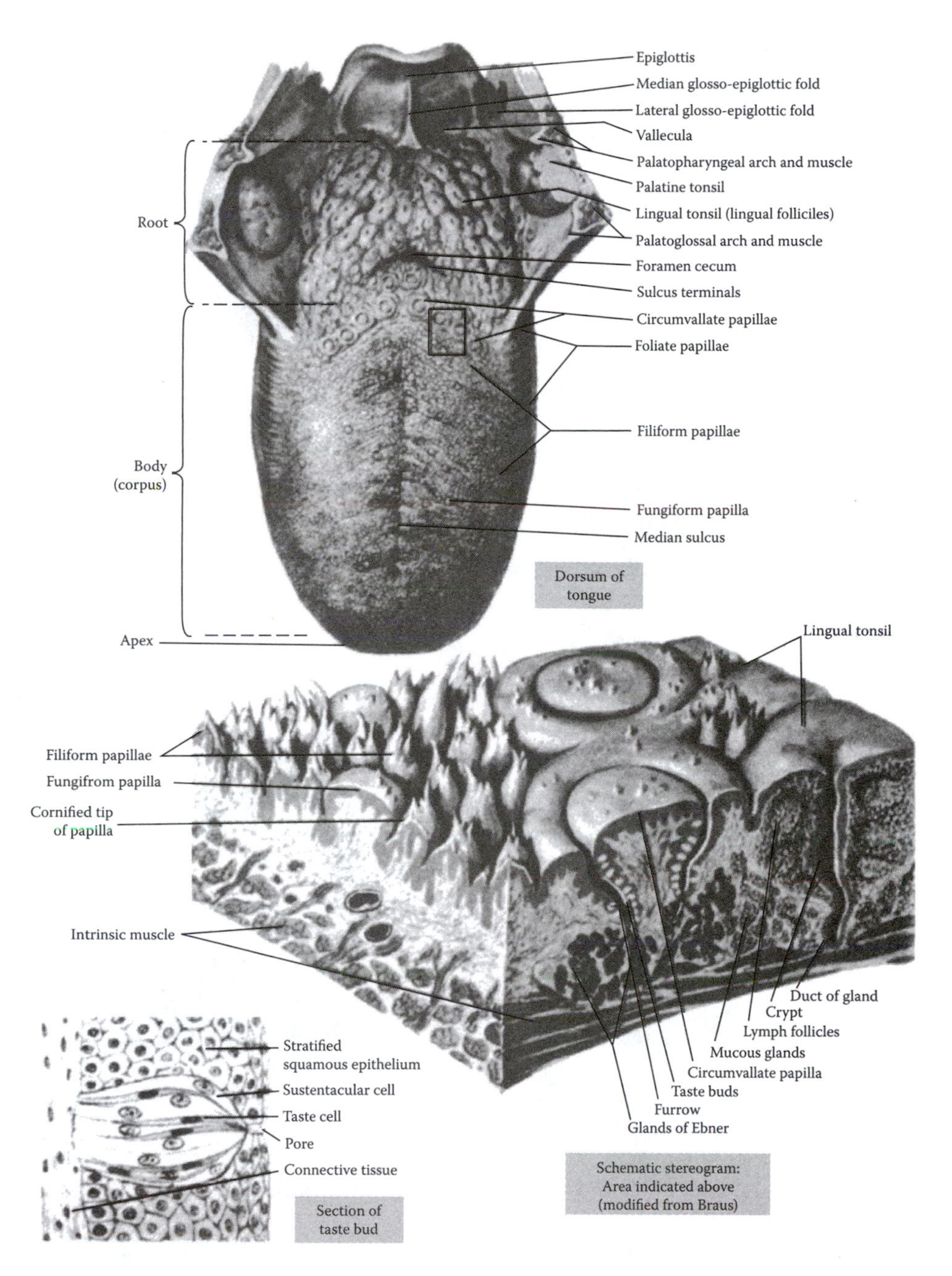

Figure 2.6 Anatomical basis of gustation, showing the tongue, a cross section of a fungiform papilla, and a section thereof showing a taste bud with receptor cells. The latter carry chemosensitive villi that protrude through the taste pore. At the opposite end, their axons continue until they make synaptic contact with cranial nerve VII, the chorda tympani. The surrounding epithelial cells will eventually differentiate into taste receptor cells that renew the current ones as often as once a week.

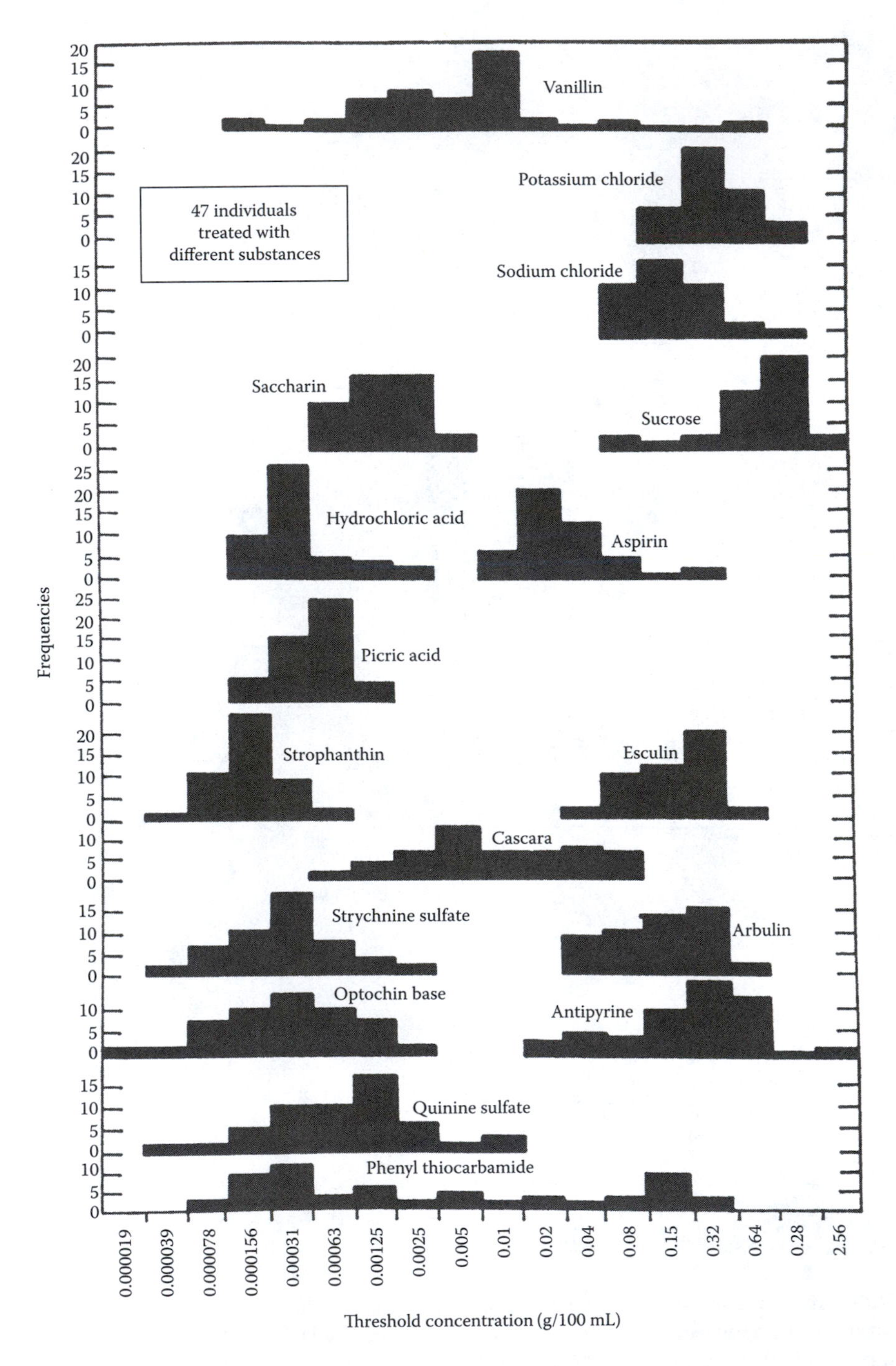

Figure 2.7 Distribution of taste thresholds for 47 individuals (From Amerine, M. A., et al., [1965]. *Principles of Sensory Evaluation of Food*, 109. New York: Academic Press. With permission.)

their role in food preferences or in odor and taste sensitivity is a subject that has not been explored (Amoore, 1977).

Genetic predisposition to perception of certain compounds (e.g., 6-n-propylthiouracil [PROP]) also exists. In the case of PROP, individuals with two recessive alleles are unable to taste bitterness in PROP. Individuals with one or two dominant alleles are PROP tasters, and those with two alleles generally find the bitter taste of PROP extremely high. One cautionary note is that the impact of taster status for one compound has not yet been shown definitively to have anything to do with the taste or odor perception of other compounds. For example, Li and Drewnoski (2000) did not find any association of PROP taster status to the liking of chocolate or sweetened caffeine solutions.

In addition to the concentration of a taste stimulus, other conditions in the mouth that affect taste perception are the temperature, viscosity, rate, duration, and area of application of the stimulus; the chemical state of the saliva; and the presence of other tastants in the solution being tasted. The incidence of ageusia, or the absence of the sense of taste, is rare. However, variability in taste sensitivity, especially for bitterness with various bitter agents, is quite common.

Researchers' understanding of the physiological mechanisms of the principal tastes has been advancing rapidly; for example, sweet (Li et al., 2001; Montmayeur et al., 2001; Nelson et al., 2001); sweet and bitter (Ruiz et al., 2001); sweet and umami (Li et al., 2002; Zhao et al., 2003); sweet, bitter, and umami (Zhang et al., 2003); and sour (Johanningsmeier, et al., 2005).

2.3.6 Sense of Hearing

Figure 2.8 shows a cross section of a human ear. Vibrations in the local medium, usually air, cause the eardrum to vibrate. The vibrations are transmitted via the small bones in the middle ear to create hydraulic motion in the fluid of the inner ear, the cochlea, which is a spiral canal covered in hair cells that, when agitated, send neural impulses to the brain. Students studying crispness and other sound aspects of products should familiarize themselves with the concepts of intensity, measured in decibels; and pitch, determined by the frequency of sound waves. A possible source of variation or error that must be controlled in such studies is the creation and/or propagation of sound inside the cranium but outside the ear; for example, by movement of the jaws or teeth and propagation via the bone structure.

Psychoacoustics is the science of building vibrational models on a sound oscilloscope to represent perceived sound stimuli such as pitch, loudness, sharpness, roughness, and so on. These models work for simple sounds but not for more complex ones. They can be used to answer questions such as "What kind of sound?" and "How loud?" However, they often fail to provide a sound that is appropriate to what the listener expects.

Recently, academics and engineers who are responsible for the sound characteristics of products have realized the need for a common vocabulary to describe sound attributes for complex sounds. This occurs because automobile, airframe, and industrial and consumer products manufacturers are concerned with the sounds that their products produce and how humans respond to those sounds. A summary of sensory methods applied to sound is given by Civille and Seltsam (2003).

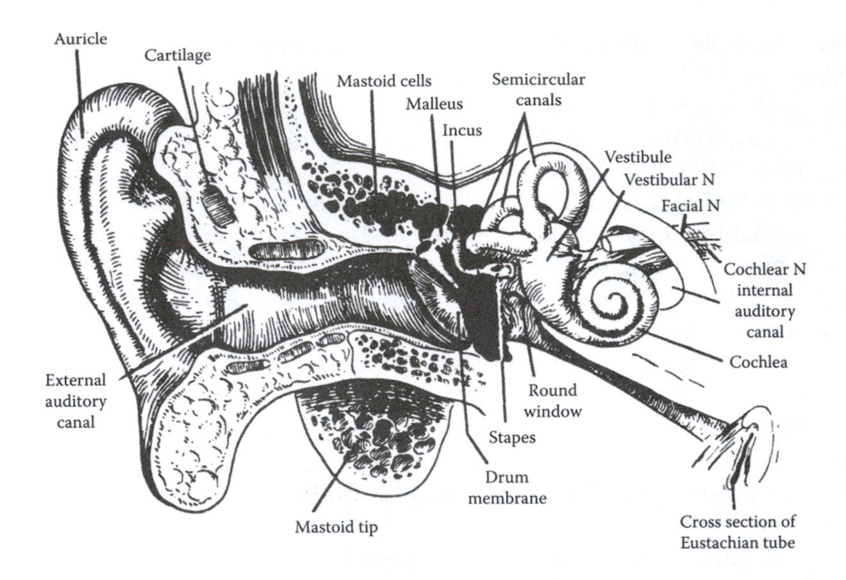

Figure 2.8 A semidiagrammatic drawing of the ear (From Kling, J. W. and L. A. Riggs [1971]. *Woodworth and Schlosberg's Experimental Psychology*, 3rd ed. New York: Holt, Rinehart & Winston. With permission.)

2.4 PERCEPTION AT THRESHOLD AND ABOVE

The reader is warned that a threshold is not a constant for a given substance but rather a constantly changing point on the sensory continuum from nonperceptible to easily perceptible (see Chapter 8). Thresholds change with moods, the time of the biorhythm, with hunger and satiety, and, clearly, from assessor to assessor. Compounds with identical thresholds can show very different rates of increase in intensity with concentration; therefore, the threshold's use as a yardstick of intensity of perception must be approached with considerable caution (Bartoshuk, 1978; Pangborn, 1984). In practical studies involving products that emit mixtures of large numbers of flavor-active substances where the purpose is to detect those compounds that play a role in the flavor of the product, the threshold has some utility, provided the range covered does not extend too far from the threshold; for example, from 0.5 × threshold to 3 × threshold. Above this range, intensity of odor or taste must be measured by scaling (see Chapter 5).

REFERENCES

Altman, L. K. (2004). 2 Americans win Nobel for demystifying sense of smell. *The New York Times*, October 4, Section A1. http://www.nytimes.com/2004/10/05/science/05nobel.html.

Amerine, M. A., R. M. Pangborn, and E. B. Roessler (1965). *Principles of Sensory Evaluation of Food*. New York: Academic Press.

Amoore, J. E. (1977). Specific anosmia and the concept of primary odors. *Chem Sens and Flav* 2: 267–81.

ASTM (1968). *Basic Principles of Sensory Evaluation*. Standard Technical Publication 433. West Conshohocken, PA: ASTM International, 110.

Axel, R. (1995). The molecular logic of smell. *Sci Am* (October): 154–9.

Bartoshuk, L. M. (1978). The psychophysics of taste. *Am J Clin Nutr* 31: 1068–77.

Beets, M. G. J. (1978). *Structure–Activity Relationships in Human Chemoreception*. London: Applied Science.

Beidler, L. M. (1960). Physiology of olfaction and gustation. *Ann Oto Rhinol and Laryn* 69: 398–409.

Boudreau, J. C. (1986). Neurophysiology and human taste sensations. *J Sens Stud* 1: 185–202.

Buck, L., and R. Axel (1991). A novel multigene family may encode odorant receptors: A molecular basis for odor reception. *Cell* 65 (1): 175–87.

Carstens, E., M. I. Carstens, J. M. Dessirier, M. O'Mahony, C. T. Simons, M. Sudo, and S. Sudo (2002). It hurts so good: Oral irritation by spices and carbonated drinks and the underlying neural mechanisms. *Food Qual Prefer* 13: 431–43.

Caul, J. F. (1957). The profile method of flavor analysis. *Adv Food Res* 7: 1–40.

Civille, G. V., and B. G. Lyon (eds) (1996). *Aroma and Flavor Lexicon for Sensory Evaluation. Terms, Definitions, References, and Examples*. ASTM Data Series Publication DS 66. West Conshohocken, PA: ASTM International.

Civille, G. V., and J. Seltsam (2003). Sensory evaluation methods applied to sound quality. *Noise Control Eng J* 51 (4): 262.

Clydesdale, F. M. (1984). Color measurement. In *Food Analysis. Principles and Techniques*, vol. 1, ed. D.W. Gruenwedel and J.R. Whitaker, 95–150. New York: Marcel Dekker.

De Man, J. M., P. W. Voisey, V. F. Rasper, and D. W. Stanley (1976). *Rheology and Texture in Food Quality*. Westport, CT: AVI Publishing.

Descoins, C., Mathlouthi, M., Le Moual, M., and J. Hennequin (2006). Carbonation monitoring of beverage in a laboratory scale unit with on-line measurement of dissolved CO_2. *Food Chem* 95: 541–53.

Doty, R. L., and D. G. Laing (2003). Psychophysical measurement of human olfactory function, including odorant mixture assessment. In *Handbook of Olfaction and Gustation*, 2nd ed., ed. R.L. Doty, 209–28. New York: Marcel Dekker.

Drewnowski, A. (2001). The science and complexity of bitter taste. *Nutr Rev* 59 (6): 163–9.

Gardner, E. (1968) *Fundamentals of Neurology*, 5th ed. Philadelphia: W.B. Saunders.

Gatchalian, M. M. (1981). *Sensory Evaluation Methods with Statistical Analysis*. University of the Philippines, Diliman: College of Home Economics.

Geldard, F. A. (1972). *The Human Senses*, 2nd ed. New York: Wiley.

Gibbons, B. (1986). The intimate sense of smell. *Natl Geogr* 170 (3): 324–61.

Gilbert, A. N., and C. J. Wysocki (1987). The National Geographic smell survey results. *Natl Geogr.* 172: 514–25.

Harper, R. (1972). *Human Senses in Action*. London: Churchill Livingston.

Hochberger, J. E. (1964) *Perception*. Englewood Cliffs, NJ: Prentice-Hall.

Howard, P. (1996). *Handbook of Physical Properties of Organic Chemicals*. Boca Raton, FL: CRC Press.

ISO (International Organization for Standardization). (2006). Initiation and training of assessors in the detection and recognition of odours, and ISO/DIS 22935- 1. Milk and milk products, sensory analysis—Part 1, General guidance for the recruitment, selection, training and monitoring of milk and milk product assessors. American National Standards Institute/ ISO. Latest revision, International Standard ISO 5496, Sensory Analysis—Methodology—General Guidance.

Johanningsmeier, S. D., R. E. McFeeters, and M. Drake (2005). A hypothesis for the chemical basis for perception of sour taste. *J Food Sci* 70 (2): R44–R48.

Jönsson, F. U., and M. J. Olsson. (2003). Olfactory metacognition. *Chem Senses* 28: 651–8.

Kling, J. W., and L. A. Riggs (eds.) (1971). *Woodworth and Schlosberg's Experimental Psychology*, 3rd ed. New York: Holt, Rinehart & Winston.

Köster, E. P., J. Degel, and D. Piper (2002). Proactive and retroactive interference in implicit odor memory. *Chem Senses* 27: 191–206.

Kramer, A., and B. A. Twigg (1973). *Quality Control for the Food Industry,* vol. 1. Westport, CT: AVI Publishing.

Laing, D. G. (1983). Natural sniffing gives optimum odor perception for humans. *Perception* 12: 99.

Lawless, H. T. (1986). Sensory interaction in mixtures. *J Sens Stud* 1 (3/4): 259–74.

Lawless, H. T., and H. Heymann (1998). *Sensory Evaluation of Food. Principles and Practices.* New York: Chapman & Hall.

Lawless, H. T., S. Schlake, J. Smythe, J. Lim, H. Yang, K. Chapman, and B. Bolton. (2004). Metallic taste and retronasal smell. *Chem Sens* 29 (1): 25–33.

Li, X., M. Inoue, D. R. Reed, T. Huque, R. B. Puchalski, M. G. Tordoff, Y. Ninomiya, G. K. Beauchamp, and A. A. Bachmanov (2001). High resolution genetic mapping of the saccharin preference locus (Sac) and putative sweet taste receptor (T1R1) gene (Gpr70) to mouse distal chromosome 4. *Mamm Genome* 12: 13–16.

Li, X., L. Staszewski, H. Xu, K. Durick, M. Zoller, and E. Adler (2002).Human receptors for sweet and umami taste. *Proc Natl Acad Sci U S A* 99: 4692–6.

Lota, M., D. D. Serra, F. Tomi, C. Jacquemond, and J. Casanova (2002). Volatile components of peel and leaf oils of lemon and lime species. *J Agric Food Chem* 50: 796–805.

Maruniak, J. A. (1988). The sense of smell. In *Sensory Analysis of Foods,* 2nd ed., ed. J. R. Piggott, 25. London: Elsevier.

McDougall, V. (1983). Assessment of the appearance of food. In *Sensory Quality in Foods and Beverages: Its Definition, Measurement and Control,* ed. A. A. Williams and R. K. Atkin, 121 ff. Chichester: Ellis Horwood.

McDougall, D. B. (1988). Color vision and appearance measurement. In *Sensory Analysis of Foods,* 2nd ed., ed. J. R. Piggott, 103 ff. London: Elsevier.

Meilgaard, M. C. (1975). Flavor chemistry of beer. II. Flavor and threshold of 239 aroma volatiles. *Tech Q Master Brew Assoc Am* 12: 151–68.

Meilgaard, M. C., D. S. Reid, and K. A.Wyborski (1982). Reference standards for beer flavor terminology system. *J Am Soc Brew Chem* 40: 119–28.

Mitchell, J. R. (1984). Rheological techniques. In *Food Analysis. Principles and Techniques,* vol. 1, ed. D. W. Gruenwedel and J. R. Whitaker. New York: Marcel Dekker.

Moncrieff, R. W. (1951). *The Chemical Senses.* London: Leonard Hill.

Montmayeur, J. P., S. D. Liberles, H. Matsunami, and L. B. Buck (2001). A candidate taste receptor gene near a sweet taste locus. *Nat Neuro Sci* 4: 492–8.

Moskowitz, H. R., A. Dravnieks, W. S. Cain, and A. Turk (1974). Standardized procedure for expressing odor intensity. *Chem Sens Flav* 1: 235–7.

Mozell, M. M., B. P. Smith, P. E. Smith, R. J. Sullivan Jr., and P. Swender (1969). Nasal chemoreception and flavor identification. *Arch Otolaryngol* 90: 131–7.

Nelson, V., M. A. Hoon, J. Chandrashekar, Y. Zhang, N. J. Ryba, and C. S. Zuker (2001). Mammalian sweet taste receptors. *Cell* 106 (3): 381–90.

Netter, F. H. (1973). *CIBA Collection of Medical Illustrations,* vols. 1 and 3. Summit, NJ: Ciba-Geigy Corporation.

Oakley, B., and D. Riddle (1992). Recepter cell regeneration and connectivity in olfaction and taste. *Exp Neurol* 115: 50–54.

Pangborn, R. M. (1984). Sensory techniques of food analysis. In *Food Analysis. Principles and Techniques,* vol. 1, ed. D.W. Gruenwedel, J.R. Whitaker. New York: Marcel Dekker.

Parr, W. V., D. Heatherbell, and K. G. White (2002). Demystifying wine expertise: Olfactory threshold, perceptual skill and semantic memory in expert and novice wine judges. *Chem Sens* 27 (8): 744–55.

Piggott, J. R. (ed.) (1998). *Sensory Analysis of Foods,* 2nd ed. London: Elsevier.

Ruiz, A. L., G. T. Wong, S. Damak, and R. E. Margolskee (2001). Dominant loss of responsiveness to sweet and bitter compounds caused by a single mutation in alpha-gusducin. *Proc Natl Acad Sci U S A* 98: 8868–73.

Sekuler, R., and R. Blake (1990). *Perception*, 2nd ed. New York: McGraw-Hill.

Siebert, K. J., L. E. Stenroos, D. S. Reid (1981). Characterization of amorphous-particle haze. *J Am Soc Brew Chem* 39: 1–11.

Smadja, J., P. Rondeau, and A. S. C. Sing (2005). Volative constituents of five *Citrus* Petitgrain essential oils from Reunion. *Flavor Frag J* 20: 399–402.

Smith II, R. F. (2010). A descriptive evaluation methodology for consumer audio equipment. *J Sens Stud* 25 (6): 804–18.

Stone, H., and J. L. Sidel (2004). *Sensory Evaluation Practices*, 3rd ed. San Diego, CA: Elsevier.

Szczesniak, A. S. (1963). Classification of textural characteristics. *J Food Sci* 28: 385–9.

Turin, L. (1996). A spectroscopic mechanism for primary olfactory reception. *Chem Sens* 21: 773–91.

Vekiari, S. A., E. E. Protopapdakis, P. Papadopoulou, D. Papanicolaou, C. Panou, and M. Vamvakias (2002). Composition and seasonal variation of the essential oil from leaves and peel of a Cretan lemon variety. *J Agric Food Chem* 50: 147–53.

Viana, F. (2011). Chemosensory properties of the trigeminal system. *ACS Chem Neurosci* 2: 38–50.

Vickers, Z. M., and M. C. Bourne (1976). Crispness in foods. A review. A psychoacoustical theory of crispness. *J Food Sci* 41: 1153–8.

Wysocki, C. J., and A. N. Gilbert (1989). National geographic smell survey. Effects of age are heterogeneous. In *Nutrition and the Chemical Senses in Aging: Recent Advances and Current Needs*, vol. 561. New York: Annals of the New York Academy of Science.

Wysocki, C. J., J. D. Pierce, and A. N. Gilbert (1991). Geographic, cross-cultural, and individual variation in human olfaction. In *Smell and Taste in Health and Disease*, ed. T.V. Getchell, 287–314. New York: Raven Press.

Zhang, V., M. A. Hoon, J. Chandrashekar, K. L. Mueller, B. Cook, D. Wu, C. S. Zuker, and J. P. Ryba (2003). Coding of sweet, bitter and umami tastes: Different receptor cells sharing similar signaling pathways. *Cell* 112 (3): 293–301.

Zhao, V., Y. Zhang, M. A. Hoon, J. Chandrashekar, I. Erlenbach, N. J. P. Ryba, and C. S. Zuker. (2003). The receptors for mammalian sweet and umami taste. *Cell* 115 (3): 255–66.

3

Controls for Test Room, Products, and Panel

3.1 INTRODUCTION

Many variables must be controlled if the results of a sensory test are to measure the true product differences under investigation. It is convenient to group these variables under three major headings:

1. Test controls: the test-room environment, the use of booths or a round table, the lighting, the room air, the preparation area, and the entry and exit areas
2. Product controls: the equipment used, the way samples are screened, prepared, numbered, coded, and served
3. Panel controls: the procedure used by a panelist evaluating the sample in question

3.2 TEST CONTROLS

The physical setting must be designed to minimize the subjects' biases, maximize their sensitivity, and eliminate variables that do not come from the products themselves. Panel tests are costly because of the high cost of panelists' time. A high level of reduction of disturbing factors is easily justified. Drop-offs in panel attendance and panel motivation are universal problems, and management must clearly show the value it places on panel tests by the care and effort expended on the test area. The test area should be centrally located, easy to reach, and free of crowding and confusion, as well as comfortable, quiet, temperature controlled, and above all, free from odors and noise.

3.2.1 Development of Test-Room Design

Since the first edition of this book (1987), test-room design has matured, as reflected in publications by national and international organizations (Eggert and Zook, 1986; European Cooperation for Accreditation of Laboratories, 1995; Chambers and Wolf, 1996;

International Organization for Standardization, 1998). A move toward requiring accreditation of sensory services under ISO 9000 has accelerated a trend toward uniformly high standards, for example, separate air exhausts from each booth.

Early test rooms made allowance for six to ten subjects and consisted of a laboratory bench or conference table on which samples were placed. The need to prevent subjects from interacting, thus introducing bias and distraction, led to the concept of the booth (see Figure 3.1).

In a parallel development, the Arthur D. Little organization (Caul, 1957) argued that panelists should interact and come to a consensus, which required a round table with a "lazy Susan" on which reference materials were placed to standardize terminology and scale values.

Current thinking often combines these two elements into (1) a booth area that is the principal room used for difference tests as well as some descriptive tests, and (2) a round-table area used for training and/or other descriptive tasks (see Figure 3.2). Convenience dictates that a sample-preparation area be located near to, but separate from, the test room. Installations above a certain size also require office area, sample storage area, and data workstation.

3.2.2 Location

The panel test area should be readily accessible to all. A good location is one that most panel members pass on their way to lunch or morning break. If panel members are drawn from the outside, the area should be near the building entrance. Test rooms should be separated by a suitable distance from congested areas because of noise and the opportunity this would provide for unwanted socializing. Test rooms should be away from other noise and from sources of odor such as machine shops, loading docks, production lines, and cafeteria kitchens.

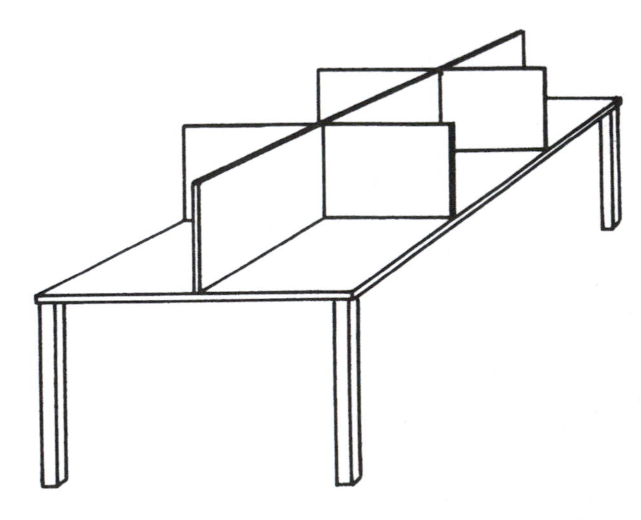

Figure 3.1 Simple booths consisting of a set of dividers placed on a table.

Figure 3.2 Top: circular table with "lazy Susan" used for consensus-type descriptive analysis. (Courtesy of Ross Products Division, Columbus, OH). Bottom: round-table discussion used for descriptive analysis ballot development. (Courtesy of NutraSweet/Kelco Inc., Mt. Pleasant, IL. With permission.)

3.2.3 Test-Room Design

3.2.3.1 Booth

It is customary for one sample-preparation area to serve six to eight booths. The booths may be arranged side by side, in an L shape, or with two sets of three to four booths facing each other across the serving area. The L shape represents the most efficient use of the "work triangle" concept in kitchen design, resulting in a minimum of time and distance covered by technicians in serving samples. One unit of six to eight booths will accommodate a moderate test volume of 300–400 sittings per year of panels up to 18–24 members. For higher volumes of testing and/or larger panels, multiple units served from one or several preparation areas are recommended. Consideration should also be given to placement of the technicians' monitor(s) and central processing unit(s) for any automated data handling system.

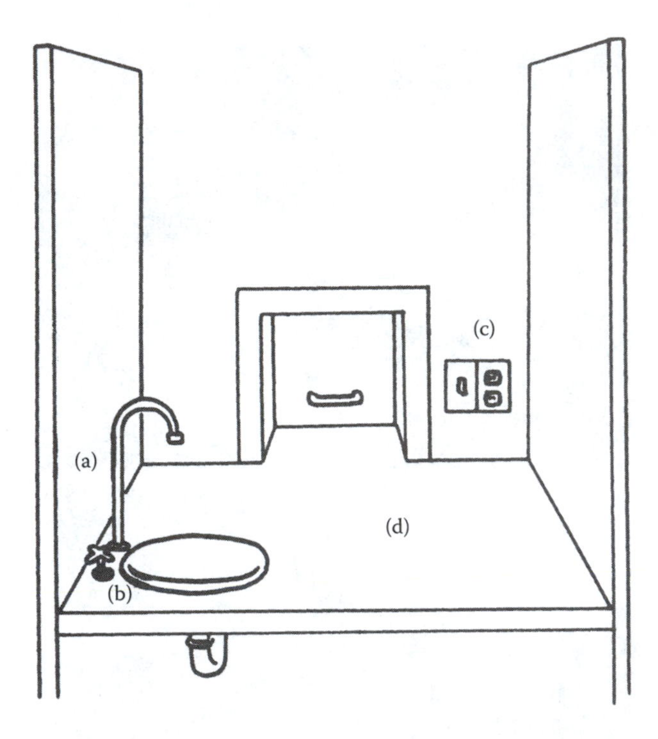

Figure 3.3 Sensory evaluation booth with hatch (in background) for receipt and return of sample tray: (a) tap water; (b) small sink; (c) electrical outlet and signal switch to panel attendant; (d) table covered with odorless Formica or another easy-to-clean surface.

Figure 3.3 shows a typical booth that is 27–32 in. wide with an 18–22 in. deep counter installed at the same height as the sample preparation table (normally 36 in.). Space can be allowed for installation of a PC monitor and a keyboard if required. The dividers should extend approximately 18 in. above the countertop to reduce visual and auditory distraction between booths. The dividers may extend from the floor to the ceiling/soffit for complete privacy (with the design allowing for adequate ventilation and/or cleaning), or it may be suspended from the wall, enclosing only the torso and head of the assessor. The latter is preferred in most cases as claustrophobia is a permanent problem, whereas assessors soon learn to refrain from looking over shoulders or uttering loud comments on the quality of samples. A minimum free distance of 4 ft is recommended as a corridor to allow easy access to the booths.

3.2.3.1.1 Special Booth Features

A small stainless-steel sink and a water faucet are usually included for rinsing. These are mandatory for the evaluation of such products as mouthwashes, toothpastes, and household items but are not recommended for solid foods that may plug the traps. Filtered water may be required if odor-free tap water is unavailable.

A signal system is sometimes included so that the panel supervisor knows when an assessor is ready for a sample or has a question. Usually this takes the form of a switch in

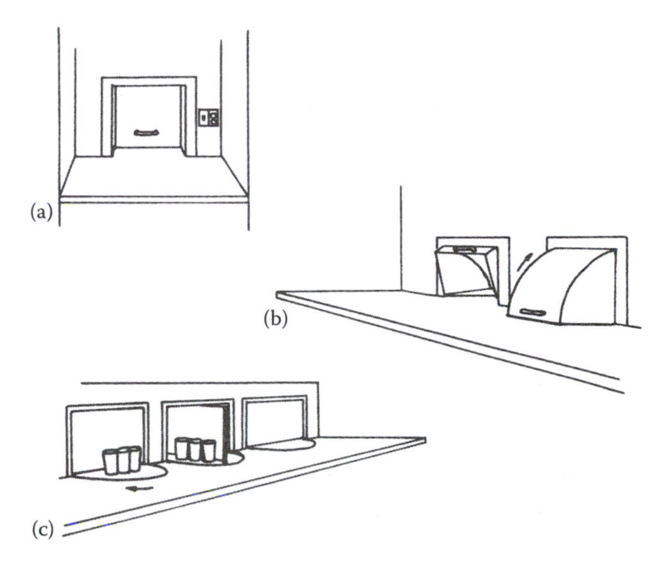

Figure 3.4 Three types of hatch for passing samples to and from the panelists: (a) sliding door; (b) breadbox; (c) carousel.

each booth that will trigger a signal light for that booth in the sample-preparation area. It may include an exterior light panel that indicates to incoming subjects which booths are available.

A direct computer entry system located in each panel booth requires space to accommodate the entry device (e.g., keyboard and tower, laptop, tablet).

Sample trays may be carried to each booth if they consist of nonodorous items that will keep their condition for 10–20 min. If these conditions are not fulfilled, the sample-preparation area must be located behind the booths and a hatch provided through which the tray can be passed once the subject is in place. Three types of pass-through are in use (see Figure 3.4). The sliding door (vertical or horizontal) requires the least space. The types known as the breadbox and the carousel are more effective in preventing the passage of odors or visible cues from the preparation area to the subject.

The materials of construction in the booths and surrounding area should be odor free and easy to clean. Formica and stainless steel are the most common surface materials.

3.2.3.2 Descriptive Evaluation and Training Area

At a minimum, this function may be filled by a table in the panel leader's office where standards may be served as a means of educating panel members. At the other extreme, if descriptive analysis is a common requirement or if needs for training and testing are large, the following equipment is recommended:

- A conference-style room with several tables that can be arranged as required by the size and objective of the group.
- Audiovisual equipment, which may include an "electronic white board" capable of making hard copies of results or a monitor/screen that projects the data from the panel leader's computer.

- Separate preparation facilities for reference samples used to illustrate the descriptors or intensities; depending on type, these may include a storage space (frozen, refrigerated, or room temperature, perhaps sealed to prevent odors from escaping) and a holding area for preparing the references (perhaps hooded).

3.2.3.3 Preparation Area

The preparation area is a laboratory that must permit preparation of all of the possible and foreseeable combinations of test samples at the maximum rate at which they are required. Each booth area and descriptive analysis area should have a separate preparation laboratory so as to maximize the technician's ability to prepare, present, and clean up each study. Typically, the preparation area includes immediate access to the following, in addition to any specialized equipment dictated by the type of samples:

- Laboratory benches flush with the hatches so that sample trays will slide through
- Benches, kitchen range, ovens, and so on, for preparation
- Refrigerator and freezer for storage of samples
- Storage for glassware, dishes, glasses, trays, and so on
- Dishwashers, disposals, trash compactors, wastebaskets, sinks, and so on
- Storage for panel member treats, if used
- Large garbage containers for quick disposal of used product, and so on
- Central computer system to keep track of products and panelists

Consideration should be given to company and local recycling policies so that appropriate receptacles are available in the preparation area.

3.2.3.4 Office Facilities

An office is usually situated within view of the panel booths, as someone must be present while testing is in progress. It may be convenient to locate records, storage space, and any computer equipment in the same area so that the panel leader's time may be effectively utilized. Equipment such as phones and printers should be at a sufficient distance to avoid distracting the subjects.

3.2.3.5 Entrance and Exit Areas

In large facilities, it is advisable to separate entrance and exit areas for assessors so as to prevent the unwanted exchange of information. The exit area commonly contains a desk where assessors can study the identity of the day's samples and where they may receive a "treat" to encourage participation. If some of the panelists are nonemployees, the entrance/exit area should contain a sufficient waiting room with comfortable seats, coat closet or coat rack, and separate restrooms.

3.2.3.6 Storage

Space must be allocated for storage of

- Samples prior to preparation, after preparation, and at the time of serving
- Reference samples and controls or standards under the appropriate temperature and humidity conditions

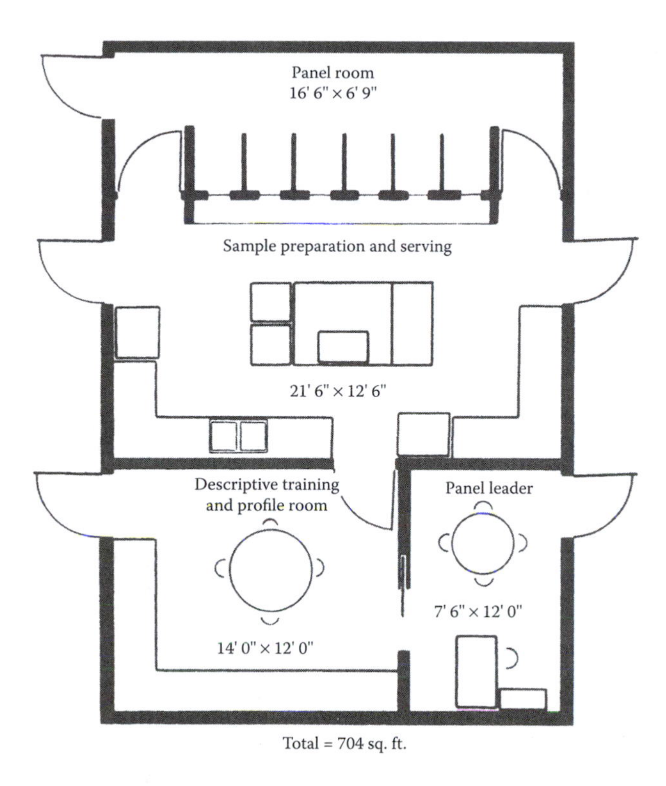

Figure 3.5 Layout for medium-sized sensory evaluation area suitable for 300–400 tests per year. (Drawn by D. Grabowski. With permission.)

- Large volumes of disposable containers and utensils
- Clean-up materials with minimal odor or fragrance
- Paper ballots or score sheets before and after use

Figures 3.5 and 3.6 show typical layouts of medium- and large-scale installations featuring various facilities that may be located around the booth area.

3.2.4 General Design Factors

3.2.4.1 Color and Lighting

The color and lighting in the booths should be planned to permit adequate viewing of samples while minimizing distractions (Amerine et al., 1965; Malek et al., 1982; Eggert and Zook, 1986; International Organization for Standardization, 1988; Poste et al., 1991; European Cooperation for Accreditation of Laboratories, 1995; Chambers and Wolf, 1996). Walls should be off-white; the absence of hues of any color will prevent unwanted differences in appearance. Booths should have even, shadow-free illumination at 70–80 footcandles (fc) (typical of an office area). If appearance is critical, rheostat control may be used to vary the light intensity up to 100 fc. Incandescent lighting allows wider variation and permits the use of colored lights (see Section 3.2.4.1.1), but more heat is generated,

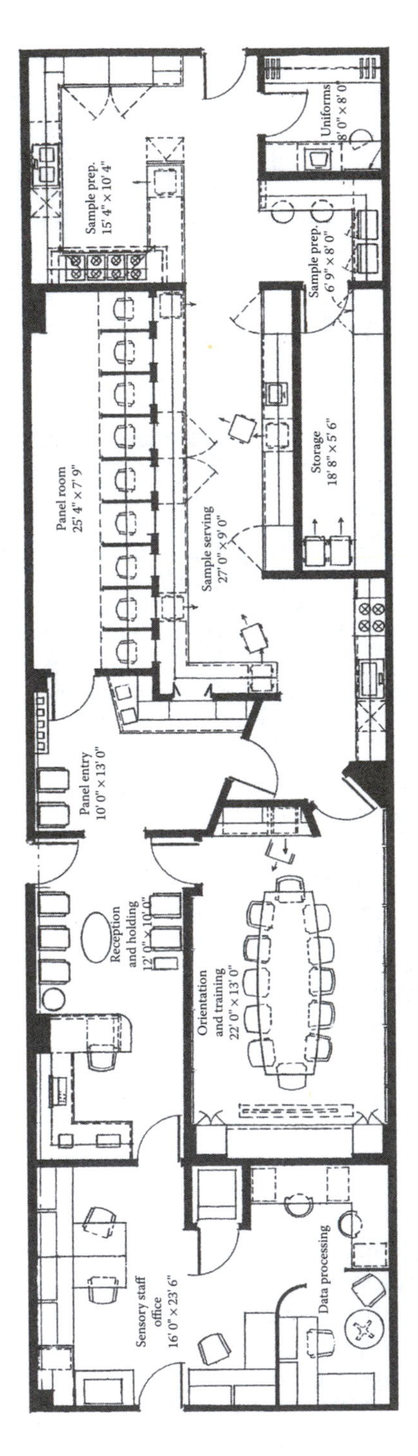

Figure 3.6 Layout for large sensory evaluation area suitable for preparation and evaluation of 600–1000 samples per year. (From Eggert, J., and Zook, K., (Eds.) [1986]. *Physical Requirements for Sensory Evaluation Laboratories*, ASTM International, West Conshohocken, PA. With permission.)

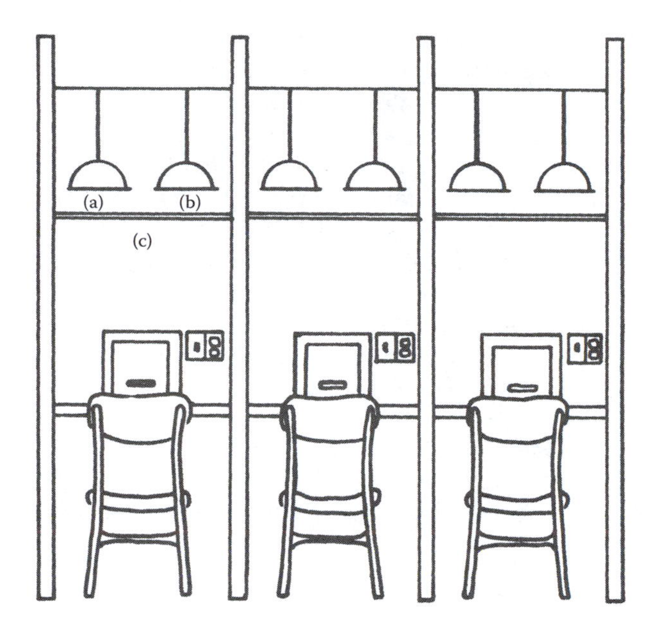

Figure 3.7 Panel booths showing arrangements for lighting: (a) incandescent; (b) fluorescent; (c) holder for sheet filters. (Courtesy Gatchalian, M. M. [1981]. *Sensory Evaluation Methods with Statistical Analysis*. University of the Philippines, College of Home Economics, Diliman. With permission.)

requiring adequate cooling. Fluorescent lighting generates less heat and allows a choice of whiteness (i.e., cool white, warm white, simulated north daylight; see Figure 3.7).

3.2.4.1.1 Colored Lights

A common feature of many panel booths is a choice of red, green, and/or blue lighting at low intensity obtained through the use of colored bulbs or special filters. The lights are used to mask visual differences between samples in difference tests calling for the subject to determine by taste (or by feel, if appropriate) which samples are similar enough to be used interchangeably.

Many colored bulbs emit sufficient white light to be ineffective in reducing color differences. Theater gel filters are quite effective and may be placed in frames over recessed spotlights. Another alternative is a low-pressure sodium lamp, which emits light at a single wavelength. Low Pressure Sodium-SOX lamps are available from Phillips and can be purchased through any NAED distributor (National Association of Electrical Distributors). Both the theater gels and color masking lamps remove colors but do not eliminate differences in color intensity. The effect is that of black-and-white television with degrees of gray still detectable.

Pangborn (1967) notes that an abnormal level of illumination may itself influence the assessor's impressions. An alternative is to choose methods other than simultaneous presentation to accommodate the presence of visual differences between samples. For example, samples may be served sequentially and scored with reference to a common standard.

3.2.4.2 Air Circulation, Temperature, and Humidity

The sensory evaluation area should be air conditioned at 72°F–75°F and 45%–55% relative humidity (RH). (For the tactile evaluation of fabrics, paper nonwovens, and skincare products, tighter humidity control may be required, e.g., 50% ± 2% or 65% ± 2% RH.) Recirculated and make-up air should pass through a bank of activated carbon canisters that are capable of removing all detectable odor. The canisters may be placed outside the testing area in a location that allows easy replacement, for example, every 2 or 3 months. Frequent monitoring is required to prevent the filters from becoming ineffective and/or becoming an odor source. A slight positive pressure should be maintained in the booth areas so as to prevent odor contamination from the sample-preparation area or from outside. If the testing of odorous materials such as sausages or cheese is a possibility, separate air exhausts must be provided from each booth.

3.2.4.3 Construction Materials

The materials used in the construction and furnishing of a sensory evaluation laboratory must be in accordance with the specific environment required for the products to be evaluated in the laboratory.

3.2.4.3.1 Nonodorous

Paper, fabric, carpeting, porous tile, and so on must be avoided because they are either odorous in themselves or may harbor dirt, molds, and so on that will emit odor. Construction materials must be smooth, easy to clean, and nonabsorbent so that they do not retain odors from previous sessions. The materials that best meet these requirements are stainless steel, Teflon, and Formica. Nonodorous vinyl laminate is suitable for ceilings, walls, and floors.

3.2.4.3.2 Color

A neutral, unobtrusive color scheme using off-white colors and few patterns provides a background that is not distracting to panelists. Especially for countertops, it is important to choose a color that does not confound or bias evaluations. A white paper or fabric on a black benchtop will show visual flaws more dramatically, thus biasing both visual and tactile evaluations.

3.2.4.3.3 Plumbing

Product trapped in pipes causes distracting and confounding odors in a sensory laboratory. It is essential that all pipes and drains open to the testing room can be cleaned and flushed. If spit sinks are necessary for some tests (toothpaste, mouthwash), thought should be given to having them detachable, that is, connected by flexible hose to water inlet and drain. When the sinks are not in use, they can be stored separately and the pipes can be closed off with caps.

3.3 PRODUCT CONTROLS

3.3.1 General Equipment

When a sensory evaluation test is conducted, the product researcher and the sensory analyst are looking for some treatment effect: effect of an ingredient change, a processing variable, a packaging change, a storage variable, and so on. One of the primary responsibilities of the sensory analyst is to control the early handling, the preparation, and the presentation of each product. These controls ensure that extraneous variables are not introduced and that no real treatment variables are obscured.

The preparation area should be situated adjacent to the test area. However, the air handling system should be structured so that the test area has positive pressure that feeds into the preparation area, which in turn contains the air return system as well as a supplementary exhaust.

3.3.2 Sample Preparation

3.3.2.1 Supplies and Equipment

In addition to the necessary major appliances, the controlled preparation of products requires adequate supplies and equipment, such as

- Scales, for weighing products and ingredients
- Glassware, for measurement and storage of products
- Timers, for monitoring of preparation procedures
- Stainless-steel and glass equipment, for mixing and storing products, and so on

3.3.2.2 Materials

Equipment used for the preparation and presentation of samples must be carefully selected to reduce the introduction of biases and new variables. Most plastic cutlery, storage containers, and wraps or bags are unsuitable for the preparation and storage of foods, beverages, or personal care products. The transfer of volatiles to and from the plastic can change the aroma and/or flavor characteristics of a product.

Wooden materials should not be used for cutting boards, bowls, mixing utensils, or pastry boards. They are porous and absorb aqueous and oil-based materials, which are then easily transferred from the wood to the next product that the wood contacts.

Containers used for storage, preparation, or serving should therefore be glass, glazed china, or stainless steel because of the reduced transfer of volatiles with these materials. Plastic, which has been pretested for low odor transfer, should be used only when the test product(s) will be held for less than 10 min in the container during and prior to the test.

3.3.2.3 Preparation Procedures

The controlled preparation of products requires careful regulation and monitoring of procedures used, with attention given to

- Amount of product to be used, measured by weight or volume using precise equipment (volumetric cylinders, gram scales, etc.)
- Amount of each added ingredient (as above)
- The process of preparation, regulation of time (stopwatch), and temperature (thermometers)
- Holding time, defined as the minimum and maximum time after preparation that a product can be used for a sensory test

3.3.3 Sample Presentation

3.3.3.1 Container, Sample Size, and Other Particulars

The equipment and procedures used for product presentation during the test must be carefully selected to reduce introduction of biases and new variables. Attention should be given to control of the following:

Serving containers. Again, these are preferably glass or glazed china, not plastic unless tested for lack of odor/flavor/fragrance transfer under the time and temperature conditions of the test.

Serving size. Extreme care must be given to regulating the precise amount of product to be given to each subject. Technicians should be carefully trained to deliver the correct amount of product with the least amount of handling. Special equipment may be advantageous for measuring precise amounts of a product for sensory testing.

Serving matrix. For most difference tests, the product under test is presented on its own, without additives. Products such as coffee, tea, peanut butter, vegetables, meats, and so on, are served without condiments or other adjuncts that may normally be used by consumers, such as milk, bread, butter, spices, and so on. In contrast, for consumer tests (preference/acceptance tests), products should be presented as normally consumed: coffee or tea with milk, sugar, or lemon, as required; peanut butter with bread or crackers; and vegetables and meat with spices, according to the consumer's preference. Products that are normally tasted in or on other products (condiments, dressings, sauces, etc.) should be evaluated in or on a uniform carrier that does not mask the product characteristics. These include a flour roux (a cooked flour-and-water base used for sauces), a fondant (sugared candy base), and sweetened milk (for vanilla and similar spices and flavorings).

Serving temperature. After the sample is distributed into each serving container, and just before serving, the product should be checked to determine if it is at the appropriate temperature. Most sensory laboratories develop standard preparation procedures that determine the needed temperature in the preparation container necessary to ensure the required temperature after delivery to the tasting/smelling container. The use of standard procedures greatly reduces the need for monitoring of each individual portion.

3.3.3.2 Order, Coding, and Number of Samples

As part of any test, the order, coding, and number of samples presented to each subject must be monitored.

The order of presentation should be "balanced" so that each sample appears in a given position an equal number of times. For example, these are the possible positions for three products, A, B, and C, to be compared in a ranking test:

$$\text{ABC—ACB—BCA—BAC—CBA—CAB} \tag{3.1}$$

Such a test should be set up with a number of subjects that is a multiple of six so as to permit presentation of the six possible combinations an equal number of times (see Chapter 4). The presentation also can be random, which may be achieved by drawing sample cards from a bag or by using a compilation of random numbers (see Table 19.1). Labels can be printed from a computer to make the sample labeling easier. Odorous tape or odorous markers should never be used to label sample containers.

The codes assigned to each product can be biasing; for example, subjects may subconsciously choose samples marked *A* over those marked with other letters. Therefore, single and double letters and digits are best avoided. In addition, letters or numbers that represent companies, area codes, and test numbers or samples should not be used. Most sensory analysts rely on the table of three-digit random numbers for product coding. Codes should not be very prominent, either on the product or on the score sheet. They can be clearly yet discreetly placed on the samples and score sheets to reduce confusion as to sample identification and to simultaneously reduce potential biases.

The number of samples that can be presented in a given session is a function of both sensory and mental fatigue in the subject. With cookies or biscuits, eight or ten may be the upper limit, while with beer, burnout may occur with six or eight samples. Products with a high carryover of flavor, such as smoked or spicy meats, bitter substances, or greasy textures, may allow only one or two samples per test. On the other hand, visual evaluations can be done on series of 20–30 samples, with mental fatigue as the limiting factor.

3.3.4 Product Sampling

The sensory analyst should determine how much of a product is required for evaluation and should know the history of the products to be tested. Information about prior handling of experimental and control samples is important in the design of the test and interpretation of the results. A log book should be kept in the sensory laboratory to record pertinent sample data:

- The source of the product: When and where it was made? Sample identification is necessary for laboratory samples (lab notebook number) as well as production samples (date and machine codes).
- The testing needs: How much product will be required for all of the tests to be run, and possibly rerun, for this evaluation? All of the product representing a sample should come from one source (same place, same line, same date, etc.). If the product is not uniform, attempts should be made to blend and repackage the different batches.
- The storage: Where has the sample been and under what conditions? If two products are to be compared for a processing or ingredient variable, it is not possible to measure the treatment effect if there are differences in age, storage temperature and humidity, shipping storage and humidity, packaging differences, and so on; these factors can cloud the measurement.

3.4 PANELIST CONTROLS

The way in which a panelist interacts with the environment, the product, and the test procedure are all potential sources of variation in the test design. Control or regulation of these interactions is essential to minimizing the extraneous variables that may potentially bias the results.

3.4.1 Panel Training or Orientation

Panelists, of course, need careful instruction with respect to the handling of samples, the use of the score sheet, and the information sought in the test. The training of panelists is discussed in detail in Chapter 10. At a minimum, panelists must be prepared to participate in a laboratory sensory test with no instruction from the sensory analysts after the test has started. They should be thoroughly familiar with

- The test procedures, such as the amount of sample to be tasted at one time; the delivery system (spoon, cup, sip, slurp); the length of time of contact with the product (sip/spit, short sniff, one bite/chew); and the disposition of the product (swallow, expectorate, leave in contact with skin or remove from skin) distance from and amount of an odor source must be predetermined and adhered to by all panelists.
- The score-sheet design, including instructions for evaluation, questions, terminology, and scales for expressing judgment must be understood and familiar to all panelists.
- The type of judgment/evaluation required (difference, description, preference, acceptance) should be understood by the panelists as part of their test orientation.

3.4.2 Product/Time of Day

With panelists who are not highly trained, it is wise to schedule the evaluation of certain product types at the time of day when that product is normally used or consumed. The tasting of highly flavored or alcoholic products in the early morning is not recommended. Product testing just after meals or coffee breaks also may introduce bias and should be avoided. Some preconditioning of the panelists' skin, hair, nose or mouth may be necessary to improve the consistency of verdicts.

3.4.3 Panelists/Environment

As discussed in Section 3.2, the test environment, as seen by the panelist, must be controlled if biases are to be avoided. Note, however, that certain controls such as colored lights, high humidity, or an enclosed testing area may cause anxiety or distraction unless panelists are given ample opportunity to become acclimated to such "different" surroundings.

Again, it is necessary to prepare panelists for what they are to expect in the actual test situation, to give them the orientation and time to feel comfortable with the test protocols, and to provide them with enough information to respond properly to the variables under study.

REFERENCES

Amerine, M. A., R. M. Pangborn, and E. B. Roessler (1965). *Principles of Sensory Evaluation of Food.* New York: Academic Press.

Caul, J. F. (1957). The profile method of flavor analysis. *Adv Food Res* 7: 1–40.

Chambers, E., and M. Baker Wolf (Eds.) (1996). *Sensory Testing Methods* (2nd edn.). ASTM Manual 26. West Conshohocken, PA: ASTM International.

European Cooperation for Accreditation of Laboratories (1995). EAL-G16, Accreditation for sensory testing laboratories. Available from national members of EAL, e.g., in the UK NAMAS, tel. 44 181 943-7068; fax 44 181 943-7134.

Gatchalian, M. M. (1981). *Sensory Evaluation Methods with Statistical Analysis*. Diliman: University of the Philippines, College of Home Economics.

International Organization for Standardization (ISO) (1998). Sensory analysis—General guidance for the design of test rooms. In *International Standard ISO 8589*. Geneva, Switzerland: International Organization for Standardization.

Kelly, F. B., and H. Heymann (1989). Contrasting the effects of ingestion and expectoration in sensory difference tests. *J Sens Stud* 3: 4, 249.

Kuesten, C. K., and L. Kruse (Eds.) (2008). *Physical Requirement Guidelines for Sensory Evaluation Laboratories* (2nd edn.). ASTM MNL 60. West Conshohocken, PA: ASTM International.

Malek, D. M., D. J. Schmitt, and J. H. Munroe (1982). A rapid system for scoring and analyzing sensory data. *J Am Soc Brew Chem* 40: 133.

Pangborn, R. M. (1967). Use and misuse of sensory methodology. *Food Qual Contr* 15: 7–12.

Poste, L. M., D. A. Mackie, G. Butler, and E. Larmond (1991). *Laboratory Methods for Sensory Analysis of Food. Publication 1864/E* (pp. 4–13). Ottawa: Agriculture Canada.

4

Factors Influencing Sensory Verdicts

4.1 INTRODUCTION

Good sensory measurements require that we look at the tasters as measuring instruments that are somewhat variable over time and among themselves and are very prone to bias. To minimize variability and bias, the experimenter must understand the basic physiological and psychological factors that may influence sensory perception. Gregson (1963) notes that perception of the real world is not a passive process but an active and selective one. An observer records only those elements of a complex situation that he can readily see and interpret as meaningful. The rest he eliminates, even if it is staring him in the face. The observer must be put in a frame of mind to understand the characteristics that he or she is to measure. This is done through training (see Chapter 10) and by avoiding a number of pitfalls (Amerine et al. 1965; Pangborn, 1979; Poste et al., 1991; Lawless and Heymann, 2010) inherent in the presentation of samples, the text of the questionnaire, and the handling of the participants.

4.2 PHYSIOLOGICAL FACTORS

4.2.1 Adaptation

Adaptation is a decrease in or change in sensitivity to a given stimulus as a result of continued exposure to that stimulus or a similar one. In sensory testing, this effect is an important unwanted source of variability of thresholds and intensity ratings.

In the following example of "cross-adaptation" (O'Mahony, 1986), the observer in condition B is likely to perceive less sweetness in the test sample because the tasting of sucrose reduces his sensitivity to sweetness:

	Adapting Stimulus	Test Stimulus
Condition A	H_2O	Aspartame
Condition B	Sucrose	Aspartame

The water used in condition A contains no sweetness and does not fatigue (or cause adaptation in the perception of sweet taste).

Condition A	H_2O	Quinine
Condition B	Sucrose	Quinine

Here, "cross-potentiation," or facilitation, is likely to occur. In condition B, the observer perceives more bitterness in the test sample because the tasting of sucrose has heightened his sensitivity to quinine. A detailed discussion of adaptation phenomena in sensory testing is given by O'Mahony (1986).

4.2.2 Enhancement or Suppression

Enhancement or suppression involves the interaction of stimuli presented simultaneously as mixtures.

Enhancement. The effect of the presence of one substance increasing the perceived intensity of a second substance.

Synergy. The effect of the presence of one substance increasing the perceived combined intensity of two substances, such that the perceived intensity of the mixture is greater than the sum of the intensities of the components.

Suppression. The effect of the presence of one substance decreasing the perceived intensity of a mixture of two or more substances.

Examples (see key below):

1. Total Perceived Intensity of Mixture

Situation	Name of Effect
MIX < A + B (each alone)	Mixture suppression
MIX > A + B (each alone)	Synergy

2. Components of Analyzable Mixture

Situation	Name of Effect
A′ < A	Mixture suppression
A′ > A	Enhancement

Key: MIX, perceived intensity of mixture; A, perceived intensity of unmixed component A; A′, perceived intensity of component A in mixture.

4.3 PSYCHOLOGICAL FACTORS

4.3.1 Expectation Error

Information given with the sample may trigger preconceived ideas. One usually finds what one expects to find. In testing, such as the classic tests for thresholds that consist of a series of ascending concentrations, the subject (through autosuggestion) anticipates the sensation and reports his response before it is applicable. A panelist who hears that an overage product has been returned to the plant will have a tendency to detect aged flavors in the samples of the day. A beer taster's verdict of bitterness will be biased if he knows the hop rate employed. Expectation errors can destroy the validity of a test and must be avoided by keeping the source of samples a secret and by not giving panelists any detailed information in advance of the test. Samples should be coded, and the order of presentation should be balanced among the participants. Theoretically, well-trained panelists would not be influenced by accidental knowledge about a sample. In practice, however, such information could influence the subject's ratings; thus, it is preferable for him/her to be ignorant of the history of the sample.

4.3.2 Error of Habituation

Humans have been described as creatures of habit. This description holds true in the sensory world and leads to an error, the error of habituation. There is a tendency to continue to give the same response when a series of slowly increasing or decreasing stimuli are presented, for example, in quality control from day to day. The panelist may be apt to repeat the same scores and hence to miss any developing trends, or even to accept an occasional defective sample. Habituation is common and can be counteracted by varying the types of product or by presenting doctored samples.

4.3.3 Stimulus Error

This error is caused when irrelevant criteria, such as the style or color of the container, influence the observer. If the criteria suggest differences, the panelist will find them even when they do not exist. For example, Amerine et al. (1965) presented an example in which tasters, knowing that wines in screw-capped bottles were, at that time, usually less expensive, may produce lower ratings when served from such bottles than if served from cork-closure bottles. Urgently-called panel sessions may trigger reports of known production defects. Samples served late in a test may be rated more flavorful because panelists know that the panel leader will present light-flavored samples first to minimize fatigue. The remedies in these cases are obvious: Avoid leaving irrelevant (as well as relevant) cues, schedule panel sessions regularly, and make frequent and irregular departures from any usual order or manner of presentation.

4.3.4 Logical Error

Logical errors occur when two or more characteristics of the samples are associated in the minds of the assessors. Knowledge that a darker beer tends to be more flavorful or that darker mayonnaise tends to be stale causes the observer to modify his verdict, thus disregarding his own perceptions. Logical errors must be minimized by keeping the samples uniform and by masking differences with the aid of colored glasses, colored lights, and so on. Certain logical errors cannot be masked but may be avoided in other ways; for example, a more bitter beer will always tend to receive a higher score for hop aroma. With trained panelists, the leader may attempt to break the logical association by occasionally doctoring a sample with quinine to produce high bitterness combined with low hop aroma.

4.3.5 Halo Effect

When more than one attribute of a sample is evaluated, the ratings will tend to influence one another (the halo effect). Simultaneous scoring of various flavor aspects along with overall acceptability can produce different results than if each characteristic is evaluated separately. For example, in a consumer test of orange juice, subjects are asked not only to rate their overall liking but also to rate specific attributes. When the product is generally well liked, all of its various aspects—sweetness, acidity, fresh orange character, flavor strength, mouthfeel—tend to be rated favorably as well. Conversely, if the product is not well liked, most of the attributes will be rated unfavorably. The remedy, when any particular variable is important, is to present separate sets of samples for evaluation of that characteristic.

4.3.6 Order of Presentation of Samples

At least five types of bias may be caused by the order of presentation.

Contrast effect. Presentation of a sample of good quality just before one of poor quality may cause the second sample to receive a lower rating than if it had been rated monadically (i.e., as a single sample). As an example, if one lives in Minneapolis in the winter and the thermometer hits 40°F, the city is having a heat wave. If one lives in Miami and the thermometer registers 40°F, the news media will report a severe cold spell. The converse is also true: A sample that follows a particularly poor one will tend to be rated higher.

Group effect. One good sample presented in a group of poor samples will tend to be rated lower than if presented on its own. This effect is the opposite of the contrast effect.

Error of central tendency. Samples placed near the center of a set tend to be preferred over those placed at the ends. In triangle tests, the odd sample is detected more often if it is in the middle position. (An error of central tendency is also found with scales and categories; see Chapter 5.)

Pattern effect. Panelists will use all available clues (this, of course, is legitimate on their part) and are quick to detect any pattern in the order of presentation.

Time error/positional bias. One's attitude undergoes subtle changes over a series of tests, from anticipation or even hunger for the first sample, to fatigue or indifference with the last. Often, the first sample is abnormally preferred (or rejected). A short-term test (sip and evaluate) will yield a bias for the sample presented first, whereas a long-term test (1-week home placement) will produce a bias for the sample presented last. Discrimination is greater with the first pair in a set than with subsequent pairs.

All of these effects must be minimized by the use of a balanced or randomized order of presentation. *Balanced* means that each of the possible combinations is presented an equal number of times. Each sample in a panel session should appear an equal number of times in first, second … and *n*th position. If there are large numbers of samples to be presented, a balanced incomplete block design can be used (see Sections 8.5.2 and 14.5.4).

Randomized means that the order in which the selected combinations appear was chosen according to the laws of chance. In practice, randomization is obtained by drawing sample cards from a bag, or it may be planned with the aid of a compilation of random numbers (see Table 19.1, also Product Controls in Section 3.3).

Computer programs for developing balanced randomized serving plans are also available, for example, from Qi Statistics (2001).

4.3.7 Mutual Suggestion

The response of a panelist can be influenced by other panelists. Because of this, panelists are separated in booths, thus preventing a judge from reacting to the facial expression registered by another judge. Vocalizing an opinion in reaction to samples is not permitted. The testing area should also be free from noise and distraction and separate from the preparation area.

4.3.8 Lack of Motivation

The degree of effort a panelist will make to discern a subtle difference, to search for the proper term for a given impression, or to be consistent in assigning scores is of decisive importance for the results. It is the responsibility of the panel leader to create an atmosphere in which assessors feel comfortable and do a good job. An interested panelist is always more efficient. Motivation is best in a well-understood, well-defined test situation. The interest of panelists can be maintained by giving them reports of their results. Panelists should be made to feel that the panels are an important activity. This can be subtly accomplished by running the tests in a controlled, efficient manner.

4.3.9 Capriciousness versus Timidity

Some people tend to use the extremes of any scale, thereby exerting more than their share of influence over the panel's results. Others tend to stick to the central part of the scale and to minimize differences between samples. To obtain reproducible, meaningful results, the panel leader should monitor new panelists' scores on a daily basis, giving guidance in the form of typical samples already evaluated by the panel and, if necessary, using doctored samples as illustrations and reference samples that represent the lower and upper ends of intensity scales.

4.4 POOR PHYSICAL CONDITION

Panelists should be excused from sessions (1) if they suffer from fever or the common cold in the case of tasters, and if they suffer from skin or nervous system disorders in the case of a tactile panel; (2) if they suffer from poor dental hygiene or gingivitis; (3) in the case of emotional upset or heavy pressure of work that prevents them from concentrating (conversely, panel work can be an oasis in a frantic day); and (4) if they currently take any medications that will interfere with their ability to taste or smell (numerous drugs have adverse effect that impacts patient's sense of taste and smell [Doty et al., 2008]). Additionally, aging affects sensory sensitivity, and elderly people can be less sensitive to different tastes and smells (Fukunaga et al., 2005; Wylie and Nebauer, 2011) (see Chapter 10). Smokers can be good tasters but should refrain from smoking for 30–60 min before a panel. Strong coffee can affect perception for up to an hour. Tasting should not take place in the first 2 h after a major meal. Therefore, the optimal time for panel work (for persons on the day shift) is between 10:00 a.m. and lunch. Generally, the best time for an individual panelist depends on his biorhythm: It is that time of the day when one is most awake and one's mental powers are at their peak. Mattes (1986) reviews the many ways in which health or nutrition disorders affect sensory function and, conversely, how sensory defects can be used in the diagnosis of health or nutrition disorders. Taste and smell dysfunctions are associated with chronic diseases (Henkin et al., 2013), and adaptation of the sensitivity of certain taste receptors may induce conditions that are related to health issues such as obesity, diabetes, and so on. (Depoortere, 2014). Thus, physical condition is very important for sensory evaluation, and panelists in poor physical condition may not produce accurate test results.

REFERENCES

Amerine, M. A., R. M. Pangborn, and E. B. Roessler (1965). *Principles of Sensory Evaluation of Food.* New York: Academic Press.

Depoortere, I. (2014). Taste receptors of the gut: Emerging roles in health and disease. *Gut* 63, no. 1: 179.

Doty, R. L., M. Shah, and S. M. Bromley. (2008). Drug-induced taste disorders. *Drug Saf* 31(3): 199–215.

Fukunaga, A., H. Uematsu, and K. Sugimoto (2005). Influences of aging on taste perception and oral somatic sensation. *J Gerontol Ser A: Biol Sci Med Sci* 60A(1): 109.

Gregson, R. A. M. (1963). The effect of psychological conditions on preference for taste mixtures. *Food Tech* 17(3): 44.

Henkin, R. I., L. M. Levy, and A. Fordyce (2013). Taste and smell function in chronic disease: A review of clinical and biochemical evaluations of taste and smell dysfunction in over 5000 patients at The Taste and Smell Clinic in Washington, DC. *Am J Otolaryngol* 34(5): 477–89.

Lawless, H. T., and H. Heymann (2010). *Sensory Evaluation of Food: Principles and Practices* (2nd edn.). New York: Chapman & Hall.

Mattes, R. D. (1986). Effects of health disorders and poor nutritional status on gustatory function. *J Sens Stud* 1(3–4): 225.

O'Mahony, M. (1986). Sensory adaptation. *J Sens Stud* 1(3–4): 237.

Pangborn, R. M. (1979). Physiological and psychological misadventures in sensory measurement or the crocodiles are coming. In M. R. Johnston (ed.), *Sensory Evaluation Methods for the Practicing Food Technologist* (pp. 2–22). Chicago: Institute of Food Technologists.

Poste, L. M., D. A. Mackie, G. Butler, and E. Larmond (1991). *Larmond Laboratory Methods for Sensory Analysis of Food, Publication 1864/E. 4–13.* Ottawa: Agriculture Canada.

Qi Statistics (2001). *Design Express Version 1.0 Reference Manual.* Reading, UK: Qi Statistics.

Wylie, K., and M. Nebauer (2011). "The food here is tasteless!" Food taste or tasteless food? Chemosensory loss and the politics of under-nutrition. *Collegian* 18: 27–35.

5

Measuring Responses

5.1 INTRODUCTION

This chapter describes the various ways in which sensory responses can be measured. The purpose is to present the principle of each method of measuring responses and to discuss its advantages and disadvantages. For a detailed critical review of this point, see Doty and Laing (2003).

In the simplest of worlds, if tasters were really measuring instruments, they could be set up with a range of 0–100 and be supplied with a couple of calibration points (doctored samples) for each attribute to be rated. Unfortunately, the real world of testing is not simple, and a much more varied approach is needed. The degree of complexity is such that the psychology departments of major universities maintain laboratories of psychophysics (Doty, 2003; Moskowitz, 2002; Lawless and Heymann, 1998; Laming, 1994; Sekuler and Blake, 1990; Cardello and Maller, 1987; Baird and Noma, 1978; Anderson, 1974; Kling and Riggs, 1971). Some of the factors to consider are outlined in this chapter.

When panelists are asked to assign numbers or labels to sensory impressions, they may do this in at least four ways (see Figure 5.1):

- Nominal data (Latin: *nomen* = name): The items examined are placed in two or more groups that differ in name but do not obey any particular order or any quantitative relationship; for example, the numbers worn by football players.
- Ordinal data (Latin: *ordinalis* = order): The panelist places the items examined into two or more groups that belong to an ordered series; for example, slight, moderate, strong.
- Interval data (Latin: *inter vallum* = space between ramparts): Panelists place the items into numbered groups separated by a constant interval; for example, three, four, five, six.
- Ratio data: Panelists use numbers that indicate by what multiple the stimulus in question is stronger (or saltier, or more irritating) than a reference stimulus presented earlier.

Nominal data contains the least information. Ordinal data carries more information and can be analyzed by most nonparametric statistical tests. Interval and ratio data are

Figure 5.1 Pictorial illustration of scales. The names of the three food items (apple, pear, banana) provide nominal data. In the example of ordinal data, three rye breads are ranked from greatest to least number of caraway seeds. The three beverages form an interval scale in that they are separated by constant intervals of one unit of sucrose. In the last example, two volatiles from three cups of coffee are measured on a GC, and it is established that the first cup contains 3/4 of the volatiles of the second cup and only 1/2 of the volatiles of the third. Note that the illustration shows physical/ chemical scales. A panelist's sensory scales may be different; for example, the sweetness of sugar increases less from 4 to 5 lumps than it does from 3 to 4 lumps. (From Cardello, A. V. and O. Maller [1987]. *Objective Methods in Food Quality Assessment,* ed. J.G. Kapsalis. Boca Raton, FL: CRC Press. With permission.)

even better because they can be analyzed by all nonparametric methods and often by parametric methods. Ratio data is preferred by some because it is free from end-of-scale distortions; however, in practice, interval data, which is easier to collect, appears to give equal results (see Section 5.6.3).

The most frequently used methods of measuring sensory response to a sample are, in order of increasing complexity:

- Classification: The items evaluated are sorted into groups that differ in a nominal manner; for example, marbles sorted by color.
- Grading: Time-honored methods used in commerce, which depend on expert graders who learn their craft from other graders; for example, "USDA Choice" grade of meat.
- Ranking: The samples (usually from three to seven) are arranged in order of intensity or degree of some specified attribute; the scale used is ordinal.
- Scaling: The subjects judge the sample by reference to a scale of numbers (often from zero to ten) that they have been trained to use. Category scaling (CS) yields ordinal data or sometimes interval data; line scales usually yield interval data; and magnitude estimation, although designed to yield ratio data, in practice seems to produce mixed interval/ratio data.

A further method, the use of odor units based on thresholds, will be discussed in Chapter 9. In choosing among these methods and training the panel to use them, the practicing panel leader needs to understand and then address the two major sources of variation in panel data: (1) the differences in the test subjects' perceptions of the stimulus and (2) the differences in the expression of those perceptions by the subjects (see Chapter 1).

Actual differences in perception are part of the considerable variability in sensory data that sensory analysts learn to live with and psychophysicists learn to measure. Sensory thresholds vary from one person to another (Pangborn, 1981; Doty and Laing, 2003; Meilgaard, 1993). In a study of difference thresholds for substances added to beer, it was found that panels of 20 trained tasters tend to contain two who exhibit a threshold four times lower than the median for the panel, and two who exhibit threshold five times higher than the median. In a study of 200+ healthy, untrained individuals (excluding anosmics), Amoore (1977) found differences in sensitivity of 1000-fold between the most and the least sensitive. It follows that the verdict of a small panel of four or seven people can be highly variant with respect to the general population, hence the tendency in this book to recommend panel sizes of 20–30, or preferably many more. A small panel is representative only of itself or the population it was specifically screened to represent.

The second source of variation, the way in which the subjects express a given sensory impression, can be many times greater again, but luckily it can be minimized by careful selection of sensitive panelists and thorough training and by careful selection of the terminology and scaling techniques provided to panelists. The literature is replete with examples of sensory verdicts that can only be explained by assuming that many panel members were quite "at sea" during the test; they probably did perceive the attribute under study. However, they did not have a clear picture in their mind of what aspect they were asked to measure, and/or they were unfamiliar with the mechanics of the test, and/or they did not understand how to express the impression.

In choosing a way of measuring responses, the sensory analyst should generally select the simplest sensory method that will measure the expected differences between the samples, thus minimizing panel training time. Occasionally, a more complex method will be employed that uses more terminology and more sophisticated scales, thus requiring more training and evaluation time. For example, there may be sample differences that were not taken into account at the planning stage and that would have been missed with the simpler method. Overall training time may end up being less because once the panel has reached the higher level of training, it can tackle many types of samples without the need for separate training sessions for each.

5.2 PSYCHOPHYSICAL THEORY

Psychophysics is a branch of experimental psychology devoted to studying the relationships between sensory stimuli and human responses, that is, to improving understanding of how the human sensory system works. University psychophysicists are constantly refining the methods by which a response can be measured, and sensory analysts need to study their techniques and cooperate in their experiments. This chapter will provide an overview of psychophysics as applied to sensory testing. Those interested in more detail should read the references listed at the beginning of Section 5.1; see also Lawless (1990).

A major focus of psychophysics is to discover the form of the psychophysical function: the relationship between a stimulus, C, and the resulting sensation, R, preferably expressed as a mathematical function, $R = f(C)$ (see Figure 5.2).

While the stimulus is either known (an added concentration) or easy to measure (a peak height, an Instron reading), it is the sensation that causes difficulty. The subject must be asked questions and given instructions such as

Judge this odor on a scale of 0–99;
Is this sensation 2× as strong or 3× as strong?
Which of these solutions has the strongest taste of quinine?

No one, however, can answer such questions reproducibly and precisely. A variety of experimental techniques are being used; for example, comparison with a second, better known sensation such as the loudness of a tone (this is called cross-modality matching, see p. 59), or direct electrical measurement of the nerve impulse generated in the chorda tympani (taste nerve) in persons undergoing inner-ear operations (Borg et al., 1967).

Over the past century, two forms of the psychophysical function have been used: Fechner's law and Stevens' law. Although neither is perfect, each (when used within its limits of validity) provides a much better guide for experiment design than simple intuition. For a thorough discussion of the two, see Lawless and Heymann (1998). Two other reviews of psychophysical theory, Laming (1994) and Norwich and Wong (1997), include worthwhile attempts at reconciling Fechner's and Stevens' laws. More recently, the Michaelis–Menten equation known from enzyme chemistry, or the Beidler equation derived from it, have been used to model the dose–response relationship (see Section 5.2.3 below and Chastrette et al., 1998).

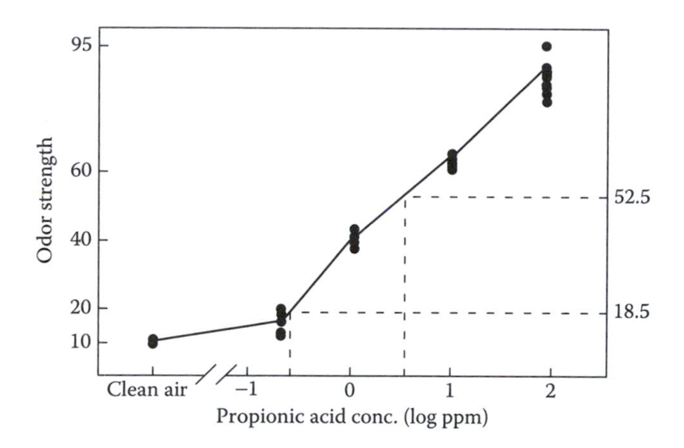

Figure 5.2 Example of a psychophysical function. Odor strength was rated 0–99 with zero = no odor or nasal irritation sensation. (From Kendal-Reed et al. [1998]. *Chem Sens*, vol. 23, Oxford: Oxford University Press, 71–82. With permission.)

5.2.1 Fechner's Law

Fechner (1860) selected as his measure of the strength of sensation the just-noticeable difference (JND; see Figure 5.3). For example, he would regard a perceived sensation of eight JNDs as twice as strong as a sensation of four JNDs. JNDs had just become accessible to measurement through difference testing, which Fechner learned from Ernst Weber at the University of Leipzig in the mid-1800s. Weber found (1834) that difference thresholds increase in proportion to the initial perceived absolute stimulus intensity at which they are measured:

$$\frac{\Delta C}{C} = k(\text{Weber's law}), \tag{5.1}$$

where C is the absolute intensity of the stimulus; for example, concentration, ΔC, is the change in intensity of the stimulus that is necessary for one JND, and k is a constant, usually between zero and one. Weber's law states, for example, that the amount of an added flavor that is just detectable depends upon the amount of that added flavor that is already present. If k has been determined, one can calculate how much extra flavorant is needed.

The actual derivation of Fechner's law,

$$R = k \log C \,(\text{Fechner's law}), \tag{5.2}$$

is complex and depends upon a number of assumptions, some of which may not hold (Norwich and Wong, 1997). Support for Fechner's law is provided by common CS. When panelists score a number of samples that vary along one dimension (e.g., sweetness) using a scale such as from zero to nine, the results plot out as a logarithmic curve similar to that of Figure 5.3. One tangible outcome of Fechner's theories was a logarithmic scale of sound intensity, the decibel (dB) scale.

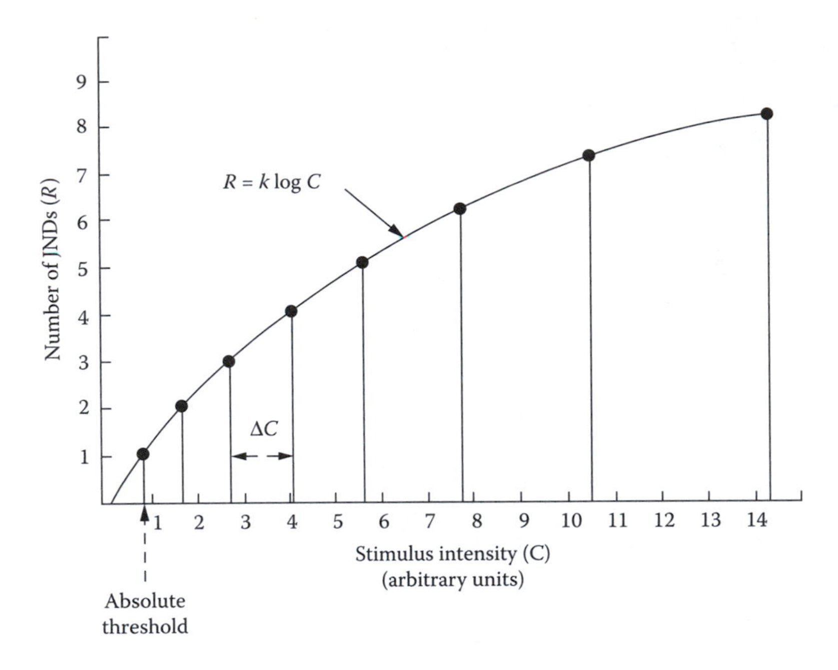

Figure 5.3 Derivation of Fechner's law by the method of summing JNDs. (Adapted from Cardello, A. V., and O. Maller [1987]. *Objective Methods in Food Quality Assessment*, ed. J.G. Kapsalis. Boca Raton, FL: CRC Press. With permission.)

5.2.2 Stevens' Law

S.S. Stevens, working at Harvard a century after Fechner, pointed out that if Equation 5.2 were correct, a tone of 100 dB should only sound twice as loud as one of 50 dB. He then showed, with the aid of magnitude estimation scaling (see p. 58), that subjects found the 100 dB tone to be 40 times as loud as the one of 50 dB (Stevens, 1970). Stevens' main contention (1957)—that perceived sensation magnitude grows as a power function of stimulus intensity—can be expressed mathematically as

$$R = k\,C^n \text{ (Stevens' power law)}, \tag{5.3}$$

where k is a constant that depends on the units in which R and C are measured, and n is the exponent of the power function, that is, n is a measure of the rate of growth of perceived intensity as a function of stimulus intensity.

Figure 5.4 shows power functions with n = 0.5, 1.0, and 1.5; and Table 5.1 lists typical exponents for a variety of sensory attributes. The finding that the exponent for visual length is 1.0, that is, simple proportionality, has led to the common use of line scales for rating sensory intensity (Einstein, 1976).

When n is larger than 1.0, the perceived sensation grows faster than the stimulus; an extreme example is electric shock (Table 5.1). Conversely, when n is smaller than 1.0, as for many odors, the sensation grows more slowly than the stimulus, and a curve results that is superficially similar to Figure 5.3.

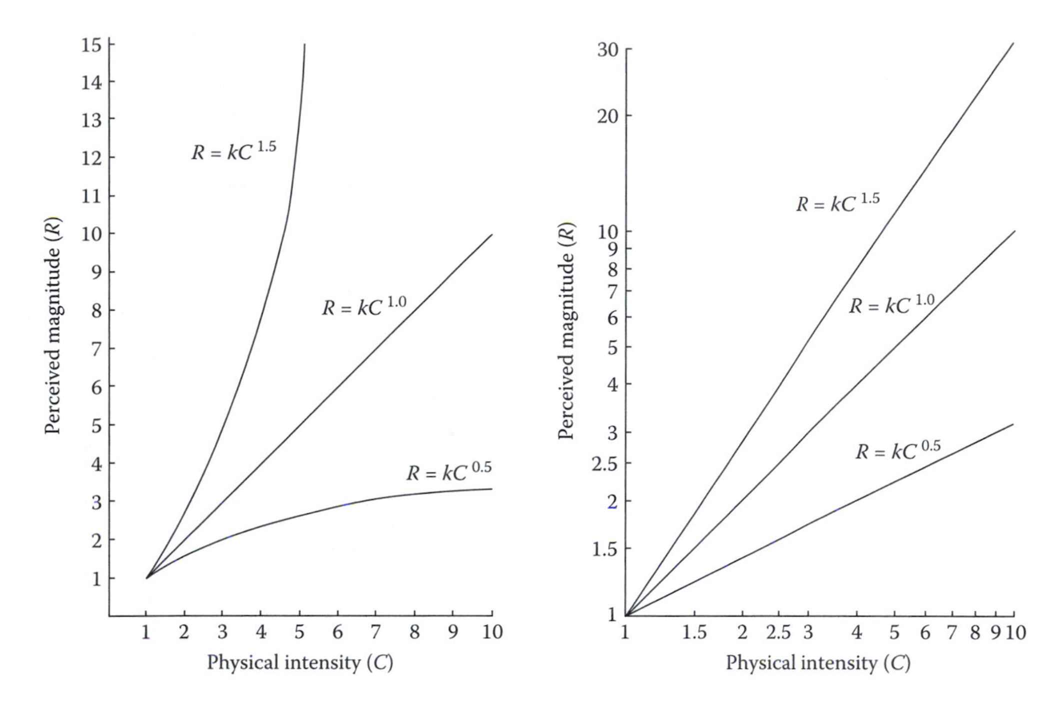

Figure 5.4 Plots of power functions with $k = 1$ and $n = 0.5, 1.0,$ and 1.5 in linear (left) and logarithmic (right) coordinates. (Adapted from Cardello, A. V., Maller, O. [1987]. *Objective Methods in Food Quality Assessment*, ed. J.G. Kapsalis. Boca Raton, FL: CRC Press. With permission.)

Stevens proposed that only ratio scales are valid for the measurement of perceived sensation, and his magnitude estimation scales are widely used in psychophysical laboratories. However, many authors have pointed out (Cardello and Maller, 1987) that for the sensory evaluation of foods and fragrances, there are serious shortcomings with these scales. The exponents vary with the range of stimuli in the test and with the modulus used; worse yet, the exponents differ greatly among investigators and among individuals because of the subjects' idiosyncratic use of numbers.

5.2.3 Beidler Model

The log function and the power function are merely mathematical equations that happen to fit observed sensory data. There is nothing physiological about them. McBride (1987) has suggested that the equation that follows, which Beidler (1954, 1974) derived from animal experiments and the Michaelis–Menten equation for the kinetics of enzyme–substrate relationships in biological systems, can be used to describe human taste response. McBride proposes that we move away from the dependence on subjects' use of numbers or scales and simply assume that human psychophysical response is proportional to the underlying neurophysiological response:

$$\frac{R}{R_{max}} = \frac{C}{k+C} \quad \text{(The Beidler equation)}, \tag{5.4}$$

Table 5.1 Representative Exponents of Power Functions for a Variety of Sensory Attributes

Attribute	Exponent	Stimulus
Bitter taste	0.65	Quinine, sipped
	0.32	Quinine, flowed
Brightness	0.33	5° field
Cold	1.0	Metal on arm
Duration	1.1	White noise
Electric shock	3.5	Current through fingers
Hardness	0.8	Squeezed rubber
Heaviness	1.45	Lifted weights
Lightness (visual)	1.20	Gray papers
Loudness	0.67	1000-Hz tone
Salt taste	1.4	NaCl, sipped
	0.78	NaCl, flowed
Smell	0.55	Coffee
	0.60	Heptane
Sour taste	1.00	HCl, sipped
Sweet taste	1.33	Sucrose, sipped
Tactual roughness	1.5	Emery cloths
Thermal pain	1.0	Radiant heat on skin
Vibration	0.95	60 Hz on finger
	0.6	250 Hz on finger
Viscosity	0.42	Stirring fluids
Visual area	0.7	Projected squares
Visual length	1.00	Projected line
Warmth	1.6	Metal on arm

Source: Cardello, A. V., Maller, O (1987). *Objective Methods in Food Quality Assessment*, ed. J.G. Kapsalis. Boca Raton, FL: CRC Press. With permission.

The equation states that the response, R, divided by the maximal response, R_{max}, shows a sigmoidal relationship to the stimulus, C (the molar concentration), when C is plotted on a logarithmic scale (see Figure 5.5). The constant, k, is the concentration at which the response is half-maximal. Beidler (1974) calls it the association constant, or binding constant, and notes that it can be seen as a measure of the affinity with which the stimulus molecule binds to the receptor. The Beidler model works best for the middle and high ranges of sensory impressions, for example, for the sweetness of sweet foods or beverages. Unlike Fechner's and Stevens' models, it assumes that the response has an upper limit, R_{max}, that is not exceeded, irrespective of the concentration of the stimulus. It is seen as that concentration when all the receptors are saturated.

McBride shows, with a number of examples for sugars, salt, citric acid, and caffeine, that the Beidler equation provides a good description of human taste response as obtained by two psychophysical methods: JND cumulation and category rating. Application of the

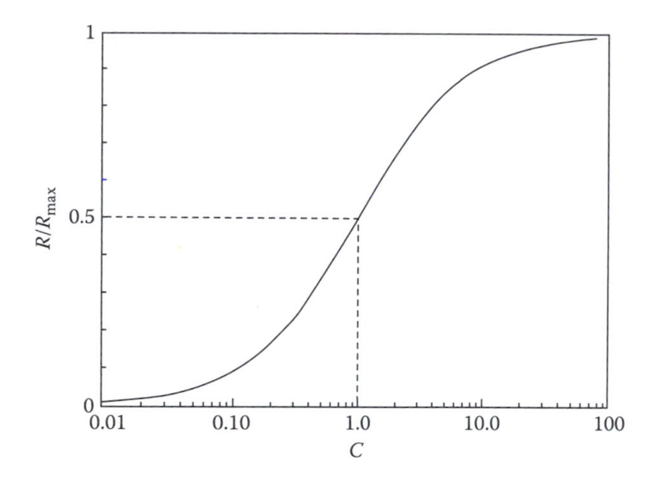

Figure 5.5 The sigmoidal relationship between taste response, R/R_{max}, and stimulus concentration, C, as specified by the Beidler equation; k is set equal to 1 for convenience. The inflexion point (maximum slope) of the curve occurs at $R = 0.5R_{max}$, when $C = k$.

Beidler equation allows the estimation of the hitherto unobtainable parameters for human taste response, R_{max} and k. Therefore, unlike the empirical Fechner and Stevens laws, the Beidler equation offers the potential for quantitative estimation of human taste response, that is, of the psychophysical function. Details of how this may be carried out for studies of the biophysics of the sensory mechanism are given by Beidler (1974), Maes (1985), and Chastrette et al. (1998).

Other techniques that are frequently used by psychophysicists to attempt to model the assessor's decision process are finding application in sensory evaluation, especially in threshold and discrimination testing. These other psychophysical models include the Thurston–Ura model and the signal detection model (see Lawless and Heymann, 1998; Chapter 5; and Section 9.3).

5.3 CLASSIFICATION

In classification tests, the subjects are asked to select an attribute or attributes that describe the stimulus. In a beverage test, for example, subjects place a mark next to the term(s) that best describe(s) the sample:

______ sweet	______ sour	______ lemony
______ blended	______ thick	______ refreshing
______ pulpy	______ natural	______ aftertaste

Alternatively, one may ask respondents to place a mark next to all the terms that describe a specific sample. The following is an example, where respondents are asked to perform such a task.

Following are some adjectives people use to describe the scent of products. Thinking about the lotion you just smelled, please indicate which of these descriptors apply to its scent. (Please check all that apply.)

The scent of this lotion is…

______ Classic	______ Feminine	______ Sporty
______ Sweet	______ Floral	______ Romantic
______ Sensual	______ Soapy	______ Light
______ Harsh	______ Heavy	______ Citrusy
______ Modern	______ Relaxing	______ Fruity
______ Fresh	______ Energizing	______ Crisp
______ Warm	______ Natural	

This is commonly referred to as check-all-that-apply or CATA information and is often used:

- When the list of descriptors is long
- As a screening tool
- When a researcher is looking for a more "in-the-moment" instinctive reaction to a product (for example, in the context of emotion research)

No attempt is made to standardize the terms, and the results are reported as the number of check marks for each term. Such data are nominal; no numbers are used, and there is no increasing or decreasing series expressed in the data. For example, the apples in a lot may be characterized by predominant color (red, green, and yellow).

The proper selection of the right terms is essential for the correct interpretation of the description of the stimulus. If panelists are not trained, as is the case with consumers, common nontechnical terms must be used. A source of confusion is that subjects often erroneously associate individual common terms with degrees of goodness or badness. The caveats below describe situations using classifications in which selection of the proper words/terms/classes is the critical first step. Selection of the best possible terminology is not only important in classification tests; it is needed in all measuring techniques that use a term or descriptor to define the perceived property being investigated.

The selection of sensory attributes and the corresponding definition of these attributes should be related closely to the real chemical and physical properties of a product that can be perceived. Adherence to an understanding of the actual rheology or chemistry of a product makes the data easier to interpret and more useful for decision making.

Caveats:

1. If a product has noticeable defects, such as staleness or rancidity, and terms to describe such defects have not been included in the list, panelists (especially if untrained) will use another term in the list to express the off note.
2. If a list of terms provided to panelists fails to mention some attribute that describes real differences between products, or which describes important characteristics in one product, panelists again will use another term from the list provided to

express what they perceive. This dumping phenomenon occurs when panelists cannot find the right or precise attribute (H. Lawless, pers. comm.).

Following are some examples of word lists that have been used for classification tasks or for subsequent rating tasks:

1. Afterfeel of skincare products, for example, soaps, lotions, creams: tacky, smooth, greasy, supple, grainy, waxy, oily, astringent, taut, dry, moist, and creamy. Note that no relationship is introduced between the attributes that may, in fact, be facets of the same parameter (moist/dry, smooth/grainy, etc.)
2. Spice notes (subjects may be asked to define which spices or herbs contribute to one overall spice complex): oregano, basil, thyme, sage, rosemary, marjoram, and/or clove, cinnamon, nutmeg, mace, cardamom.
3. Hair color/hair condition: panelists/hairdressers are asked to classify the hair color and hair condition of men and women who are to serve as subjects for half-head shampoos; such sorting may be necessary to balance all treatments.

For each subject, check the most appropriate descriptor(s) from each column:

Color of Hair	Condition of Hair
Blond	Healthy
Brown	Damaged
Red	Dull
Black	Split
Tinted/frosted	Oily scalp
	Dandruff

5.4 GRADING

Grading is a method of evaluation, used frequently in commerce, that depends on expert "graders" who learn the scale used from other graders. Scales usually have four or five steps such as "choice," "extra," "regular," and "reject." Examples of items subjected to sensory grading are coffee, tea, spices, butter, fish, and meat.

Sensory grading most often involves a process of integration of perceptions by the grader. The grader is asked to give one overall rating of the combined effect of the presence of the positive attributes, the blend or balance of those attributes, the absence of negative characteristics, and/or the comparison of the products being graded with some written or physical standard.

Grading systems can be quite elaborate and useful in commerce, where they protect the consumer against being offered low-quality products at a high price, while permitting the producer to recover the extra costs associated with the provision of a high-quality product. However, grading suffers from the considerable drawback that statistical correlation with measurable physical or chemical properties is difficult or impossible. Consequently, many of the time-honored grading scales are being replaced by the methods described in this book. Examples of good grading methods still in use are the Torry scale for freshness of fish (Sanders and Smith, 1976) and the USDA scales for butter (USDA, 1977) and meat (USDA, undated).

5.5 RANKING

In ranking, subjects receive three or more samples that are to be arranged in order of intensity or the degree of some specified attribute. For example, four samples of yogurt are to be ranked for degree of sensory acidity, or five samples of breakfast cereal may be ranked for preference. A full description of ranking tests and their statistical treatment will be found in Chapter 8.

For each subject, the sample ranked first is accorded a 1, that ranked second a 2, and so on. The rank numbers received by each sample are summed, and the resulting rank sums indicate the overall rank order of the samples. Rank orders cannot meaningfully be used as a measure of intensity, but they are amenable to significance tests such as the χ^2-test (see Chapter 14) and Friedman's test (see Chapter 8).

Ranking tests are rapid and demand relatively little training, although it should not be forgotten that the subjects must be thoroughly familiarized with the attribute under test. Ranking tests have wide application, but with sample sets above three, they do not discriminate as well as tests based on the use of scales.

5.6 SCALING

Scaling techniques involve the use of numbers or words to express the intensity of a perceived attribute (sweetness, hardness, smoothness) or a reaction to such attribute (e.g., too soft, just right, too hard). If words are used, the analyst may assign numerical values to the words (e.g., like extremely = 9, dislike extremely = 1) so that the data can be treated statistically. Methods of scaling are under intensive study around the world (ISO, 1999; Muñoz and Civille, 1998) and the subsequent paragraphs demonstrate standard protocols.

The validity and reliability of a scaling technique are highly dependent on

- The selection of a scaling technique that is broad enough to encompass the full range of parameter intensities and also has enough discrete points to pick up all the small differences in intensity between samples
- The degree to which the panel has or has not been taught to associate a particular sensation (and none other) with the attribute being scaled
- The degree to which the panel has or has not been trained to use the scale in the same way across all samples and across time (see Chapter 10 on panelist training)

Compared with difference testing, scaling is a more informative—and therefore a more useful—form of recording the intensity of perception. As with ranking, the results are critically dependent on how well the panelists have been familiarized with the attribute under test and with the scale being used. In this respect, three different philosophies have been applied (Muñoz and Civille, 1998):

- Universal scaling, in which panelists consider all products and intensities they have experienced as their highest intensity reference point (e.g., the Spectrum aromatics scale uses the cinnamon impact of Big Red chewing gum as a 12.5 in intensity on a 15-point scale)
- Product-specific scaling, in which panelists consider only their experience within the selected product category in setting their highest reference point (e.g., the vanilla impact of typical vanilla cookies was set at 10 on a 15-point product specific scale)
- Attribute-specific scaling, in which panelists consider their experience of the selected attribute across all products in setting their highest reference point (e.g., a specific toothpaste is assigned the top value of 13 for the peppermint aromatic in any product)

A common problem with scales is that panelists tend to use only the middle section. For example, if ciders are judged for intensity of "appley" flavor on a scale of between zero and nine, subjects will avoid the numbers 0, 1, and 2 because they tend to keep these in reserve for hypothetical samples of very low intensity, which may never come. Likewise, the numbers 7, 8, and 9 are avoided in anticipation of future samples of very high intensity, which may never come. The result is that the scale is distorted. For example, a cider of outstanding apple intensity may be rated 6.8 by the panel while a cider that is only just above the average may receive a 6.2.

Although the properties of data obtained from any response scale may vary with the circumstances of the test (e.g., the experience of judges in the test, the familiarity of the attribute), it is typically assumed that

- Category scaling (ISO term: *rating*) yields ordinal or interval data
- Line scaling (ISO term: *scoring*) yields interval data
- Magnitude estimation scaling (often called *ratio scaling*) sometimes, but not always, yields ratio data

5.6.1 Category Scaling

A category (or partition) scale is a method of measurement in which the subject is asked to "rate" the intensity of a particular stimulus by assigning it a value (category) on a limited, usually numerical, scale. Category scale data are generally considered to be at least ordinal-level data. They do not generally provide values that measure the degree (how much) one sample is more than another. On a 7-point category scale for hardness, a product rated a 6 is not necessarily twice as hard as a product with a 3 hardness rating. The hardness difference between 3 and 6 may not be the same as that between 6 and 9. Although attempts are made to encourage panelists to use all intervals as equal, panelists may also tend to use the categories with equal frequency, except that they usually avoid the use of the two scale endpoints so as to save them for "real extremes." Here are four examples of category scales of proven usefulness in descriptive analysis:

Number Category Scales		**Word Category Scale I**	**Word Category Scale II**
0	0	None	None at all
1	1	Threshold	Just detectable
2	2.5	Very slight	Very mild
3	5	Slight	Mild
4	7.5	Slight-moderate	Mild-distinct
5	10	Moderate	Distinct
6	12.5	Moderate-strong	Distinct-strong
7	15	Strong	Strong

Generally, even word category scales are converted to numbers. The numbers used in the above list are typical of such conversions.

The Flavor Profile® and Texture Profile® descriptive analysis methods use a numerical-type category scale anchored with words:

Numerical Value	**Word Anchor**
0	None
)(	Threshold, just detectable
½	Very slight
1	Slight
1½	Slight-moderate
2	Moderate
2½	Moderate-strong
3	Strong

Unless the scale represents a very small range of sensory perception or the number of samples to be tested is small (less than five), panel leaders should consider using at least a 10–15-point category scale. Data for category scales can be analyzed using χ^2-tests to compare the proportion of responses occurring in each category among a group of samples. Alternatively, if it is reasonable to assume that the categories are equally spaced, parametric techniques such as *t*-tests, analysis of variance, and regression can be applied to the data. Riskey (1986) discusses the use and abuse of category scales in considerable detail. James et al. (2003) review and test different ranking scales.

The practical steps involved in the construction of a scale are discussed in Chapters 11 and 12. Appendix 12.1 contains a wide selection of terms of proven usefulness as scale endpoints, and Appendix 12.2 gives reference points on a scale of 0–15 for the four basic tastes and for the intensity of selected aroma, taste, and texture characteristics of items readily available in supermarkets, such as Hellmann's mayonnaise.

5.6.2 Line Scales

With a linear or line scale, the panelist "rates" the intensity of a given stimulus by making a mark on a horizontal line that corresponds to the amount of the perceived stimulus. The lengths most used are 15 cm and 6 in. with marks ("anchors") either at the ends, or 1/2 in. or 1.25 cm from the two ends (see Figure 5.6). The use of more than two anchors tends to reduce the line scale to a category scale, which may or may not be desired. Normally, the left end of the scale corresponds to "none" or zero amount of the stimulus, while the right end of the scale represents a large amount or a very strong level of the stimulus (Anderson, 1970; Stone and Sidel, 1992). In some cases, the scale is bipolar, that is, opposite types of stimuli are used to anchor the endpoints.

Panelists use the line scale by placing a mark on the scale to represent the perceived intensity of the attribute in question. The marks from line scales are converted to numbers by manually measuring the position of each mark on each scale using a ruler, a transparent overlay, or a digitizer that is interfaced to a computer or by direct data entry by stylus on a computer screen. The digitizer converts the position of the mark to a number based on a preset program and feeds the data into the computer for analysis.

5.6.3 Magnitude Estimation Scaling

Magnitude estimation (Moskowitz, 1977; Meilgaard and Reid, 1979; ISO, 1994; Doty and Laing, 2003) or free number matching is a scaling technique based on Stevens' law (see Section 5.2.2). The first sample a panelist receives is assigned a freely chosen number (the number can be assigned by the experimenter, in which case it is referred to as a *modulus*; or the number can be chosen by the panelist). Panelists are then asked to assign all

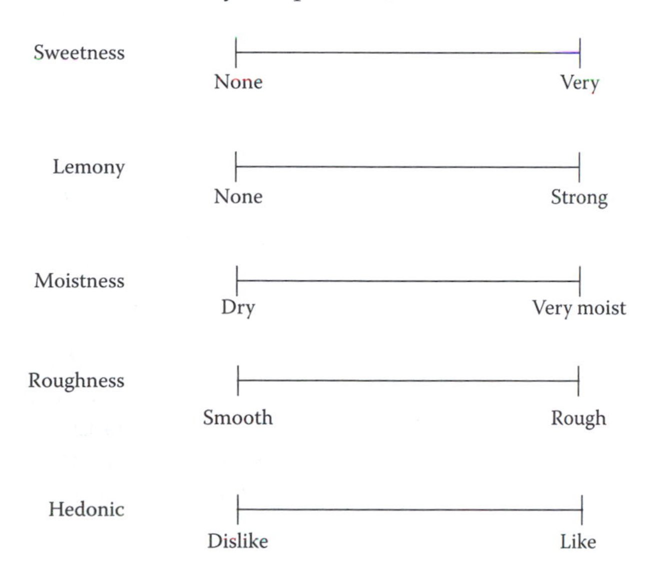

Figure 5.6 Typical line scales.

subsequent ratings of subsequent samples in proportion to the first sample rating. If the second sample appears three times as strong as the first, the assigned rating should be three times the rating assigned to the first, or reference, stimulus. Panelists are instructed to keep the number ratings in proportion to the ratios between sensations.

Examples

1. With a modulus: The first cookie that you taste has an assigned "crispness" rating of 25. Rate all other samples for crispness in proportion to that 25. If the crispness of any sample is half that of the first sample, assign it a crispness value of 12.5.
   First sample 25
   Sample 549 ______
   Sample 306 ______
2. Without a modulus: Taste the first cookie; assign any number to the "crispness" of that cookie. Rate the crispness of all other samples in proportion to the rating given the first sample.
   Sample 928 ______ (first sample)
   Sample 549 ______
   Sample 306 ______

The results are evaluated as described by Moskowitz (1977) and ISO (1994). Alternative methods of evaluation are reviewed by Butler et al. (1987) and Lawless (1989).

5.6.3.1 Magnitude Estimation versus Category Scaling

A good discussion of the advantages and disadvantages of the two methods is given by Pangborn (1984). The data produced by magnitude estimation (ME) have ratio properties, similar to the standard forms of technical measurement (length, weight, volume, etc.). ME gets around the problem that panelists avoid the ends of scales so as to leave room for another stimulus. Adherents of ME also cite the fact that users of CS must spend time and effort on the preparation of standards and on teaching the panel to use them. Those favoring CS note that ME is incapable of providing stable and reproducible values for flavor intensity. In practice, ME panelists require a good deal of training if they are to use the method with any facility; many judges rate in "nickels and dimes" using whole and half numbers or prefer the 10s or 5s in a series such as 15, 20, 25, and so on, and they have trouble thinking in pure ratio terms such as "six times stronger" or "1.3 times weaker." In a number of applications (Powers et al., 1981; Giovanni and Pangborn, 1983; Pearce et al., 1986; Lawless and Malone, 1986), ME has provided no greater discrimination than CS. Furthermore, ME is less suitable for scaling degree of liking (Pangborn et al. 1989). Where ME does offer more points of discrimination and separation is in academic applications with few judges (20 or less) studying a unidimensional system such as sucrose in water, one aromatic chemical in a diluent, or one increasing tone.

5.6.3.2 Magnitude Matching (Cross-Modality Matching)

In this technique, subjects match the intensity of attribute 1, such as the sourness of acid solutions, to the intensity of another attribute 2, such as the loudness of 1000-Hz tones. If the two intensities are governed by the functions

$$R_1 = k_1 C_1^{n1}, \text{ or } \log R_1 = \log k_1 + n_1 \log C_1$$

and

$$R_2 = k_2 C_1^{n2}, \text{ or } \log R_2 = \log k_2 + n_2 \log C_2;$$

matching the functions gives

$$\log k_1 + n_1 \log C_1 = \log k_2 + n_2 \log C_2,$$

or

$$\log C_1 = \frac{n_2}{n_1} \log C_2 \text{ plus a constant.}$$

In other words, a power function has been obtained that describes the intensity of sourness, and the exponent of the function is equal to the ratio of the original exponents (Cardello and Maller, 1987; Lawless and Heymann, 1998; Marks et al., 1988). The advantage of this approach is that no numbers are assigned, so it gets around the tendency of subjects to use numbers differently, as mentioned in Section 5.6.

5.6.4 Labelled Magnitude Scales (LMS)

Building upon Borg's (1982) early work using the category-ratio scale to study perceived exertion, Green et al. (1993) developed the LMS to study taste and oral sensations.

The LMS is a vertical line marked at various points with verbal intensity anchors, such as "weak", "moderate", and "strong." Near the bottom of the scale is the label "barely detectable" while near the top of the scale is the label "strongest imaginable". These labels are spaced along the scale in a quasi-logarithmic manner established by calibration using ratio-scaling. Individuals in studies are usually asked to mark where on the scale they perceive the intensity of a given sensation to lie (Green et al., 1993).

The scale can lend itself to the study of perceptual differences between subjects in certain cases. It provides both magnitude estimates as well as semantic descriptors of the sensations. Although it works well when the smell and taste sensations are broadly defined, e.g., "strongest imaginable taste" (Green et al., 1996), there is evidence that when the upper bound of the scale is defined more narrowly (e.g., "strongest imaginable sweetness"), the scale produces a steeper function than does magnitude estimation. It was suggested that the scale may need to be modified depending on the instructions and frame of reference used for scaling (Green et al., 1996). Modification may also be needed in cases where individuals differ greatly in sensory perception such as in PTC/PROP tasting (Bartoshuk et al., 2004).

REFERENCES

Amoore, J. E. (1977). Specific anosmia and the concept of primary odors. *Chem Sens Flav* 2: 267–81.

Anderson, N. H. (1970). Functional measurement and psychological judgment. *Psychol Rev* 77: 153–70.

Anderson, N. H. (1974). Algebraic models in perception. In *Psychophysical Judgment and Measurement, Vol. 2 of Handbook of Perception*, ed. Carterette, E. C. and M. P. Friedman, 215–98. New York: Academic Press.

Baird, J. C., and E. Noma (1978). *Fundamentals of Scaling and Psychophysics.* New York: Wiley-Interscience.

Bartoshuk, L. M., V. B. Duffy, B. G. Green, H. J. Hoffman, C. W. Ko, L. A. Lucchina, L. E. Marks, D. J. Snyder, and J. M. Weiffenbach (2004). Valid across-group comparisons with labeled scales: The gLMS versus magnitude matching. *Physiol Behav* 82: 109–114.

Beidler, L. M. (1954). A theory of taste stimulation. *J Gen Physiol* 38: 133–9.

Beidler, L. M. (1974). Biophysics of sweetness. In *Symposium: Sweeteners*, ed. G. E. Inglett, 10. Westport, CT: AVI Publishing.

Borg, G., H. Diamant, L. Strom, and Y. Zotterman (1967). The relation between neural and perceptual intensity: A comparative study on the neural and psychophysical responses to gustatory stimuli. *J Physiol* 13: 192.

Borg, G. (1982). A category scale with ratio properties for intermodal and interindividual comparisons. In *Psychophysical Judgement and the Process of Perception*, ed. Geissler, H. G. and P. Petxold, 25–34. VEB Deutxcher Verlag der Wissenschaften, Berlin.

Butler, G., L. M. Poste, M. S. Wolynetz, V. E. Ayar, and E. Larmond (1987). Alternative analyses of magnitude estimation data. *J Sens Stud* 2 (4): 243–57.

Cardello, A. V., and O. Maller (1987). Psychophysical bases for the assessment of food quality. In *Objective Methods in Food Quality Assessment*, ed. J.D. Kapsalis, 61–125. Boca Raton, FL: CRC Press.

Chastrette, M., T. Thomas-Danguin, and E. Rallet (1998). Modelling the human olfactory stimulus–response function. *Chem Sens* 23: 181–96.

Doty, R. L. (ed.) (2003). Section B: Human psychophysics and measurement of odor-induced responses. In *Handbook of Gustation and Olfaction*. New York: Marcel Dekker.

Doty, R. L., and D. G. Laing (2003). Psychophysical measurement of human olfactory function, including odorant mixture assessment. In *Handbook of Olfaction and Gustation*, 2nd ed., ed. R. L. Doty, 203–28. New York: Marcel Dekker.

Einstein, M. A. (1976). Use of linear rating scales for the evaluation of beef flavor by consumers. *J Food Sci* 41: 383.

Fechner, G. T. (1860). *Elemente der Psychophysik.* Leipzig: Breitkopf und Hartel.

Giovanni, M. E., and R. M. Pangborn (1983). Measurement of taste intensity and degree of liking of beverages by graphic scales and magnitude estimation. *J Food Sci* 48: 1175–82.

Green, B. G., G. S. Shaffer, and M. M. Gilmore (1993). Derivation and evaluation of a semantic scale of oral sensation magnitude with apparent ratio properties. *Chem Senses* 18: 683–702.

Green, B. G., P. Dalton, B. Cowart, G. Shaffer, K. Rankin, and J. Higgins (1996). Evaluating the "Labeled Magnitude Scale" for measuring sensations of taste and smell. *Chem Senses* 21 (3): 323–334.

ISO (1994). Sensory analysis—Methodology—Magnitude estimation. International Standard ISO 11056. Available from International Organization for Standardization, Case Postale 56, 1 rue Varembé, CH1211 Genève 20; or from American National Standards Institute, 11 West 42nd St., New York, NY 10036.

ISO (1999). Sensory analysis—Guidelines for the use of quantitative response scales. Draft International Standard ISO CD 4121, available from ISO, c/o AFNOR, Tour Europe, Cedex 7, 92049 Paris La Défense.

James, C. E., D. G. Laing, A. L. Jinks, N. Oram, and I. Hutchinson (2003). Taste response functions of adults and children using different rating scales. *Food Qual Prefer* 15: 77–82.

Kendal-Reed, M., J. C. Walker, W. T. Morgan, M. LaMacchio, and R. W. Lutz (1998). Human responses to propionic acid. I. Quantification of within- and between-participant variation in perception by normosmics and anosmics. *Chem Senses* 23: 71–82.

Kling, J. W., and L. A. Riggs (eds) (1971). *Woodworth & Schlosberg's Experimental Psychology*, 3rd ed. New York: Holt, Rinehart & Winston.

Laming., D. (1994). Psychophysics. In *Companion Encyclopedia of Psychology*, ed. A. M. Colman, 251–77. London: Routledge.

Lawless, H. T. (1989). Logarithmic transformation of magnitude estimation data and comparisons of scaling methods. *J Sens Stud* 4: 75–86.

Lawless, H. T. (1990). Applications of experimental psychology in sensory evaluation. In *Psychological Basis of Sensory Evaluation*, eds. R. L. McBride and H. J. H. MacFie, 69–91. London: Elsevier.

Lawless, H. T., and H. Heymann (1998). *Sensory Evaluation of Food: Principles and Practices*. New York: Chapman & Hall.

Lawless, H. T. and G. J. Malone (1986). The discriminative efficiency of common scaling methods. *J Sens Stud* 1 (1): 85–98.

Maes, F. W. (1985). Improved best-stimulus classification of taste neurons. *Chem Sens* 10: 35–44.

Marks, L. E., J. C. Stevens, L. M. Bartoshuk, J. F. Gent, B. Rifkin, and V. K. Stone (1988). Magnitude-matching: The measurement of taste and smell. *Chem Sens* 13 (1): 63–87.

McBride, R. L. (1987). Taste psychophysics and the Beidler equation. *Chem Sens* 12: 323–32.

Meilgaard, M. C. (1993). Individual differences in sensory threshold for aroma chemicals added to beer. *Food Qual Pref* 4: 153–67.

Meilgaard, M. C., and D. S. Reid (1979). Determination of personal and group thresholds and the use of magnitude estimation in beer flavour chemistry. In *Progress in Flavour Research*, ed. D. G. Land and H. E. Nursten, 67–73. London: Applied Science.

Moskowitz, H. R. (1977). Magnitude estimation: Notes on what, how and why to use it. *J Food Qual* 1: 195–228.

Moskowitz, H. R. (2002). The intertwining of psychophysics and sensory analysis: Historical perspectives and future opportunities—A personal view. *Food Qual Pref* 14: 87–98.

Muñoz, A. A., and G. V. Civille (1998). Universal, product and attribute specific scaling and the development of common lexicons in descriptive analysis *J Sens Stud* 13: 57–76.

Norwich, K. H., and W. Wong (1997). Unification of psychophysical phenomena: The complete form of Fechner's law. *Percep Psychophys* 59: 929–40.

Pangborn. R. M. (1981). Individuality in responses to sensory stimuli. In *Criteria of Food Acceptance. How Man Chooses What He Eats*, ed. J. Solms, R. L. Hall, 177–219. Zürich: Forster-Verlag.

Pangborn, R. M. (1984). Sensory techniques of food analysis. In *Physical Characterization, Vol. 1 of Food Analysis. Principles and Techniques*, ed. D. W. Gruenwedel and J. R. Whitaker, 61–8. New York: Marcel Dekker.

Pangborn, R. M., P. M. J. X. Guinard, and H. L. Meiselman (1989). Evaluation of bitterness of caffeine in hot chocolate drink by category, graphic, and ratio scaling. *J Sens Stud* 4 (1): 31–53.

Pearce, J. J., C. B. Warren, and B. Korth (1986). Evaluation of three scaling methods for hedonics. *J Sens Stud* 1 (1): 27–46.

Powers, J. J., C. B. Warren, and T. Masurat (1981). Collaborative trials involving the methods of normalizing magnitude estimations. *Lebensm Wiss Technol* 14: 86–93.

Riskey, D. R. (1986). Use and abuse of category scales in sensory measurement. *J Sens Stud* 1 (3/4): 217.

Sanders, H. R., and G. L. Smith (1976). The construction of grading schemes based on freshness assessment of fish. *J Food Technol* 11: 365.

Sekuler, R., and R. Blake (1990). *Perception*, 2nd ed. New York: McGraw-Hill.

Stevens, S. S. (1957). On the psychophysical law. *Psychol Rev* 64: 153–81.

Stevens, S. S. (1970). Neural events and the psychophysical law. *Science* 170: 1043.

Stone, H., and J. L. Sidel (1992). *Sensory Evaluation Practices*, 2nd ed. Orlando, FL: Academic Press.

USDA (n.d.). United States grading and certification standards. Meats, prepared meat and meat products. Regulations, Title 7 CFR, Part 54, updated annually or periodically. http://cfr.regstoday.com/7cfr54.aspx.

USDA (1977). United States standards for grades of butter. Regulations, Title 7 CFR, Part 58 (published in Federal Register, February 1, 1977).

Weber, E. H. (1834). *De pulsu resorptime, auditor et tache: Annotationes anatomical et physiological.* Leipzig: Koehler.

6

Guidelines for Choice of Technique

6.1 INTRODUCTION

The five tables that follow are meant as memory joggers. They are not a substitute for the study of the individual methods described in this book, but once the methods have become familiar, preferably via practical hands-on testing of most of them, the tables can be used to check whether there might be a better way to attack a given problem. Most of us tend to give preference to a few trusted favorite tests, and perhaps we bend the test objective a bit to allow their use—a dangerous habit.

To avoid this practice or to find a way out of it, the authors suggest the following practical steps.

6.2 DEFINE THE PROJECT OBJECTIVE

Read the text in Chapter 1, and then refer to Table 6.1 to classify the type of project. Review the 17 entries. Write down the project objective, and then look up the test to which the table refers.

6.3 DEFINE THE TEST OBJECTIVE

Four tables are available for this purpose:

- Table 6.2: Difference tests: Does a sensory difference exist between samples?
- Table 6.3: Attribute difference tests: How does attribute X differ between samples?
- Table 6.4: Affective tests: Which sample is preferred? How acceptable is sample X?
- Table 6.5: Descriptive tests:Rate each of the attributes listed in the score sheet.

Write down the test objective and list the tests required. Then meet with the project leader and others involved in the project and discuss and refine the design of the tests.

Table 6.1 Types of Challenges Encountered in Sensory Analysis

Type of Problem	Type of Challenge	Tests Applicable
1. How do I know what kind of products I should make?	New product development—the product development team needs information on the sensory characteristics and also on consumer acceptability of experimental products as compared with existing products in the market.	All tests in this book
2. How do I match my product to an existing product on the market?	Product matching—here the focus is on proving that no difference exists between an existing and a developmental product.	Difference tests in similarity mode, Chapter 7
3. How do I determine the direction in which I need to move my product to improve it?	Product improvement—step 1: Define exactly what sensory characteristics need improvement; step 2: determine that the experimental product is indeed different; step 3: confirm that the experimental product is liked better than the control.	All difference tests, Table 6.2; then affective tests, Table 6.4; see note
4. If my company changes the manufacturing process, how do I know if this will affect my product?	Process change—step 1: Confirm that no difference exists; step 2: If a difference does exist, determine how consumers view the difference.	Difference tests in similarity mode, Chapter 7; affective tests, Table 6.4; see note
5. If I substitute a lower-cost ingredient with a more expensive one, how do I know if my product has changed?	Cost reduction and/or selection of new source of supply—step 1: Confirm that no difference exists; step 2: If a difference does exist, determine how consumers view the difference.	Difference tests in similarity mode, Chapter 6; affective tests, Table 6.4; see note
6. How do I ensure that the quality of my product is maintained during all stages of production?	Quality control—Products sampled during production, distribution, and marketing are tested to ensure that they are as good as the standard: Descriptive tests (well-trained panel) can monitor many attributes simultaneously.	Difference tests, Table 6.2; descriptive tests, Table 6.5
7. How do I ensure that the quality of my product is being maintained throughout its shelf life?	Storage stability—testing of current and experimental products after standard aging tests; step 1: Ascertain when difference becomes noticeable; step 2: Descriptive tests (well-trained panel) can monitor many attributes simultaneously; step 3: Affective tests can determine the relative acceptance of stored products.	Difference tests, Table 6.2; descriptive tests, Table 6.5; affective tests, Table 6.4
8. How do I assign grades to my products based on their sensory quality?	Product grading or rating—used where methods of grading exist that have been accepted by agreement between producer and user, often with government supervision.	Grading, Chapter 5
9. How do I find out if consumers like my product?	Consumer acceptance and/or opinions—after laboratory screening, it may be desirable to submit product to a central-location or home placement test to determine consumer reaction; acceptance tests will indicate whether the current product can be marketed or whether improvement is needed.	Affective tests, Table 6.4; Chapter 13

10. How do I find out if consumers prefer my product?	Consumer preference—full-scale consumer preference tests are the last step before test marketing; employee preference studies cannot replace consumer tests but can reduce their number and cost whenever the desirability of key attributes of the product is known from previous consumer tests.	Affective tests, Table 6.4; Chapter 13
11. How do I ensure that my panelists are as highly trained as they need to be?	Panelist selection and training—an essential activity for any panel; may consist of (1) interview; (2) sensitivity tests; (3) difference tests; and (4) descriptive tests.	Chapter 10
12. How do my sensory results correspond with my instrumental results?	Correlation of sensory with chemical and physical tests—correlation studies are needed (1) to lessen the load of samples on the panel by replacing a part of the tests with laboratory analyses; (2) to develop background knowledge of the chemical and physical causes of each sensory attribute.	Descriptive tests, Chapter 11, Table 6.5; attribute difference tests, Table 6.3
13. How do I confirm that this taint is coming from where I suspect?	Threshold of added substances—required (1) in trouble shooting to confirm suspected source(s) of off flavor(s); (2) to develop background knowledge of the chemical cause(s) of sensory attributes and consumer preferences.	Chapter 9
14. How can I ensure that the advertising claims I make on my product are substantiated and have legal veracity?	Advertising Claims—dependent on claim. One-sided directional difference tests for preference. Descriptive tests to prove functionality.	Chapter 18
15. What is the fundamental design that this personal care product should take?	Consumer-directed product development—rapid prototyping. Use when the engineering fundamentals of the product needed to be uncovered.	Chapter 18
16. How do I obtain a detailed understanding of how consumers use my product?	Research in real context—ethnography for observing consumers during their daily lives. Sequence mapping to understand the life cycle of the product.	Chapter 18
17. How do I obtain a deep understanding of the emotional connection consumers have to my product?	Determination of how products make consumer feel—consumer emotion research.	Chapter 18

Note: In 3, 4, and 5, if the new product is different, descriptive tests (Table 6.5) may be useful in order to characterize the difference. If the difference is found to be in a single attribute, attribute difference tests (Table 6.3) are the tools to use in further work.

Table 6.2 Area of Application of Difference Tests: Does a Sensory Difference Exist between Samples? Are the Samples Similar Enough to be Used Interchangeably?

The tests in this table are suitable for applications such as

1. To determine whether product differences result from a change in ingredients, processing, packaging, or storage
2. To determine whether an overall difference exists where no specific attribute(s) can be identified as having been affected
3. To determine whether two samples are sufficiently similar to be used interchangeably
4. To select and train panelists and to monitor their ability to discriminate between test samples

Test	Areas of Application
1. Triangle test	Two samples not visibly different; one of the most used difference tests; statistically efficient, but somewhat affected by sensory fatigue and memory effects; generally 20–40 subjects, can be used with as few as 5–8 subjects; brief training required.
2. Duo–trio test	Two samples not visibly different; test has low statistical efficiency but is less affected by fatigue than the triangle test; useful where product well known to subjects can be employed as the reference; generally 30 or more subjects, can be used with as few as 12–15; brief training required.
3. Two-out-of-five test	Two samples without obvious visible differences; statistically highly efficient, but strongly affected by sensory fatigue, hence use limited to visual, auditory, and tactile applications; generally 8–12 subjects, can be used with as few as 5; brief training required.
4. Tetrad test	Two samples without obvious visible differences; based on Thurstonian theory; useful when the differences are small.
5. Same/different test (also called simple difference test)	Two samples not visibly different; test has low statistical efficiency but is suitable for samples of strong or lingering flavor, samples that need to be applied to the skin in half-face tests, and samples that are very complex stimuli and are therefore confusing to the subjects; generally 30 or more subjects, can be used with as few as 12–15; brief training required.
6. "A"–"Not A" test	As No. 5, but used where one of the samples has importance as a standard or reference product, is familiar to the subjects, or is essential to the project as the current sample against which all other samples are measured.
7. Difference-from-control test	Two samples that may show slight visual differences such as are caused by the normal heterogeneity of meats, vegetables, salads, and baked goods; test is used where the size of the difference affects a decision about the test objective, for example, in quality control and storage studies; generally 30–50 presentations of the sample pair; moderate amount of training required.
8. Sequential tests	Used with any of the above tests (1–3) to determine with a minimum of testing, at a predetermined significance level, whether the two samples are perceptibly (1) identical or (2) different.
9. Similarity mode	Used with tests 1–3 or 8, when the test objective is to prove that no perceptible difference exists between two products; used in situations such as (1) the substitution of a new ingredient for an old one that has become too expensive or unavailable or (2) a change in processing brought about by replacement of an old or inefficient piece of equipment.

Table 6.3 Area of Application of Attribute Difference Tests: How Does Attribute X Differ between Samples?

Test	Areas of Application
1. Alternative/forced choice (2-AFC test)	One of the most used attribute difference tests; used to show which of two samples has more of the attribute under test ("directional difference test") or which of two samples is preferred ("paired preference test"); test exists in one- or two-sided applications; generally 30 or more subjects, can be used with as few as 15.
2. Alternative/forced choice (3AFC, MAC)	Similar to 2-AFC but more stimuli are presented; used to show which of the samples has more of the attribute under test or which of the samples are preferred; test exists in one- or two-sided applications; number of subjects depends in part on the number of stimuli presented.
3. Pairwise ranking test	Used to rank three to six samples according to intensity of one attribute; paired ranking is simple to perform and the statistical analysis is uncomplicated, but results are not as actionable as those obtained with rating; generally 20 or more subjects, can be used with as few as 10.
4. Simple ranking test	Used to rank three to six, certainly no more than eight, samples according to one attribute; ranking is simple to perform, but results are not as actionable as those obtained by rating; two samples of small or large difference in the attribute will show the same difference in rank (i.e., one rank unit); ranking is useful to presort or screen samples for more detailed tests; generally 16 or more subjects, can be used with as few as 8.
5. Rating of several samples	Used to rate three to six, certainly no more than eight, samples on a numerical intensity scale according to one attribute; it is a requirement that all samples be compared in one large set; generally 16 or more subjects, can be used with as few as 8; may be used to compare descriptive analyses of several samples, but note (Section 13.6.4) that there will be some carryover (halo effect) between the attributes.
6. Balanced incomplete block test	As No. 5, but used when there are too many samples (e.g., 7–15) to be presented together in one sitting.
7. Rating of several samples, balanced incomplete block	As No. 6, but used when there are too many samples (e.g., 7–15) to be presented together in one sitting.

The tests in this table are used to determine whether or not, or the degree to which, two or more samples differ with respect to one defined attribute. This may be a single attribute, such as sweetness; or a combination of several related attributes, such as freshness; or an overall evaluation, such as preference. With the exception of preference, panelists must be carefully trained to recognize the selected attribute, and the results are valid only to the extent that panelists understand and obey such instructions. **A lack of difference in the selected attribute does not imply that no overall difference exists.** Samples need not be visibly identical, as only the selected attribute is evaluated.

Table 6.4 Area of Application of Affective Tests Used in Consumer Tests and Employee Acceptance Tests

Test	Questions Typically Asked	Areas of Application
Preference Tests		
1. Paired preference	Which sample do you prefer? Which sample do you like better?	Comparison of two products
2. Rank preference	Rank samples according to your preference with 1 = best, 2 = next best, etc.	Comparison of three to six products
3. Multiple paired preference	As No.1.	Comparison of three to six products
4. Multiple paired preference, selected pairs	As No.1.	Comparison of five to eight products
Acceptance Tests		
1. Simple acceptance test	Is the sample acceptable/not acceptable?	First screening in employee acceptance test
2. Hedonic rating	Figure 13.2.	One or more products to study how acceptance is distributed in the population represented by the subjects
Attribute Diagnostics		
1. Attribute-by-preference test	Which sample did you prefer for fragrance?	Comparison of two to six products to determine which attributes "drive" preference
2. Hedonic rating of individual attributes	Rate the following attributes on the hedonic scale provided.	Study of one or more products to determine which attributes, and at what level, "drive" preference
3. Intensity rating of individual attributes	Rate the following attributes on the intensity scale provided.	Study of one or more products, in cases where groups of subjects differ in their preference
4. Just about right (JAR) scale rating of individual attributes	Rate the following attributes on the JAR	Study one or more products to determine relative ratings for selected attributes

Affective tests can be divided into preference tests, in which the task is to arrange the products tested in order of preference; acceptance tests, in which the task is to rate the product or products on a scale of acceptability; and "attribute diagnostics," in which the task is to rank or rate the principal attributes that determine a product's preference or acceptance. With regard to the statistical analysis, preference and acceptance tests can be seen as a special case of attribute difference tests (Table 6.3) in which the attribute of interest is either preference or degree of acceptance. In theory, all tests listed in Table 6.3 can be used as preference tests and/or as acceptance tests. In practice, subjects in affective tests are less experienced, and complex designs such as balanced incomplete blocks are not usable. The tests in this table are equally suitable for presentation in laboratory tests, employee acceptance tests, central location consumer tests, or home use consumer tests unless otherwise indicated.

Table 6.5 Area of Application of Descriptive Tests

Tests	Areas of Application
1. Flavor profile (Arthur D. Little)	In situations where many and varied samples must be judged by a few highly trained tasters.
2. Texture profile (General Foods)	In situations where many and varied samples must be judged for texture by a few highly trained tasters.
3. Quantitative Descriptive Analysis (QDA®) method (Tragon Corp.)	In situations such as quality assurance in a large company, where large numbers of the same kind of products must be judged day in and day out by a well-trained panel; in product development in situations where reproducibility over time and place is not required.
4. Free-choice profiling	In consumer testing when it is desirable not to teach the subjects a common scale. Consumers rate the intensities of the attribute.
5. Flash profiling	In consumer testing when it is desirable not to teach the subjects a common scale. Consumers order the products by the intensities of specific attributes.
6. Napping	In testing when it is desirable not to teach the subjects a common scale. Consumers place similar products near each other and dissimilar products far from each other on a matrix. Consumers are asked to describe the samples but are instructed to avoid hedonic terms.
7. Sorting	In testing when it is desirable not to teach subjects a common scale. Subjects group samples into categories that make sense to them as individuals. Subjects may be asked to regroup samples according to different criteria.
8. Graph theoretic approach	In testing when it is desirable to select the best combinations out of a wide range of possibilities.
9. Spectrum method	A custom-design system suitable for most applications, including those under the above tests 1, 2, and 3; suitable where reproducibility over time and place is needed.
10. Modified, short-version spectrum descriptive analysis	To monitor a few critical attributes of a product through shelf-life studies; to examine possible manufacturing defects and product complaints; for routine quality assurance.
11. Time-intensity (TI)	Useful for samples in which the perceived intensity of flavor varies over time after the product is taken into the mouth, for example, bitterness of beer, sweetness of artificial sweeteners.
12. Progressive profiling	Useful for samples in which multiple attributes may change over time. A profile of up to five attributes is taken at designated intervals.
13. Temporal dominance of sensations (TDS)	Useful for samples that may change their prominent sensory attributes over time (e.g., gum).
14. Temporal order of sensations (TOS)	Useful for samples that may change their prominent sensory attributes over time (e.g., gum). Utilizes a checklist.
15. Multiple attribute time intensity (MATA)	Useful for samples that have multiple dynamic attributes. Individual attributes are measured at designated time points.

Descriptive tests are very diverse, often being designed or modified for each individual application, and they are therefore difficult to classify in a table such as this. A classification by inventor or method developer is perhaps the most helpful.

6.4 REVIEW PROJECT OBJECTIVE AND TEST OBJECTIVES: REVISE TEST DESIGN

In sensory testing, a given problem frequently requires appreciable thought before the appropriate practical tests can be selected (IFT, 1981). This is because the initial conception of the problem may require clarification. It is not unusual for the problem and test objectives to be defined and redefined several times before an acceptable design emerges. Sensory tests are expensive, and they often give results that cannot be interpreted. If this happens, the design may be at fault. Pilot tests are often useful as a means of refining a design. It would, for example, be meaningless to carry out a consumer preference test with hundreds of participants without first having shown that a perceptible difference exists; such a pilot study can be established with 10 or 20 tasters, using a difference test. In another example, islands of opposing preference may exist, invalidating a normal preference test; here, the solution may be a pilot study in which various types of customers receive single-sample acceptability tests.

REFERENCE

IFT. (1981). Guidelines for the preparation and review of papers reporting sensory evaluation data. Sensory Evaluation Division, Institute of Food Technologist. *Food Technol* 35: 11–50.

7

Overall Difference Tests

Does a Sensory Difference Exist between Samples?

7.1 INTRODUCTION

Chapter 7 and Chapter 8 contain "cookbook-style" descriptions of individual difference tests with examples. The underlying theory is found in Chapter 5, "Measuring Responses," and in Chapter 14, "Basic Statistical Methods." Guidelines for the choice of a particular test are found under "Scope and Application" for each test, and also in summary form in Chapter 6, "Guidelines for Choice of Technique."

Difference tests can be set up legitimately in hundreds of different ways, but in practice the procedures described here have acquired individual names and a history of use. There are two groups of difference tests with the following characteristics:

Overall difference tests (Chapter 7): Does a sensory difference exist between samples? These are tests, such as the triangle and the duo–trio, which are designed to show whether subjects can detect any difference at all between samples.

Attribute difference tests (Chapter 8): How does attribute X differ between samples? Subjects are asked to concentrate on a single attribute (or a few attributes), for example, "Please rank these samples according to sweetness." All other attributes are ignored. Examples are the paired-comparison tests, the n-AFC tests (alternative forced choice), and various types of multiple comparison tests. The intensity with which the selected attribute is perceived may be measured by any of the methods described in Chapter 5, for example, ranking, line scaling, or magnitude estimation (ME).

The 2- and 3-AFC tests are often used in threshold determinations (see Chapter 9). Affective tests (preference tests, e.g., consumer tests) are also attribute difference tests (see Chapter 13).

7.2 UNIFIED APPROACH TO DIFFERENCE AND SIMILARITY TESTING

Discrimination tests can be used to address a variety of practical objectives. In some cases, researchers are interested in demonstrating that two samples are perceptibly different.

In other cases, researchers want to determine if two samples are sufficiently similar to be used interchangeably. In yet another set of cases, some researchers want to demonstrate a difference, while other researchers involved in the same study want to demonstrate similarity. All of these situations can be handled in a unified approach through the selection of appropriate values for the test-sensitivity parameters α, β, and p_d. What values are appropriate depends on the specific objectives of the test.

The unified approach also applies to paired-comparison tests, such as the 2-AFC (see Section 8.2).

When testing for difference, the objective is merely to discover whether a perceptible difference exists between two samples. The statistical analysis is made under the tacit assumption that only the α-risk matters (the probability of concluding that a perceptible difference exists when one does not). The number of assessors is determined by looking at the α-risk table and taking into account material concerns, such as availability of assessors, available quantity of test samples, and so on. The β-risk (the probability of concluding that no perceptible difference exists when one does) and the proportion of distinguishers, p_d, on the panel are ignored or, more appropriately, are assumed to be unimportant. As a result, in testing for difference, the researcher selects a small value for the α-risk and accepts arbitrarily large values for the β-risk and p_d (by ignoring them) to keep the required number of assessors within reasonable limits.

In testing for similarity, the sensory analyst wants to determine that two samples are sufficiently similar to be used interchangeably. Reformulating for reduced costs and validating alternate suppliers are just two examples of this common situation. In designing a test for similarity, the analyst determines what constitutes a meaningful difference by selecting a value for p_d and then specifying a small value for β-risk to ensure that there is only a small chance of missing that difference if it really exists. The α-risk is allowed to become large to keep the number of assessors within reasonable limits.

In many cases, however, it is important to balance the risk of missing a difference that exists (β-risk) with the risk of concluding that a difference exists when it does not (α-risk). In this case, the analyst chooses values for all three parameters, α, β, and p_d, to arrive at the number of assessors required to deliver the desired sensitivity for the test (see Example 7.4).

As a rule of thumb, a statistically significant result at

- An α-risk of 10%–5% (0.10–0.05) indicates moderate evidence that a difference is apparent
- An α-risk of 5%–1% (0.05–0.01) indicates strong evidence that a difference is apparent
- An α-risk of 1%–0.1% (0.01–0.001) indicates very strong evidence that a difference is apparent
- An α-risk below 0.1% (<0.001) indicates extremely strong evidence that a difference is apparent

For β-risks, the strength of the evidence that a difference is not apparent is assessed using the same criteria as above (substituting "is not apparent" for "is apparent").

The maximum allowable proportion of distinguishers, p_d, falls into three ranges:

- $p_d < 25\%$ represent small values
- $25\% < p_d < 35\%$ represent medium-sized values
- $p_d > 35\%$ represent large values

A spreadsheet application has been developed in Microsoft Excel* to aid researchers in selecting values for α, β, and p_d that provide the best compromise between the desired test sensitivity and available resources (see Section 14.3.5). The test sensitivity analyzer allows researchers to quickly run a variety of scenarios with different combinations of the number of assessors, n; the number of correct responses, x; and the maximum allowable proportion of distinguishers, p_d, and in each case to observe the resulting impacts on α-risk and β-risk.

Recently, an alternative approach to similarity testing, called *equivalence testing*, has been adapted from the pharmaceutical industry (Westlake, 1972) and is becoming more widely used in sensory evaluation (Arents et al., 2002; Bi, 2005; Sauerhoff et al., 2005). Equivalence testing recognizes that two products can be perceptibly different and yet still be similar enough to each other to be used interchangeably. This result can occur in the unified approach when more than the minimum required number of respondents participate in the test. Equivalence testing ignores the statistically significant difference and focuses on ensuring that the maximum difference does not exceed a predetermined acceptable limit (e.g., p_d). Equivalence testing has a strong theoretical basis and is an approach worth considering when the primary test objective is to ensure that any difference that might exist does not exceed acceptable limits.

7.3 TRIANGLE TEST

The section on the triangle test (ASTM, 2004; ISO, 2004c), being the first in this book, is rather complex and includes many details that (1) all sensory analysts should know, (2) are common to many methods, and (3) are therefore omitted in subsequent methods. The application of the unified approach is described in Examples 7.3 and 7.4.

7.3.1 Scope and Application

Use this method when the test objective is to determine whether a sensory difference exists between two products. This method is particularly useful in situations where treatment effects may have produced product changes that cannot be characterized simply by one or two attributes. Although it is statistically more efficient than the paired-comparison and duo–trio methods, the triangle test has limited use with products that involve sensory fatigue, carryover, or adaptation and with subjects who find testing three samples too confusing. This method is effective in certain situations:

1. To determine whether product differences result from a change in ingredients, processing, packaging, or storage
2. To determine whether an overall difference exists where no specific attribute(s) can be identified as having been affected
3. To select and monitor panelists for ability to discriminate given differences

7.3.2 Principle of the Test

1. Present to each subject three coded samples.
2. Instruct subjects that two samples are identical and one is different (or odd).

* Available upon request in Excel as an e-mail attachment from Tom.Carr@CarrConsulting.net.

3. Ask the subjects to taste (feel, examine) each product from left to right and select the odd sample.
4. Count the number of correct replies and refer to Table 19.8 for interpretation.

7.3.3 Test Subjects

Generally, 20–40 subjects are used for triangle tests, although as few as 12 may be employed when differences are large and easy to identify. Similarity testing, on the other hand, requires 50–100 subjects. At a minimum, subjects should be familiar with the triangle test (the format, the task, the procedure for evaluation), and with the product being tested, especially because flavor memory plays a part in triangle testing.

An orientation session is recommended prior to the actual taste test to familiarize subjects with the test procedures and product characteristics. Care must be taken to supply sufficient information to be instructive and motivating, while not biasing subjects with specific information about treatment effects and product identity.

7.3.4 Test Procedure

The test controls (explained in detail in Chapter 3) should include a partitioned test area in which each subject can work independently. Control of lighting may be necessary to reduce any color variables.

Prepare and present samples under optimum conditions for the product type investigated; for example, samples should be appetizing and well presented.

Offer samples simultaneously, if possible; however, samples that are bulky, leave an aftertaste, or show slight differences in appearance may be offered sequentially without invalidating the test.

Prepare equal numbers of the six possible combinations (ABB, BAA, AAB, BBA, ABA, and BAB) and present these at random to the subjects.

Ask subjects to examine (taste, feel, smell, etc.) the samples in order from left to right, with the option of going back to repeat the evaluation of each while the test is in progress.

The score sheet, shown in Figure 7.1, could provide for more than one set of samples. However, this can only be done if sensory fatigue is minimal. Do not ask questions about preference, acceptance, degree of difference, or type of difference after the initial selection of the odd sample. This is because the subject's choice of the odd sample may bias his/her responses to these additional questions. Responses to such questions may be obtained through additional tests. See Chapter 13 for preference and acceptance tests and Chapter 8 for difference tests related to size or type (attribute) of difference.

7.3.5 Analysis and Interpretation of Results

Count the number of correct responses (correctly identified odd samples) and the number of total responses.

Determine if the number correct for the number tested is equal to or larger than the number indicated in Table 19.8.

Triangle test		
Name __________ Date ________		
Type of sample __________		
Instructions Taste samples from left to right. Two are identical; determine which is the odd sample. If no difference apparent, you must guess.		
Sets of three samples	Which is the odd sample?	Comments
___ ___ ___	______	________
___ ___ ___	______	________
___ ___ ___	______	________

Figure 7.1 Example of score sheet for three triangle tests.

Do not count "no difference" replies as valid responses. Instruct subjects to guess if the odd sample is not detectable.

Example 7.1: Triangle Difference Test: New Malt Supply

A test beer "B" is brewed using a new lot of malt, and the sensory analyst wishes to know if it can be distinguished from control beer "A" taken from current production. A 5% risk of error is accepted, and 12 trained assessors are available; 18 glasses of "A" and 18 glasses of "B" are prepared to make 12 sets that are distributed at random among the subjects, using two each of the combinations ABB, BAA, AAB, BBA, ABA, and BAB.

Eight subjects correctly identify the odd sample. In Table 19.8, the conclusion is that the two beers are different at the 5% level of significance.

Example 7.2: Detailed Example of Triangle Difference Test: Foil versus Paper Wraps for Candy Bars

Problem/situation: The director of packaging of a confection company wishes to test the effectiveness of a new foil-lined packaging material against the paper wrap currently being used for candy bars. Preliminary observation shows that paper-wrapped bars begin to show harder texture after 3 months while foil-wrapped bars remain soft. The director feels that if he can show a significant difference at 3 months, he can justify a switch in wrap for the product.

Project objective: To determine if the change in packaging causes an overall difference in flavor and/or texture after 3 months of shelf storage.

Test objective: To measure if people can differentiate between the two three-month-old products by tasting them.

Test design: a triangle difference test with 30–36 subjects. The test will be conducted under normal white lighting to allow for differences in appearance to be taken into account. The subjects will be scheduled in groups of six to ensure full randomization within groups. Significance for a difference will be determined at an α risk of 5%, that is, this test will falsely conclude a difference only 5% of the time.

Screen samples: Inspect samples initially (before packaging) to ensure that no gross sensory differences are noticeable from sample to sample. Evaluate test samples at 3 months to ensure that no gross sensory characteristics have developed that would render the test invalid.

Conduct the test: Code two groups each of 54 plates with three-digit random numbers from Table 19.1. Remove samples from package; cut-off ends of each bar and discard; cut bar into bite-size pieces and place on coded plates. Keep plates containing samples that were paper-wrapped (P) separate from those containing samples that were foil-wrapped (F). For each subject, prepare a tray marked by his/her number and containing three plates that are P or F according to the worksheet in Figure 7.2. Record the three plate codes on the subject's ballot (see Figure 7.3).

Analyze results: Of the 30 subjects who showed up for the test, 17 correctly identified the odd sample.

Date 6-2-99 Worksheet Test code 587 FF03

Post this sheet in the area where trays are prepared. Code scoresheets ahead of time. Label serving containers ahead of time.

Type of samples: Candy bars
Type of test: Triangle test

Sample identification	Code
Pkg 4736 (paper)	P
Pkg 3987 (foil)	F

Code serving containers as follows:

Panelist #	Order of presentation
1,7,13,19,25,31	P - F - F
2,8,14,20,26,32	F - P - F
3,9,15,21,27,33	F - F - P
4,10,16,22,28,34	F - P - P
5,11,17,23,29,35	P - F - P
6,12,18,24,30,36	P - P - F

1. Place stickers with panelist's number on tray.
2. Select plates "P" of "F" from those previously coded and place on tray from left to right.
3. Write codes selected on panelist's score sheet.
4. Serve samples.
5. Receive filled-in score sheet and note on it the order of presentation used, and whether reply was correct (c) or incorrect (i).

Figure 7.2 Worksheet for a triangle test. Example 7.2: foil vs. paper wraps for candy bars.

Triangle test		Test code:
Taster no. ____	Name: ________	Date: ________
Type of sample: ____________________		
Instruction Taste the samples on the tray from left to right. Two samples are identical; one is different. Select the odd/different sample and indicate by placing an X next to the code of the odd sample.		
Samples on tray	Indicate odd sample	Remarks
________	☐	________________
________	☐	________________
________	☐	________________
If you wish to comment on the reasons for your choice or if you wish to comment on the product characteristics, you may do so under Remarks.		

Figure 7.3 Score sheet for triangle test. Example 7.2: foil vs. paper wraps for candy bars. The subject places an X in one of the three boxes but may write remarks on more than one line.

Number of subjects	30
Number correct	17

Table 19.8 indicates that this difference is significant at an α-risk of 1% (probability $p \leq 0.01$).

Test report: The full report should contain the project objective, the test objective, and the test design as previously described. Examples of worksheet and score sheet may be enclosed. Any information or recommendations given to the subjects (e.g., about the origin of samples) must be reported. The tabulated results (17 correct out of 30) and the α-risk (meets the objective of 5%) follow. In the conclusion, the results are tied to the project objective: "A significant difference was found between the paper- and foil-wrapped candies. The foil does produce a perceived effect. There were 10 comments about softer texture in the foil-wrapped samples."

Example 7.3: Triangle Test for Similarity. Determining Panel Size Using α, β, and p_d: Blended Table Syrup

Problem/situation: A manufacturer of blended table syrup has learned that his supplier of corn syrup is raising the price of this ingredient. The research team has identified an alternate supplier of high-quality corn syrup whose price is more acceptable. The sensory analyst is asked to test the equivalency of two samples of blended table syrup, one formulated with the current supplier's product and the other with the less expensive corn syrup from the alternate supplier.

Project objective: To determine if the company's blended syrup can be formulated with the less expensive corn syrup from the alternate supplier without a perceptible change in flavor.

Test objective: To test for similarity of the blended table syrup produced with corn syrups from the current and alternate suppliers.

Number of assessors and choice of α, β, and p_d: The sensory analyst and the project director, looking at Table 19.7, note that to obtain maximum protection against falsely concluding similarity, for example by setting β at 0.1% (i.e., $\beta=0.001$) relative to the alternative hypothesis, that the true proportion of the population able to detect a difference between the samples is at least 20% (i.e., $p_d=0.20$). To preserve a modest α-risk of 0.10 they need to have at least 260 assessors. They decide to compromise at $\alpha=0.20$, $\beta=0.01$, and $p_d=30\%$, which requires 64 assessors.

Test design: The sensory analyst conducts a 66-response triangle test according to the established test protocol for blended table syrups. The sensory booths are prepared with red-tinted filters to mask color differences. Twelve panelists are scheduled for each of five consecutive sessions and six panelists are scheduled for the sixth and final session. Figure 7.4 shows the analyst's worksheet for a typical session.

Analyze results: Out of 66 respondents, 21 correctly picked the odd sample. Referring to Table 19.8, in the row corresponding to $n=66$ and the column corresponding to $\alpha=0.20$, one finds that the minimum number of correct responses required for significance is 26. Therefore, with only 21 correct responses, it can be concluded that any sensory difference between the two syrups is sufficiently small to be ignored; that is, the two samples are sufficiently similar to be used interchangeably.

Date 11-5-98 — Worksheet — No. 35-0032-31

Post this sheet in the area where trays are prepared. Code score sheets ahead of time. Label serving containers ahead of time.

Type of samples : Blended table syrups

Type of test : Triangle similarity test

Sample identification:	Codes used for: Sets with 2 A's	Codes used for: Sets with 2 B's
A: Lab code 47-3651	587 246	413
B: Lab code 026 (Control)	894	365 751

Code serving containers as follows:

Subject #	Codes in order	Underlying pattern*
1	587 246 894	AAB
2	413 365 751	ABB
3	751 413 365	BAB
4	246 587 894	AAB
5	751 365 413	BBA
6	587 894 246	ABA
7	413 751 365	ABB
8	246 894 587	ABA
9	894 587 246	BAA
10	365 751 413	BBA
11	894 246 587	BAA
12	365 413 751	BAB

*Each pattern is repeated twice to allow for each code in each position.

Figure 7.4 Worksheet for triangle test for similarity. Example 7.3: blended table syrup.

Interpret results: The analyst informs the project manager that the test resulted in 21 correct selections out of 66, indicating with 99% confidence that the proportion of the population who can perceive a difference is less than 30% and probably much lower. The alternate supplier's product can be accepted.

Confidence limits on p_d: If desired, analysts can calculate confidence limits on the proportion of the population that can distinguish the samples. The calculations are as follows:

$$p_c \text{ (proportion correct)} = \frac{c}{n}$$

$$p_d \text{ (proportion distinguishers)} = 1.5p_c - 0.5$$

$$s_d \text{ (standard deviation of } p_d) = 1.5\sqrt{\frac{p_c(1-p_c)}{n}}$$

$$\text{one-sided upper confidence limit} = p_d + z_\beta s_d$$

$$\text{one-sided lower confidence limit} = p_d - z_\alpha s_d,$$

where

c	is the number of correct responses
n	is the total number of assessors, and
z_α and z_β	are critical values of the standard normal distribution. Commonly used values of z for one-sided confidence limits include

Confidence Level (%)	z
75	0.674
80	0.842
85	1.036
90	1.282
95	1.645
99	2.326

For the data in the example, the upper 99% one-sided confidence limit on the proportion of distinguishers is calculated as

$$p_{\max} = p_d + z_\beta s_d = \left[1.5\left(\frac{21}{66}\right) - 0.5\right] + (2.326)(1.5)\sqrt{\frac{\left(\frac{21}{66}\right)\left(1-\left(\frac{21}{66}\right)\right)}{66}}$$

$$= [-0.023] + 2.326(1.5)(0.05733)$$

$$= 0.177 \text{ or } 18\%$$

whereas the lower 80% one-sided confidence limit falls at

$$p_{\min} = p_d - z_\alpha s_d = [-0.023] - 0.842(1.5)(0.05733)$$

$$= -0.095 \text{ (i.e., 0.0, it cannot be negative),}$$

or, in other words, the sensory analyst is 99% sure that the true proportion of the population that can distinguish the samples is no greater than 18% and may be as low as 0%.*

Example 7.4: Balancing α, β, and p_d. Setting Expiration Date for a Soft Drink Composition

Problem/situation: A producer of a soft drink composition wishes to choose a recommended expiration date to be stamped on bottled soft drinks made with it. It is known that in the cold (2°C), bottled samples can be stored for more than one year without any change in flavor, whereas at higher temperatures, the flavor shelf life is shorter. A test is carried out in which samples are stored at high ambient temperature (30°C) for 6, 8, and 12 months, then presented for difference testing.

Project objective: To choose a recommended expiration date for a bottle product made with the composition.

Test objective: To determine whether a sensory difference is apparent between the product stored cold and each of the three products stored warm.

Number of assessors and choice of α, β, and p_d: The producer would like to see the latest possible expiration date and decides he is only willing to take a 5% chance of concluding that there is a difference when there is not (i.e., $\alpha = 0.05$). The quality assurance (QA) manager, on the other hand, wishes to be reasonably certain that customers cannot detect an "aged" flavor until after the expiration date, so he agrees to accept 90% certainty (i.e., $\beta = 0.10$) that no more than 30% of the population (i.e., $p_d = 30\%$) can detect a difference. Consulting Table 19.7 in the column under $\beta = 0.10$ and the section for $p_d = 30\%$, the sensory analyst finds that a panel of 53 is needed for the tests. However, only 30 panelists can be made available for the duration of the tests. Therefore, the three of them renegotiate the test sensitivity parameters to provide the maximum possible risk protection with the number of available assessors. Consulting Table 19.7 again, they decide that a compromise of $p_d = 30\%$, $\beta = 0.20$, and $\alpha = 0.10$ provides acceptable sensitivity given the number of available assessors.

Test design: The analyst prepares and conducts triangle tests using a panel of 30.

Analyze results: The number of correct selections turns out as follows: at 6 months, 11; at 8 months, 13; at 12 months, 15. Consulting Table 19.8, the analyst concludes that, at 6 months, no proof of difference exists. At 8 months, the difference is larger. Table 19.8 shows that proof of difference would have existed had a higher $\alpha = 0.20$ been used. Finally, at 12 months, the table shows that proof of a difference exists at $\alpha = 0.05$.

Interpretation: The group decides that an expiration date of 8 months provides adequate assurance against occurrences of "aged" flavor in product that has not passed this date. As an added check on their conclusion, the 80% one-sided confidence limits are calculated for each test. It is found that they can be 80% sure that no more than

* Unified approach versus similarity tables: Notice that the unified approach used in this fifth edition does not include similarity tables such as those found in the second edition. As the present example illustrates, Table 19.8 merely shows that proof of similarity exists. To learn how strong the evidence of similarity is, that is, that "p_d is no greater than 18% and may be as low as 0%," the analyst needs to calculate the confidence limits. See Section 14.2.3 for the derivation of confidence intervals.

16% of consumers can detect a difference at 6 months, no more than 26% at 8 months, but possibly as many as 37% at 12 months. The product is safely under the $p_d = 30\%$ limit at 8 months.*

7.4 DUO–TRIO TEST

7.4.1 Scope and Application

The duo–trio test (ISO, 2004a) is statistically less efficient than the triangle test because the chance of obtaining a correct result by guessing is 1 in 2. On the other hand, the test is simple and easily understood. Compared with the paired-comparison test, it has the advantage that a reference sample is presented that avoids confusion with respect to what constitutes a difference, but a disadvantage is that three samples, rather than two, must be evaluated.

Use this method when the test objective is to determine whether a sensory difference exists between two samples. This method is particularly useful

1. To determine whether product differences result from a change in ingredients, processing, packaging, or storage
2. To determine whether an overall difference exists where no specific attributes can be identified as having been affected

The duo–trio test has general application whenever more than 15, and preferably more than 30, test subjects are available. Two forms of the test exist: the *constant reference mode*, in which the same sample, usually drawn from regular production, is always the reference; and the *balanced reference mode*, in which both of the samples being compared are used at random as the reference. Use the constant reference mode with trained subjects whenever a product well known to them can be used as the reference. Use the balanced reference mode if both samples are unknown or if untrained subjects are used.

If there are pronounced aftertastes, the duo–trio test is less suitable than the paired-comparison test (see Chapter 8.2).

7.4.2 Principle of the Test

Present to each subject an identified reference sample, followed by two coded samples, one of which matches the reference sample. Ask subjects to indicate which coded sample matches the reference. Count the number of correct replies and refer to Table 19.10 for interpretation.

7.4.3 Test Subjects

Select, train, and instruct the subjects as described under Section 7.3.3. As a general rule, the minimum is 16 subjects, but for less than 28, the β-error is high. Discrimination is much improved if 32, 40, or a larger number can be employed.

* An example of the confidence limit calculation using the 6-month results is $p_d = (1.5(11/30) - 0.5) + 0.84(1.5)\sqrt{(11/30)(1-(11/30))/30} = 0.16$.

7.4.4 Test Procedure

For test controls and product controls, (see Sections 3.2 and 3.3). Offer samples simultaneously, if possible, or else sequentially. Prepare equal numbers of the possible combinations (see examples) and allocate the sets at random among the subjects. An example of a score sheet (which is the same in the balanced reference and constant reference modes) is given in Figure 7.5. Space for several duo–trio tests may be provided on the score sheet but do not ask supplementary questions (e.g., the degree or type of difference or the subject's preference), as the subject's choice of matching sample may bias his response to these additional questions. Count the number of correct responses and the total number of responses and refer to Table 19.10. Do not count "no difference" responses; subjects must guess if in doubt. Three examples follow, all using the unified approach.

Example 7.5: Balanced Reference: Fragrance for Facial Tissue Boxes

Problem/situation: A product development fragrance chemist needs to know if two methods of fragrance delivery for boxed facial tissues—fragrance delivered directly to the tissues, or fragrance delivered to the inside of the box—will produce differences in perceived fragrance quality or quantity.

Project objective: To determine if the two methods of fragrance delivery produce any difference in the perceived fragrance of the two tissues after they have been stored for a period of time comparable to normal product age at time of use.

Test objective: To determine if a fragrance difference can be perceived between the two tissue samples after storage for 3 months.

Test No.

Duo–trio test

Taster no. ________ Name: ________ Date: ____

Type of sample ________________

Instructions: Taste samples from left to right.
The left-hand sample is a reference. Determine which of the two samples matches the reference and indicate by placing an X.

If no difference is apparent between the two unknown samples, you must guess.

Reference | Code ______ | Code ______

Comments: ________________

Figure 7.5 Score sheet for duo–trio test.

Test design: When the stimuli are complex, a duo–trio test requires less repeated sniffing of samples than triangle tests or attribute difference testing. This reduces the potential confusion caused by odor adaptation and/or the difficulty in sorting out three sample intercomparisons. The test is conducted with 40 subjects who have some experience in odor evaluation. The samples are prepared by the fragrance chemist using the same fragrance and the same tissues on the same day. The boxed tissues are then stored under identical conditions for 3 months. Test tissues are taken from the center 50% of the box; each tissue is placed in a sealed glass jar 1 h prior to evaluation. This allows for some fragrance to migrate to the headspace, and the use of the closed container reduces the amount of fragrance buildup in the testing booths. Each of the two samples is used as the reference in half (20) of the evaluations. Figure 7.6 shows the score sheet used.

Analyze results: Only 21 out of the 40 subjects chose the correct match to the designated reference. According to Table 19.10, 26 correct responses are required at an α-risk of 5%. In addition, when the data are reviewed for possible effects from the position of each sample as reference, the results show that the distribution of correct responses is even (10 and 11). This indicates that the quality and/or quantity of the two fragrances have little, if any, additional biasing effect on the results.

Interpret results: The sensory analyst informs the fragrance chemist that the odor duo–trio test failed to detect any significant odor differences between the two packing systems given the fragrance, the tissue, and the storage time used in the study.

Duo–trio test — Test no. 230S

Panelist no. 21 Name:________ Date:______

Type of Sample: Facial tissue in a glass jar

Instructions

1. Please sniff each sample, starting at the left. Remove the cap only briefly and take short, shallow sniffs.
2. The left hand sample is a reference. Determine which of the two coded samples matches the fragrance of the reference.
3. Indicate the matching sample by placing an X in the corresponding box.

If no difference is apparent between the two unknown samples, you must guess.

Reference — Code ______ ☐ — Code ______ ☐

Comments: ________________

Figure 7.6 Score sheet for duo–trio test. Example 7.5: balanced reference mode.

Sensitivity of the test: For planning future studies of this type, note that choosing 40 subjects for a duo–trio test yields the following values for the test-sensitivity parameters:

Proportion of Distinguishers (p_d) %	Probability of Detecting	
	$(1-\beta)$ @ $\alpha=0.05$	$(1-\beta)$ @ $\alpha=0.10$
10	0.13	0.21
15	0.21	0.32
20	0.32	0.44
25	0.44	0.57
30	0.57	0.69
35	0.70	0.80
40	0.81	0.88
45	0.89	0.94
50	0.95	0.97

For example, using 40 subjects and testing at the $\alpha=0.05$ level yields a test that has a 44% chance ($1-\beta=0.44$) of detecting the situation where 25% of the population can detect a difference ($p_d=25\%$). Increasing the number of subjects increases the likelihood of detecting any given value of p_d. Testing at larger values of α also increases the chances of detecting a difference at a given p_d.

Example 7 6: Constant Reference: New Can Liner

Problem/situation: A brewer is faced with two supplies of cans, A being the regular supply he has used for years and B a proposed new supply said to provide a slight advantage in shelf life. He wants to know whether any difference can be detected between the two cans. The brewer feels that it is important to balance the risk of introducing an unwanted change to his beer against the risk of passing up the extended shelf life offered by can B.

Project objective: To determine if the package change causes any perceptible difference in the beer after shelf storage as normally experienced in the trade.

Test objective: To determine if any sensory difference can be perceived between the two beers after 8 weeks of shelf storage at room temperature.

Number of assessors: The brewer knows from past experience that if no more than $p_d=30\%$ of his panel can detect a difference, then he assumes no meaningful risk in the marketplace. He is slightly more concerned with introducing an unwanted difference than he is with passing up the slightly extended shelf life offered by can B. Therefore, he decides to set the β-risk at 0.05 and his α-risk at 0.10. Referring to Table 19.9 in the section for $p_d=30\%$, the column for $\beta=0.05$ and the row for $\alpha=0.10$, he finds that 96 respondents are required for the test.

Test design: A duo–trio test in the constant reference mode is appropriate because the company's beer in can A is familiar to the tasters. A separate test is conducted at each of the brewer's three testing sites. Each test is set up with 32 subjects, with A as the reference; 64 glasses of beer A and 32 of beer B are prepared and served to the subjects in 16 combinations AAB and 16 combinations ABA, the left-hand sample being the reference.

Analyze results: 18, 20, and 19 subjects correctly identified the sample that matched the reference. According to Table 19.10, significance at the 10% level requires 21 correct.

Note: In many cases, it is permissible to combine two or more tests so as to obtain improved discrimination. In the present case, the cans were samples of the same lot, and the subjects were from the same panel, so combination is permissible. $18+20+19=57$ correct out of $3\times32=96$ trials. From Table 19.10, the critical numbers of correct replies with 96 samples are 55 at the 10% level of significance, and 57 at the 5% level.

Interpret results: Conclude that a difference exists, significant at the 5% level on the basis of combining three tests. Next, examine any notes made by panelists that describe the difference. If none is found, submit the samples to a descriptive panel. Ultimately, if the difference is neither pleasant nor unpleasant, a consumer test may be required to determine if there is preference for one can or the other.

Example 7.7: Duo–Trio Similarity Test: Replacing Coffee Blend

Problem/situation: A manufacturer of coffee has learned that one coffee bean variety, which has long been a major component of its blend, will be in short supply for the next 2 years. A team of researchers has formulated three new blends that they feel are equivalent in flavor to the current blend. The research team has asked the sensory evaluation analyst to test the equivalency of these new blends to the current product.

Project objective: To determine which of the three blends can best be used to replace the current blend.

Test objective: To test for similarity between the current blend and each of the project blends.

Test design: Preliminary tests have shown that differences are small and not particularly related to a specific attribute. Therefore, use of the duo–trio test for similarity is appropriate. To reduce the risk of missing a perceptible difference, the sensory analyst proposes the tests be run using 60 panelists each (an increase from the customary 36 used in testing for difference). Using her spreadsheet test sensitivity analyzer* (see Section 14.3.5), she has determined that a 60-respondent duo–trio test has a 90% (i.e., $\beta=0.10$) probability of detecting the situation where $p_d=25\%$ of the panelists can detect a difference, with an accompanying α-risk of approximately 0.25. The analyst accepts the large α-risk because she is much more concerned with incorrectly approving a blend that is different from the control, and she only has 60 panelists available for the tests. For each blend, the sensory analyst plans to conduct one 60-response coffee test spaced over one week. As the preparation and holding time of the product is a critical factor that influences flavor, subjects must be carefully scheduled to arrive within 10 min after preparation of the products. Using the 12 booths in the sensory lab, prepared with brown-tinted filters on the lights, the analyst schedules 12 different subjects for each cell of each test. The use of 12 panelists per session permits a balanced presentation of each sample as the reference sample as well as a balanced order of presentation of the two test samples within the cell. Figure 7.7 shows the analyst's worksheet.

Samples are presented without cream and sugar. The pots are kept at 175°F and poured into heated (130°F) ceramic cups that are coded as per the worksheet and placed

* Available upon request in Excel as an e-mail attachment from Tom.Carr@CarrConsulting.net.

Date 3-4-99 Cell no. 3 Worksheet No. 2803–30

Post this sheet in the area where trays are prepared. Code score sheets ahead of time. Label serving containers ahead.

Type of samples: cups of coffee

Type of test: Duo-trio similarity test (balanced reference)

Samples served A = Control B = Blend 62-A C = Blend 223B D = Blend 211

Codes Used:

	For B Sets w/ 2 A'S	versus A Sets w/ 2 B'S	For C Sets w/ 2 A'S	versus A Sets w/ 2 C'S	For D Sets w/ 2 A'S	versus A Sets w/ 2 D'S
Sample A	317 543	986	866 581	541	121 225	965
Sample B	314	393 737				
Sample C			674	373 158		
Sample D					221	499 134

Code serving containers as follows:

Subject No.	Pattern	Codes in order	Pattern	Codes in order	Pattern	Codes in order
37	ABA	R – 314–543	AAC	R – 581–674	DAD	R – 965–134
38	BBA	R – 737–986	ACA	R – 674–866	AAD	R – 225–221
39	BAB	R – 986–393	CCA	R – 158–541	ADA	R – 221–121
40	AAB	R – 317–314	CAC	R – 541–373	DDA	R – 499–965
41	ABA	R – 314–317	AAC	R – 866–674	DAD	R – 965–499
42	BBA	R – 393–986	ACA	R – 674–581	AAD	R – 121–221
43	BAB	R – 986–737	CCA	R – 373–541	ADA	R – 221–225
44	AAB	R – 543–314	CAC	R – 541–158	DDA	R – 134–965
45	ABA	R – 314–543	AAC	R – 581–674	DAD	R – 965–134
46	BBA	R – 737–986	ACA	R – 674–866	AAD	R – 225–221
47	BAB	R – 986–393	CCA	R – 158–541	ADA	R – 221–121
48	AAB	R – 317–314	CAC	R – 541–373	DDA	R – 499–965

Figure 7.7 Worksheet for duo–trio similarity test. Example 7.7: replacing coffee blend.

in the order that it indicates. Score sheets (see Figure 7.8) are prepared in advance to save time, and samples are poured when the subject is already sitting in the booth.

Analyze results: The number of correct responses for the three test blends were

Cell No. (of 12 Subjects)	Blend B	Blend C	Blend D
1	3	6	8
2	4	5	8
3	5	7	5
4	7	7	7
5	5	5	7
Total	24	30	35

From her spreadsheet test sensitivity analyzer, the analyst knows that 33 correct responses are necessary to conclude that a significant difference exists at the α-risk chosen for the test (approximately 0.25), so 32 or fewer correct responses from the 60-respondent test is evidence of adequate similarity.

	Duo–trio test	Test no. 28 03-03
Taster no. ________	Name:________________	Date:____
Type of sample: cups of coffee, freshly brewed		

Instructions: Taste sample from left to right
The left-hand sample is a reference. Determine which
of the two axis samples matches the reference and indicate
by placing an X

If no difference is apparent between the two unknown samples,
you must guess

Reference	Code ______	Code ______
▨	☐	☐

Comments: ______________________________

Figure 7.8 Score sheet for duo–trio similarity test. Example 7.7: replacing coffee blend.

Output from Test Sensitivity Analyzer

Inputs				**Output**			
Number of Respondents	**Number of Correct Responses**	**Probability of Correct Guess**	**Proportion Distinguishers**	**Probability of a Correct Response @ p_d**	**Type I Error**	**Type II Error**	**Power**
(n)	**(X)**	**(p_0)**	**(p_d)**	**(p_{max})**	**(α-risk)**	**(β-risk)**	**(1 − β)**
60	33	0.50	0.25	0.625	0.2595	0.0923	0.9077

Interpretation:

33 or more correct responses is evidence of a difference at the $a = 0.26$ level of significance.

32 or fewer correct responses indicates that you can be 91% sure that no more than 25% of the panelists can detect a difference—that is, evidence of similarity relative to $p_d = 25\%$ at the $\beta = 0.09$ level of significance.

Therefore, it is concluded that test blends B and C are sufficiently similar to the control to warrant further consideration, but that test blend D, with 35 correct answers, is not. The 90% upper one-tailed confidence interval on the true proportion of distinguishers for test blend D (based on the duo–trio test method) is

$$p_{\max(90\%)} = \left[2\left(\frac{x}{n}\right) - 1\right] + z_\beta \sqrt{\left[4\left(\frac{x}{n}\right)\left(1 - \left(\frac{x}{n}\right)\right)\right] n}$$

$$= \left[2\left(\frac{35}{60}\right) - 1 \right] + 1.282 \frac{\sqrt{\left[4\left(\frac{35}{60}\right)\left(1 - \left(\frac{35}{60}\right)\right) \right]}}{60}$$

$$= [0.1667] + 1.282(0.1273)$$

$$= 0.33, \text{ or } 33\%.$$

The sensory analyst concludes with 90% confidence that the true proportion of the population that can distinguish test blend D from the control may be as large as 33%, thus exceeding the prespecified critical limit (p_d) of 25% by as much as 8%.

The sensory analyst may have an additional concern. Only 24 of the 60 respondents correctly identified test blend B. In a duo–trio test involving 60 respondents, the expected number of correct selections when all of the respondents are guessing ($p_d = 0$) is $n/2 = 30$. The less-than-expected number of correct responses may indicate that some extraneous factor was active during the testing of blend B that biased the respondents away from making the correct selection, for example, mislabeled samples or poor preparation or handling of the samples before serving. The sensory analyst tests the hypothesis that the true probability of a correct response is at least 50% (H_0: $p \geq 0.5$) against the alternative that it is less than 50% (H_a: $p < 0.5$) using the normal approximation to the binomial with the one-tailed confidence level set at 95% (i.e., $\alpha = 0.05$, lower tail). The test statistic is

$$z = \frac{\left[\left(\frac{x}{n}\right) - p_0 \right]}{\sqrt{\frac{p_0(1 - p_0)}{n}}}$$

$$= \frac{\left[\left(\frac{24}{60}\right) - 0.05 \right]}{\sqrt{\frac{0.50(1 - 0.50)}{60}}}$$

$$= \frac{[-0.10]}{(0.06455)}$$

$$= -1.55.$$

Using Table 19.2 (noting that $\Pr[z < -1.55] = \Pr[z > 1.55]$), the sensory analyst finds that the probability of observing a value of the test statistic no larger than −1.55 is (0.5–0.4394) = 0.0606. This probability is greater than the value of $\alpha = 0.05$, and the analyst concludes that there is not sufficient evidence to reject the null hypothesis at the 95% level. The 24 correct responses were not sufficiently off the mark (of 30) for the analyst to conclude that an extraneous factor was active.

7.5 TWO-OUT-OF-FIVE TEST

7.5.1 Scope and Application

This method is statistically very efficient because the chances of correctly guessing two out of five samples are 1 in 10 as compared with 1 in 3 for the triangle test. By the same token, the test is so strongly affected by sensory fatigue and by memory effects that its principal use has been in visual, auditory, and tactile applications, not in flavor testing.

Use this method when the test objective is to determine whether a sensory difference exists between two samples and particularly when only a small number of subjects is available (e.g., ten).

As with the triangle test, the two-out-of-five test is effective in certain situations:

1. To determine whether product differences result from a change in ingredients, processing, packaging, or storage
2. To determine whether an overall difference exists, where no specific attribute(s) can be identified as having been affected
3. To select and monitor panelists for ability to discriminate given differences in test situations where sensory fatigue effects are small

7.5.2 Principle of the Test

Present to each subject five coded samples. Instruct subjects that two samples belong to one type and three to another. Ask the subjects to taste (feel, view, examine) each product from left to right and select the two samples that are different from the other three. Count the number of correct replies and refer to Table 19.14 for interpretation.

7.5.3 Test Subjects

Select, train, and instruct the subjects as described in Section 3.3. Generally, 10–20 subjects are used. As few as five to six may be used when differences are large and easy to identify. Use only trained subjects.

7.5.4 Test Procedure

For test controls and product controls, Sections 3.2 and 3.3. Offer samples simultaneously if possible; however, samples that are bulky or show slight differences in appearance may be offered sequentially without invalidating the test. If the number of subjects is other than 20, select the combinations at random from the following, taking equal numbers of combinations with three *A*s and three *B*s:

AAABB	ABABA	BBBAA	BABAB
AABAB	BAABA	BBABA	ABBAB
ABAAB	ABBAA	BABBA	BAABB
BAAAB	BABAA	ABBBA	ABABB
AABBA	BBAAA	BBAAB	AABBB

Two-out-of-five test		
Name: ______________ Date: ________		
Type of sample: ______________		
Instructions 1. Examine the samples from left to right. Two are of one type, and the other three of another. 2. Identify the group of two samples by placing an X in the corresponding boxes.		
Test 1	Test 2	Test 3
Left ______	______	______
______	______	______
______	______	______
______	______	______
Right ______	______	______
Comments		
Left ______	______	______
______	______	______
______	______	______
______	______	______
Right ______	______	______

Figure 7.9 Score sheet for three two-out-of-five tests.

An example of a score sheet is given in Figure 7.9. Count the number of correct responses and the number of total responses and refer to Table 19.14. Do not count "no difference" responses; subjects must guess if in doubt.

Example 7.8: Comparing Textiles for Roughness

Problem/situation: A textile manufacturer wishes to replace an existing polyester fabric with a polyester/nylon blend. He has received a complaint that the polyester/nylon blend has a rougher and scratchier surface.

Project objective: To determine whether the polyester/nylon blend needs to be modified because it is too rough.

Test objective: To obtain a measure of the relative difference in surface feel between the two fabrics.

Test design: As sensory fatigue is not a large factor, the two-out-of-five test is the most efficient for assessing differences. A small panel of 12 will be able to detect quite small differences. Choose, at random, 12 combinations of the two fabrics from the table of 20 combinations previously presented. Ask the panelists "Which two samples feel the same and different from the other three?"

Conduct the test: Place each of the anchored or loosely mounted fabric swatches inside a cardboard tent in a straight line in front of each panelist (see Figure 7.10), who

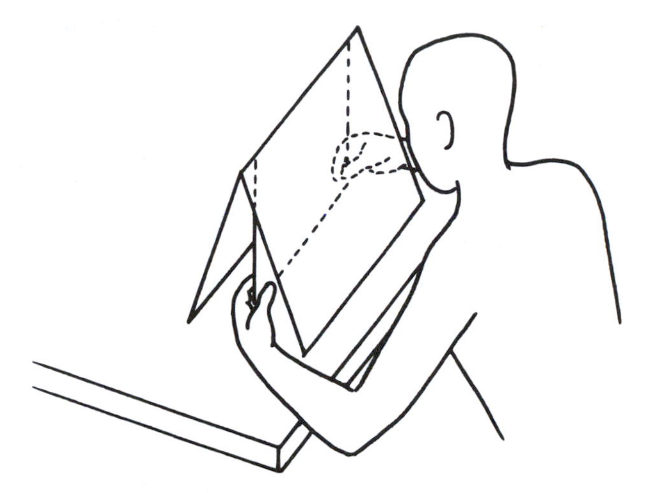

Figure 7.10 Two-out-of-five test. Example 7.8: arrangement of fabric samples in front of panelist.

must be able to feel the fabrics but not see them. Assign sample codes from a list of random three-digit numbers (see Table 19.1). Use the score sheet in Figure 7.11.

Analyze results: Of the 12 subjects, 9 were able to correctly group the fabric samples. Reference to Table 19.14 shows that the difference in surface feel was detectable at a level of significance of $\alpha = 0.001$.

Two-out-of-five test — Test code

Name: ______________ Date: ________

Type of sample: ______________

Type of difference: ______________

____

Instructions

1. Examine the samples in the order listed below. Two are of one type and the other three of another. Feel the surface gently with fingers or palm of hand.

2. Identify the two samples that feel the same by placing an X in the corresponding boxes.

Codes	x	Comments
________	☐	________________
________	☐	________________
________	☐	________________
________	☐	________________
________	☐	________________

Figure 7.11 Score sheet for two-out-of-five test. Example 7.8: comparing textiles for roughness.

Interpret results: The fabric manufacturer is informed that a difference in surface feel between the two fabric types is easily detectable.

Example 7.9: Emollient in Face Cream

Problem/situation: The substitution of one emollient for another in the formula for a face cream is desirable because of a significant saving in cost of production. The substitution appears to reduce the surface gloss of the product.

Project objective: The marketing group wishes to determine whether a visually detectable difference exists between the two formulas before going to consumers to determine any effect on acceptance.

Test objective: To determine whether a statistically significant difference in appearance exists between the two formulas of face cream.

Test design/screen samples: Use ten subjects who have been screened for color blindness and impaired vision. Test 2 mL of product under white incandescent light on a watch glass against a white background. Pretest samples to be sure that surfaces do not change (crust, weep, discolor) within 30 min after exposure, the maximum length of one test cell.

Conduct test: Arrange samples in a straight line from left to right according to the plan shown on the worksheet (see Figure 7.12); use a score sheet similar to the one in Figure 7.11. Ask the subjects to "identify the two samples that are the same in appearance and different from the other three."

Analyze results: Five subjects group the samples correctly. According to Table 19.14, this corresponds to 1% significance for a difference.

Date 3-05-99 Worksheet Test code TO-AF88

Post this sheet in the area where trays are prepared. Code score sheets ahead of time. Label serving containers ahead of time.

Type of samples : Face cream for viewing
Type of test : Two-out-of-five test

Sample identification	Code
Px-2316 (control)	A
Px-2602 (new emollient)	B

Arrange samples as follows in the front of each subject:

Judge no.	Order of samples
1	A A B B B
2	A B B A B
3	B A A B B
4	B A B B A
5	B B A B A
6	B B A A A
7	B A A B A
8	A B B A A
9	A B A A B
10	A A B A B

Figure 7.12 Worksheet for two-out-of-five test. Example 7.9: emollient in face cream. Arrangement of samples for viewing.

7.6 SAME/DIFFERENT TEST (OR SIMPLE DIFFERENCE TEST)

7.6.1 Scope and Application

Use this method when the test objective is to determine whether a sensory difference exists between two products, particularly when these are unsuitable for triple or multiple presentation, for example, when the triangle and duo–trio tests cannot be used. Examples of such situations are comparisons between samples of strong or lingering flavor, samples that need to be applied to the skin in half-face tests, and samples that are very complex stimuli and are mentally confusing to the panelists.

As with other overall difference tests, the same/different test is effective

1. To determine whether product differences result from a change in ingredients, processing, packaging, or storage
2. To determine whether an overall difference exists where no specific attribute(s) can be identified as having been affected

This test is somewhat time consuming because the information on possible product differences is obtained by comparing responses obtained from different pairs (A/B and B/A) with those obtained from matched pairs (A/A and B/B). The presentation of the matched pair enables the sensory analyst to evaluate the magnitude of the "placebo effect" of simply asking a difference question.

7.6.2 Principle of the Test

Present each subject with two samples, asking whether the samples are the same or different. In half of the pairs, present the two different samples; in half of the pairs, present a matched pair (the same sample, twice). Analyze results by comparing the number of "different" responses for the matched pairs to the number of "different" responses for the different pairs, using the χ^2-test.

7.6.3 Test Subjects

Generally, 20–50 presentations of each of the four sample combinations (A/A, B/B, A/B, B/A) are required to determine differences. Up to 200 different subjects can be used, or 100 subjects may receive two of the pairs. If the same/different test has been chosen because of the complexity of the stimuli, then no more than one pair should be presented to any one subject at a time. Subjects may be trained or untrained, but panels should not consist of mixtures of the two.

7.6.4 Test Procedure

For test controls and product controls, see Sections 3.2 and 3.3. Offer samples simultaneously, if possible, or else successively. Prepare equal numbers of the four pairs and present them at random to the subjects, if each is to evaluate only one pair. If the test is designed such that each subject is to evaluate more than one pair (one matched and one different or all four combinations), then records of each subject's test scores must be kept. Typical worksheets and score sheets are given in Example 7.10.

7.6.5 Analysis and Interpretation of Results

See Example 7.10.

Example 7.10: Replacing a Processing Cooker for Barbecue Sauce

Problem/situation: In an attempt to modernize a condiment plant, a manufacturer must replace an old cooker used to process barbecue sauce. The plant manager would like to know if the product produced in the new cooker tastes the same as that made in the old cooker.

Project objective: To determine if the new cooker can be put into service in the plant in place of the old cooker.

Test objective: To determine if the two barbecue sauce products, produced in different cookers, can be distinguished by taste.

Test design: The products are spicy and will cause carryover effects when tested. Therefore, the same/different test with a bland carrier, such as white bread, is an appropriate test. A total of 60 responses, 30 matched and 30 unmatched pairs, are collected from 60 subjects. Each subject evaluates either a matched pair (A/A or B/B) or an unmatched pair (A/B or B/A) in a single session. The worksheet and the score sheet for the test are shown in Figures 7.13 and 7.14. The test is conducted in the booth area under red lights to mask any color differences.

Screen samples: Preliminary tests are made with five experienced tasters to determine if the samples are easier to taste plain or on a carrier, such as white bread. The carrier is used to make comparison easier without introducing extraneous sensory factors.

Date 2-26-99 Worksheet Test code 84-46F09

Post this sheet in the area where trays are prepared. Code score sheets ahead of time. Label serving containers ahead of time.

Type of samples : Barbecue sauce on white bread pieces
Type of test : Same/Different test

Sample identification	Code
5-117-36 (old cooker)	36
5-117-39 (new cooker)	39

Code serving containers with 3-digit random numbers and divide into two lots, one lot to receive sample 36, the other sample 39.

When preparing panelists' trays, place samples from left to right in the following order :

Panelist code	Sample order
1–15	36–36
16–30	36–39
31–45	39–36
46–60	39–39

Figure 7.13 Worksheet for same/different test. Example 7.10: replacing a processing cooker for barbecue sauce.

Same/different test	Test no. 84-4639
Taster no. ________ Name: ____________ Date: ____	
Type of sample: Barbecue sauce on white bread pieces	
Instructions 1. Taste the two samples from left to right. 2. Determine if samples are the same/identical or different. 3. Mark your response below. Note that some of the sets consist of two identical samples.	
______ Products are the same ______ Products are different	
Comments: ________________________________	

Figure 7.14 Score sheet for same/different test. Example 7.10: replacing a processing cooker for barbecue sauce.

The pretest is also helpful in determining the appropriate amount of product (by weight or volume) relative to bread (by size) for the test.

Conduct test: Just before each subject is to taste, add the premeasured sauce to the precut bread pieces that had been stored cold in an airtight container. Place samples on labeled plates in the order indicated on the worksheet for each panelist.

Analyze results: In the table below, the columns indicate the samples that were tested; the rows indicate how they were identified by the subjects:

	Subjects Received		
	Matched Pair AA or BB	**Unmatched Pair AB or BA**	**Total**
Subjects said			
Same	17	9	26
Different	13	21	34
Total	30	30	60

The χ^2-analysis (see Section 14.3, Example 14.6) is used to compare the placebo effect (17/13) with the treatment effect (9/21). The χ^2-statistic is calculated as

$$x^2 = \sum \frac{(O-E)^2}{E},$$

where:

- O is the observed number and
- E is the expected number, in each of the four boxes same/matched, same/unmatched, different/matched, and different/unmatched. For example, for the box same/matched,

$$E=\frac{(26\times 30)}{60}=13\text{, and,}$$

$$x^2=\frac{(17-13)^2}{13}+\frac{(9-13)^2}{13}+\frac{(13-17)^2}{17}+\frac{(21-17)^2}{17}=4.34,$$

which is greater than the value in Table 19.5 (df = 1, probability = 0.05, $\chi^2=3.84$); that is, a significant difference exists.

Interpret results: The results show a significant difference between the barbecue sauces prepared in the two different cookers. The sensory analyst informs the plant manager that the equipment supplier's claim is not true. A difference has been detected between the two products. The analyst suggests that if the substitution of the new cooker remains an important cost/efficiency item in the plant, the two barbecue sauces should be tested for preference among users. A consumer test resulting in parity for the two sauces or in preference for the sauce from the new cooker would permit the plant to implement the process.

Note: If Example 7.10 had been run with 30 subjects rather than 60, and with each of the 30 receiving both a matched and an unmatched pair in separate sessions, the results could have been the same as above, but the χ^2-test would have been inappropriate and a McNemar test would be indicated (Conover, 1980). To perform the McNemar procedure, the analyst must keep track of both responses from each panelist and tally them in the following format:

		Subject Received A/B or B/A and Responded	
		Same	**Different**
Subject received A/A or B/B and responded	Same	$a=2$	$b=15$
	Different	$c=7$	$d=6$

The test statistic is

$$\text{McNemar's } T=\frac{(b-c)^2}{b+c}.$$

For $(b+c)\geq 20$, the assumption of no difference is rejected if T is greater than the critical value of a χ^2 with one degree of freedom from Table 19.5. For $(b+c)<20$, a binomial procedure is applied (see Conover, 1980). For the present example,

$$\text{McNemar's } T=\frac{(15-7)^2}{15+7}=2.91,$$

which is less than $\chi^2_{1,0.05}=3.84$. Therefore, one cannot conclude that the samples are different.

If the paired data from the 30 panelists had been treated as if they were individual observations from 60 panelists, one would have obtained the data as presented under "Analyze results," Example 7.6. The standard χ^2-analysis would have led to the incorrect conclusion that a statistically significant difference existed between the samples.

7.7 "A"–"NOT A" TEST

7.7.1 Scope and Application

Use this method (ISO, 1987) when the test objective is to determine whether a sensory difference exists between two products, particularly when these are unsuitable for dual or triple presentation, that is, when the duo–trio and triangle tests cannot be used. Examples of such situations are comparisons of products with a strong and/or lingering flavor, samples that need to be applied to the skin in half-face tests, products that differ slightly in appearance, and samples that are very complex stimuli and are mentally confusing to the panelists. Use the "A"–"not A" test in preference to the same/different test (Section 7.6) when one of the two products has importance as a standard or reference product, is familiar to the subjects, or is essential to the project as the current sample against which all others are measured.

As with other overall difference tests, the "A"–"not A" test is effective

1. To determine whether product differences result from a change in ingredients, processing, packaging, or storage
2. To determine whether an overall difference exists where no specific attribute(s) can be identified as having been affected

The test is also useful for screening of panelists, for example, in determining whether a test subject (or group of subjects) recognizes a particular sweetener relative to other sweeteners, and it can be used for determining sensory thresholds by a signal detection method.

7.7.2 Principle of the Test

Familiarize the panelists with samples "A" and "not A." Present each panelist with samples, some of which are product "A" while others are product "not A"; for each sample, the subject judges whether it is "A" or "not A." Determine the subjects' ability to discriminate by comparing the correct identifications with the incorrect ones using the χ^2-test.

7.7.3 Test Subjects

Train 10–50 subjects to recognize the "A" and the "not A" samples. Use 20–50 presentations of each sample in the study. Each subject may receive only one sample ("A" or "not A") or two samples (one "A" and one "not A"), or each subject may test up to ten samples in a series. The number of samples allowed is determined by the degree of physical and/or mental fatigue they produce in the subjects.

Note: A variant of this method, in which subjects are not familiarized with the "not A" sample, is not recommended. This is because subjects, lacking a frame of reference, may guess wildly and produce biased results.

7.7.4 Test Procedure

For test controls and product controls, see Sections 3.2 and 3.3. Present samples with a score sheet one at a time. Code all samples with random numbers and present them in random order so that the subjects do not detect a pattern of "A" versus "not A" samples in any series. Do not disclose the identity of samples until after the subject has completed the test series.

Note: In the standard version of the procedure, the following protocol is observed:

1. Products "A" and "not A" are available to subjects only until the start of the test.
2. Only one "not A" sample exists for each test.
3. Equal numbers of "A" and "not A" are presented in each test.

These protocols may be changed for any given test, but the subjects must be informed before the test is initiated. Under no. 2, if more than one "not A" samples exist, each must be shown to the subjects before the test.

7.7.5 Analysis and Interpretation of Results

The analysis of the data with four different combinations of sample versus response is somewhat complex and can best be understood by referring to Example 7.11.

Example 7.11: New Sweetener Compared with Sucrose

Problem/situation: A product development chemist is researching alternate sweeteners for a beverage that uses sucrose as 5% of the current formula. Preliminary taste tests have established 0.1% of the new sweetener as the level equivalent to 5% sucrose but have also shown that if more than one sample is presented at a time, discrimination suffers because of carryover of the sweetness and other taste and mouthfeel factors. The chemist wishes to know whether the two beverages are distinguishable by taste.

Project objective: To determine if the alternate sweetener at 0.1% can be used in place of 5% sucrose.

Test objective: To compare the two sweeteners directly while reducing carryover and fatigue effects.

Test design: The "A"–"not A" test allows the samples to be indirectly compared, and it permits the subjects to develop a clear recognition of the flavors to be expected with the new sweetener. Solutions of the sweetener at 0.1% are shown repeatedly to the subjects as "A," and 5% sucrose solutions are shown as "not A"; 20 subjects each receive 10 samples to evaluate in one 20 min test session. Subjects are required to taste each sample once, record the response ("A" or "not A"), rinse with plain water, and wait 1 min before tasting the next sample. Figure 7.15 shows the test worksheet and Figure 7.16 shows the score sheet.

Analyze results: In the table below, the columns show how the samples were presented and the rows show how the subjects identified them:

Date 1-15-99	Worksheet	Test code 612A83
Post this sheet in the area where trays are prepared. Code score sheets ahead of time. Label serving containers ahead of time.		
Type of samples : Sweetened beverage		
Type of test : "A" – "Not A" test		
Sample identification		Code
Beverage with 0.1% sweetener	("A")	A
Beverage with 5% sucrose	("Not A")	B

Code 200 6-oz cups with random three-digit numbers and divide into two lots of 100 each. Use sample "A" for the first 100 cups and sample "Not A" for the second 100 cups.
When preparing panelists' trays, place samples from left to right in the following order:

Panelist	Sample order									
1 – 5	A	A	B	B	A	B	A	B	B	A
6 – 10	B	A	B	A	A	B	A	A	B	B
11 – 15	A	B	A	B	B	A	B	B	A	A
16 – 20	B	B	A	A	B	A	B	A	A	B

Figure 7.15 Worksheet for "A"–"Not A" test. Example 7.11: new sweetener compared with sucrose.

		Subject Received		
		A	Not A	Total
Subject said	A	60	35	95
	Not A	40	65	105
	Total	100	100	200

The χ^2-statistic is calculated as in Section 7.6:

$$x^2 = \frac{(60-47.5)^2}{47.5} + \frac{(35-47.5)^2}{47.5} + \frac{(40-52.5)^2}{52.5} + \frac{(65-52.5)^2}{52.5} = 12.53,$$

which is greater than the value in Table 19.5 (df = 1, α-risk = 0.05, χ^2 = 3.84); that is, a significant difference exists.

Note: The χ^2-analysis just presented is not entirely appropriate because of the multiple evaluations performed by each respondent. However, no computationally convenient alternative method is currently available. The levels of significance obtained from this test should be considered approximate values.

Interpret results: The results indicate that the 0.1% sweetener solution is significantly different from the 5% sucrose solution. The sensory analyst informs the development chemist that the particular sweetener is likely to cause a detectable change in flavor of the beverage. The next logical step may be a descriptive analysis to characterize the difference.

"A"–"Not A" Test

Test code

Taster no:______ Name:____________ Date:_______

Type of sample: Sweetened beverage

Instructions

1. Before taking this test, familiarize yourself with the flavor of the samples "A" and "Not A" which are available from the attendant.
2. Taste the test samples from left to right. After each sample, record your response below, rinse your palate with water, and wait one full minute between samples.

Note: You have received approximately equal numbers of "A" and "Not A" samples.

Sample No.	Code	The sample is: "A"	"Not A"	Sample No.	Code	The sample is: "A"	"Not A"
1	____	☐	☐	6	____	☐	☐
2	____	☐	☐	7	____	☐	☐
3	____	☐	☐	8	____	☐	☐
4	____	☐	☐	9	____	☐	☐
5	____	☐	☐	10	____	☐	☐

Comments: ______________________________

Figure 7.16 Score sheet for "A"–"Not A" test. Example 7.11: new sweetener compared with sucrose.

One might ask "What would it take for the difference to be nonsignificant?" This would be the case if results had been:

60	50
40	50

for which χ^2 equals 2.02, a value less than 3.84. See ISO (1987) for a number of similar examples.

7.8 DIFFERENCE-FROM-CONTROL TEST

7.8.1 Scope and Application

Use this test when the project or test objective is twofold, both (1) to determine whether a difference exists between one or more samples and a control, and (2) to estimate the size of any such differences. Generally one sample is designated the "control," "reference," or "standard," and all other samples are evaluated with respect to *how different* each is from that control.

The difference-from-control test is useful in situations in which a difference may be detectable, but the size of the difference affects the decision about the test objective. Quality assurance/quality control and storage studies are cases in which the relative size of a difference from a control is important for decision making. The difference-from-control test is appropriate where the duo–trio and triangle tests cannot be used because of the normal heterogeneity of products such as meats, salads, and baked goods.

The difference-from-control test can be used as a two-sample test in situations where multiple sample tests are inappropriate because of fatigue or carryover effects. The difference-from-control test is essentially a simple difference test with an added assessment of the size of the difference.

7.8.2 Principle of the Test

Present to each subject a control sample plus one or more test samples. Ask subjects to rate the size of the difference between each sample and the control and provide a scale for this purpose. Indicate to the subject that some of the test samples may be the same as the control. Evaluate the resulting mean difference-from-control estimates by comparing them to the difference-from-control obtained with the blind controls.*

7.8.3 Test Subjects

Generally 20–50 presentations of each of the samples and the blind control with the labeled control are required to determine a degree of difference. If the difference-from-control test is chosen because of a complex comparison or fatigue factor, then no more than one pair should be given to any one subject at a time. Subjects may be trained or untrained, but panels should not consist of a mixture of the two. All subjects should be familiar with the test format, the meaning of the scale, and the fact that a proportion of test samples will be blind controls.

7.8.4 Test Procedure

For test controls and product controls, see Sections 3.2 and 3.3. When possible, offer the samples simultaneously with the labeled control evaluated first. Prepare one labeled control sample for each subject plus additional controls to be labeled as test samples. If the test is designed to have all subjects eventually test all samples but this cannot be done in one test session, a record of subjects by sample must be kept to ensure that remaining samples are presented in subsequent sessions.

---

* The use of the estimate obtained with the blind controls amounts to obtaining a measure of the placebo effect. This estimate represents the numerical effect of simply asking the difference question, when in fact no difference exists.

The scale used may be any of those discussed in Section 5.6. For example,

Verbal Category Scale	Numerical Category Scale
No difference	0 = No difference
Very slight difference	1
Slight/moderate difference	2
Moderate difference	3
Moderate/large difference	4
Large difference	5
Very large difference	6
	7
	8
	9 = Very large difference

(When calculating results with the verbal category scale, convert each verdict to the number placed opposite, e.g., large difference = 5.)

7.8.5 Analysis and Interpretation of Results

Calculate the mean difference-from-control for each sample and for the blind controls, and evaluate the results by analysis of variance (or paired *t*-test if only one sample is compared with the control), as shown in the examples.

Example 7.12: Analgesic Cream—Increase of Viscosity

Problem/situation: The home healthcare division of a pharmaceutical company plans to increase the viscosity of its analgesic cream base. The two proposed prototypes are instrumentally thicker in texture than the control. Sample F requires more force to initiate flow/movement, while sample N initially flows easily but has higher overall viscosity. The product researchers wish to know how different the samples are from the control. As this type of test is best done on the back of the hands, evaluation is limited to two samples at a time.

Project objective: To decide whether sample F or sample N is closest overall to the current product.

Test objective: To measure the perceived overall sensory difference between the two prototypes and the regular analgesic cream.

Test design: A preweighed amount of each product is placed on a coded watch glass. The same amount (the weight of product that is normally used on a 10 cm^2 area) is weighed out for each sample. A 10 cm^2 area is traced on the back of the subjects' hands. The test uses 42 subjects and requires 3 subsequent days for each. On each of the 3 days, a subject sees one pair, which may be

- Control versus product F
- Control versus product N
- Control versus blind control

(See worksheet, Figure 7.17.) All subjects receive the labeled control first and the test sample second. Subjects are seated in individual booths that are well ventilated to reduce odor buildup and well lighted to permit visual cues to contribute to the assessment.

Date 10-2-98	Worksheet	No. 13-625

Post this sheet in the area where trays are prepared. Code score sheets ahead of time. Label serving containers ahead of time.

Type of samples: Analgesic cream
Type of test: Difference from control test

Sample	Description	Sample code
C	Control	C
F	Experimental 10A3 (thixotropic)	Random #s under 500
N	Experimental 2-6X (high viscosity)	Random #s over 500

Serve in the following order:

Subject #	Day 1	Day 2	Day 3
1 – 7	C - F	C - N	C - C
8 – 14	C - N	C - F	C - C
15 – 21	C - F	C - C	C - N
22 – 28	C - N	C - C	C - F
29 – 35	C - C	C - N	C - F
36 – 42	C - C	C - F	C - N

Hour	Subject #
9:00	1,8,15,22,29,36
9:45	2,9,16,23,30,37
10:30	3,10,17,24,31,38
11:15	4,11,18,25,32,39
1:00	5,12,19,26,33,40
1:45	6,13,20,27,34,41
2:30	7,14,21,28,35,42

Figure 7.17 Worksheet for difference-from-control test. Example 7.12: analgesic cream.

Conduct test: Weigh out samples within 15 min of each test. Label the two samples to be presented with a three-digit code. Using easily removed marks, trace the 10 cm^2 area on the backs of the hands of each subject. Instruct subjects to follow directions on the score sheet (see Figure 7.18) carefully.

Analyze results: The results obtained are shown in Table 7.1, and an analysis of variance (ANOVA or AOV) procedure appropriate for a randomized (complete) block design is used to analyze the data. The 42 judges are the "blocks" in the design. The three samples are the "treatments" (or, more appropriately, are the three levels of the treatment). (See Section 14.5 for a general discussion of ANOVA and block designs.)

Table 7.2 summarizes the statistical results of the test. The total variability is partitioned into three independent sources of variability, that is, variability due to the difference among the panelists (i.e., the block effect), variability due to the differences among the samples (i.e., the treatment effect of interest), and the unexplained variability that remains after the other two sources of variability have been accounted for (i.e., the experimental error).

The F-statistic for samples is highly significant (Table 19.6); $F_{2,82} = 127.0$, $p < 0.0001$. The F-statistic is a ratio: the mean square for samples divided by the mean square for error. The appropriate degrees of freedom are those associated with the mean squares in the numerator and denominator of the F-statistic (2 and 82, respectively). A Dunnett's test (Dunnett, 1955; 1984) for multiple comparisons with a control was applied to the sample

Difference-from-control test

Name: ______________________ Date: ______________

Type of sample: ______________________________

______________________ Code of test sample _____

Instructions

1. You have received two samples, a control sample labeled C and a test sample labeled with a three-digit number.
2. Remove all of the control sample from the watch glass using your right Index and middle fingers.
3. Using the index and middle fingers, spread the control product around the area traced on the back of your left hand.
4. Wipe finger tips with cloth on tray.
5. Pick up all of the test sample from the labeled watch glass using your left index and middle fingers.
6. Using the index and middle fingers, spread the product across the area traced on your right hand.
7. Indicate the size of the difference in skinfeel of the sample, relative to the control, on the scale below.

_______ 0 = no difference
_______ 1 =
_______ 2 =
_______ 3 =
_______ 4 =
_______ 5 =
_______ 6 =
_______ 7 =
_______ 8 =
_______ 9 =
_______ 10 = extreme difference

Remember that a duplicate control is the sample some of the time.

Comments : ______________________________

____

Figure 7.18 Worksheet for difference-from-control test. Example 7.12: analgesic cream.

means and revealed that both of the test samples were significantly different from the blind control. It could also be concluded that product N is significantly ($p < 0.05$) more different from the control than product F based on a least significant difference (LSD) multiple comparison (LSD = 0.4).

Interpretation: Significant differences were detected for both samples, and it is concluded that the two formulas are sufficiently different from the control to make it worthwhile to conduct attribute difference tests (see Chapter 6, Table 6.3) or descriptive tests (see Chapter 12, Appendix 12.2 E, pp. 202–5) for viscosity/thickness, skin heat, skin cool, and afterfeel.

Table 7.1 Results from Example 7.12: Difference-from-Control Test—Analgesic Cream

Judge	Blind Control	Product F	Product N	Judge	Blind Control	Product F	Product N
1	1	4	5	22	3	6	7
2	4	6	6	23	3	5	6
3	1	4	6	24	4	6	6
4	4	8	7	25	0	3	3
5	2	4	3	26	2	5	1
6	1	4	5	27	2	5	5
7	3	3	6	28	2	6	4
8	0	2	4	29	3	5	6
9	6	8	9	30	1	4	7
10	7	7	9	31	4	6	7
11	0	1	2	32	1	4	5
12	1	5	6	33	3	5	5
13	4	5	7	34	1	4	4
14	1	6	5	35	4	6	5
15	4	7	6	36	2	3	6
16	2	2	5	37	3	4	6
17	2	6	7	38	0	4	4
18	4	5	7	39	4	8	7
19	0	3	4	40	0	5	6
20	5	4	5	41	1	5	5
21	2	3	3	42	3	4	4

Table 7.2 Analysis of Variance Table for Example 7.12: Difference-from-Control Test—Analgesic Cream

Source	Degrees of Freedom	Sum of Squares	Mean Square	*F*	*p*
Total	125	545.78			
Judges	41	247.11	6.03	6.8	0.0001
Samples	2	225.78	112.89	127	0.0001
Error	82	72.89	0.89		

Sample Means with Dunnett's Multiple Comparisons

Sample	Blind control	Product F
Mean response	2.4a	4.8b
Sample	Blind control	Product N
Mean response	2.4a	5.5b

Note: Within a row, means not followed by the same letter are significantly different at the 95% confidence level. Dunnett's $d_{0.05}=0.46$. Product N is significantly more different from the control than product F ($LSD_{0.05}=0.4$).

Example 7.13: Flavored Peanut Snacks

Problem/situation: The quality assurance manager of a large snack processing plant needs to monitor the sensory variation in a line of flavored peanut snacks and to set specifications for production of the snacks. The innate variations among batches of each of the added flavors (honey, spicy, barbecue, etc.) preclude the use of the triangle, duo–trio, or same/different tests. In most overall difference tests such as these, if subjects can detect variations within a batch, then this severely reduces the chances of a test detecting batch-to-batch differences. What is needed is a test that allows for separation of the variation within batches from the variation between batches.

Project objective: To develop a test method suitable for monitoring batch-to-batch variations in the production of flavored peanut snacks. Ultimately to set QA/quality control (QC) sensory specifications.

Test objective: To measure the perceived difference within batches and between batches of flavored peanuts of known origin.

Test design: Samples from a recent control batch (normal production) are pulled from the warehouse. Jars from each of two lines are sampled and labeled control A and control B. These samples represent the variation within a batch. Samples are also pulled from a lot of production in which a different batch of peanuts served as the raw material. The sample is marked "test." A difference-from-control test design is set up in which three pairs are tested:

- Control A versus control A (the blind control)
- Control A versus control B (the within batch measure)
- Control A versus test (the between batch measure)

Fifty subjects are scheduled to participate in three separate tests (C_A vs. C_A; C_A vs. C_B; C_A vs. test) over a 3-day period. The pairs are randomized across subjects. In all pairs, C_A is given first as the control, and subjects rate the difference between the members of the pair on a scale of 0–10. The results are analyzed by the procedure of Aust et al. (1985), according to which the difference between the score for the blind control and that for the within batch measure is subtracted from the between batch measure to determine statistical significance for a difference.

Screen samples: The samples are prescreened for flavor, texture, and appearance by individuals from production, QA, marketing, and research and development (R&D) who are familiar with the product to determine that each sample is representative of the within and between batch variations for the product. Along with the sensory analyst, the group decides that for the test, only whole peanuts will be sampled and tested.

Conduct test: Count out 15 whole peanuts for each sample and place in a labeled cup. Control A when in first position is labeled "control"; all other samples have three-digit codes:

Pair 1: Control A versus Control A
Labels: "Control" versus [three-digit code]
Pair 2: Control A versus Control B
Labels: "Control" versus [three-digit code]
Pair 3: Control A versus Test Sample
Labels: "Control" versus [three-digit code]
The score sheet is shown in Figure 7.19.

Analyze results: The data from the evaluations (see Table 7.3) were analyzed according to the procedure described by Aust et al. (1985). This procedure tests whether the

Difference-from-control test

Name: ______________ Date: 8-7-98 Test # 1103-6B

Type of sample: Flavored peanut snacks.

Instructions

1. Taste the sample marked "Control" first.
2. Taste the sample marked with the three digit code.
3. Assess the overall sensory difference between the two samples using the scale below.
4. Mark the scale to indicate the size of the overall difference.

	Scale	Mark to indicate difference
No difference	0	______
	1	______
	2	______
	3	______
	4	______
	5	______
	6	______
	7	______
	8	______
	9	______
Extremely different	10	______

Remember that a duplicate control is the sample some of the time.

Comments : ______________________________

Figure 7.19 Score sheet for difference-from-control test. Example 7.13: flavored peanut snacks.

score for the test sample is significantly different from the average of the two control samples. The null and alternate hypotheses are

$$H_0 : \mu_T = \frac{\mu_{C_A} + \mu_{C_B}}{2} \text{ vs. } H_a : \mu_T > \frac{\mu_{C_A} + \mu_{C_B}}{2}$$

The error term used to test this hypothesis, called *pure error mean square* (see Table 7.4), is calculated by summing the squared differences between the two control samples over all the panelists, then dividing by twice the number of panelists. The resulting ANOVA in Table 7.4 shows that the *F*-test ($F_{1,24} = MS_{\text{T vs. R}}/MS_{\text{pure error}} = 326.54$) for differences between the test and control samples is highly significant.

Interpretation: The analyst concludes that, even in the presence of variability among the control samples, the test sample is significantly different from the average of the

Table 7.3 Results from Example 7.13: Difference-from-Control Test—Flavored Peanut Snacks

Judge	Control A	Control B	Test
1	2	1	6
2	0	3	7
3	1	2	5
4	1	3	7
5	0	3	6
6	2	2	6
7	3	1	6
8	2	3	6
9	2	2	6
10	3	4	6
11	1	2	7
12	0	1	7
13	3	1	4
14	0	2	8
15	0	0	6
16	0	1	8
17	1	1	7
18	3	4	6
19	1	1	9
20	0	3	6
21	0	1	7
22	1	2	6
23	2	1	4
24	1	1	6

Table 7.4 Analysis of Variance Table According to the Difference-from-Control Test of Aust et al. (1985) for the Data of Example 7.13: Flavored Peanut Snacks

Source	Degrees-of-Freedom	Sum of Squares	Mean Squares	F	p
Total	71	456.61			
Test vs. references	1	367.36	367.36	326.54	< 0.0001
Pure error	24	27	1.13		
Residual	46	62.25			

controls. He suggests, as a next step, to determine with consumers whether the test batch is different in *preference* or *acceptance*. Such determination allows the company to determine the degree to which the difference perceived by the panel is meaningful to consumers. Further study with the difference-from-control test paired with consumer tests permits the establishment of realistic specifications for QA.

7.9 SEQUENTIAL TESTS

7.9.1 Scope and Application

Sequential tests are a means to economize the number of evaluations required to draw a conclusion, for example, acceptance versus rejection of a trainee on a panel or shipment versus destruction of a lot of produced goods. Unlike the preceding tests in this chapter, where the size of the type II error (β) is minimized for a fixed α and number of judgments, n, in sequential tests the values of α and β are decided on beforehand, and n is determined by evaluating the outcome of each sensory evaluation as it occurs. Also, because α and β are determined beforehand, sequential tests provide a direct approach to simultaneously testing for either the difference or the similarity (see Section 7.2) between the two samples.

Sequential tests are very practical and efficient because they take into consideration the possibility that the evidence derived from the first few evaluations may be quite sufficient (for fixed values of α and β) to draw a conclusion. Any further testing would be a waste of time and money. In fact, sequential tests can reduce the number of evaluations required by as much as 50%.

The sequential approach may be used with those existence-of-difference tests in which there is a correct and an incorrect answer, for example, the triangle, two-out-of-five, and duo–trio tests.

7.9.2 Principle of the Test

Conduct a sequence of evaluations according to the procedure appropriate for the chosen method and enter the results of each completed test into a graph, such as that in Figure 7.20, in which three regions are identified: the acceptance region, the rejection region, and the continue-testing region. In Figure 7.20, the number of trials is plotted on the horizontal (x) axis and the total number of correct responses is plotted on the vertical (y) axis. Enter the result of the first test, if correct, as $(x,y)=(1,1)$ and, if incorrect, as $(x,y)=(1,0)$. For each succeeding test, increase x by 1 and increase y by 1 for a correct reply and by 0 for an incorrect reply. Continue testing until a point touches or crosses one of the lines bordering the region of indecision. The indicated conclusion (i.e., accept or reject) is then drawn.

7.9.3 Analysis and Interpretation of Results: Parameters of the Test

The version of the sequential test used here is that of the ISO (2004b). The test itself is due to Wald (1947), and an alternative test is presented by Rao (1950). Both tests are clearly explained by Bradley (1953), who gives methods for calculating the expected number of

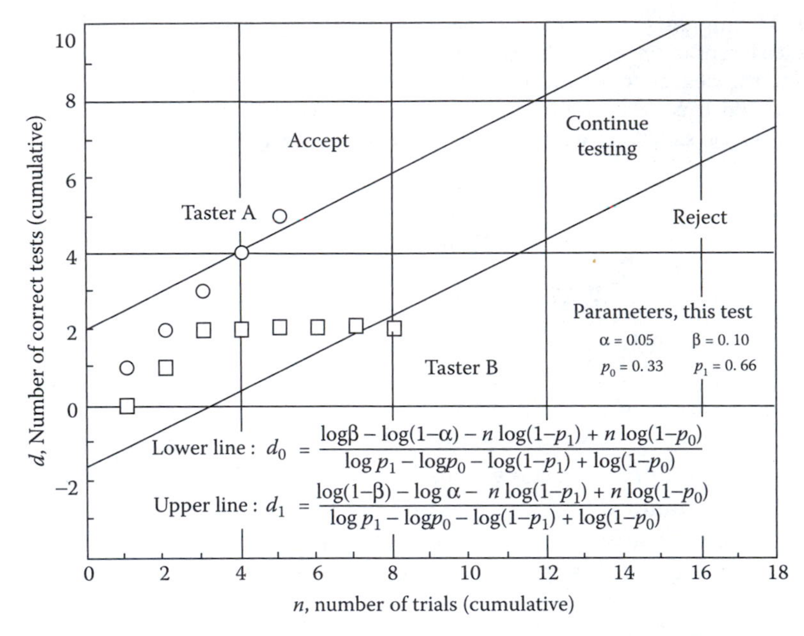

α is the probability of stating that a difference occurs when it does not
β is the probability of stating that no difference occurs when it does
p_0 is the expected proportion of correct decisions when the samples are identical
p_1 is the expected proportion of correct decisions when the odd sample is detected (other than by guess) on half the total number of occasions

Figure 7.20 Example of sequential approach for selection of panel trainees by triangle tests.

evaluations needed to reach a decision, as well as rules for choosing the parameters associated with the method, as shown in Examples 7.14 and 7.15.

Example 7.14: Acceptance versus Rejection of Two Trainees on a Panel

Project objective: To select or reject the trainees on the basis of their sensitivity to the differences in a series of test samples.

Test objective: To determine for each trainee whether his/her long-term proportion, p, of correct answers is sufficiently large for admittance onto the panel.

Test design: The sample pairs are submitted one at a time in the form of triangle tests. Intervals between tests are kept long enough to avoid fatigue. As each triangle is completed, the result is entered in Figure 7.20. The tests series continue until the trainee is either accepted or rejected.

Analyze results: Test parameters—values for four parameters are assigned by the panel leader:

- α is the probability of selecting an unacceptable trainee
- β is the probability of rejecting an acceptable trainee
- p_0 is the maximum unacceptable ability (measured as the proportion of correct answers)
- p_1 is the minimum acceptable ability (measured as the proportion of correct answers)

As can be seen in Figure 7.20, the equations for the lines dividing the graph into regions for acceptance, and so on, depend on α, β, p_0, and p_1. In the present example, trainee A is correct in all tests and is accepted after five triangles. Trainee B fails in the first triangle, succeeds in triangles two and three, but then fails on every subsequent triangle and is rejected after number eight.

Various values of the four parameters may be used. As p_0 approaches p_1, the number of required trials increases. There are several methods for reducing the average number of trials required. First, using the triangle test example, the minimum acceptable probability of detecting a difference can be set higher, for example, increased from 50% in our present example to 67%, which would make $p_1=0.78$ [from $p_1=0.67+(1-0.67)(1/3)$].* Second, if many trainees are available, α and β could be assigned larger values (e.g., $\alpha>0.05$ and/or $\beta>0.10$).

Example 7.15: Sequential Duo–Trio Tests—Warmed-Over Flavor in Beef Patties

Project objective: The routine QC panel at an army food engineering station has detected warmed-over flavor (WOF) in beef patties refrigerated for 5 days and then reheated. The project leader, knowing that "an army marches on its stomach," wishes to set a realistic maximum for the number of days beef patties can be refrigerated.

Test objective: To determine, for samples stored for 1, 3, and 5 days, whether a difference can be detected versus a freshly grilled control.

Test design: Preliminary tests show that in duo–trio tests, 5-day patties show strong WOF and 1-day patties none, hence a sequential test design is appropriate; a decision for these two samples could occur with few responses.

The three sample pairs (control vs. 1-day; control vs. 3-day; control vs. 5-day) are presented in separate duo–trio tests, in which the control and storage samples are presented as the reference for every other subject. As each subject completes one test, the result is added to previous responses, and the cumulative results are plotted (see later). The test series continues until the storage sample is declared similar to or different from the control.

Analyze results: The results obtained are shown in Table 7.5. Here α is the probability of declaring a sample different from the control, when no difference exists; β is the probability of declaring a sample similar to the control, when it is really different.

The sensory analyst and the project leader decide to set both $\alpha=0.10$ and $\beta=0.10$. They set $p_0=0.50$, the null hypothesis p-value of a duo–trio test. Further, they decide that the maximum proportion of the population that can distinguish the fresh and stored samples should not exceed 40%. Therefore, the value of p_1 is

$$p_1 = (0.04)(1.0) + (0.06)(0.05) = 0.70,$$

(from: p_1 = Pr[distinguisher]Pr[correct response given by a distinguisher] + Pr [nondistinguisher]Pr[correct response given by a nondistinguisher]).

The equations of the two lines that form the boundaries of the acceptance, rejection, and continue-testing regions are

$$d_0 = 2.59 - 0.60\,n,$$

$$d_1 = 2.59 + 0.60\,n.$$

* See Chapter 14 for the derivation of this equation.

Table 7.5 Results Obtained in Example 7.15: Sequential Duo–Trio Tests—Warmed-Over Flavor in Beef Patties

Subject No.	Test A Control vs. 1 Day		Test B Control vs. 3 Day		Test C Control vs. 5 Day	
1	I	0	I	0	C	1
2	I	0	C	1	C	2
3	I	0	I	1	C	3
4	C	1	C	2	C	4
5	I	1	I	2	I	4
6	C	2	C	3	C	5
7	I	2	I	3	C	6
8	C	3	C	4	C	7
9	I	3	C	5	I	7
10	C	4	C	6	C	8
11	I	4	C	7	C	9
12			I	7	C	10
13			C	8		
14			C	9		
15			C	10		
16			C	11		
17			I	11		
18			I	11		
19			C	12		
20			C	13		
21			I	13		
22			I	13		
23			I	13		
24			C	14		
25			I	14		
26			C	15		
27			C	16		
28			C	17		
29			C	18		
30			C	19		

Note: Column 1: I, incorrect; C, correct; Column 2: cumulative correct.

These lines are plotted in Figure 7.21 along with the cumulative number of correct duo–trio responses for each of the three stored samples (see Table 7.5). The sample stored for 1 day is declared similar to the control. The sample stored for 5 days is declared significantly different from the control. The sample stored for 3 days had not been declared significantly similar to or different from the control after 30 trials.

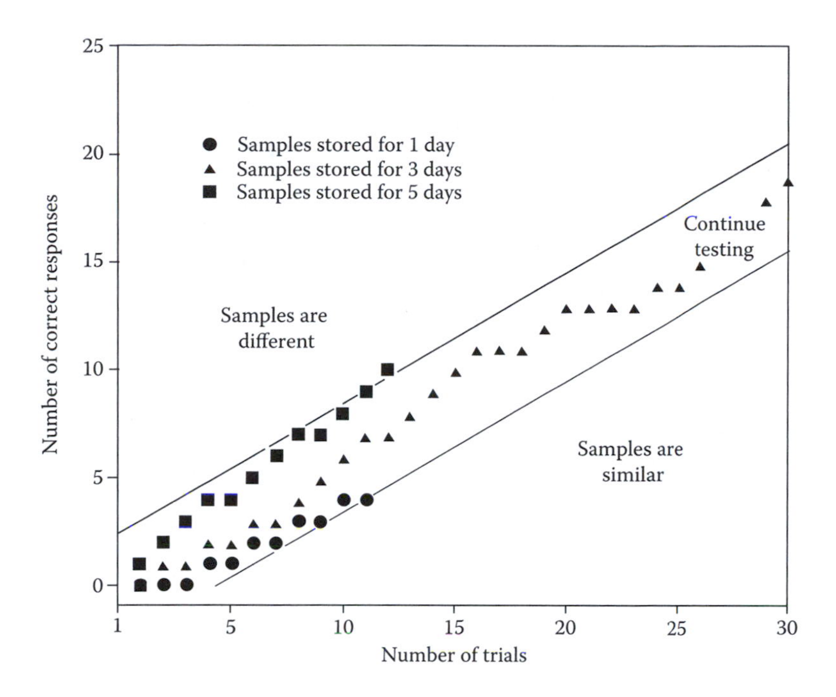

Figure 7.21 Test plot of results from Example 7.15: sequential duo–trio tests, warmed-over flavor in beef patties.

Interpret results: The project leader receives the decisive results for 1-day and 5-day samples and is informed that the result for the 3-day samples is indecisive after 30 tests. He can accept 3 days as the specification or choose to continue testing until a firm decision results.

REFERENCES

Arents, P. C., A. A. Duineveld, and B. King (2002). Sensory equivalence testing—The reversed null hypothesis and the size of a difference that matters. 6th Sensometrics Meeting in Dortmund, Germany.

ASTM (2004). Standard test method E1885-04 (2011). In *Standard Test Method for Sensory Analysis—Triangle Test.* West Conshohocken, PA: ASTM International.

Aust, L. B., M. C. Gacula, S. A. Beard, and R. W. Washam (1985). Degree of difference test method in sensory evaluation of heterogeneous product types. *J Food Sci* 50: 511–3.

Bi, J. (2005). Similarity testing in sensory and consumer research. *Food Qual Prefer* 16: 139–49.

Bradley, R. A. (1953). Some statistical methods in taste testing and quality evaluation. *Biometrics* 9: 22–38.

Conover, W. J. (1980). *Practical Nonparametric Statistics.* New York: Wiley.

Dunnett, C. W. (1955). A multiple comparison procedure for comparing several treatments with a control. *J Am Stat Assoc* 50:1096–121.

Dunnett, C. W. (1984). New tables for multiple comparisons with a control. *Biometrics* 20: 482–91.

ISO (1987). *Sensory Analysis—Methodology—"A"—"Not A" Test. ISO Standard 8588.* Available New York: International Organization for Standardization.

ISO (2004a). *Sensory Analysis—Methodology—Duo–Trio Test. ISO Standard 10399.* New York: International Organization for Standardization.

ISO (2004b). *Sensory Analysis—Methodology—Sequential Tests. ISO Standard 16820.* New York: International Organization for Standardization.

ISO (2004c). *Sensory Analysis—Methodology—Triangle Test. ISO Standard 4120.* New York: International Organization for Standardization.

Rao, C. R. (1950). Sequential tests of null hypothesis. *Sankhya* 10: 361.

Sauerhoff, K., T. Gualtieri, K. Brumbaugh, and D. Craig-Petsinger (2005). The application of equivalence testing to consumer research where the objective is parity. 6th Pangborn Sensory Science Symposium in Harrogate, UK.

Wald, A. (1947). *Sequential Analysis.* New York: Wiley.

Westlake, W. J. (1972). Use of confidence intervals in analysis of comparative bioavailability trials. *J Pharm Sci* 61: 1340–41.

8

Attribute Difference Tests

How Does Attribute X Differ between Samples?

Attribute difference tests occupy a middle ground between the overall difference tests, discussed in the previous chapter, and descriptive analysis, discussed in several chapters later in the book. Attribute difference tests focus on one or a small number of attributes as opposed to assessing overall similarity or difference between products. They do not attempt to provide a complete characterization of the sensory properties of a product, which may be obtained from a descriptive analysis.

8.1 INTRODUCTION: PAIRED COMPARISON DESIGNS

Attribute difference tests measure a single attribute, for example, sweetness, comparing one sample with one or several others. The lack of a difference between samples with regard to one attribute does not imply that no overall difference exists. Attribute difference tests involving two samples (Section 8.2) are simple regarding test design and statistical treatment; the main difficulty is that of determining whether test situations are one sided or two sided (see Section 8.2.1, Example 8.1 and Example 8.2).

With more than two samples, some designs can be analyzed by the analysis of variance, whereas others require specialized statistics. The degree of complexity increases rapidly with sample numbers, as does the economy of testing, which is possible by improved test designs. A description of the various multiple pair tests follows; multisample tests and their designs are discussed in Section 8.4.

In Sections 8.3 and 8.4, the subjects are asked to compare each sample with every other sample. Such paired comparisons provide good measures for the intensity of the attribute of interest for each sample on a meaningful scale, and they have the advantage that a measure is obtained of the relative intensity of the attribute within each pair that can be formed. However, the number of possible pairs increases geometrically with the number of samples:

Number of samples, t	3	4	5	6	7	8	9
Number of possible pairs, $N = t(t-1)/2$	3	6	10	15	21	28	36

In Sections 8.3 and 8.4, the question "Which sample is sweeter (fresher, preferred)?" is asked. This approach is based on rank data (e.g., the sweeter sample is assigned rank 2 and the other sample rank 1), which introduces a degree of artificiality; no measure of the degree of difference is obtained directly from each respondent. In return, the statistics are simpler. With rating data, specialized statistics become necessary.

8.2 DIRECTIONAL DIFFERENCE TEST: COMPARING TWO SAMPLES

8.2.1 Scope and Application

Use this method when the test objective is to determine in which way a particular sensory characteristic differs between two samples (e.g., which sample is sweeter). In this mode, the method is also called the paired-comparison test or the 2-alternative forced choice (2-AFC) test. It is one of the simplest and most used sensory tests and is often used first to determine if other more sophisticated tests should be applied. Other forms of paired comparisons of two samples are the same/different test (see Section 7.6) and the paired preference test (see Section 13.5.3.1).

When using a paired comparison test, it is necessary from the outset to distinguish between two-sided applications (*bilateral*, the most common) and one-sided applications (*unilateral*, when only one reply is of interest or only one reply is correct). (See Section 14.3.2 and the note in Example 8.2).

The unified approach also applies to the paired-comparison test. The number of respondents required for the test is affected by (1) whether the test is one sided (use Table 19.9) or two sided (use Table 19.11); and (2) by the values chosen for the test-sensitivity parameters α, β, and p_{max}. In paired-comparison tests, the parameter p_{max} replaces the parameter p_d from the overall difference methods discussed in Chapter 7. p_{max} is the departure from equal intensity (i.e., a 50:50 split of opinion among respondents) that represents a meaningful difference to the researcher. For example, if the researcher considers a 60:40 split in the population of respondents to be a meaningfully large departure from equal intensity, then $p_{max} = 0.60$ and the researcher finds the number of respondents in that section of the appropriate Table 19.9 or Table 19.11 for the chosen values of α and β. As a rule of thumb:

- $p_{max} < 55\%$ represents small departures from equal intensity.
- $55\% \leq p_{max} \leq 65\%$ represents medium departures.
- $p_{max} > 65\%$ represents large departures.

8.2.2 Principle

Present to each subject two coded samples. Prepare equal numbers of the combinations AB and BA and allot them at random among the subjects. Ask the subject to taste the products from left to right and fill in the score sheet. Clearly inform the subject whether "no difference" verdicts are permitted.

Only the "forced choice technique" is amenable to formal statistical analysis. However, in some cases, subjects may object quite strenuously to inventing a difference when none is perceived. The sensory analyst must then decide whether to (1) divide their scores evenly over the two samples or (2) ignore them. Procedure (1) decreases the probability of finding a difference, while procedure (2) increases it; hence, the analyst must face the temptation to influence the results one way or the other. In practice, about one half of analysts prohibit "no difference" verdicts. The other half, having found that a happy panel is a better panel, most frequently use procedure (1).

8.2.3 Test Subjects

Because of the simplicity of the test, it can be conducted with subjects who have received a minimum of training; it is sufficient that subjects are completely familiar with the attribute under test. Or, if a test is of particular importance (e.g., an off flavor in a product already on the market), highly trained subjects may be selected who have shown special acuity for the attribute.

Because the chance of guessing is 50%, fairly large numbers of test subjects are required. Table 19.12 shows that with 15 presentations, for example, 13 must agree if a significance level of $\alpha = 0.01$ is to be obtained, while with 50 presentations, the same significance can be obtained with 35 agreeing verdicts.

8.2.4 Test Procedure

For test controls and product controls, see Sections 3.2 and 3.3. Offer samples simultaneously if possible, or else sequentially. Prepare equal numbers of the combinations AB and BA and allocate the sets at random among the subjects. Refer to Section 7.3.4 for details of procedure. A typical score sheet is shown in Figure 8.1. Note that the score sheet is the same whether the test is one- or two sided, but the score sheet must show whether "no difference" verdicts are permitted (or the subjects must know this). Space for several successive paired comparisons may be provided on a single score sheet, but do not add supplemental questions because these may introduce bias.

Count the number of responses of interest. In a one-sided test, count the number of correct responses or the responses in the direction of interest, and refer to Table 19.10. In a two-sided test, count the number of agreeing responses citing one sample more frequently, and refer to Table 19.12.

Example 8.1: Directional Difference (Two-Sided): Crystal Mix Lemonade

Problem/situation: Consumer research on lemonades indicates that consumers are most interested in a lemon/lemonade flavor most like "fresh-squeezed lemonade." The company has developed two promising flavor systems for a powdered mix. The developers wish to get some measure of whether one of these has more fresh-squeezed lemon character than the other.

Project objective: To develop a product that is high in fresh-squeezed lemon character.

Test objective: To determine which, if either, of the two flavor systems tastes more like fresh-squeezed lemonade.

Directional Difference Test		
Name: ______________	Date: ______________	
Type of sample: ______________		
Characteristic studied: ______________		
Instructions: Taste each pair from left to right and enter your verdict below. If no difference is apparent, enter your best guess, however uncertain. "No difference" verdicts are permitted, but only as a last resort.		
Test pairs		Which sample is more ______
______	______	______
______	______	______
______	______	______
Comments : ______________		

Figure 8.1 Example of score sheet for directional difference test. Presentation: paired comparisons. "No difference" verdicts permitted.

Test design: As different people may have different ideas of what constitutes a fresh-squeezed flavor, a large panel is needed, but training is not a strong requirement. A paired-comparison test with 40 subjects and an α-error of 5%, that is, $\alpha=0.05$, is deemed suitable. The null hypothesis is H_0: Freshness A = Freshness B. The alternative hypothesis is H_a: Freshness A≠Freshness B; either outcome is of interest, hence the test is two sided. The samples are coded "691" and "812," and the score sheet shown in Figure 8.1 is used to collect the data.

Screen samples: Taste samples in advance to confirm that the intensity of lemon flavors is similar in the two samples.

Analyze results: Sample 812 is chosen by 26 subjects as having more fresh-squeezed lemon flavor. Four subjects report "no difference" and are divided between the two samples. From Table 19.12, conclude that, with 28 out of 40 choosing 812, the number is sufficient to constitute a significant difference.

Interpret results: Suggest that formulation 812 is used in the future, as it has significantly more fresh-squeezed lemon character in the opinion of this test panel.

Example 8.2: Directional Difference (One-Sided): Beer Bitterness

Problem/situation: A brewer receives reports from the market that his beer "A" is deemed insufficiently bitter, and a test brew "B" is made using a higher hop rate.

Project objective: To produce a beer that is perceptibly more bitter but not excessively so.

Test objective: To compare beers A and B to determine whether a small but significant increase in bitterness has been attained.

Test design: A paired-comparison/directional difference test is chosen because the point of interest is the increase in bitterness and nothing else. The project leader opts for a high degree of certainty, that is, $\alpha = 0.01$. The sensory analyst codes the beers "452" and "603" and offers them to a panel of 30 subjects of proven ability to detect small changes in bitterness. The score sheet asks "Which sample is more bitter?" (not "Is 603 more bitter than 452?") so as not to bias the subjects.

Screen samples: The samples are tasted by a small panel of six to make certain that differences other than bitterness are minimal.

Analyze results: Sample B is selected by 22 subjects. The null hypothesis is H_0: Bitterness A = Bitterness B, but the alternate hypothesis is H_a: Bitterness B > Bitterness A, making the test one-sided. The analyst concludes from Table 19.10 that a difference in bitterness was perceived at $\alpha = 0.01$. The test brew was successful.

Note: The important point in deciding whether a paired-comparison test is one- or two sided is whether the alternative hypothesis is one- or two sided, not whether the question asked of the subjects has one or two replies. One-sided test situations occur mainly where the test objective is to confirm a definite "improvement" or treatment effect (see also Section 14.3.1). Some examples of one- and two-sided test situations are:

One-Sided	Two-Sided
Confirm that test brew is more bitter	Decide which test brew is more bitter
Confirm that test product is preferred (as we had prior reason to expect)	Decide which product is preferred
In training tasters: which sample is more fruity (doctored samples used)	Most other test situations—whenever the alternative hypothesis is that the samples are different, rather than "one is more than the other"

8.3 SPECIFIED METHOD OF TETRADS: COMPARING TWO SAMPLES ON A SPECIFIED ATTRIBUTE USING THE METHOD OF TETRADS

8.3.1 Scope and Application

Use this method when the test objective is to compare two products on a single attribute (e.g., sweetness). Research has shown that the specified tetrad test may be more sensitive than the directional difference test just discussed (Garcia et al., 2013). All of the same considerations regarding one-sided and two-sided applications and the number of assessments required to deliver the desired level of sensitivity that apply to the directional difference test also apply to the specified tetrad test.

The number of assessments required to deliver the desired level of sensitivity from the specified tetrad test must be determined using a Thurstonian model (see Section 14.4) as opposed to a guessing model, because the guessing model does not accurately reflect the higher level of sensitivity provided by the tetrad test as compared, for example, to the triangle test, which has the same guessing probability.

8.3.2 Principle of the Test

Present each assessor with four samples. Ask each assessor to select the two of the four samples that have the highest (or lowest) intensities of a prespecified attribute. Count the number of correct replies and refer to Table 19.15 for interpretation.

8.3.3 Test Assessors

Select, train, and instruct the assessors as described in Section 10.3. As a general rule, the minimum is 12 assessors, but for less than 18, the β-error is high. Discrimination is much improved if 24, 30, or a larger number can be employed.

8.3.4 Test Procedure

For test controls and product controls, see Sections 3.2 and 3.3. Offer samples simultaneously, if possible, or else sequentially. Prepare equal numbers of the six possible combinations (AABB, ABAB, ABBA, BAAB, BABA, and BBAA) and allocate the sets at random among the assessors. There is an example of a score sheet for the specified Tetrad test in Figure 8.2. Space may be provided on the score sheet for the valuation of more than one set of samples. However, this can only be done if sensory fatigue is minimal. Do not ask questions about preference, acceptance, degree of difference, or type of difference after the initial selection of the two samples. This is because the assessor's choice may bias his or her responses to these additional questions. Count the number of correct responses and the total number of responses and refer to Table 19.15. Do not count "no difference" responses; assessors must guess if in doubt.

Example 8.3: Specified Tetrad Test—New Bulking Agent for Sweetener Packets

Problem/situation: A developer wants to identify an array of bulking agents that can be blended with her company's high-potency sweetener for use as a table-top sweetener. All candidate bulking agents have been prescreened to be acceptably bland in taste with no undesirable appearance of mouthfeel properties when dissolved in water, tea, or coffee. The key characteristic of interest to the developer is that the finished blend of bulking agent and sweetener will be equally sweet to the company's gold-standard reference product.

Project objective: To determine if the candidate bulking agent yields a finished product that is equally sweet to the gold standard.

Test objective: To determine if a difference in sweetness can be perceived between the product made with the candidate bulking agent and the gold-standard product in lemonade, tea, and coffee.

Test design: Separate specified Tetrad tests are conducted for each beverage type using the score sheet in Figure 8.2. Twenty-four assessors participate in each test. Each assessor receives four test samples, two with the candidate bulking agent and two with the gold-standard product. Four assessors each are randomly assigned one of the six possible orders of the four samples. Each assessor is asked to select the two samples from among the four that are sweetest.

Tetrad Test
Name: ____________ Date: ____________ Type of sample: ________________________
Instructions: Taste the sample on the tray from left to right. There are two groups of 2 similar samples. Group samples into 2 groups of 2 based on similarity.
Samples on tray _____ _____ _____ _____ Write the sample codes for Group 1. ____ ____ Write the sample codes for Group 2. ____ ____ Remarks: ________________________ ________________________________ ________________________________
If you wish to comment on the reasons for your choice or on the product characteristics, you may do so under "Remarks."

Figure 8.2 Score sheet for a specified tetrad test. Example 8.3: New bulking agent for sweetener packets.

Analyze results: The results of the three tests are

	Both Candidate	Both Gold Standard	
Liquid	**Samples Sweeter**	**Samples Sweeter**	**Maximum**
Lemonade	1	5	5
Tea	3	4	4
Coffee	6	4	6

This is a two-sided application because any difference in sweet taste intensity (higher or lower) would constitute a failure of the candidate bulking agent to match the gold standard. From Table 19.15, it is observed that the maximum number of matched pairs being perceived as sweeter is never more than the 5% α-risk cutoff of 8 pairings.

Interpret results: The sensory analyst informs the developer that there is no significant difference in perceived sweetness between the product made with the candidate bulking agent and the gold-standard product at the 95% confidence level.

8.4 PAIRWISE RANKING TEST: FRIEDMAN ANALYSIS—COMPARING SEVERAL SAMPLES IN ALL POSSIBLE PAIRS

8.4.1 Scope and Application

Use this method when the test objective is to compare several samples for a single attribute, for example, sweetness, freshness, or preference. The test is particularly useful for sets of three to six samples that are to be evaluated by a relatively inexperienced panel. It arranges the samples on a scale of intensity of the chosen attribute and provides a numerical indication of the differences between samples and the significance of such differences.

8.4.2 Principle of the Test

Present to each subject one pair at a time in random order, with the question "Which sample is sweeter?" (fresher, preferred, etc.) Continue until each subject has evaluated all possible pairs that can be formed from the samples. Evaluate the results by a Friedman-type statistical analysis.

8.4.3 Test Subjects

Select, train, and instruct subjects as described in Section 7.3.3. Use no fewer than 10 subjects; discrimination is much improved if 20 or more can be used. Ascertain that subjects can recognize the attribute of interest, for example, by training with various pairs of known intensity difference in the attribute. Depending on the test objective, subjects may be required who have proven ability to detect small differences in the attribute.

8.4.4 Test Procedure

For test controls and product controls, see Sections 3.2 and 3.3. Offer samples simultaneously, if possible, or else sequentially. Refer to Section 7.3.4 for details of procedure. Make certain that the order of presentation is truly random; subjects must not be led to expect a regular pattern, as this will influence verdicts.

Randomize presentation within pairs, between pairs, and among subjects. Ask only one question: "Which sample is more ...?" Do not permit "no difference" verdicts; if they nevertheless occur, distribute the votes evenly among the samples.

Example 8.4: Mouthfeel of Corn Syrup

Problem/situation: A manufacturer of blended table syrups wishes to market a product with low thickness at a given solids content. Four unflavored corn syrup blends, A, B, C, and D, have been prepared for evaluation (Carr, 1985).

Project objective: To evaluate the suitability of the four syrup blends.

Test objective: To establish the positions of the four blends on a subjective scale of perceived mouthfeel thickness.

Test design: The pairwise ranking test with Friedman analysis is chosen because (1) paired presentation is less affected by fatigue with these samples, and (2) this test establishes a meaningful scale. Twelve subjects of proven ability evaluate the six possible pairs AB, AC, AD, BC, BD, and CD. The worksheet and the score sheet are shown in Figures 8.3 and 8.4, respectively.

Analyze results: The table below shows the number of times (out of 12) each "row" sample was chosen as being thicker than each "column" sample. For example, when Sample B was presented with Sample D, it was perceived thicker by 2 of the 12 subjects.

WORKSHEET

Date 11-6-98 No. 78

CODE	SAMPLE	CODE	SAMPLE
A	Blend 4238	C	CCSA Blend III
B	Blend 133.8B	D	Test Sample 11.3A

Each panelist receives the six possible pairs in balanced random order. Each sample is coded with a random number.

Panelist No.	Order of presentation and serving code											
	1st		2nd		3rd		4th		5th		6th	
1	A 119	D 634	B 128	D 824	B 316	C 967	C 242	D 659	A 978	C 643	A 224	B 681
2	B 293	D 781	A 637	D 945	A 661	B 153	A 837	C 131	C 442	D 839	B 659	B 718
3	A 926	C 563	B 873	C 611	C 194	D 228	A 798	B 478	A 184	D 278	B 478	D 924
4	B 455	C 857	C 764	B 452	A 975	C 815	B 523	D 824	A 556	B 982	A 737	D 539
5	C 834	D 245	A 764	B 299	B 782	D 679	A 114	D 966	B 713	C 561	A 393	C 495
6	A 662	B 196	A 516	C 777	A 843	D 581	B 375	C 313	B 327	D 415	C 881	D 242
7	A 341	D 918	B 949	D 188	B 428	C 742	C 486	D 585	A 635	C 154	A 545	B 363
8	A 787	B 479	A 491	C 563	A 259	D 396	B 659	C 797	B 899	D 727	C 112	D 157
9	C 578	D 322	A 352	B 336	B 537	D 434	A 961	D 242	B 261	C 396	A 966	C 876
10	A 814	C 952	B 378	C 381	C 148	D 297	A 848	B 383	A 679	D 165	B 448	D 781
11	B 498	D 383	A 131	D 919	A 466	B 866	A 794	C 898	C 526	D 851	B 721	D 122
12	B 675	C 536	C 495	D 778	A 622	C 159	B 263	D 751	A 953	B 779	B 296	D 956

Figure 8.3 Worksheet for pairwise ranking test: Friedman analysis. Example 8.4: mouthfeel of corn syrup.

Row Samples (Thicker)	**Column Samples (Thinner)**			
	A	**B**	**C**	**D**
A	—	0	1	0
B	12	—	6	2
C	11	6	—	7
D	12	10	5	—

The first step in the Friedman analysis (Friedman, 1937; Hollander and Wolfe, 1973) is to compute the rank sum for each sample. In the present example, the rank of one is assigned to the "thicker" and the rank of two to the "thinner" sample. The rank sums

Multiple Paired-Comparisons Test

Name: ____________________ Date: ____________________

Type of sample: Unflavored table syrup

and difference: thickness (mouthfeel)

Instructions:

1. Receive the sample tray and note each sample code below according to its position on the tray.
2. Taste the first sample pair from left to right and note which sample is thicker (more viscous). Indicate by placing an X next to the code.
3. Continue until all 6 pairs have been evaluated. Rinse with water as needed to clear your palate.

Pair no.	Left sample	Right sample	Remarks
6	________	________	________
5	________	________	________
4	________	________	________
3	________	________	________
2	________	________	________
1	________	________	________

If you perceive no difference, please make a best guess. Comments regarding reasons for your choice or the characteristics of the samples may be made under Remarks.

Figure 8.4 Score sheet for pairwise ranking test: Friedman analysis. Example 8.4: mouthfeel of corn syrup.

are then obtained by adding the sum of the row frequencies to twice the sum of the column frequencies, for example, for Sample B, (12 + 6+2) +2(0 + 6+10) =52:

Sample	A	B	C	D
Rank sum	71	52	48	45

The test statistic, Friedman's T, is computed as follows:

$$T = \left(\frac{4}{pt}\right)\sum_{i=1}^{t} R_i^2 - 9p(t-1)^2 = \left[\frac{4}{(12)(4)}\right]\left[71^2 + 52^2 + 48^2 + 45^2\right] - \left[9(12)(3)^2\right] = 34.17$$

where p is the number of times the basic design is repeated (here = 12), t is the number of treatments (here = 4), R_i = the rank sum for the ith treatment, and $\sum R^2$ = sum of all R's squared, from R_1 to R_t.

Critical values of T have been tabulated (Skillings and Mack, 1981) for t = 3, 4, and 5 and small values of p; for experimental designs not in the tables, the value of T is compared to the critical value of χ^2 with (t–1) degrees of freedom (see Table 19.5). In the present case, the critical Ts are

Level of significance, α	0.10	0.05	0.01
Critical T	6.25	7.81	11.3

The results can be shown on a rank sum scale of thick versus thin:

On the same scale, the HSD value (see Section 14.5.7) for comparing two rank sums (α = 0.05) looks like this:

$$HSD = q_{\alpha,t,\infty}\sqrt{\frac{pt}{4}} = 3.63\sqrt{\frac{(12)(4)}{4}} = 12.6$$

where the value $q_{\alpha,t,\infty}$ is found in Table 19.4. The difference between A and B is much larger than 12.6, that is, A is significantly thinner and thus more desirable than the group formed by B, C, and D.

8.5 INTRODUCTION: MULTISAMPLE DIFFERENCE TESTS—BLOCK DESIGNS

The tests described in Sections 8.1 through 8.4 dealt with pairwise comparison of samples according to one selected attribute. The tests in the next four sections are based on groups of more than two samples, again compared according to one selected attribute (such as sweetness, freshness, or preference) and using the blocking designs discussed in Section 14.5.2.

8.5.1 Complete Block Designs

The simplest design is to rank all of the samples simultaneously (see Section 8.7), but the results are not as precise or actionable as those of more complex tests. The next simplest is to compare all samples together using a rating scale. We can compare all samples in one complete block (Section 8.7, multisample difference test), or we can limit the load on the taste buds (or other sensory organs) and the short-term memory of the panelists by splitting the comparison into several smaller blocks (balanced incomplete block [BIB] designs, Sections 8.8 and 8.9).

8.5.2 Balanced Incomplete Block (BIB) Designs

In the complete block designs, the size of each block (row) equals the number of treatments being studied. A block in the present context is identified by the set of samples served to one panelist. Generally, the panelist cannot evaluate more than four to six samples in a single sitting. If the number of samples (treatments) to be compared is larger, for example, 7–12, a balanced incomplete block (BIB) design can be used. Instead of presenting all the t samples in one large block, the experimenter presents them in b smaller blocks, each of which contains $k<t$ samples. The k samples that form each block must be selected so that all the samples are evaluated an equal number of times and so that all pairs of samples appear together in the b blocks an equal number of times. Cochran and Cox (1957) present an extensive list of BIB designs that can be used in most test situations. Computer programs, such as *Design Express* (2003), also can be used to generate BIB designs.

8.6 SIMPLE RANKING TEST: FRIEDMAN ANALYSIS: RANDOMIZED (COMPLETE) BLOCK DESIGN

8.6.1 Scope and Application

Use this method when the test objective is to compare several samples according to a single attribute, for example, sweetness, freshness, or preference. Ranking is the simplest way to perform such comparisons, but the data are merely ordinal, and no measure of the degree of difference is obtained from each respondent. Consecutive samples that differ widely, as well as those that differ slightly, will be separated by one rank unit. A good, detailed discussion of the virtues and limitations of rank data is given by Pangborn (1984). Ranking is less time consuming than other methods and is particularly useful when samples are to be presorted or screened for later analysis.

8.6.2 Principle of the Test

Present the set of samples to each subject in balanced, random order. Ask subjects to rank them according to the attribute of interest. Calculate the rank sums and evaluate them statistically with the aid of Friedman's test as described in Section 14.5.3.2.

8.6.3 Test Subjects

Select, train, and instruct the subjects as described in Section 7.3.3. Use no fewer than 8 subjects; discrimination is much improved if 16 or more can be used. Subjects may require special instruction or training to enable them to recognize the attribute of interest reproducibly (see Section 10.3.1.2). Depending on the test objective, subjects may be selected on the basis of proven ability to detect small differences in the attribute.

8.6.4 Test Procedure

For test controls and product controls, see Sections 3.2 and 3.3. Offer samples simultaneously, if possible, or else sequentially. The subject receives the set of t samples in balanced random order; the task is to rearrange them in rank order. The set may be presented once or several times with different coding. Accuracy is much improved if the set can be presented two or more times. In preference tests, instruct subjects to assign rank 1 to the preferred sample, rank 2 to the next preferred, and so on. For intensity tests, instruct subjects to assign rank 1 to the lowest intensity, rank 2 to the next lowest, and so on.

Recommend that subjects arrange the samples in a provisional order based upon a first trial of each and then verify or change the order based on further testing. Instruct subjects to make a "best guess" about adjacent samples, even if they appear to be the same; however, if a subject declines to guess, he or she should indicate under "comments" the samples considered identical. Assign the average rank to each of the identical samples for statistical analysis. For example, in a four-sample test, if a panelist cannot differentiate the two middle samples, assign the average rank of 2.5 to each, that is, $(2+3)/2$.

If a rank order for more than one attribute of the same set of samples is needed, carry out the procedure separately for each attribute, using new samples coded differently so that one evaluation does not affect the next. A score sheet is shown in Figure 8.5.

8.6.5 Analysis and Interpretation of Results

Analysis by Friedman's test (Friedman, 1937; Hollander and Wolfe, 1973) is preferred to the use of Kramer's tables (Kramer et al., 1974), as the latter provides inaccurate evaluation of samples of intermediate rank. Tabulate the scores as shown in Example 8.5 and calculate the rank sums for each sample (column sums). Then use Equation 14.14 to calculate the value of the test statistic, T. If the value of T exceeds the upper-α critical value of a χ^2 random variable with $(t - 1)$ degrees of freedom, then conclude that significant differences exist among the samples. Use the multiple comparison procedure appropriate for rank data presented in Equations 14.15 or 14.24 to determine which samples are different.

Example 8.5: Comparison of Four Sweeteners for Persistence

Problem/situation: A laboratory of psychophysics wishes to compare four artificial sweeteners—A, B, C, or D—for the degree of persistence of sweet taste.

Project/Test objective: To determine whether there is a significant difference among the sweeteners in the persistence of sweetness in the mouth after swallowing.

Test design: The feeling of persistence may show large person-to-person variations, so it is desirable to work with a large panel. The ranking test is suitable because it is simple to carry out and does not require much training. The four samples are tested with a panel of 48 students. Each subject receives the four samples coded with three-digit numbers and served in balanced, random order. The score sheet is shown in Figure 8.5.

Screen samples: This test requires very careful preparation to ensure that there are no other differences between the four compounds than those intended, that is, those resulting from different chemical composition. Four experienced tasters evaluate and adjust the samples to ensure that they are equally sweet to the average observer and that any differences in temperature, viscosity, or appearance (color, turbidity, and remains of foam, etc.) are absent or masked so as to preclude recognition by means other than taste and smell.

Analyze results: Table 8.1 shows how the results are compiled and the rank sums calculated. The value of the test statistic T in Equation 14.14 is

Ranking Test

Name: ______________________ Date: ______________________

Type of sample: Artificial sweeteners

____

Characteristic studied: Persistence of sweet taste

Instructions

1. Receive the sample tray and note each sample code below according to its position on the tray.
2. Taste the samples from left to right and note the *degree of persistence of the sweetness*

   Wait at least 30 seconds between samples and rinse palate as required.
3. Write "1" in the box of the sample which you find *least persistent*

   Write "2" for the next, "3" for the next, and "4" for the *most persistent*

   You may find it expedient to first arrange the samples in a provisional order, and then resolve the positions of adjacent samples by more careful tasting.
4. If two samples appear the same, make a "best guess" as to their rank order.

Code	______	______	______	______
Rank	☐	☐	☐	☐

Comments: ____

____

____

Figure 8.5 Score sheet for simple ranking test. Example 8.5: comparison of four sweeteners for persistence.

$$T = ([12/(48)(4)(5)][1352 + 1032 + 1372 + 1052]) - 3(48)(5) = 12.85$$

Use Table 19.5 to find that the upper 5% critical value of a χ^2 with three degrees of freedom is 7.81. Because the value of $T = 12.85$ is greater than 7.81, the samples are significantly different at the 5% level in their persistence of sweet taste. To determine which samples are significantly different, calculate the critical value of the multiple comparison in Equation 14.15 as

$$\text{LSD}_{\text{rank}} = 1.96\sqrt{48(4)(5)/6} = 24.8$$

Table 8.1 Table of Results for Example 8.5: Comparison of Four Sweeteners for Persistence

Subject No.	Sample A	Sample B	Sample C	Sample D
1	3	1	4	2
2	3	2	4	1
3	3	1	2	4
4	3	1	4	2
5	1	3	2	4
—	—	—	—	—
—	—	—	—	—
—	—	—	—	—
44	4	2	3	1
45	3	1	4	2
46	3	4	1	2
47	4	1	2	3
48	4	2	3	1
Rank sum	135	103	137	105

Any two samples whose rank sums differ by more than $LSD_{rank}=24.8$ are significantly different at the 5% level. Therefore, samples B and D both show significantly less persistence of sweet taste than samples A and C. Sample B is not significantly different from D, nor A from C.

Example 8.6: Bitterness in Beer Not Agreeing with Analysis

Problem/situation: A manager of quality control at a brewery knows that the company's brand P reads the same as the competition's brand by the standard analysis method for hop bitter substances, yet he hears reports that it tastes more bitter. Before commencing an investigation into possible contamination by non-hop-bitter substances, he wishes to confirm that there is a difference in perceivable bitterness.

Project/Test objective: To taste beer P for bitterness against the competitive brands A, B, and C.

Test design: The four samples are ranked by 12 subjects of proven ability to detect small differences in bitterness. The null hypothesis is H_0: Bitterness P = Bitterness A, B, or C; and the alternative hypothesis is H_a: Bitterness P $\neq$ Bitterness A, B, or C, there being no advance information about any systematic difference between A, B, and C. The score sheet used is patterned on Figure 8.5.

Analyze results: See Table 8.2. Note that the experienced panelists were permitted to assign equal ranks or "ties" to the samples. The alternate form of the test statistic T' in Equation 14.16 must be used when ties are present in the data. To calculate the value of T', the number of tied groups (g_i) in each block (i) and the size of each tied group ($t_{i,j}$) must be determined (each nontied sample is considered as a separate group of size $t_{i,j}=1$). Only blocks in which ties occur need to be considered, because only these blocks affect the calculation of T'. According to Table 8.2, ties occur in blocks 1, 3, 8, and 10. The values of g_i and $t_{i,j}$ for these blocks are

Table 8.2 Table of Results for Example 8.6: Bitterness of Beer Not Agreeing with Analysis

Subject No.	Sample A	Sample B	Sample C	Sample P
1	1	2.5	2.5	4
2	2	1	4	3
3	1	3	3	3
4	2	1	3	4
5	2	3	1	4
6	2	1	4	3
7	3	1	2	4
8	1	2	3.5	3.5
9	2	3	4	1
10	2	1	3.5	3.5
11	2	3	1	4
12	2	1	4	3
Rank sum	22	22.5	35.5	40

$$g_1 = 3, \quad t_{1,1} = 1 \quad t_{1,2} = 2 \quad t_{1,3} = 1 \quad g_3 = 2, \quad t_{3,1} = 1 \quad t_{3,2} = 3$$

$$g_8 = 3, \quad t_{8,1} = 1 \quad t_{8,2} = 1 \quad t_{8,3} = 2 \quad g_{10} = 3, \quad t_{10,1} = 1 \quad t_{10,2} = 1 \quad t_{10,2} = 1$$

These values are used to calculate the second term in the denominator of T' in Equation 14.16 as

$$T' = \left[12 \sum_{j=1}^{t} (x_j - G/t)^2 \right] / \left[bt(t+1) - (1/(t-1)) \sum_{i=1}^{b} \left(\left(\sum_{j=1}^{g_i} t^3{}_{i,j} \right) - t \right) \right]$$

$$\frac{12\left[(22-30)^2 + (22.5-30)^2 + (35.5-30)^2 + (40-30)^2\right]}{(12)(4)(5) - (1/3)(6+24+6+6)} = 13.3$$

The value of $T' = 13.3$ exceeds the upper 5% critical value of a χ^2 with three degrees of freedom ($\chi^2_{0.05,3} = 7.81$); therefore, differences exist among the samples.

Only comparisons of samples A, B, and C versus sample P are of interest. Therefore, the multiple comparison procedure for comparing test samples to a control or standard sample, appropriate for rank data, is used (see Hollander and Wolfe, 1973). The upper 5% (one-sided) critical value of the multiple comparison is 13.1. The rank sum of sample P is more than 13.1 units higher than the rank sums of samples A and B.

Test report: The QA manager concludes that the company's sample P is significantly more bitter than the competition's beers A and B; he therefore commences an investigation of the possible contamination of P with extraneous bitter-tasting substances.

8.7 MULTISAMPLE DIFFERENCE TEST: RATING APPROACH—EVALUATION BY ANALYSIS OF VARIANCE (ANOVA)

8.7.1 Scope and Application

Use this method when the test objective is to determine in which way a particular sensory attribute varies over a number of t samples, where t may vary from 3 to 6 or, at most, 8, and it is possible to compare all t samples as one large set.

Note: In descriptive analysis (see Chapter 11), when several samples are compared, the present method may be applied to each attribute.

8.7.2 Principle of the Test

Subjects rate the intensity of the selected attribute on a numerical intensity scale, for example, a category scale (see Section 5.6.1). Specify the scale to be used. Evaluate the results by the analysis of variance.

8.7.3 Test Subjects

Select, train, and instruct the subjects as described in Section 7.3.3. Use no fewer than 8 subjects; discrimination is much improved if 16 or more can be used. Subjects may require special instruction to enable them to recognize the attribute of interest reproducibly (see Section 10.3). Depending on the test objective, subjects may be selected who show high discriminating ability in the attribute.

8.7.4 Test Procedure

For test controls and product controls, see Sections 3.2 and 3.3. Offer samples simultaneously, if possible, or else sequentially. The subject receives the set of t samples in balanced randomized order; the task is to rate each sample using the specified scale. The set may be presented once only or several times with different coding. Accuracy is much improved if the set can be presented two or more times.

If more than one attribute is to be rated, theoretically the sample should be presented separately for each attribute. In practical descriptive analysis, this can become impossible because of the number of attributes to be rated in a given sample (typically from 6 to 25). *In dispensing with the requirement to rate each attribute separately, the sensory analyst accepts that there will be some interdependence between the attributes.* For example, if in a shelf-life study, the product can go stale microbiologically (e.g., sourness) or oxidatively (e.g., rancidity), high ratings on one will raise the rating on the other, even if it is absent. The effect must be counteracted by making subjects aware of it and by vigorous training that will enable them to recognize each attribute independently.

8.7.5 Analysis and Interpretation of Results

The results are analyzed using analysis of variance; see Section 14.5.2.1.

Example 8.7: Popularity of Course in Sensory Analysis. Randomized Complete Block Design

Problem/test objective: A department of food science routinely asks the students at the end of each semester to rate the courses they have taken on a scale of −3 to+3, where −3 is very poor, 0 is indifferent, and+3 is excellent. Thirty students complete the score sheet with the results shown in Table 8.3. The objective of the evaluation is to identify courses that require improvement.

Analyze results: The data lend themselves to analysis of variance for a randomized (complete) block design. The students are treated as "blocks"; the courses evaluated are the "treatments." The *F*-statistic for "courses evaluated" in Table 8.4 is highly significant ($F_{3,87}=12.91$, $p<0.0001$). Therefore, the course evaluator concludes that there are differences among the average responses for the courses. The course evaluator performs an LSD multiple comparison procedure to determine which of the course means are significantly different from each other (see Table 8.4, bottom). The results of the LSD procedure reveal that the nutrition course has a significantly lower (poorer) average rating than the other three. There are no other significant differences among the mean ratings of the other three courses. The course evaluator communicates these results to the professor and the department for further action.

Example 8.8: Hop Character in Five Beers: Split-Plot Design

Problem/situation: A brewer is producing a new brand of beer that is to have a high level of hop character. He is brewing with five alternative lots of hops that cost $1.00, $1.20, $1.40, $1.60, and $1.80/lb.

Project objective: To choose the lot that gives the most hop character for the money.

Test objective: To compare the resulting five beers for degree of hop character; to obtain a measure of the reliability of the results.

Test design: The logical way is to line up the five beers in front of a large enough number of capable tasters. This is therefore a typical multisample difference test; 20 subjects evaluate the samples on a scale of 0–9 using the score sheet in Figure 8.6. The order of presentation is randomized, and the samples are presented on three separate occasions with different coding.

Screen samples: Two experienced tasters evaluate the samples to make certain that they are representative of the type of beer to be produced and that there are no disturbing sensory differences in attributes other than hop character.

Analyze results: The results of the evaluations are shown in Table 8.5 and the corresponding split-plot ANOVA in Table 8.6. The subject-by-sample interaction was not significant:

$$F_{\text{interaction}} = 0.97, \Pr[F_{76,190} \geq 0.97] = 0.56 > 0.05.$$

The sample effect and the subject effect were both highly significant:

$$F_{\text{sample}} = 41.88 \Pr[F_{4,8} \geq 41.88] < 0.01.$$

$$F_{\text{subject}} = 17.79 \Pr[F_{19,190} \geq 17.79] < 0.01.$$

Because the interaction was not significant, it may be assumed that the subjects were consistent in their ratings of the samples. However, the significance of the subject effect suggests that the subjects used different parts of the scale to express their perceptions. This is not uncommon. Furthermore, when there is no interaction, subject-to-subject

Table 8.3 Results Obtained in Example 8.7: Multisample Difference Test (Rating)

Student	Courses Evaluated			
No.	**Biology**	**Nutrition**	**Sensory**	**Statistics**
1	2	−2	1	1
2	3	0	2	1
3	1	−3	0	0
4	2	0	1	0
5	0	1	0	0
6	−3	−3	−3	−3
7	1	3	1	1
8	−1	−1	−1	−1
9	2	−2	1	1
10	0	−3	−1	−1
11	2	0	2	2
12	−1	−2	0	1
13	3	−3	3	3
14	0	0	0	0
15	−2	2	−1	−1
16	2	−2	1	1
17	1	−1	0	0
18	0	−1	0	−1
19	3	3	3	3
20	1	−2	1	0
21	−2	−2	−2	−2
22	2	−1	1	1
23	1	0	1	1
24	3	−3	3	3
25	1	1	1	1
26	0	−1	1	−1
27	1	0	2	−1
28	2	−2	0	0
29	−2	−3	−1	−2
30	2	2	2	2

Note: Scale used: −3 to +3, where −3 = very poor, 0 = indifferent, +3 = excellent.

differences are normally of secondary interest. The differences among the samples are of primary concern. To determine which samples differ significantly in average hop character, compare the sample means using an HSD multiple comparison procedure:

Sample	4	2	5	1	3
Mean	3.9	3.0	2.9	2.1	1.4

Table 8.4 Randomized (Complete) Block ANOVA of Results in Table 8.3: Popularity of Courses in Food Science

Source of Variation	Degrees of Freedom	Sum of Squares	Mean Square	F	p
Total	119	344.37			
Students (blocks)	29	188.87			
Courses evaluated	3	47.9	15.97	12.91	<0.0001
Error	87	107.6	1.24		

Average ratings for the items evaluated with the 95% LSD multiple comparison results.

Courses Evaluated	Biology	Nutrition	Sensory	Statistics
Mean rating	0.80a	–0.83b	0.60a	0.30a

Note: Mean ratings not followed by the same letter are significantly different at the 95% confidence level—$LSD_{95\%} = 0.57$.

Note: Means not connected by a common underscore are significantly different at the 5% significance level. $HSD_{5\%} = q_{0.05,5,8}\sqrt{MS_{Error(A)} / n} = 4.89\sqrt{1.32 / 60} = 0.7$ (q-value from Table 19.4).

Sample 4 had a significantly greater average rating than all of the other samples. Samples 2 and 5, with nearly identical average ratings, had significantly less hop character than sample 4 and significantly more than samples 1 and 3. Samples 1 and 3 showed significantly less hop character than samples 2, 4, and 5.

Interpret and report results: The sensory analyst's report to the brewer contains the table of sample means and the ANOVA table, and it concludes that, of the five samples tested, sample 4 produced a significantly higher level of hop character. Sample 2, of a less expensive variety, also merits consideration.

8.8 MULTISAMPLE DIFFERENCE TEST: BIB RANKING TEST (BALANCED INCOMPLETE BLOCK DESIGN)—FRIEDMAN ANALYSIS

8.8.1 Scope and Application

Use this method when the test objective is to determine in which way a particular sensory attribute varies over a number of samples, and there are too many samples to evaluate at any one time. Typically, the method is used when the number of samples to be compared is from 6 to 12 or, at most, 16.

Choose the present method (ranking) when the panelists are relatively untrained for the type of sample and/or a relatively simple statistical analysis is preferred. Use the method described in Section 8.7 (rating) when panelists trained to use a rating scale are available.

8.8.2 Principle of the Test

Instead of presenting all t samples as one large block, present them in a number of smaller blocks according to one of the designs of Cochran and Cox (1957) or, for example, *Design Express* (2003). Ask subjects to rank the samples according to the attribute of interest.

Multisample Comparisons Test

Name: ____________ Date: ____________

Type of sample: Beer

Characteristic studied: Hop character

Instructions

Taste the samples from left to right and note the intensity of the characteristic studied. Rate each sample on the following scale:

0 1	Imperceptible
2 3	Slightly perceptible
4 5	Moderately perceptible
6 7	Strongly perceptible
8 9	Extremely perceptible

Sample Code: ______ ______ ______ ______

Rating: ______ ______ ______ ______

Comments: ____________

Figure 8.6 Score sheet for multisample difference test (rating). Example 8.8: hop character in five beers.

8.8.3 Test Subjects

Select, train, and instruct the subjects as described in Section 7.3.3. Ascertain that subjects can recognize the attribute of interest, for example, by training with sets of known intensity levels in the attribute (see Section 10.3 and Appendix 12.2).

8.8.4 Test Procedure

For test controls and product controls, see Sections 3.2 and 3.3. Offer samples simultaneously, if possible, or else sequentially. Refer to Section 7.3.4 for details of the procedure.

Table 8.5 Results Obtained in Example 8.8 Multisample Difference Test (Rating): Hop Character in Five Beers

Sample No.	1	2	3	4	5
1	2,2,1	3,4,5	1,0,2	5,4,3	3,2,4
2	0,0,1	1,2,1	0,0,0	2,1,2	2,1,1
3	0,2,1	2,0,2	0,2,0	2,3,2	0,2,2
4	3,3,3	4,5,6	2,3,1	5,8,4	5,6,4
5	2,4,3	4,3,1	3,0,3	3,5,6	1,4,3
6	2,4,1	3,2,4	3,2,1	4,6,7	3,4,2
7	0,0,1	1,2,1	0,0,0	0,2,1	2,1,1
8	6,4,3	4,6,3	3,4,6	4,6,3	3,4,6
9	2,2,2	3,3,5	0,1,1	4,6,5	3,5,3
10	1,4,3	2,5,3	2,0,2	5,4,5	5,2,3
11	3,4,2	1,3,4	3,0,3	6,5,3	3,4,1
12	1,0,0	1,2,1	0,0,0	1,2,1	1,1,2
13	1,0,0	1,2,1	0,0,0	2,1,2	1,1,2
14	3,3,3	6,5,4	1,3,2	4,8,5	4,6,5
15	2,2,2	5,3,3	1,1,0	5,6,4	3,5,3
16	1,4,2	4,2,3	1,2,3	7,6,4	2,4,3
17	3,4,1	3,5,2	2,0,2	5,4,5	3,2,5
18	1,2,0	2,0,2	0,2,0	2,3,2	2,2,0
19	1,2,2	5,4,3	2,0,1	3,4,5	4,2,3
20	3,4,6	3,6,4	6,4,3	3,6,4	6,4,3

Explanation: For example, Subject no. 20 rated sample no. 1 a 3 the first time, a 4 the second time, and a 6 the third time.

Table 8.6 Split-Plot ANOVA of Results in Table 8.5: Hop Character in Five Beers

Source of Variation	Degrees of Freedom	Sum of Squares	Mean Squares	*F*
Total	299	975.64		
Replications	2	8.89		
Samples	4	221.52	55.38	41.88[a]
Error(A)	8	10.58	1.32	
Subjects	19	412.30	21.70	17.79[a]
Sample × Subject	76	89.81	1.18	0.97
Error(B)	190	232.53	1.22	

Note: Error (A) is calculated as would be the Rep × Sample interaction. Error (B) is calculated by subtraction.

[a] Significant at the 1% level.

Make certain that order of presentation is truly random; subjects must not be led to suspect a regular pattern, as this will influence verdicts. For example, state only to "rank the samples according to sweetness, giving rank 1 to the sample of lowest sweetness, rank 2 to the next lowest, and so on."

Example 8.9: Species of Fish

Problem/situation: Military field ration XPQ-6 (fish fingers in aspic) has been prepared in the past from 15 different species of fish. Serious complaints of "fishy" flavor have been traced to the use of some of these species. Those in command want to specify a limited number of species so as to be able to weigh availability and price against the probability of food riots.

Project objective: To compare the 15 species such that quantitative information on the degree of fishy flavor is obtained that can be applied to the problem at hand.

Test objective: To compare fish fingers produced from the 15 species for degree of fishy flavor.

Test design: The multisample difference test with balanced incomplete design is chosen because it permits comparison of the 15 test products in groups of three. A randomly selected group of 105 enlisted personnel are randomly divided into 35 groups of three subjects each. Each group of three subjects is randomly assigned one of the 35 groups of three samples according to the design in Table 8.7. The score sheet asks the subject to rank his three samples according to fishy flavor, from least (=1) to most (=3).

Screen samples: The help of the cook is enlisted in preparing samples as uniformly as possible regarding texture, appearance, and flavor, minimizing the differences attributable to species by suitable changes in cooking methods and secondary ingredients. The pieces prepared for each serving are screened for appearance, and any that contain coarse fragments or show other visible deviations are discarded.

Analyze results: To make the results easier to analyze, the rank data from the study are arranged as shown in Table 8.8. The rank sum for a given species of fish is simply the sum of all the numbers in the column corresponding to that species. The value of Friedman's test statistic T (see Equation 14.18) is computed to determine if there are any differences among the species in the intensity of fishy flavor. The value of $T = 68.53$ exceeds the upper 5% critical value of a χ^2 with $(t-1) = 14$ degrees of freedom ($\chi^2_{14,\,0.05} = 23.69$), and it is concluded that there are indeed significant differences in the data set. Next, Equation 14.19 is used to calculate the value of a 95% LSD multiple comparison to determine which of the species are significantly different (see Table 8.9).

Interpret and report results: The military leadership concludes from Table 8.9 that the species identified as samples 5, 15, 13, 1, 6, and 9 should be retained for price and availability consideration, as these produce the least degree of fishy flavor and are not significantly different from each other. The species denoted as samples 14, 8, and 4 are provisionally retained if too many of the species in the first group are eliminated because of high cost or unavailability. This is done in recognition of the fact that samples 14, 8, and 4 have rank sums for the intensity of fishy flavor that are significantly greater than only samples 5 and 15 and are not significantly different from the remaining samples in the first group. The remaining species in Table 8.9 (2, 11, 10, 12, 3, and 7) are eliminated from use in field ration XPQ-6.

Table 8.7 Multisample Difference Test: BIB Design for Example 8.9—Fish Fingers in Aspic ($t=15, k=3, r=7, b=35, l=1, E=0.71$)

Block			
(1)	1	2	3
(2)	4	8	12
(3)	5	10	15
(4)	6	11	13
(5)	7	9	14
(6)	1	4	5
(7)	2	8	10
(8)	3	13	14
(9)	6	9	15
(10)	7	11	12
(11)	1	6	7
(12)	2	9	11
(13)	3	12	15
(14)	4	10	14
(15)	5	8	13
(16)	1	8	9
(17)	2	13	15
(18)	3	4	7
(19)	5	11	14
(20)	6	10	12
(21)	1	10	11
(22)	2	12	14
(23)	3	5	6
(24)	4	9	13
(25)	7	8	15
(26)	1	12	13
(27)	2	5	7
(28)	3	9	10
(29)	4	11	15
(30)	6	8	14
(31)	1	14	15
(32)	2	4	6
(33)	3	8	11
(34)	5	9	12
(35)	7	10	13

Source: From Cochran, W. G., and G. M. Cox. *Experimental Designs*, Wiley, New York, 1957. With permission.

Table 8.8 Results Obtained in Example 8.9, Multisample Difference Test: BIB Design with Rank Data—Fish Fingers in Aspic

	Sample/Species														
Block/ Subject	**1**	**2**	**3**	**4**	**5**	**6**	**7**	**8**	**9**	**10**	**11**	**12**	**13**	**14**	**15**
1	1	2	3												
2				3			1			2					
3					1					3					2
4						3					2		1		
5							3		1					2	
6	3			2	1										
7		2						3		1					
8			3										2	1	
9						3			2						1
10							2				1	3			
—	—	—	—	—	—	—	—	—	—	—	—	—	—	—	—
—	—	—	—	—	—	—	—	—	—	—	—	—	—	—	—
—	—	—	—	—	—	—	—	—	—	—	—	—	—	—	—
101	2													3	1
102		1		3		2									
103			1					2			3				
104					1				2			3			
105							3			2			1		
Rank sum	35	45	54	43	28	37	55	42	37	50	49	50	34	42	29

Note: Response: 1, least fishy; 2, intermediate; 3, most fishy.

8.9 MULTISAMPLE DIFFERENCE TEST: BIB RATING TEST—EVALUATION BY ANALYSIS OF VARIANCE

8.9.1 Scope and Application

Use this method when the test objective is to determine in which way a particular sensory attribute varies over a number of samples, and there are too many samples to evaluate at any one time. Typically, the method is used when the number of samples to be compared is from 6 to 12 or, at most, 16.

Choose the present method (rating) when panelists trained to use a rating scale are available and results need to be as precise and actionable as possible. Use the method described in Section 8.8 (ranking) when panelists have less training and/or the ranking test gives sufficient information.

Note: In descriptive analysis (see Chapter 11), when the number of samples to be compared is large, the present method may be applied to each attribute.

Table 8.9 Summary of Results and Statistical Analysis of the Data in Table 8.9: Fish Fingers in Aspic

Sample/ Species	Rank Sum					
5	28	a				
15	29	a				
13	34	a	b			
1	35	a	b	c		
6	37	a	b	c		
9	37	a	b	c		
14	42		b	c	d	
8	42		b	c	d	
4	43		b	c	d	
2	45			c	d	e
11	49				d	e
10	50				d	e
12	50				d	e
3	54					e
7	55					e

Note: Means followed by the same letter are not significantly different at the 5% significance level ($LSD_{rank} = 10.74$).

8.9.2 Principle of the Test

Instead of presenting all t samples as one large block, present them in a number of smaller blocks according to one of the designs of Cochran and Cox (1957) or, for example, *Design Express* (2003). Ask subjects to rate the intensity of the attribute of interest on a numerical intensity scale (see Section 5.6). Specify the scale to be used. Evaluate the results by the analysis of variance.

8.9.3 Test Subjects

Select, train, and instruct the subjects as described in Section 7.3.3. Ascertain that subjects can recognize the attribute of interest, for example, by training with sets of known intensity levels in the attribute. Use no fewer than 8 subjects; discrimination is much improved if 16 or more are used.

Subjects may require special instruction to enable them to recognize the attributes of interest reproducibly (see Section 10.3.1.2). Depending on the test objective, subjects may be selected who show high discriminating ability in the attribute(s) of interest.

8.9.4 Test Procedure

For test controls and product controls, see Sections 3.2 and 3.3. Offer samples simultaneously, if possible, or else sequentially. Refer to Section 7.3.4 for details of the procedure.

Make certain that the order of presentation is truly random; subjects must not be led to suspect a regular pattern, as this will influence verdicts.

Note: If more than one attribute is to be rated, unavoidably there will be some interdependence in the resulting ratings (see Section 4.3.5).

8.9.5 Analysis and Interpretation of Results

The results are analyzed by the analysis of variance (see Section 14.5.4.1 and Example 8.10).

Example 8.10: Reference Samples of Ice Cream

Problem/situation: As part of an ongoing program, the quality control (QC) manager of an ice cream plant routinely screens samples of finished product to select lots that will be added to the pool of quality reference samples for use in the main QC testing program. New reference samples are needed at regular intervals because the older samples will have changed with time and are no longer appropriate. The procedure is also used to eliminate from the pool any current reference samples that may have deteriorated.

Project objective: To maintain a sufficient inventory of reference samples of finished ice cream for QC testing purposes.

Test objective: To rate the inventory of six lots each day for overall off flavor and to discard any lot that may not be suitable as a reference.

Test design: Samples of the six lots are evaluated for overall off flavor by 15 well-trained panelists who use a 10-point category scale from 0 (no off flavor) to 9 (extreme off flavor). The panelists cannot evaluate more than four samples in one sitting. Therefore, the sensory analyst chooses a BIB design from Cochran and Cox (1957) (see Table 8.10).

Table 8.10 BIB Design for Example 8.10: Reference Samples of Ice Cream ($t=6, k=4, r=10, b=15, l=6, E=0.90$)

Block				
(1)	1	2	3	4
(2)	1	4	5	6
(3)	2	3	5	6
(4)	1	2	3	5
(5)	1	2	4	6
(6)	3	4	5	6
(7)	1	2	3	6
(8)	1	3	4	5
(9)	2	4	5	6
(10)	1	2	4	5
(11)	1	3	5	6
(12)	2	3	4	6
(13)	1	2	5	6
(14)	1	3	4	6
(15)	2	3	4	5

Source: Cochran, W. G. and G. M. Cox. *Experimental Designs*, Wiley, New York, 1957. With permission.

Each of the 15 panelists is randomly assigned one block of four samples from the design. The order of presentation of the samples within each block is randomized.

Analyze results: The ratings data for the overall off-taste attribute are presented in Table 8.11. The data are analyzed by a computer program capable of performing a BIB ANOVA (see Section 14.5.4.1). The resulting BIB ANOVA table is presented in Table 8.12. The *F*-statistic for "treatments" (i.e., samples of ice cream), when compared to the upper 5% critical value of an *F*-distribution with $(t-1)=5$ and $(tpr\text{-}t\text{-}pb+1)=40$ degrees of freedom, is found to be significant ($F=9.33>F_{0.05;5,40}=2.45$). An LSD multiple comparison procedure is applied to the average ratings of the samples to determine which samples have significantly different overall off flavor (see note 3 at the foot of Table 8.11).

Interpret and report results: The average off-taste rating of sample 1 is significantly greater than the average ratings of the remaining samples. There are no other significant differences among the mean ratings of the other samples. The sensory analyst reports the results to the QC manager with the recommendation that the lot from which sample 1 was taken be discarded from the pool of reference samples.

Table 8.11 Table of Results for Example 8.10: Reference Samples of Ice Cream

	Sample					
Block/ Subject	**1**	**2**	**3**	**4**	**5**	**6**
1	6	1	1	2		
2	6			1	3	3
3		4	2		5	2
4	7	2	3		2	
5	3	5		1		1
6			1	1	3	2
7	7	4	4			3
8	2		1	1	1	
9		2		2	2	3
10	4	2		2	5	
11	5		3		1	1
12		3	2	1		2
13	4	2			1	1
14	5		2	2		1
15		2	4	5	3	
Adjusted means	5.0	2.5	2.2	2.0	2.6	1.9

Note: 1: BIB design with rating. 2: Response—10-point category scale with 0 = no off flavor, 9 = extreme off flavor. 3: Adjusted means that are not connected by a common underscore are significantly different at the 5% significance level ($LSD_{5\%}=1.1$).

Table 8.12 Balanced Incomplete Block ANOVA Table for Example 8.10: Reference Samples of Ice Cream

Source of Variation	Degrees of Freedom	Sum of Squares	Mean Square	*F*	*P*
Total	59	150.98			
Judges (blocks)	14	39.73			
Samples (treatments, adjusted for blocks)	5	59.89	11.98	9.33	<0.0001
Error	40	51.36	1.28		

REFERENCES

Carr, B. T. (1985). Statistical models for paired-comparison data. In *American Society for Quality Control 39th Congress Transactions*. Baltimore, MD: American Society for Quality, 295–300.

Cochran, W. G. and G. M. Cox (1957). *Experimental Designs*. New York: Wiley.

Design Express (2003). *Presentation Orders for Consumer Trials*. Berkshire, UK: Qi Statistics.

Friedman, M. (1937). The use of ranks to avoid the assumption of normality implicit in the analysis of variance. *J Am Stat Assoc* 32: 675–701.

Garcia, K., J. M. Ennis, and W. Prinyawiwatkul (2013). Reconsidering the specified tetrad test. *J Sens Stud* 28(6): 445–9.

Hollander, M. and D. A. Wolfe (1973). *Nonparametric Statistical Methods*. New York: Wiley.

Kramer, A., G. Kahan, D. Cooper, and A. Papavasiliou (1974). A non-parametric method for the statistical evaluation of sensory data. *Chem Sens Flav* 1: 121–3.

Pangborn, R. M. (1984). Sensory techniques of food analysis. In D. W. Gruenwedel and J. R. Whitaker (eds.), *Food Analysis. Principles and Techniques* (vol. 1, p. 59). New York: Marcel Dekker.

Skillings, J. H. and A. G. Mack (1981). On the use of a Friedman-type statistic in balanced and unbalanced block designs. *Technometrics* 23: 171–7.

9

Determining Threshold

9.1 INTRODUCTION

Sensory thresholds are ill defined in theory (Lawless and Heymann, 1998; Morrison, 1982). A good determination requires hundreds of comparisons with a control, and results do not reproduce well (Brown et al., 1978; Marin et al., 1988; Stevens et al., 1988). Published group thresholds (Fazzalari, 1978; Van Gemert and Nettenbreijer, 1984; Devos et al., 1990) vary by a factor of 100 for quinine sulfate in water and by much more in complex systems. Swets (1964) doubts even the existence of a sensory threshold. A first reaction is that it is futile to invest time and money in threshold studies; however, in situations such as those described in the next paragraph, the threshold approach is still the best available.

Thresholds in air, determined by automated flow olfactometry, are used to determine degrees of air pollution (CEN, 1997) and to set legal limits for polluters. Thresholds of added substances are used with water supplies, foods, beverages, cosmetics, paints, solvents, and so on to determine the point at which known contaminants begin to reduce acceptability. These are the most important uses, and testing may be done with hundreds of panelists to map the distribution of relative sensitivity in the population. Thresholds may also be used as a means of selecting or testing panelists, but this should not be the principal basis for selection (see Chapter 10) unless the test objective requires detection of the stimulus at very low levels. The threshold of added desirable substances may be used as a research tool in the formulation of foods, beverages, and so on.

It should be kept in mind that a low detection threshold for a given compound corresponds to a high sensitivity for the flavor in question. The concepts of the odor unit (O.U.) (Guadagni et al., 1966) or flavor unit (F.U.) (Meilgaard, 1975) use the threshold as a measure of flavor intensity. For example, if H_2S escapes from a leaking bottle into a room, when the level reaches the threshold of detection, the odor intensity is at 1 O.U.; at double that level of H_2S, the intensity is at 2 O.U., and so on. This use of thresholds requires caution and is not applicable at intensities above 3–6 O.U. (Chapter 2). Procedures for estimating sensory intensity at levels above threshold are discussed in Chapter 5.

The methods used to determine olfactory thresholds can have a profound influence on the results. Hangartner and Paduch (1988) show that odorant flows below the usual sniffing volume of 1–2 L/sec will give rise to thresholds severalfold too high. Doty et al. (1986) found that the use of a larger sniff bottle resulted in 10–20-fold lower thresholds because panelists were able to raise the sniffing volumes. Training can lower thresholds as much as 1000-fold (Powers and Shinholser, 1988). For a detailed review of the history and an evaluation of current practices of odor measurement, the reader is referred to Doty and Laing (2003).

Experience shows that with practice and training (Brown et al., 1978), it is possible to obtain reproducibility levels of ±20% for a given panel and ±50% between one large panel (>25) and another. The important factors, in addition to repeated training with the actual substance under test, are those described in Chapter 4: Subjects will pride themselves and hope to please the experimenter by finding the lowest threshold, and this must be counteracted by meticulous attention to the details of sample preparation and sample presentation so as to not leave clues to their identity.

9.2 DEFINITIONS

Thresholds are the limits of sensory capacities. It is convenient to distinguish between the absolute threshold, the recognition threshold, the difference threshold, and the terminal threshold.

The absolute threshold (detection threshold) is the lowest stimulus capable of producing a sensation: the dimmest light, the softest sound, the lightest weight, the weakest taste. The recognition threshold is the level of a stimulus at which the specific stimulus can be recognized and identified. The recognition threshold is usually higher than the absolute threshold. If a person tastes water containing increasing levels of added sucrose a transition in sensation will occur in at some point from "water taste or pure water" to "a very mild taste." As the concentration of sucrose increases, a further transition will occur from "a very mild taste" to "mild sweet." The level at which this second transition occurs is called the recognition threshold.

The difference threshold is the extent of change in the stimulus necessary to produce a noticeable difference. It is usually determined by presenting a standard stimulus that is then compared to a variable stimulus. The term *just-noticeable difference* (JND) is used when the difference threshold is determined by changing the variable stimulus by small amounts above and below the standard until the subject notices a difference. Chapter 5 addresses this subject directly.

The terminal threshold is that magnitude of a stimulus above which there is no increase in the perceived intensity of the appropriate quality for that stimulus. Above this level, pain often occurs.

JNDs increase as one proceeds up the scale of concentration, and they have been used as scale steps of sensory intensity. Hainer et al. (1954) calculated that their subjects could distinguish some 29 JNDs between the absolute and the terminal thresholds. However, thresholds vary too much from person to person, and from group to group, for the JND to have gained practical application as a measure of perceived intensity.

The conventional notion of a threshold (e.g., for diesel exhaust in air) is that shown in Figure 9.1. Above 5 ppm, the exhaust can be detected; below 5 ppm, it cannot be detected. However, an observer making repeated tests using a dilution olfactometer will produce a set of results such as those shown in Figure 9.2. The observer's sensitivity will vary with

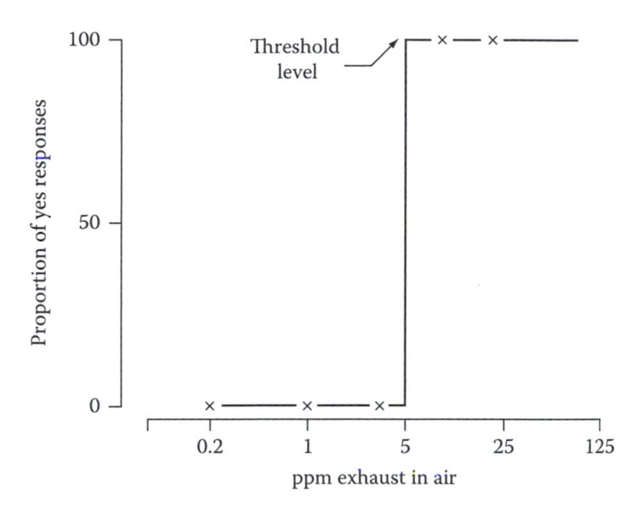

Figure 9.1 Conventional notion of the absolute threshold (for diesel exhaust in air).

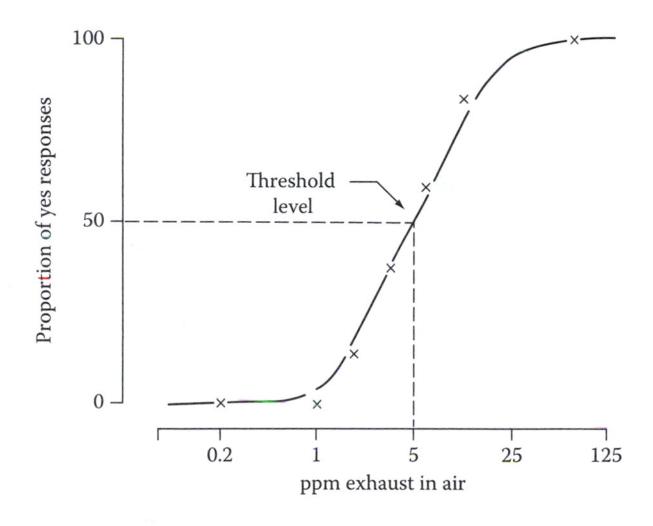

Figure 9.2 Typical data from determination of personal threshold (for diesel exhaust in air).

chance air currents over the olfactory membrane and with momentary or biorhythmic variations in the sensitivity of his nervous system. The ticking of a watch held at a certain distance can be heard one moment, and can be inaudible the next, and then audible again, and so on. The threshold is not a fixed point but a value on a stimulus continuum. By convention, the observer's personal threshold is that concentration that can be detected 50% of the time and not the concentration that can be detected at *X*% significance, an error frequently committed (Laing, 1987). As a rule, one finds a typical Gaussian dose–response curve from which the 50% point can be accurately determined after transformation of the experimental percentage points by one of the methods described in Example 9.2.

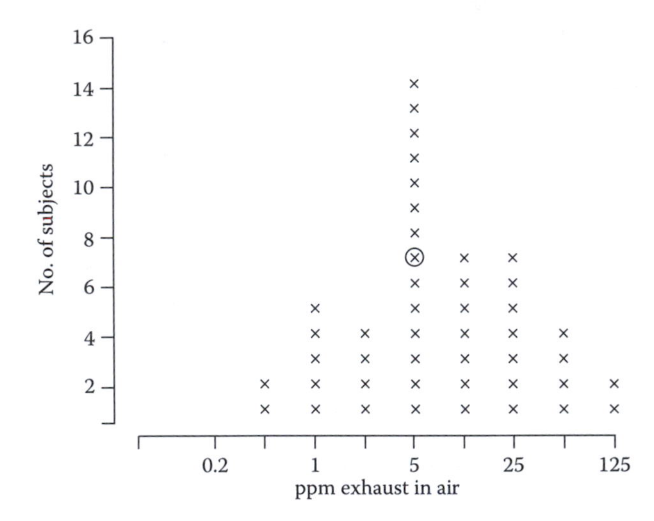

Figure 9.3 Typical histogram of threshold for a group of 45 subjects; ⊗ = subject from Figure 9.2.

To get from a collection of personal thresholds to a group threshold, it is noted that the frequency distribution tends to be bell shaped for the majority (Meilgaard, 1993). However, the curve's right-hand tail tends to be longer than the left (see Figure 9.3) because most groups contain a proportion of individuals who show very low sensitivity to the stimulus in question. The measure of central tendency that makes most sense for such a group of observers may be the geometric mean, as it gives less weight to the highest thresholds. A rank probability graph (Figure 9.4) is a useful tool for testing if a set of individual thresholds are normally distributed. This is determined to be true if a good straight line can be drawn through the points. In this case, the graph can serve to locate not only the group threshold as the 50% point, but also the concentrations that can be detected by 5% or 90%, for example, of the corresponding population.

9.3 APPLICATIONS OF THRESHOLD DETERMINATIONS

Thresholds can be measured by a variety of the classical psychophysical designs based on, for example, the method of limits, the method of average error, or the frequency method (Kling and Riggs, 1971). In recent years, a tendency among psychophysicists has been to choose a different route by applying the signal detection theory (SDT) (Swets, 1964; Macmillan and Creelman, 1991; Doty and Laing, 2003). SDT is a system of methods based on the idea that the point of interest is not the threshold as such but rather "the size of the psychological difference between the two stimuli," which has the name d'. The advantage of SDT is that the subject's decision process becomes more explicit and can be statistically modeled. However, SDT procedures are more time consuming than the classical threshold designs, and it has been shown (Frijters, 1980) that for forced-choice methods of sample presentation, there is a 1:1 relationship between d' and the classical threshold.

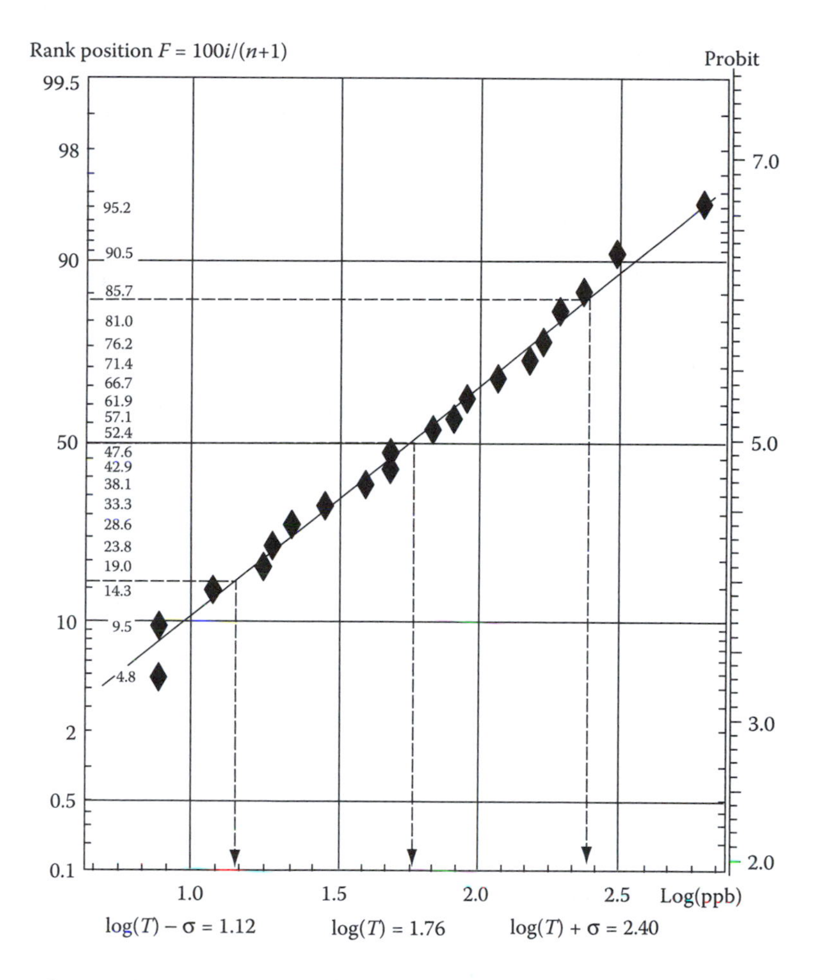

Figure 9.4 Rank probability graph for the 20 panelists in Table 9.2. Result: a straight line can be drawn through the points; consequently, the panelists are normally distributed with $\log(T) = 1.76$ ($T = 58$ ppb); $\log(T) - \sigma = 1.12$ (13 ppb); $\log(T) + \sigma = 2.40$ (255 ppb).

For these reasons, both the American Society for Testing and Materials (ASTM, 1997a, 1997b) and the International Organization for Standardization (ISO, 2002) have decided to stick with the method of limits and what is known as the three-alternative forced-choice (3-AFC) method of sample presentation, in which three samples are presented: Two are controls, and one contains the substance under test. The ASTM's rapid method (E679, see Example 9.1) aims to determine a practical value close to the threshold based on a minimum of testing effort (e.g., 50–150 3-AFC presentations). It makes a very approximate (e.g., ±200%) best estimate determination of each panelist's threshold. In return, the panel can be larger, and the resulting group threshold and distribution become more reliable because the variation between individuals is much greater (up to 100-fold) than the variation between tests by a single individual (up to 5-fold). The result is slightly biased at best and can be very biased if subjects falling on the upper or lower limits of the range under test are not reexamined (see Example 9.1).

The ASTM's intermediate method (E1432) proceeds to determine individual thresholds according to Figure 9.2 and then, in a second step, it determines the group threshold according to Figure 9.3. For this, it requires approximately five times as many sample presentations per panelist as the rapid method. In return, the group threshold and distribution of individual thresholds are both bias free.

The ISO Standard 13301 (ISO, 2002) is, in effect, a combination of both of the above. For the curve-fitting step, the intermediate method uses nonlinear least-squares regression (see Example 9.2). (The ISO procedure permits logistic regression and a maximum likelihood procedure for which a procedure of calculation using computer spreadsheets has been introduced.) If results more precise than can be obtained with these methods are desired, one enters the field of research projects as such, and any of a number of designs may be appropriate, for example, Powers' multiple pairs test (Powers and Ware, 1976; Kelly and Heymann, 1989) or SDT (Macmillan and Creelman, 1991). Bi and Ennis (1998) provide a review of these methods and propose an additional procedure for population thresholds based on the beta-binomial distribution that takes account of the fact that data for one individual tends to have a much narrower distribution than data for a group of individuals.

Example 9.1: Threshold of Sunstruck Flavor Compound Added to Beer

Problem/situation: A brewer, aware that beer exposed to UV light develops sunstruck flavor (3-methyl-2-butene-l-thiol, a compound not otherwise present), wishes to test the protection offered by various types of packaging material.

Project objective: To choose packaging that offers acceptable protection at minimum cost, using as criterion the amount of the sunstruck compound formed during irradiation compared with the threshold amount.

Test objective: To determine the threshold of purified 3-methyl-2-butene-1-thiol added to the company's beer.

Test design: The E679 rapid test is suitable as the need is for good coverage of the variability among people; 25 panelists each receive six 3-AFC tests with concentrations spaced by a factor of three. Limit bias by (1) choosing the range of concentrations offered with the aid of a preliminary test using five panelists and (2) retesting those panelists who are correct at the lowest or fail at the highest level.

Screen samples: In the preliminary test, ascertain that the base beer is free of sunstruck flavor and that the 3-methyl-2-butene-1-thiol confers a pure sunstruck character at the chosen test concentrations.

Conduct the test: Test each panelist at the six concentrations. Test any panelist who is correct at the lowest level once or twice more at that level and include sets at one or two lower levels. Likewise, test any panelist who fails at the highest level twice more at that level and at one or two higher levels. Record and analyze results, as shown in Table 9.1. The best estimate threshold (BET) for each subject is the geometric mean of the highest concentration missed and the next higher concentration. The group BET is the geometric mean of the individual BETs. Repeat the test series at least once on a different day, using the same observers. Note that thresholds often decrease as panelists become accustomed to the flavor of the substance and the mechanics of the test. If the threshold decreases more than 20%, repeat the test series until the values stabilize.

Test report: Include the complete Table 9.1 and give demographics of the panelists.

Table 9.1 Sensory Threshold of the Sunstruck Flavor Compound Added to Beer

	Concentrations Presented (ppb)								Best Estimate Threshold	
Panelist	**0.27**	**0.80**	**2.41**	**7.28**	**21.7**	**65.2**	**195**	**Over**	**ppb**	**Log(10)**
01		0	0	+	+	+	+		4.19	0.622
02	0	+	+	+	+	+	+		0.46	−0.337
03		0	+	+	+	+	+		1.39	0.143
04	0	+	+	+	+	+	+		0.46	−0.337
05		0	+	0	+	+	+		12.6	1.100
06		0	+	+	+	+	+		1.39	0.143
07		+	0	+	+	+	+		4.19	0.622
08	0	+	+	+	+	+	+		0.46	−0.337
09	0	+	+	+	+	+	+		0.46	−0.337
10		0	+	0	0	+	0	+	338	2.529
11		0	+	+	+	+	+		1.39	0.143
12		0	+	+	+	+	+		1.39	0.143
13		+	0	+	+	+	+		4.19	0.622
14		0	0	+	+	+	+		4.19	0.622
15	0	+	+	+	+	+	+		0.46	−0.337
16		0	+	0	+	+	+		12.6	1.100
17		0	+	+	+	+	+		1.39	0.143
18		+	+	0	0	+	+		37.7	1.576
19	0	+	+	+	+	+	+		0.46	−0.337
20		+	0	+	+	+	+		4.19	0.622
21		0	+	+	+	+	+		1.39	0.143
22	0	+	+	+	+	+	+		0.46	−0.337
23		+	0	0	+	+	+		12.6	1.100
24	0	+	+	+	+	+	+		0.46	−0.337
25		0	0	+	+	+	+		4.19	0.622
									Sum →	9.299
Group BET, geometric mean (ppb)									2.35 ←	0.3720
Log standard deviation =										0.719

(Continued)

Table 9.1 (Continued) Sensory Threshold of the Sunstruck Flavor Compound Added to Beer

Histogram of Individual BE Thresholds
Geometric Mean = 2.35 ppb

24						
22						
19	21	25				
15	17	20				
09	12	14				
08	11	13	23			
04	06	07	16			
02	03	01	05	18		10
0.46	1.39	4.19	12.6	37.7	113	338

Procedure: ASTM E679 Ascending concentration series method of limits.
Equipment: Colorless beer glasses, 250 mL; 50 mL beer "A" per glass.
Sample: 3-Methyl-2-butene-1-thiol (Aldrich).
Purification: By preparative gas chromatography on two columns.
Number of scale steps: 6; concentration factor per step: 3.0.
Number of subjects: 25.
High and low results confirmed? Yes.
3-Methyl-2-butene-1-thiol, ppb in beer "A".

Example 9.2: Threshold of Isovaleric Acid in Air

Problem/situation: A rendering plant produces air emissions containing isovaleric acid as the most flavor-active component. The neighbors complain, and an ordinance is passed requiring a reduction below threshold.

Project objective: To choose between various process alternatives and a higher chimney.

Test objective: To determine the threshold of isovaleric acid in air.

Test design: A fairly thorough method such as the ASTM Intermediate Method (E1432) or the second example of ISO Standard 13301 is suitable because of the economic consequences of the issue. Use a dynamic olfactometer (CEN, 1997) and 20 panelists. Give each panelist 3-AFC tests six times at each of five or more concentrations spaced twofold apart and chosen in advance (see *Conduct the test*, below). The apparatus contains three sniff ports; the panelist knows that two produce odor-free air and must choose the one that he believes to contain added isovaleric acid. The added concentration is at the lowest level in the first test and increases by a constant factor in each subsequent test. From the percentage of correct results at each concentration, calculate each panelist's threshold and, from these, the group threshold.

Screen samples: Ascertain that the air supply is free from detectable odors and that the isovaleric acid is of sensory purity and free from foreign odors. Check the reliability of the olfactometer by chemical analysis.

Conduct the test: Test each panelist in turn at the chosen concentrations. Make this choice, in advance, by a single test (or a few tests) at each of a set of widely spaced

```
** Purpose: Fit logistic models P = (1/3 + EXP[K])/(1 + EXP[K]),
**           where K = B(T - LOG[X]),
**           P is the proportion of correct identifications,
**           B is the slope,
**           X is the acutal concentration (ppb) of Isovaleric
**           Acid in air,
**       and T is the threshold value in log(ppb).

PROC NLIN    Method=DUD    Data=Input;     by panelist;
              PARMS B=-4   T=2
                  K    =  B*(T - LOG10(K));
                  K    =  EXP(K);
                  K    =  (1/3 + E);
                  D    =  (1 + E);
              MODEL    P =  N/D
          TITLE2  'Logistic Regression Modles';
RUN;

Output for panelist 13:
    Logistic Regression of Threshold Data Using SAS PROC NLIN
                     Logistic Regression Models
            Non-linear Least Squares Iterative Phase
          Dependent Variable: P         Method:  DUD
Iteration              B              T              Residual  SS
   -3           -4             2.000000000        0.025885700365
   -2           -4.4           2.000000000        0.02054415598
   -1           -4             2.200000000        0.084958944779
    0           -4.4           2.000000000        0.02054415598
    1           -5.852958      1.961443385        0.010812277188
    2           -6.259745      1.967823308        0.010766524899
    3           -6.189164      1.951938036        0.010504941622
    4           -6.283542      1.954261395        0.010481402394
    5           -6.280162      1.954257276        0.010481361251
    6           -6.281544      1.954068199        0.010481219887
    7           -6.277816      1.953905805        0.010481193047
    8           -6.280506      1.953919346        0.010481176612
    9           -6.281737      1.953896400        0.010481179219
   10           -6.281715      1.953899496        0.010481176193
Convergence criterion met.

Non-linear Least Squares Summary Statistics  Dependent Variable P
    Source            DF   Sum of squares      Mean Square
    Regression         2    2.3500748238       1.1750374119
    Residual           3    0.0104811762       0.0034937254
    Uncorrected Total  5    2.3605560000
    (Corrected Total)  4    0.3550844800

                                           Asymptotic 95%
                           Asymptotic    Confidence interval
    Parameter  Estimate    STD. Error    Lower              Upper
    B      -6.281714751 1.6824126163-11.635992903-0.9274366000
    T       1.953899496 0.0473965533  1.803059965 2.1047390264
```

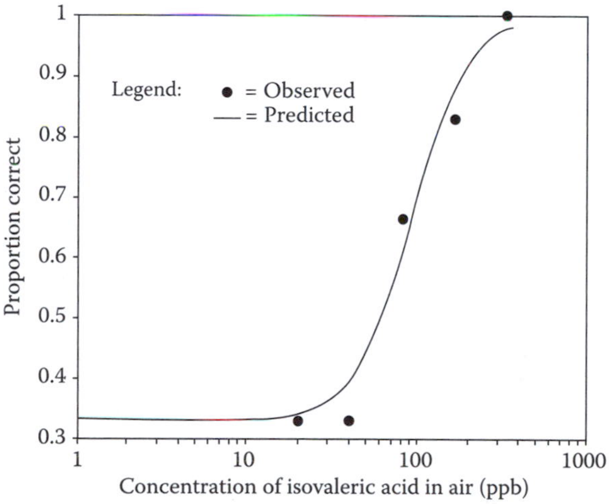

```
Predicted values
  P     C    LOG[X]   X
0.967 0.95  2.4226  265
0.933 0.90  2.3037  201
0.833 0.75  2.1288  135
0.667 0.50  1.9539   90
0.500 0.25  1.7790   60
0.400 0.10  1.6041   40
0.367 0.05  1.4770   30
```

Figure 9.5 Fitting of a dose–response curve to the data in Table 9.2 using "SAS® PROC NLIN" and the logistic method. The estimated value of $T=1.954$ is the threshold concentration in log (ppb) for Panelist 13.

Table 9.2 Determination of Olfactory Thresholds to Isovaleric Acid in Air by ASTM Intermediate Method E1432

Concentrations	**Panelist**					
Presented (ppb)	**2**	**11**	**13**	**15**	**18**	**19**
Example of results, showing six panelists: Number of correct tests (out of six)						
640				6	6	5
320			6	5	6	4
160	6	6	5	4	2	3
80	5	4	4	2	2	2
40	4	4	2	3	3	1
20	6	0	2	2	1	2
10	5	3				
5	3					
2.5	2					
Converted to proportion correct = C/N where C is the above number and $N = 6$						
640				1.000	1.000	0.833
320			1.000	0.833	1.000	0.667
160	1.000	1.000	0.833	0.667	0.333	0.500
80	0.833	0.667	0.667	0.333	0.333	0.333
40	0.667	0.667	0.333	0.500	0.500	0.167
20	1.000	0.000	0.333	0.333	0.167	0.333
10	0.833	0.500				
5	0.500					
2.5	0.333					

Using Logistic Regression (Computer Package SAS® PROC NLIN, See Figure 9.5) the Individual Thresholds are

Panelist No.	**1**	**2**	**3**	**4**	**5**	**6**	**7**	**8**	**9**	**10**
Log (threshold)	0.84	0.84	1.04	1.20	1.26	1.32	1.43	1.58	1.67	1.67
Threshold (ppb)	7	7	11	16	18	21	27	38	47	47
Panelist No.	**11**	**12**	**13**	**14**	**15**	**16**	**17**	**18**	**19**	**20**
Log (threshold)	1.81	1.91	1.95	2.07	2.19	2.25	2.29	2.40	2.52	2.82
Threshold (ppb)	64	81	90	118	154	178	196	249	330	665

Procedure: ASTM Intermediate Method E1432.
Equipment: Dynamic triangle olfactometer, after A. Dravnieks.
Sample: Isovaleric acid (Sigma).
Purification: Recrystallization as calcium salt.
Number of panelists: 20.
Number of scale steps presented to each: min 5.
Concentration factor per scale step: 2.0.

concentrations (e.g., 2.5, 10, 40, 160, and 640 ppb). In the test, if a panelist should score 100% correct at the lowest concentration, reschedule the concentration series with this as the highest. Likewise, if a panelist scores less than 80% correct at the highest concentration, continue the series by presenting higher concentrations until this no longer happens.

Analyze the results: Plot the data as shown in Figure 9.5, in which the abscissa is the concentration, x (or log concentration), and the ordinate is the proportion distinguishers (or percent correct above chance), p_d. p_d is obtained from the proportion correct, p_c, as follows:

Test	Formula	Chance Level
Triangle or 3-AFC	$p_d = 1.5 \times p_c - 0.5$	0.333
Paired comparison or 2-AFC	$p_d = 2.0 \times p_c - 1.0$	0.500
Two-out-of-five	$p_d = 1.111 \times p_c - 0.111$	0.100

Calculate the individual thresholds by one of the six curve-fitting methods allowed by ASTM E1432 or ISO 13301 (e.g., by logistic regression using a computer package as shown in Figure 9.5). Plot the individual thresholds in a rank/probability graph, as shown in Figure 9.4, and obtain the group threshold as the 50% point. If a straight line can be drawn through the points, conclude that the panelists represent a normal distribution and that other points of interest can be read from the graph, for example, the concentration that 10% of a population similar to the panel can detect.

Test report: Include the information in Table 9.2 and Figure 9.4 and give demographics of the panelists.

Group threshold: Obtain the group threshold T by rank probability graph as shown in Figure 9.4. The result is $\log(T) = 1.76$; $T = 58$ ppb. Alternatively, calculate T as the geometric mean of the individual thresholds:

$$\log T = \frac{0.84 + 0.84 + 1.04 + L + 2.40 + 2.52 + 2.82}{20} = \frac{36.06}{20} = 1.753;\ T = 56.6\ \text{ppb}.$$

REFERENCES

ASTM (1997a). Determination of odor and taste thresholds by a forced-choice ascending concentration series method of limits. *Standard Practice E679–97,* West Conshohocken, PA, ASTM International.

ASTM (1997b). Defining and calculating sensory thresholds from forced-choice data sets of intermediate size. *Standard Practice E1432–97,* West Conshohocken, PA, ASTM International.

Bi, J. and D. M. Ennis (1998). Sensory thresholds: Concepts and methods. *J Sens Stud* 13: 133–48.

Brown, D. G. W., J. F. Clapperton, M. C. Meilgaard, and M. Moll (1978). Flavor thresholds of added substances. *J Am Soc Brew Chem* 36: 73–80.

CEN (1997). Determination of odour concentration by dynamic olfactometry. In *Air Quality.* Brussels, Belgium: European Committee for Standardisation.

Devos, M., F. Patte, J. Rouault, P. Laffort, and L. J. Van Gemert (1990). *Standardized Human Olfactory Thresholds.* Oxford: IRL Press.

Doty, R. L. and D. G. Laing (2003). Psychophysical measurement of human olfactory function, including odorant mixture assessment. In R. L. Doty (ed.), *Handbook of Olfaction and Gustation* (2nd ed., pp. 203–28). New York: Marcel Dekker.

Doty, R. L., T. P. Gregor, and R. G. Settle (1986). Influence of intertrial interval and sniff-bottle volume on phenyl alcohol detection thresholds. *Chem Sens* 11: 2, 259–64.

Fazzalari, F. A. (ed) (1978). *Compilation of Odor and Taste Threshold Values Data.* ASTM data series publications, West Conshohocken, PA: ASTM International.

Frijters, J. E. R. (1980). Three-stimulus procedures in olfactory psychophysics. An experimental comparison of Thurstone–Ura and three-alternative forced-choice models of signal detection theory. *Percept Psychophys* 28: 5, 390–7.

Guadagni, D. G., S. Okano, R. G. Buttery, and H. K. Burr. (1966). Correlation of sensory and gas–liquid chromatographic measurement of apple volatiles. *Food Tech* 30: 518.

Hainer, R. M., A. G. Emslie, and A. Jacobson (1954). Basic odor research correlation. *Ann NY Acad Sci* 58: 158.

Hangartner, M. and M. Paduch (1988). Interface human nose—olfactometer. In *Measurement of Odor Emissions, Proceedings of Workshop,* Annex V, 53, Commission of the EEC.

ISO (2002). *Sensory Analysis—Methodology—General Guidance for Measuring Odour, Flavour, and Taste Detection Thresholds by a Three-alternative Forced-choice (3-AFC) Procedure.* International Organization for Standardization, International Standard ISO 13301:2002, Switzerland: ISO.

Kelly, F. B. and H. Heymann (1989). Contrasting the effect of ingestion and expectoration in sensory difference tests. *J Sens Stud* 3: 4, 249–55.

Kling, J. W. and L. A. Riggs. (eds) (1971). *Woodworth & Schlosberg's Experimental Psychology* (3rd ed., Chapter 2). New York: Holt, Rinehart and Winston.

Laing, G. G. (1987). *Optimum Perception of Odours by Humans* (vol. 8). Australia: CSIRO Division of Food Research.

Lawless, H. T. and H. Heymann (1998). *Sensory Evaluation of Food. Principles and Practices* (Chapter 6). New York: Chapman and Hall.

Macmillan, N. A. and C. D. Creelman (1991). *Detection Theory, A User's Guide* (vol. 391). Cambridge: Cambridge University Press.

Marin, A. B., T. E. Acree, and J. Barnard (1988). Variation in odor detection thresholds determined by charm analysis. *Chem Sens* 13: 3, 435–44.

Meilgaard, M. C. (1975). Flavor chemistry of beer. Part I. Flavor interaction between principal volatiles. *Tech Quart Mastr Brew Assoc Am* 12: 107–17.

Meilgaard, M. C. (1993). Individual differences in sensory threshold for aroma chemicals added to beer. *Food Qual Prefer* 4: 153–67.

Morrison, G. R. (1982). Measurement of flavor threshold. *J Inst Brew* 88: 170–4.

Powers, J. J. and K. Shinholser (1988). Flavor thresholds for vanillin and predictions of higher or lower thresholds. *J Sens Stud* 3: 1, 49–61.

Powers, J. J. and G. O. Ware (1976). Comparison of sigmplot. Probit and extreme-value methods for the analysis of threshold data. *Chem Sens Flav* 2: 2, 241–53.

Stevens, J. C., W. W. Cain, and R. J. Burke (1988). Variability of olfactory thresholds. *Chem Sens* 13:4, 643–53.

Swets, J. A. (1964). Is there a sensory threshold? In J. A. Swets (ed.), *Signal Detection and Recognition by Human Observers.* New York: Wiley.

Van Gemert, L. J. and A. H. Nettenbreijer (1984). In V. Zeist (eds.), *Compilation of Odour Threshold Values in Air and Water.* The Netherlands: Central Institute for Nutrition and Food Research TNO, Supplement.

10

Selection and Training of Panel Members

10.1 INTRODUCTION

This section is partly based on ASTM Special Technical Publication 758, *Guidelines for the Selection and Training of Sensory Panel Members* (1981), and on the ISO *Guide for Selection and Training of Assessors* (1993). The development of a sensory panel deserves thought and planning with respect to the inherent need for the panel, the support from the organization and its management, the availability and interest of panel candidates, the need for the screening of training samples and references, and the availability and condition of the panel room and booths. In the food, fragrance, and cosmetic industries, the sensory panel is the company's single most important tool in research and development (R&D) and in quality control (QC). The success or failure of the panel development process depends on the strict criteria and procedures used to select and train the panel.

The project objective of any given sensory problem or situation determines the criteria for selection and training of the subjects. Too often in the past (ISO, 1991), the sole criterion was a low threshold for one or more of the basic tastes. Today, sensory analysts use a wide selection of tests specifically selected to correspond to the proposed training regimen and end use of the panel. Taste acuity is only one aspect; much more important is the ability to discern and describe a particular sensory characteristic in a "sea" or "fog" of other sensory impressions.

This chapter describes specific procedures for the decision to establish a panel, the selection and training of both difference and descriptive panels, and ways to monitor and motivate panels. This chapter does not apply to consumer testing (see Chapter 13), which uses naïve subjects representative of the population for whom the product is intended. Although the text uses the language of a commercial organization that exists to develop, manufacture, and sell a product and has its upper management, middle management, and reward structure, the system described can be easily modified to fit the needs of other types of organizations such as universities, hospitals, civil or military service organizations, and so on.

10.2 PANEL DEVELOPMENT

Before a panel can be selected and trained, the sensory analyst must establish that a need exists in the organization and that commitment can be obtained to expend the required time and money to develop a sensory tool. Upper management and the project group (R&D or quality assurance [QA]/QC) must see the need to make decisions based on sound sensory data with respect to overall differences and attribute differences (difference panels) or full descriptions of product standards, product changes over time, or ingredient and processing manipulation; and for the construction and interpretation of consumer questionnaires (descriptive panels). The sensory analyst must also define the resources required to develop and maintain a sensory panel system.

10.2.1 Personnel

Heading the list of resources required is (1) a large enough pool of available candidates from which the panel can be selected (see Appendix 10.2A for possible sources for recruiting panel candidates); (2) a sensory staff to implement the selection, training, and maintenance procedures, including a panel leader and technician; and (3) a qualified person to conduct the training process. Ideally, panelists should come from within the organization, as they are located at the site where the samples are prepared (e.g., R&D facility or plant). Before a descriptive panel is trained, consideration is given to the choice of a panel leader. An effective panel leader is a person who is able to serve as the connection between the panel and product developers or other panel clients. The panel leader works with the panel to ensure that the panel has a clear understanding of attributes and scales as well as the ability to translate the panel data into actionable information. A successful panel leader is also a person who (1) has knowledge of sensory attributes, (2) has good group dynamic skills, (3) has listening and/or attending skills, (4) is creatively alert, and (5) is patient. A panel leader may come from the panel itself. If this is the case, the panel leader should be additionally trained to manage the panel and communicate with the research team so that the information provided to product developers and other scientists is reliable, valid, and useful. If a panel is large, a panel technician may also be required to be responsible for all sample procurement, preparation, and presentation, as well as for completing all the necessary documentation of the panel protocol and data output.

Some companies choose to test products at a different site, which may be another company facility. With reduced laboratory staffing, many companies have opted to use residents recruited from the local community as panelists rather than bench chemists and support staff from the labs. Outside panelists may be available for more hours per week and may be cheaper and more focused for longer panel sessions. The primary drawbacks of using external panelists are that they often require more time and effort to train in the technical aspects of panel work, and they do not provide the inherent proprietary security of internal employees.

Panel candidates and management must understand, in advance, the amount of time required (personnel hours) for the selection and training of the particular panel in question. An assessment of the number of hours needed for panelists, technicians, and a panel leader should be presented and accepted before the development process is initiated. The individual designated to select and train the panel is often a member of the sensory staff who is experienced and trained in the specific selection and training techniques needed for the challenge at hand.

10.2.1.1 Special Considerations for a Quality Control/Quality Assurance (QC/QA) Panel

While the criteria for selecting panelists for QC and shelf-life panels is similar to the criteria for selecting descriptive panelists, a few key differences are recognized. Typically, QC and shelf-life panelists are internal plant employees who dedicate time during their shifts to evaluate products made at that plant. Evaluations are streamlined (e.g., short ballots, fewer panelists) to minimize panelists' time away from their primary jobs while efficiently evaluating as many products as possible. However, as it is not unusual to have two or more evaluations during a shift, it is necessary to recruit, screen, and train a pool of panelists from which to draw. This allows for panelist attendance flexibility while enabling the intended number of evaluations to be executed each day/week.

10.2.2 Facilities

The physical area for the selection, training, and ongoing work of a panel must be defined before development of the panel begins. A training room and panel testing facilities (booths and/or round table, conference room, etc.) must have the proper environmental controls (see Chapter 3), be of sufficient size to handle all of the panelists and products projected, and be located near the product preparation area and panelist pool.

10.2.3 Data Collection and Handling

This is another resource to be defined: the personnel, hardware, and software required to collect and treat the data generated by the panel. Topics such as the use of personal computers with specific software versus the company server should be addressed before the data begin to accumulate on the sensory analyst's desk. The specific ways in which the data are generated and used (i.e., frequency data, scalar data—category, linear, magnitude estimation), the number of attributes, the number of replications, and the need for statistical analysis all contribute to the requirements for data collection and handling.

10.2.4 Projected Costs

After upper management and the project group understand the need to have a panel and the time and costs required for its development and use, the costs and benefits can be assessed from a business and investment perspective. This phase is essential so that the support from management is based on a full understanding of the panel development process. After management and the project team are "on board," the sensory analyst can expect the support that is needed to satisfy the requirements for personnel (both panelists and staff), facilities, and data handling. Management can then, through circulars, letters, and/or seminars, communicate its support for the development of and participation in sensory testing. As the reader will have gathered by now, public recognition by management of the importance of the sensory program and of the involvement of employees as panelists is essential for the operation of the system. If participation in sensory tests is not seen by upper and middle management as a worthwhile expenditure of time, the sensory analyst will find the recruiting task to be difficult, if not impossible, and test participation will dry up more quickly than new recruits can be enrolled.

After management support has been communicated through the organization and has been demonstrated in terms of facilities and personnel for the panel, the sensory analyst can use presentations, questionnaires, and personal contacts to reach potential panel members. The time commitment and qualifications must be clearly iterated so that candidates understand what is required of them. General requirements include an interest in the test program, availability, promptness, and general good health (no allergies or health problems affecting participation), articulateness, and absence of aversions to the product classes involved. Other specific criteria are listed for individual tasks in Sections 10.3 and 10.4.

10.3 SELECTION AND TRAINING FOR DIFFERENCE TESTS

10.3.1 Selection

Assume that the early recruitment procedure has provided a group of candidates free of obvious drawbacks, such as heavy travel or work schedules or health problems that would make participation impossible or sporadic. The sensory analyst must now devise a set of screening tests that teach the candidates the test process while weeding out unsuitable nondiscriminators as early as possible. Such screening tests should use the products to be studied and the sensory methods to be used in the study. It follows that they should be patterned on those described in this chapter rather than being used directly. The screening tests aim to determine differences among candidates in the ability to (1) discriminate (and describe, if attribute difference tests are to be used) character differences among products, and (2) discriminate (and describe with a scale for attribute difference tests) differences in the intensity or strength of the characteristic.

Suggested rules for evaluating the results are given at the end of each section. The analyst should consider that although candidates with high success rates may, on the whole, be satisfactory, the best panel will result if selection can be based on potential rather than on current performance.

10.3.1.1 Matching Tests

Matching tests are used to determine a candidate's ability to discriminate (and describe, if asked) differences among several stimuli presented at intensities well above threshold level. Familiarize candidates with an initial set of four to six coded, but unidentified, products. Then present a randomly numbered set of eight to ten samples, of which a subset is identical to the initial set. Ask candidates to identify on the score sheet the familiar samples in the second set and to label them with the corresponding codes from the first set.

Table 10.1 contains a selection of samples suitable for matching tests. These may be common flavor substances in water, common fragrances, lotions with different fat/oil

Table 10.1 Suggested Samples for Matching Tests

Tastes, Chemical Feeling Factors		
Flavor	**Stimulus**	**Concentration (g/L)**[a]
Sweet	Sucrose	20
Sour	Tartaric acid	0.5
Bitter	Caffeine	1.0
Salty	Sodium chloride	2.0
Astringent	Alum	10

Aroma, Fragrances, Odorants[b]	
Aroma Descriptors	**Stimulus**
Peppermint, minty	Peppermint oil
Anise, anethole, licorice	Anise oil
Almond, cherry	Amaretto, benzaldehyde, oil of bitter almond
Orange, orange peel	Orange oil
Floral	Linalool
Ginger	Ginger oil
Jasmine	Jasmine-74-d-10%
Green	*cis*-3-Hexenol
Vanilla	Vanilla extract
Cinnamon	Cinnamaldehyde, cassia oil
Clove, dentist's office	Eugenol, oil of clove
Wintergreen	BenGay, methyl salicylate, oil of wintergreen

[a] In tasteless and odorless water at room temperature.
[b] Perfume blotters dipped in odorant, dried in hood 30 min, placed in wide-mouthed jar with tight cap.

systems, products made with pigments of different colors, fabrics of similar composition but differing in basis weight, and so on. Care should be taken to avoid carryover effects; for example, samples must not be too strong. Table 10.2 shows an example of a score sheet for matching fragrances at above threshold levels in a nonodorous diluent.

10.3.1.2 Detection/Discrimination Tests

This selection test is used to determine a candidate's ability to detect differences among similar products with ingredient or processing variables. Present candidates with a series of three or more triangle tests (Rainey, 1979; Zook and Wesmann, 1977) with differences ranging from easy to moderately difficult (see, e.g., Bressan and Behling, 1977). Duo–trio tests (Section 6.3) may also be used. Table 10.3 lists some common flavor standards and the levels at which they may be used. "Doctored" samples, such as beers spiked (Meilgaard et al., 1982) with substances imitating common flavors and off notes, may also be used. Arrange preliminary tests with experienced tasters to determine the optimal order of the test series and to control stimulus levels such that they are appropriate and detectable but not overpowering. Use standard triangle or duo–trio score sheets

Table 10.2 Score Sheet for Fragrance Matching Test

First Set	Second Set Match		Descriptor[a]
079	________		________
318	________		________
992	________		________
467	________		________
134	________		________
723	________		________
Floral	Peppermint	Vanilla	Wintergreen
Green	Cinnamon	Ginger	Clove
Jasmine	Orange	Cherry, almond	Anise/licorice

Note: Instructions: Sniff the first set of fragrances; allow time to rest after each sample. Sniff the second set of fragrances and determine which samples in the second set correspond to each sample in the first set. Write down the code of the fragrance in the second set next to its match from the first set. Optional: Determine which descriptor from the list below best describes the fragrance pair.

[a] A list of descriptors, similar to the one given below, may be given at the bottom of the score sheet. The ability to select and use descriptors should be determined if the candidates will be participating in attribute difference tests.

Table 10.3 Suggested Materials for Detection Tests

Substance	Concentration (g/L)[a]	
Caffeine	0.2[b]	0.4[c]
Tartaric acid	0.4[b]	0.8[c]
Sucrose	7.0[b]	14.0[c]
γ-Decalactone	0.002[b]	0.004[c]

[a] Amount of substances added to tasteless and odorless water.
[b] 3 × threshold level.
[c] 6 × threshold level.

when suitable. If desired, use sequential triangle tests (Chapter 7, Section 7.9) to decide acceptance or rejection of candidates. However, as already mentioned, do not rely too much on taste acuity.

10.3.1.3 Ranking/Rating Tests for Intensity

These tests are used to determine candidates' ability to discriminate graded levels of intensity of a given attribute. Ask candidates to rate on an appropriate scale, if this is the method the test panelist will eventually use; otherwise use ranking (Chapter 8, Section 8.4). Present a series of samples in random order, in which one parameter is present at different levels that cover the range present in the product(s) of interest. Ask candidates to rank the samples in ascending order (or rate them using the prescribed scale) according to the level of the stated attribute (sweetness, oiliness, stiffness, surface smoothness, etc.); see suggested materials in Table 10.4.

Typical score sheets are shown in Tables 10.5 and 10.6. The selection sequence may make use of more than one attribute ranking/rating test, especially if the ultimate panel will need to cover several sense modalities, for example, color, visual surface oiliness, stiffness, and surface smoothness.

Table 10.4 Suggested Materials for Ranking/Rating Tests

Taste	Sensory Stimuli	Concentration			
Sour	Citric acid/water, g/L	0.25	0.5	1.0	1.5
Sweet	Sucrose/water, g/L	10	20	50	100
Bitter	Caffeine/water, g/L	0.3	0.6	1.3	2.6
Salty	Sodium chloride/water, g/L	1.0	2.0	5.0	10
Odor					
Alcoholic	3-Methylbutanol/water, mg/L	10	30	80	180
Texture					
Hardness	Cream cheese,[a] American cheese,[a] peanuts, carrot slices[a]				
Fracturability	Corn muffin,[a] graham cracker, Finn crisp bread, Life Saver				

[a] At 1/4-inch thickness.

Table 10.5 Score Sheet, Ranking Test for Intensity

	Code
Least salty	________
	________
	________
Most salty	________

Rank the salty taste solutions in the coded cups in ascending order of saltiness.

Table 10.6 Score Sheet, Rating Test for Intensity

Code		
463	None________________________________	Strong
318	None________________________________	Strong
941	None________________________________	Strong
502	None________________________________	Strong

Rate the saltiness of each coded solution for intensity/strength of saltiness using the line scale for each.

10.3.1.4 Interpretation of Results of Screening Tests

Matching Tests: Reject candidates scoring less than 75% correct matches. Reject candidates for attribute tests who score less than 60% in choosing the correct descriptor.

Detection/Discrimination Tests: When using triangle tests, reject candidates scoring less than 60% on the "easy" tests (6 × threshold) or less than 40% on the "moderately difficult" tests (3 × threshold). When using duo–trio tests, reject candidates scoring less than 75% on the easy tests or less than 60% on the moderately difficult tests. Alternatively, use the sequential tests procedure as described in Chapter 7, Section 7.9.

Ranking/Rating Tests: Accept candidates ranking samples correctly or inverting only adjacent pairs. In the case of rating, use the same rank-order criteria and expect candidates to use a large portion of the prescribed scale when the stimulus covers a wide range of intensity.

10.3.2 Training

To ensure development of a professional attitude to sensory analysis on the part of panelists, conduct the training in a controlled professional sensory facility. Instruct subjects how to precondition the sensory modality in question, for example, not to use perfumed cosmetics and to avoid exposure to foods or fragrances for 30 min before sessions; how to prepare skin or hands for fabric and skinfeel evaluations; and how to notify the panel leader of allergic reactions that affect the test modality. On any day, excuse subjects suffering from colds, headaches, lack of sleep, and so on.

From the outset, teach subjects the correct protocols for handling the samples before and during evaluation. Stress the importance of adhering to the prescribed test procedures, reading all instructions, and following them scrupulously. Demonstrate ways to eliminate or reduce sensory adaptation, for example, by taking shallow sniffs of fragrances and leaving several tens of seconds between sample evaluations. Stress the importance of disregarding personal preferences and concentrating on the detection of differences or descriptions.

Begin by presenting samples of the product(s) under study that represent large, easily perceived sensory differences. Concentrate initially on helping panelists to understand the scope of the project and to gain confidence. Repeat the test method using somewhat smaller but still easily perceived sample differences. Allow the panel to learn through repetition until full confidence is achieved.

For attribute difference tests, carefully introduce panelists to the attributes, the terminology used to describe them, and the scale method used to indicate intensity. Present a range of products showing representative intensity differences for each attribute.

Continue to train "on the job" by using the new panelists in regular discrimination tests. Occasionally, introduce training samples to simulate off notes or other key product differences to keep the panel on track and attentive.

Be aware of changes in attitude or behavior on the part of one or more panelists who may be confused, losing interest, or distracted by other problems. The history of sensory testing is full of incredible results that could have come only from panelists who were "lost" during the test with the sensory analyst failing to anticipate and detect a failure in the "test instrument."

10.4 SELECTION AND TRAINING OF PANELISTS FOR DESCRIPTIVE TESTING

10.4.1 Recruiting Descriptive Panelists

Panelist recruiting is a key element in creating a successful descriptive panel (Appendix 10.2A). A descriptive panel describes products using attributes and intensities, so panelists must be capable of using both terms and expressions of magnitude to "tell the story" of the products. Even though a descriptive panelist should be a discriminator, it is important that the panelist also has proven abilities to think and to communicate.

Step one in panel building is recruiting as many interested, potential panelists as possible. They must be informed of some of the details surrounding the study and what the benefits are for them (money, knowledge, expertise, etc.) Postings, advertisements (print or online), and announcements on radio or during public events are only a few possible pathways. The postings and advertisements should be placed where they are most likely to be seen by people who are interested in food, beauty, and home (Stoer et al., 2002). Figure 10.1 is an example of such an ad. Some companies use temp agencies to recruit panelists who, if selected, are managed and paid by the temp agency.

Figure 10.1 Example of a descriptive panel recruiting advertisement. (From Stoer et al., *J Sens Stud.*, 17, 2002.)

10.4.2 Selection for Descriptive Testing

When selecting panelists for descriptive analysis, the panel leader or panel trainer should determine each candidate's capabilities in three major areas:

1. For each of the sensory properties under investigation (such as fragrance odor, flavor, oral texture, handfeel, or skinfeel), the ability to detect differences in characteristics present and in their intensities
2. The ability to describe those characteristics using (a) verbal descriptors for the characteristics and (b) scaling methods for the different levels of intensity
3. The capacity for abstract reasoning, as descriptive analysis depends heavily on the use of references when characteristics must be quickly recalled and applied to other products

In addition to screening panelists for these descriptive capabilities, panel leaders must prescreen candidates for the following personal criteria:

1. Interest in full participation in the rigors of the training, practice, and ongoing work phases of a descriptive panel
2. Availability to participate in 80% or more of all phases of the panel's work; whether conflict with home life, work load, travel, or even the candidate's supervisor may eventually cause the panelist to drop off the panel during or after training, thus losing one panelist from an already small number of 10–15
3. General good health and no illnesses related to the sensory properties being measured, such as
   a. Diabetes, hypoglycemia, hypertension, dentures, chronic colds or sinusitis, or food allergies in those candidates for flavor and/or texture analysis of foods, beverages, pharmaceuticals, or other products for internal use
   b. Chronic colds or sinusitis, for aroma analysis of foods, fragrances, beverages, personal care products, pharmaceuticals, or household products
   c. Central nervous system disorders or reduced nerve sensitivity due to the use of drugs affecting the central nervous system, for tactile analysis of personal care skin products, fabrics, or household products

The ability to detect and describe differences, the ability to apply abstract concepts, and the degree of positive attitude and predilection for the tasks of descriptive analysis can all be determined through a series of tests that include

- A set of prescreening questionnaires
- A set of acuity tests
- A set of ranking/rating tests
- A personal interview

The investment in a descriptive panel is large in terms of time and human resources, and it is wise to conduct an exhaustive screening process rather than train unqualified subjects.

Lists of screening criteria for three descriptive methods (the flavor profile, quantitative descriptive analysis, and texture profile) can be found in ASTM Special Technical

Publication 758 (1981). The following criteria listed are those used to select subjects for training in the Spectrum™ method of descriptive analysis as described in Chapter 12. These can be applied to the screening of employees or to external screening in cases where recruiting from the local community is preferred due to the amount of time necessary (20–50 h per person per week). The additional prescreening questionnaires are used to select individuals who can verbalize and think conceptually. This reduces the risk of selecting outside panelists who have sensory acuity but cannot acquire the technical orientation of panels recruited from inside the company.

10.4.2.1 Prescreening Questionnaires

For a panel of 15, typically 40–50 candidates may be prescreened using questionnaires such as those shown in Appendix 10.1. Appendix 10.1A applies to a tactile panel (skin-feel or fabric feel), Appendix 10.1B to a flavor panel, Appendix 10.1C to an oral texture panel, and Appendix 10.1D to a fragrance panel. Appendix 10.1E evaluates the candidate's potential to learn scaling and can be used with any of the preceding questionnaires in Appendix 10.1. Of the 40–50 original candidates, generally 20–30 qualify and proceed to the acuity tests.

10.4.2.2 Acuity Tests

To qualify for this stage, candidates should

- Indicate no medical or pharmaceutical causes of limited perception
- Be available for the training sessions
- Answer 80% of the verbal questions in the prescreening questionnaires in Appendix 10.1A through Appendix 10.1D correctly and clearly
- In the questionnaire Appendix 10.1E, assign scalar ratings that are within 10%–20% of the correct value for all figures

Candidates should demonstrate ability to

- Detect and describe characteristics present in a qualitative sense
- Detect and describe intensity differences in a quantitative sense

Therefore, although detection tests (e.g., triangle or duo–trio tests using variations in formulation or processing of the product to be evaluated) may yield a group of subjects who can detect small product variables, detection alone is not enough for a descriptive panelist. To qualify, subjects must be able to adequately discriminate and describe some key sensory attributes within the modalities used within the product class under test, and they also must show the ability to use a rating scale correctly to describe differences in intensity.

Detection. The panel trainer presents a series of samples representing key variables within the product class, in the form of triangle or duo–trio tests (Zook and Wesmann, 1977). Differences in process time or temperature (roast, bake, etc.), ingredient level (50% or 150% of normal), or packaging can be used as sample pairs to determine acuity in detection. Attempt to present the easier pairs of samples first and follow with pairs of increasing difficulty. Select subjects who achieve 50%–60% correct replies in triangle tests or 70%–80% in duo–trio tests, depending on the degree of difficulty of each test.

Description. Present a series of products showing distinct attribute characteristics (fragrance/flavor oils, geometrical texture properties [Civille and Szczesniak, 1973]) and ask candidates to describe the sensory impression. Use the fragrance list in Table 10.1 without a list of descriptors from which to choose. The candidate must describe each fragrance using his/her own words. These may include chemical terms (e.g., cinnamic aldehyde), common flavor terms (e.g., cinnamon), or related terms (e.g., Red Hots candy, Big Red gum, and Dentyne). Candidates should be able to describe 80% of the stimuli using chemical, common, or related terms and should at least attempt to describe the remainder with less specific terms (e.g., sweet, brown spice, hot spice).

10.4.2.3 Ranking/Rating Screening Tests for Descriptive Analysis

Having passed the prescreening tests and acuity tests, the candidate is ready for screening with the actual product class and/or sensory attribute for which the panel is being selected. A good example for a Camembert cheese panel is given by Issanchou et al. (1995). Candidates should rank or rate a number of products on a selection of key attributes using the technique of the future panel. These tests can be supplemented with a series of samples that demonstrate increasing intensity of certain attributes, such as tastes and odors (see Table 10.4), or oral texture properties (Appendix 12.2, Texture Section D, Scale 5 is suitable, containing hardness standards from cream cheese = 1.0 to hard candy = 14.5; also Scale 10, which contains standards for crispness from granola bar at 2.0 to cornflakes at 14.0). A questionnaire such as Table 10.7 is suitable. For certain skinfeel and fabric-feel properties, use Appendix 12.2E or 12.2F, or reference samples may need to be selected from among commercial products and laboratory prototypes that represent increasing intensity levels of selected attributes. Choose candidates who can rate all samples in the correct order for 80% of the attributes scaled. Allow for reversal of adjacent samples only, and check that candidates use most of the scale for at least 50% of the attributes tested.

10.4.2.4 Personal Interview

Especially for descriptive panels, a personal interview is necessary to determine whether candidates are well suited to the group dynamics and analytical approach. Generally, candidates who have passed the prescreening questionnaire and all of the acuity tests are interviewed individually by the panel trainer or panel leader. The objective of the interview is to confirm the candidate's interest in the training and work phases of the panel, including his/her availability with respect to workload, supervisor, and travel; and also his/her communication skills and general personality. Candidates who express little interest in the sensory programs as a whole, and in the descriptive panel in particular, should be excused. Individuals with very hostile or very timid personalities may also be excluded, as they may detract from the needed positive input of each panelist.

10.4.2.5 Mock Panel

Some companies further screen panelist candidates by inviting them to a mock panel, at which they are asked to evaluate and comment on two or more products. Candidates are

Table 10.7 Scoresheet Containing Two Ranking Tests Used to Screen Candidates for a Texture Panel

Descriptive Texture Panel Screening	
1. Place one piece of each product between molars; bite through once; evaluate for hardness. Rank the samples from least hard to most hard	
Least hard	______________
	______________
	______________
	______________
Most hard	______________
2. Place one piece of each product between molars; bite down once and evaluate for crispness (crunchiness)	
Least crisp	______________
	______________
	______________
	______________
Most crisp	______________

presented with the products, write down their perceptions (sensory parameters described by the session panel leader, for example, "The flavor and texture of these crackers"). The panel leader then directs a discussion of the results that provides each panelist with a time to express his or her perceptions. Observation of the panelists' behavior is helpful in deciding which candidates work best in a group, express concepts clearly, and participate in discussions of different perceptions.

10.4.3 Training for Descriptive Testing

The important aspect of any training sequence is to provide a structured framework for learning based on demonstrated facts and to allow the students, in this case panelists, to grow in both skills and confidence. Most descriptive panel training programs require between 40 and 120 h of training. The amount of time needed depends on the complexity of the product (wine, beer, and coffee panels require far more time than those evaluating lotions, creams, or breakfast cereals); on the number of attributes to be covered (a short-version descriptive technique for QC or storage studies in Chapter 12 requires fewer and simpler attributes); and on the requirements for validity and reliability (a more experienced panel will provide greater detail and nuance with greater reproducibility).

10.4.3.1 Terminology Development

The panel leader or panel trainer, in conjunction with the project team, must identify key product variables to be demonstrated to the panel during the initial stages of training. The project team should prepare a prototype or collect an array of products from commercially available samples as a frame of reference that represents as many of the attribute differences likely to be encountered in the product category as possible. The panel is first introduced to the chemical (olfaction, taste, chemical feeling factors) and physical principles (rheological, geometrical, etc.) that govern or influence the perception of each product attribute. With these concepts and terms as a foundation, the panel then develops procedures for evaluation, terminology with definitions and references for the product class.

Examples of this process are discussed by Szczesniak and Kleyn (1963) for oral texture, Schwartz (1975) and Civille and Dus (1991) for skincare products, McDaniel et al. (1987) for wines, Meilgaard and Muller (1987) for beer, Lyon (1987) for chicken, Johnsen et al. (1988) for peanuts, Johnsen and Civille (1986) for beef, and Johnsen et al. (1987) for catfish. Typically, the first stage of training may require 15–20 h as panelists begin to develop an understanding of the broad array of descriptors that fall into the category being studied (appearance, flavor, oral texture, etc.). This first phase is designed to provide them with a firm background in the underlying modality and for them to begin to perceive the different characteristics as they manifest in different product types.

10.4.3.2 Introduction to Descriptive Scaling

The scaling method of choice may be introduced during the first 10–20 h of training. By using a set of products or references that represent three to five different levels of each attribute, the panel leader reinforces both the sensory characteristic and the scaling method by demonstrating different levels or intensities across several attributes. Appendix 12.2 provides examples of different intensity levels of several sensory attributes for several sensory descriptive categories: flavor (aromatics, tastes, feeling factors), solid and semisolid texture (Muñoz, 1986) (hardness, adhesiveness, springiness, etc.), skinfeel (ASTM, 1997; Civille and Dus, 1991) (wetness, slipperiness, oiliness, etc.), and fabric feel (Civille and Dus, 1990) (slipperiness, grittiness, fuzziness, etc.).

The continued use of intensity reference scales during practice is meant to provide ongoing reinforcement of both attributes and intensities so that the panel begins to see the descriptive process as a use of terms and numbers (characteristics and intensities) to define or document any product in the category learned.

10.4.3.3 Initial Practice

The development of a precise lexicon for a given product category is often a three-step process. In the first step, a full array of products, prototypes, or examples of product characteristics are presented to the panel as a frame of reference. It is critical to include examples of all or most of the anticipated attributes for the product category. From this frame of reference, the panel generates an original long list of descriptors to which all panelists are invited to contribute. In the second stage, the original list, containing many overlapping terms, is rearranged and reduced into a working list in which the descriptors are comprehensive (they describe the product category completely) and yet discrete (overlapping is minimized). The third and last stage consists of choosing products, prototypes, and external references that can serve to represent good examples of the selected terms.

After the panel has a grasp of the terminology and a general understanding of the use of each scale, the panel trainer or leader presents a series of samples to be evaluated, one at a time, two or more of which represent a very wide spread in qualitative (attributes) and quantitative (intensity) differences. At this early stage of development, which lasts

15–40 h, the panel gains basic skills and confidence. The disparate samples allow the panel to see that the terms and scales are effective as descriptors and discriminators and help the members to gain confidence both as individuals and as a group.

10.4.3.4 Small Product Differences

With the help of the project/product team, the panel leader collects samples that represent smaller differences within the product class, including variations in production variables and/or bench modifications of the product. The panel is encouraged to refine the procedures for evaluation and the terminology with definitions and references to meet the needs of detecting and describing product differences. Care must be taken to reduce variations between supposedly identical samples; panelists in training tend to see variability in results as a reflection of their own lack of skill. Sample consistency contributes to panel confidence. This stage represents 10–15 h of panel time.

10.4.3.5 Final Practice

The panel should continue to test and describe several products during the final practice stage of training (15–40 h). The earlier samples should be fairly different, and the final products tested should approach the real-world testing situations for which the panel will be used.

During all five stages of the training program, panelists should meet after each session and discuss results, resolve problems or controversies, and ask for additional qualitative or quantitative references for review. This interaction is essential for developing the common terminology, procedures for evaluation, and scaling techniques that characterize a finely tuned sensory instrument.

10.5 PANEL PERFORMANCE AND MOTIVATION

Any good measuring tool needs to be checked regularly to determine its ability to perform validly and consistently. In the case of a sensory panel, the individuals, as well as the panel as a whole, need to be monitored. Panels are comprised of human subjects who have other jobs and responsibilities in addition to their participation in the sensory program; it is necessary to find ways to maintain the panelists' interest and motivation over long periods of product testing.

10.5.1 Performance

For both difference and descriptive panels, the sensory analyst needs to have a measure of the performance of each panelist, and of the panel as a whole, in terms of validity and reproducibility. Validity is the correctness of the response. In certain difference tests, such as the triangle and duo–trio, and in some directional attribute tests, the analyst knows the correct answer (the odd sample, the coded reference, the sweeter sample) and can assess the number of correct responses over time. The percent of correct responses can be computed for each panelist on a regular monthly or bimonthly basis. Weighted scores can also be calculated based on the difficulty of each test in which the panelist participated (Aust, 1984).

For the panel as a whole, validity can be measured by comparing panel results to other sensory test data, instrumental data, or the known variation in the stimulus, such as increased heat treatment, addition of a chemical, and so on.

Reliability, or the ability to reproduce results, can be easily assessed for the individual panelists and for the panel as a whole by replicating the test, using duplicate test samples, or using blind controls.

For descriptive data that are analyzed statistically by the analysis of variance, the panelists' performance can be assessed across each attribute as part of the data analysis (see ASTM, 1981 or Lea et al., 1997 for a detailed description of this analysis applied to a set of Quantitative Descriptive Analysis [QDA] results). It is recognized and accepted in QDA that panelists will use different parts of the scale to express their perceptions of the same sample. It is the relative differences in their ratings and not their absolute values that are considered important. In other descriptive methods, such as Spectrum (see Chapter 12), panelists are calibrated through the use of references to use the same part of the scale when evaluating the same sample. A descriptive panel of this type is equivalent to an analytical instrument that requires regular calibration checks. Several approaches, in addition to the ASTM guideline just mentioned, are appropriate for monitoring the individual and combined performance of "calibrated" panelists. Two aspects of performance that require monitoring are the panel's accuracy (bias) and its precision (variability) (see also Nielsen et al. [2005]).

Bias

To assess a panelist's ability to be "on target," the panel leader can determine the panelist's ability to match the accepted intensity of the attributes of a control or reference. The statistical measure of difference from the target or control rating, called *bias*, is defined as

$$\text{panelist bias, } d = x - m\,, \tag{10.1}$$

where:

- d is the deviation or bias
- x is the observed panelist value
- μ is the value for the control or target attribute

Variability

With several evaluations of a blind control or reference, the panelist's variability about his/her own mean rating is calculated using the panelist's standard deviation as follows:

$$\text{panelist SD, } s = \sqrt{\frac{\sum_{i=1}^{n}(x_i - \bar{x})^2}{(n-1)}}\,. \tag{10.2}$$

Good panelists have both low bias and low variability. The bias formula may be modified by removing the sign; this produces the absolute bias, calculated as

$$\text{panelist bias, } |d| = |x - \mu|, \tag{10.3}$$

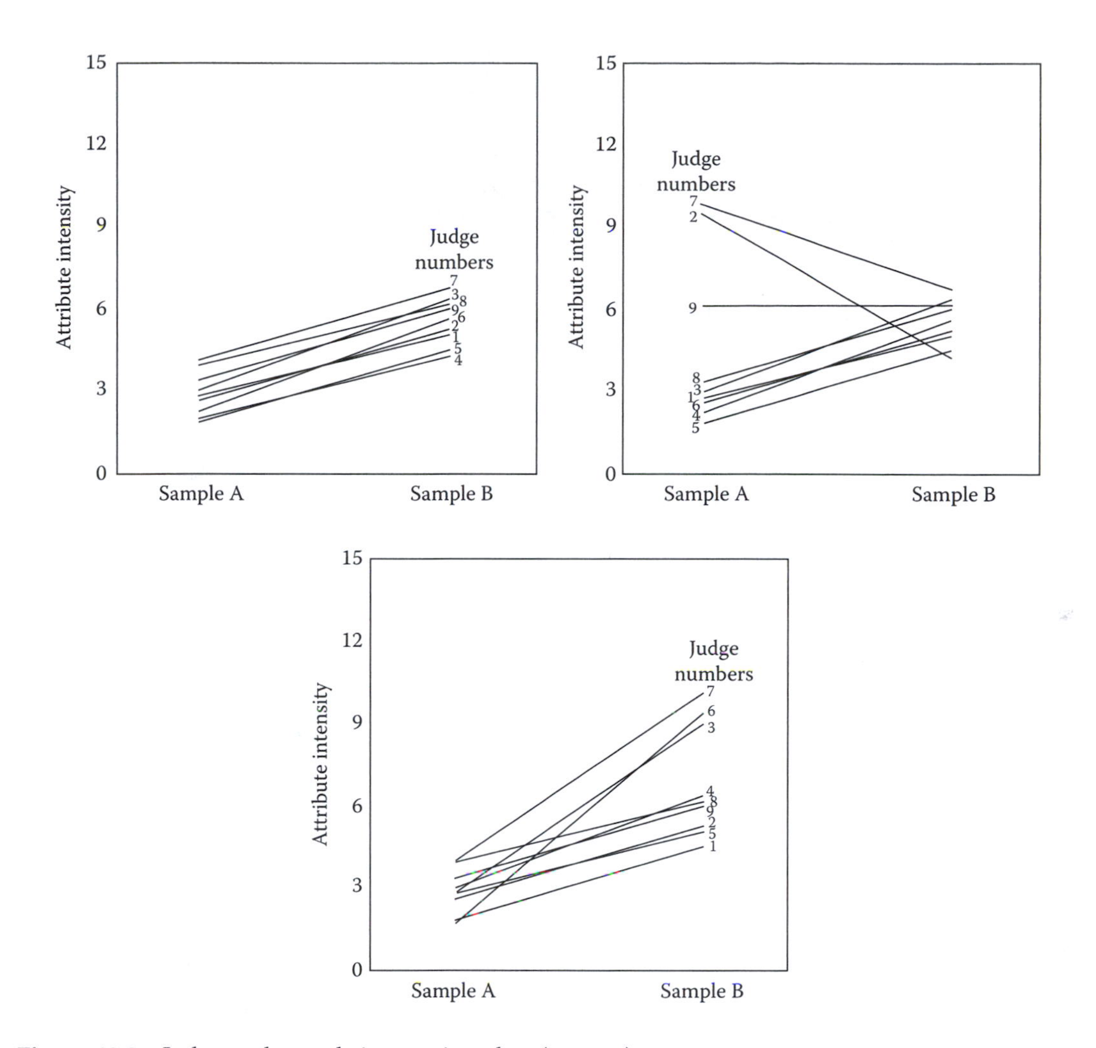

Figure 10.2 Judge and sample interaction plots (see text).

so that large positive and negative deviations do not offset each other. Small values of absolute bias are desirable. The panelists' statistics should be plotted over time to identify those panelists who need retraining or calibration.

When split-plot analysis of variance is used for descriptive data analysis, the judge-by-sample interaction is part of the results. When this interaction is significant, it is necessary to look at plots of the data to determine the source(s). Figure 10.2 shows three plots of judge-by-sample interactions. In each graph, each line represents one panelist's average ratings for two samples. In the first plot (A), the judge-by-sample interaction is not significant. All judges tend to rate the samples in the same direction and with the same relative degrees of intensity. Thus, the lines are in the same direction and similar in slope. The second plot (B) shows an extreme case of judge-by-sample interaction: Several samples are rated quite differently by some of the judges. Consequently, the lines run in different directions and have different slopes. The third plot (C) shows a few judges whose slopes differ from the rest. In this case, although the judge-by-sample interaction is statistically significant, the problem is less extreme; it is one of slight differences in the use of scales rather than total reversals, as in plot B. Generally, a judge-by-sample interaction indicates the need for more training, more frequent use of reference scales, or a review of terminology.

Additionally, individual panelist performance may be measured by assessing the panelist's ability to discriminate (sensitivity) and repeat their ratings (reproducibility). *Sensitivity* refers to the ability of each individual panelist to detect a small difference when it exists. Heightened sensitivity exists when panelists are both accurate and precise.

Measuring sensitivity for a given attribute implies that known small differences exist between two samples. Because the size of difference is often unknown, panelist ability to discriminate is often used as an alternative to sensitivity. The primary diagnostic used to measure discrimination is an ANOVA performed for each panelist separately where the product p-value indicates the panelist's ability to discriminate.

This can be performed for all attributes or for a subset of attributes (descriptors for which an overall product effect exists at the panel level). An acceptable success criteria can be defined at p-values <0.05, <0.1, and <0.2 and should be determined based on the set of products tested.

Overall panelist ability to discriminate can then be assessed by averaging diagnostics across all attributes or computing the frequency at which the panelist passed/failed to meet the established criteria at the attribute level. A criteria for overall performance can then be established (e.g., pass >50% of attributes for which overall panel found a difference).

Reproducibility is consistency within oneself. This measure should be computed by panelist for each product and each attribute. For individual panelists, precision can be measured by computing standard deviation (across replicates). Overall panelist precision can be assessed by averaging diagnostics across all attributes or computing the frequency at which the panelist passed/failed to meet the established criteria at the attribute level.

A criteria for overall performance can then be established (e.g., pass >75% of attributes).

A summary of both measures can be presented graphically by placing the average of all standard deviation and p-values across all attributes as the x and y axes of a scatter plot as in Figure 10.3.

10.5.2 Panelist Maintenance, Feedback, Rewards, and Motivation

One of the major sources of motivation for panelists is a sense of doing meaningful work. After a project is completed, panelists should be informed by letter or a posted circular of the project and test objectives, the test results, and the contribution made by the sensory results to the decision taken regarding the product. Immediate feedback after each test also tends to give the individual panelist a sense of "How am I doing?" The fears of some project leaders that panelists might become discouraged in tests with a low probability of success (a triangle test often has fewer than 50% correct responses) have proven groundless. Panelists do take into account the complexity of the sample, the difficulty of the test, and the probability of success. Panelists do want to know about the test and can indeed learn from past performance. Discussion of results after a descriptive

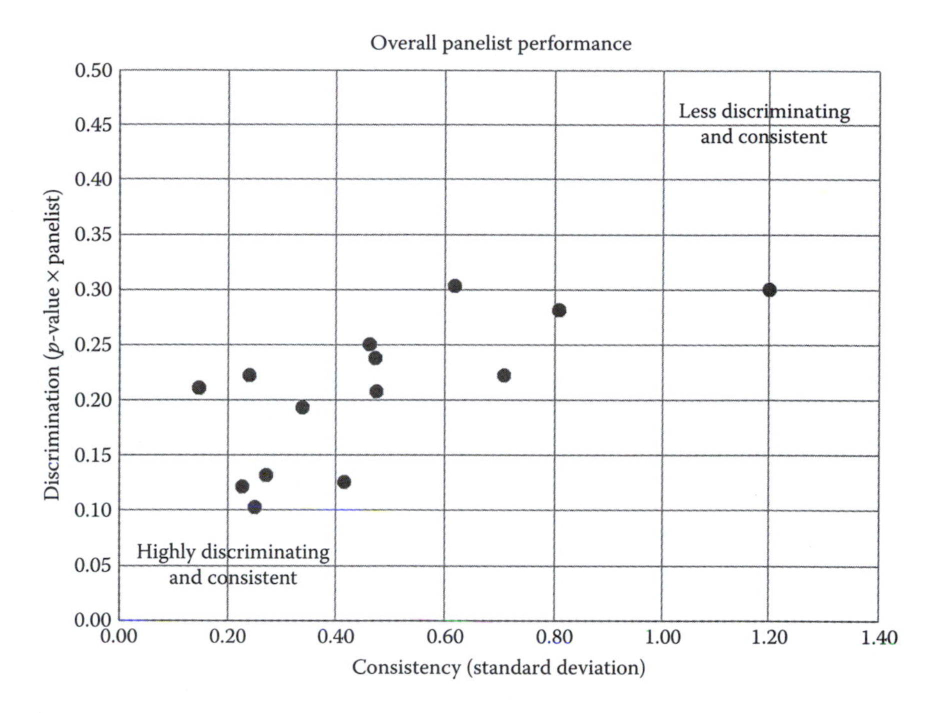

Figure 10.3 Summary of panelist performance. Consistency (standard deviation) by discrimination (*p*-value by panelist) for a panel that evaluates hair products.

panel session is highly recommended. The need to constantly refine the terms, procedures, and definitions is best served by regular panel interaction after all the data have been collected.

Feedback to panelists on performance can be provided in the form of data regarding their individual performance over three to five repeat evaluations of the same product vis-à-vis the panel as a whole. The data in Table 10.8 for a given sample indicates the mean and standard deviation for each panelist (numbers) for each attribute (letters), as well as the panel mean and standard deviation. Panelists can then determine how well the individual means agree with that of the panel as a whole (bias). In addition, the panelist's standard deviation provides an indication of that panelist's reliability (variability) on that attribute. Data for two or three products or samples over three to five evaluations should be shown to panelists on a regular basis, for example, every 3–4 months. Plots of judge-by-sample interaction, such as those shown in Figure 10.2, may also be shown to panelists to demonstrate both the general agreement among all the panelists and the performance of each panelist relative to the others.

In addition to the psychological rewards derived from feedback, panelists also respond positively and are further motivated to participate enthusiastically by a recognition and/or reward system. The presentation of certificates of achievement for

- High panel attendance
- High panel performance

Table 10.8 Panel Performance Summary

	Panelist						
Attributes	**1**	**2**	**3**	**4**	**5...**	**14**	**Panel X±SD**
A	7.5±02[a]	7.0±2	6.8±2	6.9±1	7.9±2.5	6.2±1.9	6.9±05[b]
B	4.2±1.4	4.8±2	5.5±1.6	5.0±0	4.2±1.2	4.6±1.6	4.8±0.4
C	1.4±1	3±1.3	1.5±1.2	1.0±0.9	1.1±0.8	3±1.3	1.8±0.8
D	9.0±0.5	8.0±0.7	9.0±1.0	6.4±1.2	12±1.1	10±1.3	9.4±1.6
E	4.0±0.7	4.2±0.8	3.5±1	1.9±1.2	4.4±0.9	3.8±2	3.9±1.1

Note: The 14 panelists evaluated the same sample in between other samples over a period of 3 weeks. The panel grand mean for attribute A was 6.9 and the SD over the 14 panelist means was 0.5 or 7.2%, showing satisfactory agreement between panelists for this attribute. Panelist 5 rated the attributes A and D much higher than the panel means and showed a high SD for attribute A.

[a] Panelist mean±standard deviation.

[b] Panelist grand mean±grand standard deviation.

- Improved performance
- Completion of a training program
- Completion of a special project

stimulates panel performance and communicates to panelists that the evaluation is recognized as worthwhile. Short-term rewards, such as snacks, tokens for company products, and raffle tickets for larger prizes, are often given to subjects daily. Over the longer term, sensory analysts often sponsor parties, outings, luncheons, or dinners for panelists, if possible, with talks by project or company management describing how the results were used. Publicity for panel work in the company newspaper or the local community media serves to recognize the current panel members and stimulates inquiry from potential candidates. Being a panelist is about discovering all of the sensory nuances the samples display. The ability to discover is strengthened by encouraging the panelists to become more sensory aware. Activities designed to increase sensory awareness are also motivating to the panel. The activities allow the panelists to learn new information while having a bit of fun, and they further stimulate the mind (Appendix 10.2B). Panel breakdown can occur if the panel leader does not set clear boundaries on what is acceptable and unacceptable behavior. It is a good idea to establish guidelines for expected behavior with the panel early on. Written guidelines that are reviewed and signed by the panelists serve as the foundation for panel operations (Appendix 10.2C). The underlying support by management for the full sensory program and for the active participation by panelists is a key factor in recruiting and maintaining an active pool of highly qualified members.

APPENDIX 10.1 PRESCREENING QUESTIONNAIRES

Each of the prescreening questionnaires is designed to enable the panel leader or trainer to select from a large group of candidates those individuals who are both verbal with respect to sensory properties to be evaluated and capable of expressing perceived amounts. For each type of panel to be trained (tactile, flavor, oral, texture, or fragrance), use the prescreener for that category plus the scaling exercises in Appendix 10.1E. The dates and times of training sessions and practice sessions should be provided to the applicant at the start of the prescreening process.

A. Prescreening Questionnaire for a Tactile Panel (Skinfeel or Fabric Feel)

Health Statement
Please be advised that as a tactile care product tester the essential functions of your position will be to apply products made with ingredients that may include but are not limited to fragrances, silicones, soaps, cleansers, abrasives, alcohol and petroleum products to the feet, hands, face and body. In addition, it is essential that personal care product testers be able to distinguish between different visual and tactile attributes of the products. Applicants with skin allergies, medical conditions or other conditions, such as circulatory issues, calluses or hypersensitivity that may be exacerbated by the application of certain products or ingredients are encouraged to check with their medical provider about whether this type of position is safe for them and if they will be able to perform the essential functions of the position with or without reasonable accommodations.

History
Name:
Phone Number:
Address:
E-Mail Address:
From what group or organization did you hear about this program?

Time

1. Are you available to attend **all** Training sessions listed in the Introduction?
   If not, please indicate days and times you are not available.
2. Are you available to attend all of the Practice sessions listed in the Introduction?
   If not, please indicate days you are not available.
3. After training, are you available to work (***specify HOURS, DAYS***)?
   If not, please indicate days/times you are not available.
4. How many weeks of vacation, on average, do you take every year?
5. Barring unforeseen circumstances, are you willing and able to make a two year commitment to this job?

Use Habits and General Information

1. Are you open to trying new products? (This includes but is not limited to creams and lotions for feet, hands, face and body; cosmetics; body wash; hair and anti-aging skin products.)
   If not please specify:
2. Do you apply a moisturizer daily that contains an SPF in it?
   If not how often do you apply an SPF?

3. How many products do you apply to your face regularly?
4. Think about the soap/body wash you most recently used. Describe what you like or dislike about it in detail.
5. Describe a memory related to touch/feel.
6. Does anyone in your immediate family work for a company that makes food, paper, or skincare products?
   If yes, explain:
7. Does anyone in your immediate family work for an advertising or market research agency?
   If yes, explain:

Tactile/Touch Quiz

1. What tactile characteristics of a lotion would make you think it is an expensive or high end product?
2. Which is thicker, an oily or greasy film?
3. When you rub something oily on your skin, how do your fingers move on your skin?
4. What specific appearance characteristics (visual) of a hand cream might influence the way you think it might feel during use?
5. When your skin feels moist, what other words or properties could describe it?
6. Name some things that are slippery, and what makes it slippery.
7. Name something that is smooth but not very slippery.
8. Briefly, how would you describe absorbent in a lotion?
9. What makes a deodorant feel sticky on your skin?
10. Describe the tactile characteristics (feel, touch) of soft hair.

B. Prescreening Questionnaire for a Flavor Panel

Health Statement

Please be advised that as a taste tester the essential functions of your position will be to taste products made with ingredients that include, but are not limited to, nuts, dairy products, wheat products, sugar, soy, eggs, meats and salt. In addition, it is essential that taste testers be able to distinguish between different flavors and textures. Applicants with food allergies, medical conditions or other oral conditions that may be exacerbated by the ingestion and chewing of certain foods are encouraged to check with their medical provider about whether this type of position is safe for them and if they will be able to perform the essential functions of the position with or without reasonable accommodations.

History

Name:
Phone Number:
Address:
E-Mail Address:
From what group or organization did you hear about this program?

Time

1. Are you available to attend **all** Training sessions listed in the Introduction? If not, please indicate days and times you are not available.
2. Are you available to attend all of the Practice sessions listed in the Introduction? If not, please indicate days you are not available.
3. After training, are you available to work (***specify HOURS, DAYS***)? If not, please indicate days/times you are not available.
4. How many weeks of vacation, on average, do you take every year?
5. Barring unforeseen circumstances, are you willing and able to make a two year commitment to this job?

Dynamics

1. Do you enjoy working with a group of people to accomplish a common goal? Why or why not?
2. Describe an important characteristic of an individual who makes a good team member.
3. Describe a time where you worked with others on a project or activity.

Food Habits and General Information

1. How often do you eat out in a month?
2. How often do you eat out fast foods in a month?
3. How often in a month do you eat a complete frozen meal?
4. What are your favorite foods?
5. Are there any foods that you are allergic to, or have an intolerance for, or simply choose not to eat? (Please list)
6. What foods do you not like to eat?
7. Think about what you had for dinner last night. Please describe the meal–how the food tasted and what you liked about it and what you disliked about in as much detail as possible.
8. Describe an event or instance where you were able to distinguish smells or tastes in products.
9. Thinking back in your life, describe a memory related to smell.
10. What do you like to read? For pleasure? For information?
11. What styles of restaurant do you prefer to frequent on a regular basis?
12. What are your hobbies? What makes them enjoyable?
13. Where have you traveled in the past 5 years? Where was the most fascinating and why?
14. Does anyone in your immediate family work for a food company?
15. Does anyone in your immediate family work for an advertising company or marketing research agency?

Flavor and Texture Quiz

1. If a recipe calls for thyme and there is none available, what would you substitute?
2. What are some other foods that taste like plain yogurt?
3. How would you describe the difference between flavor and aroma/smell?

4. How would you describe the difference between flavor and texture?
5. How would you describe the flavor of grated Italian cheese (Parmesan or Romano)?
6. How would describe the noticeable flavors in mayonnaise?
7. How would you describe the noticeable flavors in cola?
8. What is the difference between crispy and crunchy?
9. How would you describe the texture of an Oreo cookie when you eat it?
10. How would you describe the texture differences between meatloaf and steak?
11. When a cracker becomes stale, how does it change in flavor? In texture?
12. What are foods that come to mind when you think of "fried flavor"?

C. Prescreening Questionnaire for an Oral Texture Panel

History
Name:
Phone Number:
Address:
E-Mail Address:
From what group or organization did you hear about this program?

Time

1. Are you available to attend **all** Training sessions listed in the Introduction? If not, please indicate days and times you are not available.
2. Are you available to attend all of the Practice sessions listed in the Introduction? If not, please indicate days you are not available.
3. After training, are you available to work (***specify HOURS, DAYS***)? If not, please indicate days/times you are not available.
4. How many weeks of vacation, on average, do you take every year?
5. Barring unforeseen circumstances, are you willing and able to make a two year commitment to this job?

Texture Quiz

1. How would you describe the difference between flavor and texture?
2. Describe some of the textural properties of foods in general.
3. Describe some of the particles one finds in foods.
4. Describe some of the properties which are apparent when one chews on a food.
5. Describe the differences between crispy and crunchy.
6. What are some textural properties of potato chips?
7. What are some textural properties of peanut butter?
8. What are some textural properties of oatmeal?
9. What are some textural properties of bread?
10. For what type of products is texture important?

D. Prescreening Questionnaire for a Fragrance Panel

Health Statement
Please be advised that as a fragrance evaluator the essential functions of your position will be to smell products made with ingredients that include, but are not limited to, the

components of the fragrance such as essential oils, benzaldehyde, camphor, ethanol, propylene glycol and sodium lauryl sulfate as well as the ingredients of the products themselves. Products which have fragrance added include, but are not limited to, detergents, fabric softeners, skin care products, soap, shampoo, hairspray and sunscreens. In addition, it is essential that fragrance evaluators be able to distinguish between the various fragrance notes. Applicants with asthma, allergies or other conditions that may be exacerbated by the inhalation of fragrances in products are encouraged to check with their medical provider about whether this type of position is safe for them and if they will be able to perform the essential functions of the position with or without reasonable accommodations.

History
Name:
Phone Number:
Address:
E-Mail Address:
From what group or organization did you hear about this program?

Time

1. Are you available to attend **all** Training sessions listed in the Introduction?
   If not, please indicate days and times you are not available.
2. Are you available to attend all of the Practice sessions listed in the Introduction?
   If not, please indicate days you are not available.
3. After training, are you available to work (***specify HOURS, DAYS***)?
   If not, please indicate days/times you are not available.
4. How many weeks of vacation, on average, do you take every year?
5. Barring unforeseen circumstances, are you willing and able to make a two year commitment to this job?

Use Habits and General Information

1. Do you regularly wear a fragrance or an aftershave/cologne?
   If yes, what brands:
2. Do you prefer scented or unscented soap, detergents, fabric softeners and so on?
   Why?
3. What are some fragranced products that you like? Types or brands:
4. What are some fragranced products that you dislike? Types or brands:
5. Describe a memory related to smell.
6. Are there odors or smells that make you feel ill?
   In what way do they make you feel ill?
7. Does anyone in your immediate family work for a company that makes food, paper, or skincare products?
   If yes, explain:
8. Does anyone in your immediate family work for an advertising or market research agency?
   If yes, explain:

9. Members of a trained panel should not use perfumes/colognes on evaluation days, nor should they smoke an hour before the panel meets. Would you be willing to comply with this requirement if you are chosen as a panelist?

Fragrance Quiz

1. If a perfume is "floral" in type, what other words could be used to describe it?
2. What are some products that have an herbal smell?
3. What are some products that have a sweet smell?
4. What types of odors/smells are associated with cleanliness and freshness?
5. How would you describe the difference between fruity and lemony?
6. Briefly, how would you describe the difference between a feminine fragrance and a masculine fragrance?
7. What are some words that would describe the smell of a hamper full of clothes?
8. Describe some of the smells in a bakery.
9. Describe some of the smells in a liquid dish detergent.
10. Describe some smells in a McDonald's restaurant.
11. Describe some smells in a basement.

E. Scaling Exercises

(To be included with each of the prescreening questionnaires)

Instructions: Mark on the line at the right to indicate the proportion of the area that is shaded.

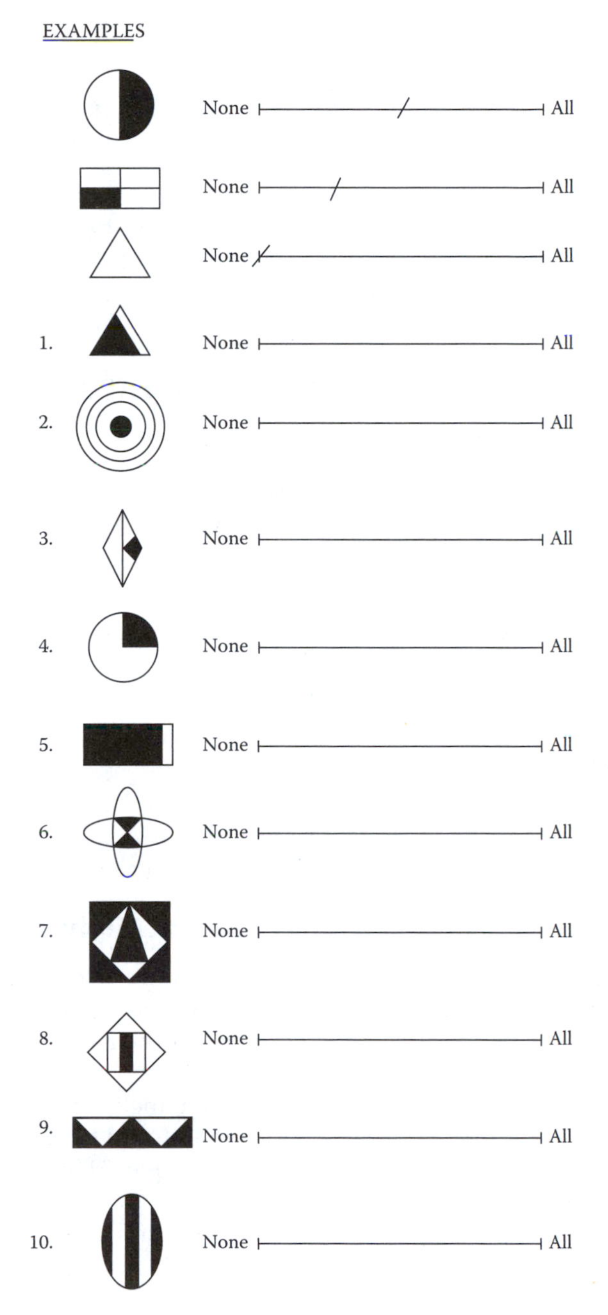

Prescreening questionnaire: scaling exercise. The answers are

1	7/8
2	1/8
3	1/6
4	1/4
5	7/8
6	1/8
7	3/4
8	1/8
9	1/2
10	1/2

APPENDIX 10.2 PANEL LEADERSHIP ADVICE

A. Panelist Recruiting Hints

Recruiting panelists requires creativity and perseverance. Creativity plays a role in the design and placement of advertising. Below is a list of possible places to advertise:

- Local newspapers within the home, food, or weekend sections
- Coupons in home mailers
- Community bulletin boards at grocery stores, health clubs, community pools, and so on
- Online jobsites like Craigslist (http://www.craigslist.com)
- Referrals from existing panelists or employees
- Local radio stations
- Community cable stations
- Colleges and adult schools
- Laundromats

B. Panel Activities for Sensory Awareness and Motivation (Dus, 2004)

- Share a sensory memory. Ask your panelists to share a sensory memory—write the words "I remember when …" in large letters on easel paper or the white board and encourage your panelists to tell a story about themselves. Be ready to share one of your sensory memories.
- Institute a sensory "show and tell" day. Invite your panelists to bring a sensory experience to share with each other. Make it once a month and have panelists sign up.
- Create a top 10 list, à la David Letterman; for example, "The top 10 things we are glad we do not have to evaluate."
- Do a Pepsi versus Coke triangle test or some other test (Puffs vs. Kleenex, etc.) then discuss.

- Create a wall (or bulletin board) of sensory related cartoons/comics or highlight one comic/cartoon per month. Ask your panelists to bring them in. Note: *New Yorker Magazine* is a great resource for this.
- Take a smell walk. Pair up your panelists and ask them to take a 10-min stroll outside and record all the smells they notice from the moment they leave the room to the moment they return.
- Draw a sound contest. Play a sound effect and ask panelists to illustrate what the sound looks like (encourage them not to draw what they think created the sound)—what the sound would look like if it had a shape/form and color. Then ask what it would smell like, taste like, and/or feel like.
- Blindfold test. Blindfold your panelists and present them with sensory stimuli and ask them to not only guess what it is but also to describe the sensory characteristics.
- Develop a sensory experience wish list with your panelists. What sensory stimuli do they wish they could experience? Brainstorm 100 and then choose 12 (one per month) and do them. Remember to defer judgment during the brainstorming—you will have plenty of opportunity to apply criteria after the list is generated. Of course some will be far-fetched (wallow in fur), but that is part of the fun.
- Ask your panelists to come up with one sensory fact that they think is cool and have them share it with the whole panel. Allow them to share facts about other creatures—do not confine yourself to human sensory perception. Have a coolest (or most obscure) fact contest.
- Have a "food that scares me" tasting day. Ask each panelist to bring in a food that they are reluctant to taste (like pig's feet) and then taste and describe. Then ask some questions: Which food tasted the best? Which one surprised you?
- Aromatherapy smells experience. Choose an aromatherapy category (i.e., lavender) and explore various products that are using that scent. Do a smell compare and contrast among lavender scented products. In what way are they similar? In what way are they different?
- Institute a weekly (or monthly) "sensory reading." Ask panelists to bring in and read aloud some sort of sensory related written piece. This should take no more than 5 min. We all know about Marcel Proust and his descriptions of madeleines—encourage your panelists to bring others. Create a recommended sensory reading list.
- Optical illusion day: Pass around different optical illusions just for fun.
- Go to http://puzzlemaker.school.discovery.com/and create a sensory crossword puzzle for your panel. See who can fill it out the fastest.
- Collect sensory scenes from movies. Play the scene at the end of a session. Ask you panelists to bring in other scenes. If you do not know where to start, use the scene from *French Kiss* where Kevin Kline teaches Meg Ryan about the flavor nuances in wine; he even pulls out a reference kit.
- Have a touchy-feely day. Ask each panelist to bring in something that they like the feel of. Pass them around and describe. You can focus on one type of feel—things that are soft, things that are sticky, and so on.

- Meet your panelists at the local mall and have a sensory treasure hunt. Pair up your panelists or have them work in teams.
- Do a sensory mad-lib with your panelists (the bookstore has *Mad-Lib* books). If you cannot find one that will fit your situation, then create your own. If you do not know what a mad-lib is, find an 8-year-old—they know. Laughter is guaranteed.
- Ask your panelists to create a collage that illustrates what a complex sensory experience such as creamy, fresh, refreshing, moisturized, or soft means to them. Ask each panelist to explain his or her collage and then hang them on the wall to create one big collage. What sensory characteristics do they see? What sensory insights do you see?
- Invite an expert to talk to your panel. Experts include perfumers, chefs, wine sommeliers, floral designers, fashion designers, sound mixers, and so on.
- Play a game of sensory charades.
- Have a "Design the Worst_______________" (product) contest. Have your panelists work in teams to create a worst sensory experience on a product. Give them 10 min.
- Bring in paint chips (many different colors) and ask your panelists to come up with their own names. Give them a theme: food names only, vacation place names, and so on.
- Just for fun, play "The Name Game." Prepare nametags with the panelists' names written on them. Place a nametag on their backs (do not put their own name on their back). The object of game is to find out whose name they have on their back by asking only yes or no questions.

C. Panel Guidelines Company

Company L provides panelists with a safe, pleasant working environment and offers panelists the flexibility to choose the studies in which they wish to participate.

Arrival Time

- Panelists are asked to arrive for panel 10 min before session is scheduled.
- Coats and sweaters need to be hung up in the hall closet.
- Prepare sites with templates.
- Panelists are responsible for signing in and out on the time sheet. Because payment is based on timesheets, it is imperative for panelists to record their participation in order to be paid.

Scheduling of Practice and Product Orientation

- These sessions are designed to review the attributes and provide feedback regarding the issues that may come about during a study.
- Practice sessions will be scheduled as needed.
- Full attendance at product orientation sessions prior to studies is required in order to participate in the study. If you cannot make all of the scheduled orientation sessions, you will be unable to participate in the study.

Scheduling of Study

- Client evaluations are scheduled with the panel leader.
- The panel leader informs the panel of any upcoming studies and the requirements. Typically, a study requires a panel of 10 panelists to provide data about the samples.
- The potential panelist must be able to make all of the scheduled sessions in order to participate in the study.
- To meet our client deadlines, we will generally be unable to schedule individual makeup days. Exceptions are at the panel leader's discretion.
- If the entire panel is canceled due to weather or illness, a makeup day will be scheduled.

Vacations and Down Time

- The panel leader needs to know about panelists' vacation plans. This is necessary to meet our clients' project time lines and to avoid scheduling conflicts.
- Panel is not scheduled during the week of Thanksgiving and the week between Christmas and New Year's Day.

Example of Panel Guidelines for a Skinfeel Panel

Salary

- Panelists will be paid once a month, with the pay periods ending on or near the 20th of the month. When possible, checks will be given to panelists at the end of the scheduled session. Some checks may have to be mailed.
- Pay scale is as follows:
- During training: $X per hour. (Note: Training includes an intensive series of sessions followed by biweekly sessions for approximately 3 months.)
- First year after training: $1.2X per hour.
- After second year: $1.35X per hour.
- Panelist hours are calculated to the nearest half hour with the quarter hour being the determining factor.
- Two hours guaranteed pay based on arrival time for panel for each scheduled session, unless otherwise agreed on for special studies.
- Panelists must participate in all scheduled sessions to receive payment. Emergency situations will be reviewed on an individual basis based on the nature of the circumstance.
- Punctuality is important so as not to delay or disrupt panel sessions. It is unfair to delay the session for panelists who arrive on time.
- If panelists are more than 5 min late, payment will commence with the next quarter hour.
- Panelists should avoid making their personal appointments during panel sessions.
- If a panelist needs to make an early departure, the panel leader should be informed as soon as possible. Payment will be based only on the time present at panel, and the 2-hour minimum does not apply.

Bonus

- Once a year, panelists will be reviewed for a potential bonus.
- Criteria for the bonus will include performance, following proper panel protocol, attendance, and data validity.
- Attendance includes on-time arrival to scheduled panel sessions and participation in scheduled studies (80%).
- Panel protocol includes site preparation, minimal talking, organized work space, awareness to detail, and so on.
- Data validity will be determined during client studies and validation studies.
- Data reliability is very important.
- Note: Panelists are measured and documented on their performance in evaluating the attributes required to review skinfeel products. Nonperforming panelists do not receive a bonus and are subject to review and reorientation without pay.

Ballot Completeness

- Ballots will be checked for completeness before turning them in to the panel leader.
- Ballot completeness includes name, panel ID number, sample code, all attribute scores, and so on. Data should be proofed by a panelist partner.
- Data cannot be reconstructed after the fact. Missing data cannot be used and may cause a study to have to be conducted again at the expense of Company L and the panel.
- If a study needs to be conducted again due to incomplete ballots, those panelists who had incomplete ballots will not be paid to repeat the study.

Validation

- Validation studies are used to document the panelists' mean and standard deviations in relationship to the panel as a whole and individually by panelists.
- This includes measuring the reliability of sample repetition by panelists.
- These studies are conducted for all areas of skinfeel and odor evaluations, including lotions, liquid soap, bar soaps, and so on.
- The results are used to document the integrity of the panel and are provided to our clients as requested.

Study Design

- Studies will use a panel pool composed of available panelists.
- A minimum of 10 (ten) panelists is needed for each study. Panels greater than 10 are encouraged.
- All panelists are expected to attend practice sessions.
- Panelists must be able to attend all sessions to be eligible for a given study.

Talk

- Discussions within panel room are limited to those directed by the panel leader.

- Panelists need to concentrate on their evaluations. Talking distracts other panelists from the task at hand, leaving opportunities to make mistakes and to forget to record data points.
- In between evaluations, panelists are welcome to bring reading material or quiet work.
- Some studies may not allow free time to spend doing other things other than waiting to complete the next time evaluation.

Panel Room

- No refreshments are allowed in the panel room.
- Smoking is prohibited.
- Panel areas must be cleaned up after each session.
- All magazines and newspapers are to be stored after use.

Panel Cancellation

- During inclement weather, a decision will be made as soon as possible and the panel leader will contact the panelists about cancellation.
- Winter weather cancellations usually follow the local school policy. If the roads are clear for the schools to open and panel is scheduled, panel will take place unless the panel leader decides differently. The panel leader will start the call chain notifying panelists of the change.
- Call chain is used to notify panelists of changes in the schedule.
- Panel leader starts chain and calls person on list.
- That person then calls the next person on the list.
- If there is no live person on the phone, leave a message and continue on the list until you reach a person.
- The process continues until the last person calls the panel leader to indicate that the chain is complete.
- The list is updated as necessary. Panelists should alert the panel leader to any phone number and address changes that occur during the year.

Emergency Calls

- If you cannot make a scheduled panel session please call and leave a message.
- There is an answering machine during nonbusiness hours.
- Please limit personal calls to the office. The phone should be used for emergency contacts only.

Outside Preparation

- For skinfeel panel:
    - It is recommended that panelists review pertinent protocols prior to a study and practice sessions.
    - Treat skin with care during the seasons. The weather may damage your skin surfaces for evaluations.

- Do not apply lotions or creams to the skin surfaces the day of evaluation prior to panel. This also includes items with strong lingering fragrances such as shampoos, hair sprays, perfumes, and so on.
- Use rubber gloves when working with detergents and dish washing.
- Use gloves when gardening to protect from calluses and blisters.
- Use sunscreen when outside to prevent sunburn.
- Be aware of changes in your skin's texture and surface before and after panel sessions.
- Panelists should report any allergic reactions to the panel leader.

REFERENCES

ASTM (1981). Committee E-18, guidelines for the selection and training of sensory panel members. In *ASTM Special Technical Publication 758*, West Conshohocken, PA: ASTM International.

ASTM (1997). *E1490–92 Standard Practice for Descriptive Skinfeel Analysis of Creams and Lotions*. West Conshohocken, PA. Available from http://www.astm.org.

Aust, L. B. (1984). Computers as an aid in discrimination testing. *Food Technol* 38:9, 71–3.

Bressan, L. P. and R. W. Behling (1977). The selection and training of judges for discrimination testing. *Food Technol* 31:11, 62–7.

Civille, G. V. and C. A. Dus (1990). Development of terminology to describe the handfeel properties of paper and fabrics. *J Sens Stud* 5: 19–32.

Civille, G. V. and C. A. Dus (1991). Evaluating tactile properties of skincare products: A descriptive analysis technique. *Cosmet Toiletries* 106:5, 83–8.

Civille, G. V. and A. S. Szczesniak (1973). Guide to training a texture panel. *J Text Stud* 4: 204–23.

Dus, C. A. (2004). 25 Activities for creative descriptive panel sessions. *Spectrum Sens* 7: 1.

ISO (1991). *Sensory Analysis—Methodology—Method of Investigating Sensitivity of Taste. ISO 3972: 1979.* New York, NY: International Organization for Standardization.

ISO (1993). *Sensory Analysis—General Guidance for the Selection, Training, and Monitoring of Assessors—Part I: Selected Assessors. ISO 8586-1:1993.* New York: International Organization for Standardization.

Issanchou, S., I. Lesschaeve, and E. P. Köster (1995). Screening individual ability to perform descriptive analysis of food products: Basic statements and application to a Camembert cheese descriptive panel. *J Sens Stud* 10: 349–68.

Johnsen, P. B. and G. V. Civille (1986). A standardized lexicon of meat W.O.F. descriptors. *J Sens Stud* 1:1, 99–104.

Johnsen, P. B., G. V. Civille, and J. R. Vercellotti (1987). A lexicon of pond-raised catfish flavor descriptors. *J Sens Stud* 2:2, 85–91.

Johnsen, P. B., G. V. Civille, J. R. Vercellotti, T. H. Sanders, and C. A. Dus (1988). Development of a lexicon for the description of peanut flavor. *J Sens Stud* 3:1, 9–17.

Lea, P., T. Næs, and M. Rødbotten (1997). *Analysis of Variance for Sensory Data*. Chichester: Wiley.

Lyon, B. G. (1987). Development of chicken flavor descriptive attribute terms by multivariate statistical procedures. *J Sens Stud* 2:1, 55–67.

McDaniel, M., L. A. Henderson, B. T. Watson Jr., and D. Heatherbill (1987). Sensory panel training and screening for descriptive analysis of the aroma of pinot noir wine fermented by several strains of malolactic bacteria. *J Sens Stud* 2:3, 149–67.

Meilgaard, M. C. and J. E. Muller (1987). Progress in descriptive analysis of beer and brewing products. *Tech Q Master Brew Assoc Am* 24:3, 79–85.

Meilgaard, M. C., D. S. Reid, and K. A. Wyborski (1982). Reference standards for beer flavor terminology system. *J Am Soc Brew Chem* 40: 119–28.

Muñoz, A. (1986). Development and application of texture reference scales. *J Sens Stud* 1:1, 55–83.

Nielsen, D., G. Hyldig, and R. Sørensen (2005). Performance of a sensory panel during long-term projects: A case study from a project on herring quality. *J Sens Stud* 20:1, 35–47.

Rainey, B. (1979). Selection and training of panelists for sensory panels. In *IFT Shortcourse: Sensory Evaluation Methods for the Practicing Food Technologist*. Chicago, IL: Institute of Food Technologists.

Schwartz, N. (1975). Adaptation of the sensory texture profile methods to skin care products. *J Text Stud* 6: 33–42.

Stoer, N., M. Rodriguez, and G. V. Civille (2002). New method for recruitment of descriptive analysis panelists. *J Sens Stud* 17: 77–88.

Szczesniak, A. and D. Kleyn (1963). Consumer awareness of texture and other food attributes. *Food Technol* 17:1, 74–7.

Zook, K. and C. Wesmann (1977). The selection and use of judges for descriptive panels. *Food Technol* 31:11, 56–61.

11

Descriptive Analysis Techniques

11.1 DEFINITION

All descriptive analysis methods involve the detection (discrimination) and the description of both the qualitative and quantitative sensory aspects of a product by trained panels of 5–100 judges (subjects). Smaller panels of 5–10 subjects are used for the typical product on the grocery shelf, whereas larger panels are used for products of mass production where small differences can be very important, for example, beers and soft drinks.

Panelists must be able to detect and describe the perceived sensory attributes of a sample. These qualitative aspects of a product combine to define the product and include all of the appearance, aroma, flavor, texture, or sound properties of a product that characterize or differentiate it from others. In addition, panelists must learn to rate the quantitative (i.e., intensity) aspects of a sample and to define to what degree each characteristic or qualitative note is present in that sample. Two products may contain the same qualitative descriptors, but they may markedly differ in the intensity of each, which results in quite different and easily distinctive sensory profiles (i.e., pictures) of each product. The two samples below have the same qualitative descriptors, but they substantially differ in the amount of each characteristic (quantitatively). The numbers used represent intensity ratings on a 15 cm line scale, where a zero means no detectable amount of the attribute and a 15 means a very large amount (Civille, 1979).

The two samples (385 and 408) below are commercially available potato chips.

Characteristic	**385**	**408**
Fried potato	7.5	4.8
Raw potato	1.1	3.7
Heated vegetable oil	3.6	1.1
Salty	6.2	13.5
Sweet	2.2	1.0

Although these two samples of chips have the same attribute descriptors, they markedly differ in the intensity of each flavor note. Sample 385 has distinct fried potato character with underlying oil, sweet, and raw-potato notes. Sample 408 is dominated by saltiness, with the potato, oil, and sweet notes being of lower impact.

11.2 FIELD OF APPLICATION

Use descriptive tests to obtain detailed description of the aroma, flavor, and oral texture of foods and beverages, the skinfeel of personal care products, the handfeel of fabrics and paper products, and the appearance and sound of any product. These sensory profiles are used in research and development (Meilgaard and Muller, 1987) and in manufacturing to:

- Document a control product before a product development project designed to change an ingredient or a process; the control profile is compared to one or more prototypes to determine which is a match or closest to the original control profile
- Define the sensory properties of a target product for new product development (Szczesniak et al., 1975)
- Define the characteristics/specifications for a control or standard for QA/QC and R&D applications (Muñoz et al., 1992)
- Document product attributes before a consumer test to help in the selection of attributes to be included in the consumer questionnaire and to help in an explanation of the results of the consumer test
- Track a product's sensory changes over time with respect to understanding shelf life, packaging, and so on
- Map perceived product attributes for the purpose of relating them to instrumental, chemical, or physical properties (Bargmann et al., 1976; Moskowitz, 1979)
- Measure short-term changes in the intensity of specific attributes over time (time–intensity analysis)

11.3 COMPONENTS OF DESCRIPTIVE ANALYSIS

11.3.1 Characteristics: The Qualitative Aspect

Those perceived sensory parameters that define the product are referred to by various terms such as attributes, characteristics, character notes, descriptive terms, descriptors, or terminology (Johnsen et al., 1988). The attributes are collectively referred to as a lexicon (Lawless and Civille, 2013).

These qualitative factors (which are the same as the parameters discussed under classification, Section 5.3) include terms that define the sensory profile or picture or thumbprint of the sample. An important aspect is that panelists, unless well trained, may have very different concepts of what a term means. The question of concept formation is reviewed in detail by Lawless and Heymann (1998). The selection of sensory attributes and the corresponding definition of these attributes should be related to the underlying chemical and physical properties of a product that can be perceived (Civille and Lawless, 1986). Adherence to an understanding of a product's actual rheology or chemistry makes the descriptive data easier to interpret and more useful for decision making. Statistical methods such as ANOVA and multivariate analysis can be used to select the more discriminating terms (Jeltema and Southwick, 1986; ISO, 1994).

The components of a number of different descriptive profiles are given below (examples of each are shown in parentheses). Note that this general list is also the key to a more complete list of descriptive terms given in Appendixes 12.1 through 12.4. The repeat appearance of certain properties and examples is intentional.

1. Appearance characteristics
   a. Color (hue, chroma, uniformity, depth)
   b. Surface texture (shine, smoothness/roughness)
   c. Size and shape (dimensions and geometry)
   d. Interactions among pieces or particles (stickiness, agglomeration, loose particles)
2. Aroma characteristics
   a. Olfactory sensations (vanilla, fruity, floral, skunky)
   b. Nasal feeling factors (cool, pungent)
3. Flavor characteristics
   a. Olfactory sensations (vanilla, fruity, floral, chocolate, skunky, rancid)
   b. Taste sensations (salty, sweet, sour, bitter, umami)
   c. Oral feeling factors (heat, cool, burn, astringent, metallic)
4. Oral texture characteristics (Brandt et al., 1963; Szczesniak, 1963; Szczesniak et al., 1963)
   a. Mechanical parameters; reaction of the product to stress (hardness, viscosity, deformation/fracturability)
   b. Geometrical parameters, that is, size, shape, and orientation of particles in the product (gritty, grainy, flaky, stringy)
   c. Fat/moisture parameters, that is, presence, release, and adsorption of fat, oil, or water (oily, greasy, juicy, moist, wet)
5. Skinfeel characteristics (Schwartz, 1975; Civille and Dus, 1991; ASTM, 2011)
   a. Mechanical parameters; reaction of the product to stress (thickness, ease to spread, slipperiness, denseness)
   b. Geometrical parameters, for example, size, shape, and orientation of particles in product or on skin after use (gritty, foamy, flaky)
   c. Fat/moisture parameters, for example, presence, release, and absorption of fat, oil, or water (greasy, oily, dry, wet)
   d. Appearance parameters; visual changes during product use (gloss, whitening, peaking)
6. Texture/handfeel of woven and nonwoven fabrics (Civille and Dus, 1990)
   a. Mechanical properties; reaction to stress (stiffness, force to compress or stretch, resilience)
   b. Geometrical properties, that is, size, shape, and orientation of particles (gritty, bumpy, grainy, ribbed, fuzzy)
   c. Moisture properties; presence and absorption of moisture (dry, wet, oily, absorbent)

Again, the keys to the validity and reliability of descriptive analysis terminology are

- Terms based on a thorough understanding of the technical and physiological principles of flavor or texture, sound, or appearance
- Thorough training of all panelists to fully understand the terms in the same way and to apply them in the same way
- Use of terminology (see Appendixes 12.1 and 12.2) to ensure consistent application of descriptive terms to a perception

11.3.2 Intensity: The Quantitative Aspect

Intensity, the quantitative aspect of descriptive analysis, expresses the degree to which each of the characteristics (i.e., terms, qualitative components) is present. This is expressed by the assignment of some value along a measurement scale.

As with the validity and reliability of terminology, the validity and reliability of intensity measurements are highly dependent upon

- The selection of a scaling technique that is broad enough to encompass the full range of parameter intensities and that has enough discrete points to pick up all the small differences in intensity between samples
- The thorough training of the panelists to use the scale in a similar way across all samples and across time (see Chapter 10 on panelist training)
- The use of reference scales for intensities of different properties (see Appendix 12.2) to ensure consistent use of scales across panelists and repeated evaluations

Three types of scales are commonly used in descriptive analysis (see also Lawless and Heymann, 1998):

1. Category scales are limited sets of words or numbers, constructed (as best as one can) to maintain equal intervals between categories. A full description can be found in Chapter 5. A category scale from 0 to 9 is perhaps the most used in descriptive analysis, but longer scales are often justified. A good rule of thumb is to evaluate how many steps a panelist can meaningfully employ and to adopt a scale twice that length. Sometimes a 100-point scale is justified, for example, in visual and auditory studies.
2. Line scales utilize a line 6 in. (15 cm) long, on which the panelist makes a mark; they are described in Chapter 5. Line scales are almost as popular as category scales. Their advantage is that the intensity can be more accurately graded because there are no steps or "favorite numbers." The chief disadvantage to using line scales is that it is harder for a panelist to be consistent because a position on a line is not as easily remembered as a number.
3. Magnitude estimation (ME) scales are based on free assignment of the first number; after that, all subsequent numbers are assigned in proportion (see Chapter 5). ME is mostly used in academic studies where the focus is on a single attribute that can vary over a wide range of sensory intensities (Moskowitz, 1975, 1978).

Appendix 12.2 contains sets of reference samples useful for the establishment of scales for various odors and tastes and also for the mechanical, geometrical, and moisture properties of oral texture. All the scales in Appendix 12.2 are based on a 15 cm line scale; however, the same standards can be distributed along a line or scale of any length or numerical value. The scales employ standard aqueous solutions such as sucrose, sodium chloride, citric acid, and caffeine as well as certain widely available supermarket items that have shown adequate consistency, for example, Hellmann's® mayonnaise and Welch's® grape juice.

11.3.3 Order of Appearance: The Time Aspect

In addition to accounting for the attributes (qualitative) of a sample and the intensity of each attribute (quantitative), panels can often detect differences among products in the order in which certain parameters manifest themselves. The order of appearance of physical properties, related to oral, skin, and fabric textures, are generally predetermined by the way the product is handled (the input of forces by the panelist). By controlling the manipulation (one chew, one manual squeeze), the subject induces the manifestation of only a limited number of attributes (hardness, denseness, deformation) at a time (Civille and Liska, 1975).

However, with the chemical senses (aroma and flavor), the chemical composition (e.g., amount of fat or oil) of the sample and some of its physical properties (temperature, volume, concentration) may alter the order in which certain attributes are detected (IFT, 1988). In some products such as beverages, the order of appearance of the characteristics is often as indicative of the product profile as the individual aroma and flavor notes and their respective intensities.

Included as part of the treatment of the order of appearance of attributes is aftertaste or after feel, which are those attributes that can still be perceived after the product or sample has been used or consumed. A complete picture of a product requires that all characteristics that are perceived after use of the product should be individually mentioned and rated for intensity.

Attributes described and rated for aftertaste or after feel do not necessarily imply a defect or negative note. For example, the cool aftertaste of a mouthwash or breath mint is a necessary and desirable property. On the other hand, a cola beverage's metallic aftertaste may indicate a packaging contamination or a problem with a particular sweetener.

When the intensity of one or more sensory properties is repeatedly tracked across a designated time span, techniques called time–intensity analysis are used. More detailed descriptions of these techniques are given in subsequent sections of this chapter.

11.3.4 Overall Impression: The Integrated Aspect

In addition to the detection and description of the qualitative, quantitative, and time factors that define the sensory characteristics of a product, panelists are capable of, and management is often interested in, some integrated assessment of the product properties. Ways such integration has been attempted include the following:

Total intensity of aroma or flavor: A measure of the overall impact (intensity) of all the aroma components (perceived orthonasally in nose) or a measure of the overall flavor impact of the aromatics (perceived retronasally in mouth). Such an evaluation can be important in determining the general fragrance or flavor impact that a product delivers to the consumer who does not normally understand all of the nuances of the contributing odors or tastes that the panel describes. The components of texture are more functionally discrete, and "total texture" is not a property that can be determined.

Balance/blend (amplitude): A well-trained descriptive panel is often asked to assess the degree to which various flavor or aroma characteristics fit together in the product. Such an evaluation involves a sophisticated understanding, half learned and half intuitive, of the appropriateness of the various attributes, their relative intensity in the complex, and the way(s) they harmonize in the complex. Specifically, blend refers to the melding of individual sensory notes such that the products present a unified overall sensory experience as opposed to spikes or individual notes. Balance, however, is the extent to which the sensory notes are in proportion and complementary to each other. Balance and blend are often evaluated jointly on a ten-point scale.

0 --	-- 10
Not balanced/blended; easy to distinguish components	Well balanced/blended; difficult to distinguish components

In practice, the aromatics in a flavor system may be well blended yet inappropriate for the product category. Sometimes these aromatics are "off notes" present at low intensities, such as cardboard, paint, or metal. However, it is possible for the inappropriate aromatic to be an "on" note. For example, garlic is wonderful in tomato sauce but inappropriate in vanilla pudding. Alternatively, roasted garlic may be well blended in a crème brûlée, but it is still inappropriate. Evaluation of balance or blend (or *amplitude* as it is called in the flavor profile method [Cairncross and Sjöstrom, 1950; Caul, 1957; Keane, 1992]) is difficult even for the highly trained panelist and should not be attempted with naïve or less sophisticated subjects. In addition, care must be taken in the use of data on balance or blend. Often a product is not intended to be blended or balanced; a preponderance of spicy aromatics or toasted notes may be essential to the full character of a product. In some products, the consumer may not appreciate a balanced composition, despite its well-proportioned notes, as determined by a trained panel. Therefore, it is important to understand the relative importance of blend or balance among consumers for the product in question before measuring and/or using such data.

Fidelity: Fidelity is the total sensory experience of the trueness of the product in the stated context or, in other words, its believability. In practice, fidelity cannot be evaluated and rated if testing is blind, because part of what fidelity captures is delivery against a concept. Fidelity may not be appropriate for innovative prototypes because often the context in which these prototypes fit is yet to be determined. Furthermore, since the context of products may differ based on regional or cultural identities, fidelity of the same product may change in different cultures (Chambers et al., 2012).

Integrity: Integrity is the integration of balance, blend, complexity (the existence of multiple sensory attributes or sensory layers that make up a singular sensory experience), longevity (the time that the full integrated sensory experience sustains itself in the month and after swallowing), and amplitude with fidelity, which presents a unified, full sensory experience that compliments the stated expectation. Integrity is not measureable unless fidelity is measured (Chambers et al., 2012).

Overall difference: In certain product situations, the key decisions involve determination of the relative differences between samples and some control or standard product. Although the statistical analysis of differences between products on individual attributes is possible with many descriptive techniques, project leaders are often concerned with just how different a sample or prototype is from the standard. The determination of an overall difference (see difference-from-control test, Section 7.8) allows the project management to make decisions regarding disposition of a sample based on its relative distance from the control; the accompanying descriptive information provides insight into the source and size of the relative attributes of the control and the sample (Aust et al., 1985; Pecore et al., 2006).

Hedonic ratings: It is a temptation to ask the descriptive panel, once the description has been completed, to rate the overall acceptance of the product. In most cases, this temptation is to be resisted, as the panel, through its training process, has been removed from the world of consumers and is no longer representative of any section of the general public. Training tends to change the personal preferences of panelists. As they become more aware of the various attributes of a product, panelists tend to weigh attributes differently to the way a regular consumer would in terms of each attribute's contribution to the overall quality, blend, or balance.

11.4 COMMONLY USED DESCRIPTIVE TEST METHODS WITH TRAINED PANELS

Over the last 40 years, many descriptive analysis methods have been developed, and some have gained and maintained popularity as standard methods (ASTM, 1992, 1996). The fact that these methods are described below is a reflection of their popularity, but it does not constitute a recommendation for use. On the contrary, a sensory analyst who needs to develop a descriptive system for a specific product and project application should study the literature on descriptive methods and should review several methods and combinations of methods before selecting the descriptive analysis system that can provide the most comprehensive, accurate, and reproducible description of each product and the best discrimination between products (see ASTM 1992, which also contains case studies of four methods). Modifications of existing descriptive methods are common and can result in precise, customized descriptive methods to precisely document the sensory attributes of a specific product category.

11.4.1 Flavor Profile Method

The flavor profile method was developed by Arthur D. Little, Inc., in the late 1940s (Caul, 1957; Keane, 1992). It involves the analysis of a product's perceived aroma and flavor characteristics, their intensities, order of appearance, and aftertaste by a panel of four to eight trained judges. An amplitude rating (see page 202) is generally included as part of the profile.

Panelists are selected on the basis of a physiological test for taste discrimination, taste intensity discrimination, and olfactory discrimination and description. A personal interview is conducted to determine interest, availability, and potential for working in a group situation. For training, panelists are provided with a broad selection of reference samples representing the product range as well as examples of ingredient and processing variables for the product type. Panelists, with the panel leader's help in providing and maintaining reference samples, develop and define the common terminology to be used by the entire panel. The panel also develops a common frame of reference for the use of the seven-point flavor profile intensity scale shown in Section 5.6.1.

The panelists, seated at a round or hexagonal table, individually evaluate one sample at a time for both aroma and flavor, and they record the attributes (called *character notes*), their intensities, order of appearance, and aftertaste. Additional samples can be subsequently evaluated in the same session, but samples are not tasted back and forth. The results are reported to the panel leader who then leads a general discussion of the panel to arrive at a consensus profile for each sample. The data is generally reported in tabular form, although a graphic representation is possible.

The flavor profile method may be applied when a panel must evaluate many different products with none that are a major producer's major line. The main advantage, but also a major limitation, of the flavor profile method is that it only uses four to eight panelists. The lack of consistency and reproducibility that this limitation entails is somewhat overcome by training and by the consensus method. However, the latter has been criticized for one-sidedness. The panel's opinion may become dominated by that of a senior member or a dominant personality, and equal input from other panel members is not always obtained. Other points of criticism of the flavor profile are that screening methods do not include tests for the ability to discriminate specific aroma or flavor differences that may be important in specific product applications, and that the seven-point scale limits the degree of discrimination among products showing small, but important, differences.

11.4.2 Texture Profile Method

Based somewhat on the principles of the flavor profile method, the texture profile method was developed by the Product Evaluation and Texture Technology groups at General Foods Corp. to define the textural parameters of foods (Skinner, 1988). Later, the method was expanded by Civille and Szczesniak (1973) and Civille and Liska (1975) to include specific attribute descriptors for specific products, including semisolid foods, beverages, skinfeel products (Schwartz, 1975; ASTM, 1997a), and fabric and paper goods (Civille and Dus, 1990). In all cases, the terminology is specific for each product type, but it is based on the underlying rheological properties expressed in the first texture profile publications (Brandt et al., 1963; Szczesniak, 1963; Szczesniak et al., 1963).

Panelists are selected on the basis of their ability to discriminate known textural differences in the specific product application for which the panel is to be trained (solid foods, beverages, semisolids, skincare products, fabrics, paper, etc.). As with most other descriptive analysis techniques, panelists are interviewed to determine interest, availability, and attitude. Panelists selected for training are exposed to a wide range of products from the category under investigation to provide a wide frame of reference. In addition, panelists are introduced to the underlying textural principles involved in the structure of the products under study. This learning experience provides panelists with an understanding of the concepts of input mechanical forces and the resulting strain on the product. In turn, panelists are able to avoid lengthy discussions about redundant terms and to select the most technically appropriate and descriptive terms for the evaluation of products. Panelists also define all terms and all procedures for evaluation, thereby reducing some of the variability encountered in most descriptive testing. The reference scales used in the training of panelists can later serve as references for approximate scale values that further reduce panel variability.

Each panelist, using one of the scaling techniques previously discussed, independently evaluates samples. The original texture profile method used an expanded 13-point version of the flavor profile scale. In the last several years, however, texture profile panels have been trained using category, line, and ME scales (see Appendix 12.2, for food texture references for use with a 15-point or 15 cm line scale). Depending on the type of scale used by the panel and on the way the data is to be treated, the panel verdicts may be derived by group consensus, as with the flavor profile method, or by statistical analysis of the data. For final reports, the data may be displayed in tabular or graphic form.

11.4.3 Quantitative Descriptive Analysis (QDA®) Method

In response to dissatisfaction among sensory analysts with the lack of statistical treatment of data obtained with the flavor profile or related methods, the Tragon Corp. developed the Tragon QDA® method of descriptive analysis (Stone et al., 1974; Stone and Sidel, 1992). This method relies heavily on statistical analysis to determine the appropriate terms, procedures, and panelists to be used for the analysis of a specific product.

Panelists are selected from a large pool of candidates according to their ability to discriminate differences in sensory properties among samples of the specific product type for which they are to be trained. The training of QDA panels requires the use of product and ingredient references, as with other descriptive methods, to stimulate the generation of terminology. The panel leader acts as a facilitator, rather than as an instructor, and refrains from influencing the group. Attention is given to development of consistent terminology, but panelists are free to develop their own approach to scoring, using the 15 cm (6 in.) line scale that the method provides.

QDA panelists evaluate products one at a time in separate booths so as to reduce distraction and panelist interaction. Panelists enter the data into a computer, or the score sheets are individually collected from the panelists as they are completed, and the data are entered for computation usually with a card reader directly from the score sheets or by a direct data entry system. Panelists do not discuss data, terminology, or samples after each taste session, and they must depend on the discretion of the panel leader for any information on their performance relative to other members of the panel and to any known differences between samples.

The results of a QDA test are statistically analyzed, and the report generally contains a graphic representation of the data in the form of a spider web with a branch or spoke from a central point for each attribute.

The QDA method was developed in partial collaboration with the Department of Food Science at the University of California at Davis. It represents a large step toward the ideal of this book, the intelligent use of human subjects as measuring instruments, as discussed in Chapter 1. In particular, the use of a graphic scale (visual analog scale) that reduces that part of the bias in scaling resulting from the use of numbers; the statistical treatment of the data; the separation of panelists during evaluation; and the graphic approach to presentation of data have done much to change the way that sensory scientists and their clients view descriptive methodology. The following areas could benefit from change or further development:

1. The panel, because of a lack of formal instruction, may develop erroneous terms. For example, the difference between natural vanilla and vanillin should be easily detected and described by a well-trained panel; however, an unguided panel would choose the term *vanilla* to describe the flavor of vanillin. Lack of direction also may allow a senior panelist or stronger personality to dominate the proceedings in all or part of the panel population in the development of terminology.
2. The free approach to scaling can lead to inconsistency of results, partly because of particular panelists' evaluating a product on a given day and not on another, and partly because of the context effects of one product seen after the other with no external scale references for intensity.
3. The lack of immediate feedback to panelists on a regular basis reduces the opportunity for learning and the expansion of terminology for greater capacity to discriminate and describe differences.
4. On a minor point, the practice of connecting the spokes of the spider web can be misleading to some users, who, because of their technical training, expect the area under a curve to have some meaning. In reality, the sensory dimensions shown in the web may be either unrelated to each other or related in ways that cannot be represented in this manner.

11.4.4 Spectrum™ Descriptive Analysis Method

This method, designed by Civille, is described in detail in Chapter 12 and other literature (Civille and Seltsam, 2014; Dus et al., 2014). Its principal characteristic is that the panelist scores the perceived intensities with reference to prelearned, absolute intensity scales. The purpose is to make the resulting profiles universally understandable and usable, not only at a later date, but also at any laboratory outside the originating one. The method provides for this purpose an array of standard attribute names (lexicons), each with its set of standards that define a scale of intensity, usually from 0 to 15, which can be measured on a 15 cm line scale or simply recorded as a straight number. Protocols for creating lexicons and a list of published lexicons are covered in Chapter 12 and other literature (Lawless and Civille, 2013; Drake and Civille, 2002).

11.4.5 Time–Intensity Descriptive Analysis

Every product has a temporal component to its sensorial experience. Determining which temporal measuring tool is appropriate for a particular test should be part of the sensory scientist's expertise. Temporal methods fall within two main categories: fixed-time-point methods and continuous measurement methods.

11.4.5.1 Fixed-Time-Point Methods

11.4.5.1.1 Progressive Profiling

In progressive profiling, a series of line-scales are presented on the computer screen at predetermined intervals. The intensities of selected attributes (typically up to 5) are then recorded as if the panelist is creating a descriptive profile (Jack et al., 1994; Findlay, 2011). The advantages of this approach include the facilitation of multiple product exposures and longer time intervals as well as the ability to set time intervals to vary in duration (Pecore et al., 2013a). However, in this approach, the number of attributes should be limited, or panelist performance may suffer due to time pressure (Wilkinson et al., 2000).

11.4.5.1.2 Sequential Profiling

Sequential profiling is an expansion of progressive profiling that includes evaluations at multiple consecutive tastings. The aftertaste of each serving is profiled at designated intervals, and designated intervals are enforced between each evaluation (Findlay, 2011). For example, Methven et al. (2010) applied this method to the evaluation of oral nutrition supplements. Panelists were asked to evaluate consecutive servings of the same product, and after each taste, panelists rated the intensity of sweetness, metallic flavor, soya milk flavor, mouthcoating, and mouth drying. Panelists were served eight consecutive servings of each supplement, and they evaluated all five attributes after each tasting. The aftertaste of each serving was profiled at 30 and 60 s intervals. Data was analyzed with three-way analysis of variance with sample, assessors, and time as variables.

11.4.5.1.3 Multiple Attribute Time Intensity (MATI)

MATI is similar to progressive and sequential profiling in that attributes are measured at designated intervals. In the case of MATI, the intensities of all attributes are measured, and the attributes are cycled. In other words, the attribute that the panelist is currently rating shifts based on the pacing pattern of the computer software (Kuesten et al., 2013).

11.4.5.2 Continuous Measurement Methods

11.4.5.2.1 Single Attribute Time Intensity

Traditional time intensity involves the continuous evaluation of one attribute. An attribute's time–intensity curve may be a key aspect defining the product (Larson-Powers and Pangborn, 1978; Lee and Pangborn, 1986; Overbosch, 1986; Overbosch et al., 1986; ASTM, 1997b). Long-term time–intensity studies measure the reduction of skin dryness periodically over several days of a skin lotion's application. A lipstick's color intensity can be periodically evaluated over several hours. Shorter term time–intensity studies track certain flavor and/or texture attributes of chewing gum over several minutes. In the shortest term studies, completed within 1–3 min, the response can be continuously recorded.

Examples include the bitterness and pungency of olive oil (Sinesio et al., 2005), the sweetness of sweeteners (IFT, 1988; Shamil et al., 1988), the bitterness of beer (Pangborn et al., 1983; Schmitt et al., 1984; Leach and Noble, 1986), and the effects of topical analgesics. The response may be recorded using pencil and paper, a scrolling chart recorder (Larson-Powers and Pangborn, 1978), or a commercially available computer system (Guinard et al., 1985; IFT, 1988; Le Révérend et al., 2008; Sokolowsky and Fischer, 2012). The panelist should not see the evolving response curve being traced because this may result in bias from preconceived notions of its form.

Time–intensity methodology has been comprehensively reviewed by Lee and Pangborn (1986), by Cliff and Heymann (1993), and, for sweeteners in particular, by Booth (1989). There are some important variables to consider:

Protocols for evaluation—type of delivery, amount of product, time to hold in the mouth, type of manipulation, expectoration, or swallow—need to be clearly defined.

Protocols for coordinating product evaluation (sample holding) and response recording (data entry) need to be worked out in advance to reduce bias from the mode of presentation.

Panelists may require several training sessions to develop and learn all of the protocols necessary for a well-controlled time–intensity study. Figure 11.1 is an example of the parameters that can be recorded in a time–intensity study; a more detailed example is given by Lee and Pangborn (1986).

Table 11.1 shows an example of responses obtained with three sweeteners.

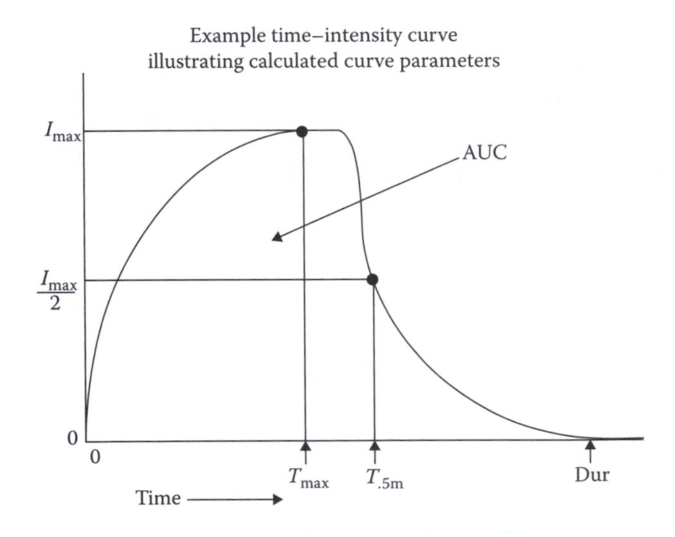

Figure 11.1 Example of a time–intensity curve illustrating calculated curve parameters. I_{max}, the maximum observed intensity; T_{max}, the time when the maximum intensity occurs; AUC, the area under the curve; Dur, the intensity duration (the time until the intensity drops back to zero); $T_{.5\,m}$, the time (after T_{max}) when intensity has fallen to half of I_{max}.

Table 11.1 Time–Intensity Data for Three Sweeteners

Parameter	7.5% Sucrose (Conditioning Sample)	0.05% Aspartame	0.4% Acesulfam-K	7.5% Sucrose
Area under the curve, cm^2	121.2	153.7	98.6	154.2
Maximum intensity, I_{max}	7.2	7.6	7.8	7.6
Time of maximum intensity, t_{max}, s	7.4	8.2	4.8	6.2
Duration, s	28.3	33.3	24.7	33.4

11.4.5.2.2 Dual Attribute Time Intensity (DATI)

Like single attribute time intensity, in DATI, attributes are measured continuously throughout the time curve, although dual attribute TI entails the measurement of two attributes simultaneously. For example, in the original publications, the sweetness and peppermint flavor of gum were rated (Duizer et al., 1996; Duizer et al., 1997). Later, DATI was used to evaluate juiciness and force to chew of meat (Zimoch and Findlay, 1998). Authors have reported comparable results between single and dual attribute TI; however, there has been some controversy since its introduction due to the number of tasks that panelists are asked to perform simultaneously (Duizer et al., 1997; Findlay, 2011).

11.4.5.2.3 Temporal Order of Sensations (TOS)

Temporal order of sensations is a method used to determine the order of appearance of key attributes throughout a sensory experience, even through multiple product exposures (Pecore et al., 2013a). The primary difference between TOS and temporal dominance of sensations (TDS) is that TOS employs a checklist approach. The advantages of this method are that compared to other temporal methods, the amount of training required for panelists is reduced, and the need for specialized software is eliminated (Pecore et al., 2013b).

11.4.5.2.4 Temporal Dominance of Sensations (TDS)

Pineau et al. (2003) introduced TDS, in which panelists are instructed to record the most dominant attribute throughout a designated time span (Pineau et al., 2009). When the first attribute ceases to be dominant, the panelist marks the new dominant attribute, and this continues over the time duration. The dominance scale itself is a proportional scale. With advanced training, panelists can also indicate the intensity of the attributes (Pecore et al., 2013a). Best practices include limiting the number of attributes in the list to 10 and balancing the order that the attributes appear across panelists (Pineau et al., 2012).

TDS has been applied to the evaluation of texture attributes during reformulation of cookies; the information was presented alongside consumer liking scores, which gave direction as to which attributes were most important (Laguna et al., 2013). Other applications include evaluating the flavor of blackcurrant squashes (Ng et al., 2012) and the texture of fish sticks (Albert et al., 2012).

11.5 COMMONLY USED DESCRIPTIVE TEST METHODS WITH UNTRAINED PANELS

11.5.1 Free-Choice Profiling

Free-choice profiling was developed by Williams and Arnold (1984) at the Agricultural and Food Council (United Kingdom) as a solution to the problem of consumers' using different terms for a given attribute. Free-choice profiling allows the panelist to invent and use as many terms as he or she needs to describe the sensory characteristics of a set of samples (Marshall and Kirby, 1988; Guy et al., 1989; Oreskovich et al., 1991). The samples are all from the same category of products, and the panelist develops his or her own score sheet. The data are analyzed by generalized procrustes analysis (Gower, 1975), a multivariate technique that adjusts for the use of different parts of the scale by different panelists and then manipulates the data to combine terms that appear to measure the same characteristic. These combined terms provide a single product profile.

The main advantage of the new technique is that it saves time, not requiring any training of the panelists other than an hour's instruction in the use of the chosen scale. A disadvantage is that the panelists, not having been trained, can still be regarded as representing naive consumers. However, questions regarding the ability of the sensory analyst to interpret the resulting terms, combined from all panelists, need to be addressed. To provide reliable guidance for product researchers, the experimenter/sensory analyst must decide what each combined term actually means. Therefore, the words or terms for each resulting parameter come from the experimenter or sensory analyst, not the panelists. The results may be colored more by the perspective of the analyst than by the combined weight of the panelists' verdicts.

11.5.2 Flash Profiling

Flash profiling is adapted from free-choice profiling. The primary difference between the two is that in flash profiling, participants rank the attributes instead of measuring their intensities. Dairou and Sieffermann (2002) first published the method in the context of jams, and since then, its use has been expanded (Tarea et al., 2007). The advantages and disadvantages of the method are similar to that of free-choice profiling. Minimal training of the panelists is required; however, panelists do not share a common vocabulary, which may allow inconsistent use of terms.

11.5.3 Projective Mapping (Napping)

Projective mapping refers to a process in which panelists group samples on a two-dimensional plane according to their similarity (Risvik et al., 1994; Risvik et al., 1997).

Attributes that describe products may be added to the plane. Subjects can be naïve consumers or trained panelists (Ross et al., 2012; Torri et al., 2013); however, using trained or expert panelists may help improve product differentiation. Data may be analyzed with multiple factor analysis or general procrustes analysis (Torri et al., 2013; Nestrud and Lawless, 2010; Pagès and Husson, 2001).

11.5.4 Sorting

Several authors have adapted sorting techniques to describe products (Steinberg, 1967; Gains, 1994; Dehlholm et al., 2012). In sorting methods, panelists group samples in a way that makes sense to them as individuals. They may be asked to regroup the products as many times as necessary until all possible groupings have been made. Data may be analyzed with frequency analysis.

11.6 APPLICATION OF DESCRIPTIVE ANALYSIS PANEL DATA

Descriptive analysis data is a versatile source of product information and understanding for both research and marketing professionals in corporate, government, and academic settings. The descriptive analysis results provide guidelines for professionals seeking to identify all of the sensory properties that can be perceived in a given product or set of products (for comparison). These results are used for

1. The documentation of the sensory properties of products: This is the primary use of descriptive analysis data. The output of a panel session, the description of each product in terms of the detailed attributes and the attribute intensity, provides a thumbprint or profile of the product in words and numbers that characterizes the aroma, flavor, appearance, texture, and/or sound of a product or set of products. Each description is unique for each product and can be considered a blueprint for that sample. In Table 11.2, the descriptions of three orange juices are shown side by side to demonstrate the detail and the relationship of the data to the actual products. Table 11.3 provides a complete description of one commercial orange juice. Several attributes are rated zero, and those attributes are shown with the zero ratings. The same product profile is also graphically displayed in Figure 11.2.

   The following attributes were not present in this sample and had intensities of zero: fruity/floral aromatics; caramelized aromatics; hydrolyzed oil aromatics; distilled orange oil; paper/gelatin; fruity/floral; smokey/phenol; fermented; hydrolyzed oil; vitamin; green; sweet aromatics; metallic prickle; chemical (stabilizer); salt.
2. The comparison of product attributes: This provides documentation of perceived characteristics for making business decisions such as setting QC sensory specifications based on a range of consumer acceptance of specific attributes; predicting market success based on comparison with highly accepted

Table 11.2 Comparison of Fresh-Squeezed, Frozen Concentrate, and Canned Orange Juice Descriptive Flavor Profiles

	Fresh-Squeezed OJ	Frozen Minute Maid® OJ	Kroger Canned OJ
Aromatics			
Orange complex	9.5	7.5	4.0
Raw	6.0	1.0	0.0
Cooked	0.0	5.0	4.0
Distilled orange oil	0.0	0.0	2.0
Expressed orange oil	3.5	2.0	0.0
Fruity/floral	4.0	0.0	0.0
Other citrus	2.0	1.0	2.0
Type:	Tangerine (1.0) Terpene (1.0)		Grapefruit
Other fruit	0.0	0.0	2.0
Type:			Pineapple/banana
Sweet aromatic (caramelized/maltol)	0.0	0.0	0.0
Green	1.0	0.0	0.0
Vitamin	0.0	0.0	0.0
Cardboard/oxidized	0.0	0.0	0.0
Hydrolyzed oil	0.0	0.0	6.0
Fermented	0.0	0.0	0.0
Smokey/phenol	0.0	0.0	0.0
Paper/gelatin	0.0	0.0	0.0
Basic Tastes			
Sweet	8.0	8.0	7.0
Salt	0.0	0.0	0.0

products (Table 11.2); or making advertising claims based on an increase or decrease in position or negative attributes that are seen as an opportunity to market product benefits.

3. Benchmarking products and prototypes alongside the current market players: This is a critical component in the different types of category appraisal projects that relate descriptive benchmarking with consumer acceptance to define the product attributes that are key drivers of acceptance, performance, benefits, or defects. Figure 11.3 is a principal component analysis map of commercial orange juice products in various packaging and storage conditions compared to two sources of fresh-squeezed orange juice.

Table 11.3 Minute Maid® Frozen Concentrate Orange Juice Complete Flavor and Chemical Feeling Factor Profile

Attribute	Intensity
Aromatics	
Orange complex	7.5
Raw	1.0
Cooked	5.0
Distilled orange oil	0.0
Expressed orange oil	2.0
Fruity/floral	0.0
Other citrus	1.0
Other fruit	0.0
Sweet aromatics (caramelized/maltol)	0.0
Green	0.0
Vitamin	0.0
Cardboard/oxidized	0.0
Hydrolyzed oil	0.0
Fermented	0.0
Smokey/phenol	0.0
Paper/gelatin	0.0
Basic Tastes	
Sweet	8.0
Salt	0.0
Sour	3.8
Bitter	1.5
Chemical feeling factors	
Astringent	2.5
Burn	2.0
Chemical (stabilizer)	0.0
Prickle	0.0
Aftertaste	
Metallic	0.0

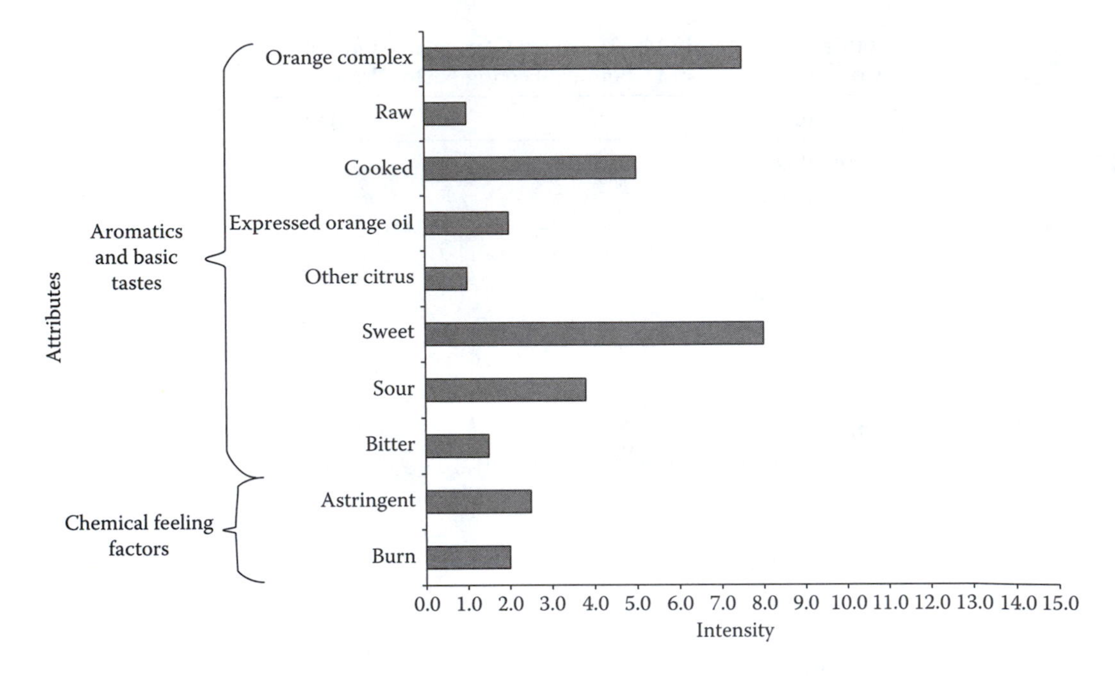

Figure 11.2 Minute Maid® frozen concentrate complete descriptive profile.

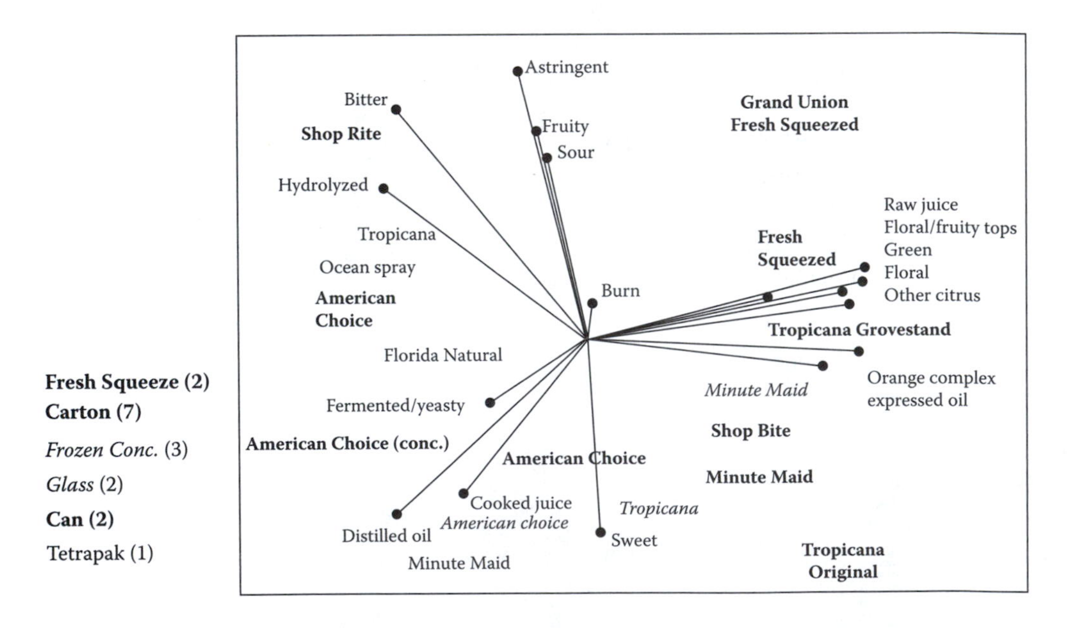

Figure 11.3 Principal component analysis map of orange juice products.

REFERENCES

Albert, A., A. Salvador, P. Schlich, and S. Fiszman (2012). Comparison between temporal dominance of sensations (TDS) and key-attribute sensory profiling for evaluating solid food with contrasting textural layers: Fish sticks. *Food Qual Prefer* 24: 111–8.

ASTM (1992). Manual on descriptive analysis testing for sensory evaluation. In R.C. Hootman (ed.), *ASTM Manual 13.*West Conshohocken, PA: ASTM International.

ASTM (1996). Sensory testing methods. In E. Chambers and M. Baker Wolf (eds.), *ASTM Manual 26*, 2nd Ed. West Conshohocken, PA: ASTM International.

ASTM (1997a). Standard practice for descriptive skinfeel analysis of creams and lotions. In *ASTM Standard Practice E1490–92*, West Conshohocken, PA, ASTM International.

ASTM (1997b). Standard guide for time–intensity evaluation of sensory attribute. In *ASTM Standard Guide E1909–97.* West Conshohocken, PA, ASTM International.

ASTM (2011). Standard guide for two sensory descriptive analysis approaches for skin creams and lotions. *ASTM Standard Practice E1490–11.* West Conshohocken, PA, ASTM International.

Aust, L. B., M. C. Gacula, S.A. Beard Jr., and R. W. Washam II (1985). Degree of difference test method in sensory evaluation of heterogeneous product types. *J Food Sci* 50: 511–3.

Bargmann, R. E., L. Wu, and J. J. Powers (1976). Search for the determiners of food quality ratings—description of methodology with application to blueberries. In J. E. Powers and H. R. Moskowitz (eds.), *Correlating Sensory Objective Measurements—New Methods for Answering Old Problems* (pp. 56–72). West Conshohocken, PA: ASTM International.

Booth, B. (1989). Time–intensity parameters considered in sweetener research at the NutraSweet Co., presentation to American Society for Testing and Materials (ASTM). *Subcommittee E18 on Sensory Evaluation*, Kansas City, MO.

Brandt, M. A., E. Z. Skinner, and J. A. Coleman (1963). Texture profile method. *J Food Sci* 28(4): 404.

Cairncross, S. E. and L. B. Sjöstrom (1950). Flavor profiles—a new approach to flavor problems. *Food Technology* 4: 308–11.

Caul, J. F. (1957). The profile method of flavor analysis. *Advances in Food Research* 7: 1–40.

Civille, G. V. (1979). Course notes for IFT short course in sensory analysis. In *Descriptive Analysis* (Chapter 6). Chicago: Institute of Food Technology.

Civille, G. V. and C. A. Dus (1990). Development of terminology to describe the handfeel properties of paper and fabrics. *J Sens Stud* 5: 19–32.

Civille, G. V. and C. A. Dus (1991). Evaluating tactile properties of skincare products: A descriptive analysis technique. *Cosmet Toiletries* 106(5): 83–8.

Civille, G. V. and H. T. Lawless (1986). The importance of language in describing perceptions. *J Sens Stud* 1(3/4): 203–15.

Civille, G. V. and I. H. Liska (1975). Modifications and applications to foods of the general foods sensory texture profile technique. *J Text Stud* 6: 19–31.

Civille, G. V. and J. Seltsam (2014). Descriptive analysis of food texture: Advances in the sensory characterizations of food textures. In Y. Dar and J. Light (eds.), *Food Texture Design and Optimization* (pp. 321–41). Hoboken: John Wiley & Sons.

Civille, G. V. and A. S. Szczesniak (1973). Guide to training a texture profile panel. *J Text Stud* 4: 204–23.

Chambers, E. IV, M. Wolf, and G. Civille (2012). Understanding flavor complexity: What individual attributes can't tell. In *Society of Sensory Professionals Conference*, October 10–12, Jersey City, NJ.

Cliff, M. and H. Heymann (1993). Development and use of time-intensity methodology for sensory evaluation: A review. *Food Res Int* 26: 375–85.

Dairou, V. and J. M. Sieffermann (2002). A comparison of 14 jams characterized by conventional profile and a quick original method, the flash profile. *J Food Sci* 67:826–34.

Dehlholm, C., P. B. Brockhoff, L. Meinert, M. D. Aaslyng, and W. L. P. Bredie (2012). Rapid descriptive sensory methods—Comparison of free multiple sorting, Partial Napping, Napping, Flash Profiling and conventional profiling. *Food Qual Prefer* 26(2): 267–77.

Drake, M. A., and G. V. Civille (2002). Flavor lexicon. *Compr Rev Food Sci Food Saf* 2:33–40.

Duizer, L. M., K. Bloom, and C. J. Findlay (1997). Dual-attribute time-intensity sensory evaluation: A new method for temporal measurement of sensory perceptions. *Food Qual Prefer* 8(4): 261–69.

Duizer, L. M., K. Bloom, and C. J. Findlay (1996). Dual-attribute time-intensity measurement of sweetness and peppermint perception of chewing gum. *J Food Sci* 61:636–8.

Dus, C., L. Stapleton, A. Trail, A. Krogmann, and G. Civille (2015). The Spectrum™ method. In S. E. Kemp, J. Hort, and T. Hollowood (eds.), *Descriptive Analysis in Sensory Evaluation*. West Sussex: Wiley.

Findlay, C. (2011). Measuring multiple-attribute time intensity. *The Compusense Blog*. https://compusense.com/en/research-and-community/blogs/2011/december/measuring-multiple-attribute-time-intensity (accessed August 10, 2015).

Gains, N. (1994). The repertory grid approach. In H. J. MacFie and D. M. H. Thomson (eds.), *Measurement of Food Preferences*. Springer.

Gower, J. C. (1975). Generalized procrustes analysis. *Psychometrika* 40: 33–51.

Guinard, J. X., R. M. Pangborn, and C. F. Shoemaker (1985). Computerized procedure for time–intensity sensory measurements. *J Food Sci* 50: 543–4.

Guy, C., J. R. Piggott, and S. Marie (1989). Consumer free-choice profiling of whisky. In J. R. Piggott and A. Paterson (eds.), *Distilled Beverage Flavour: Recent Developments* (pp. 41–55). Chinchester, UK: Ellis Horwood/VCH.

IFT. 1988. Computers tell "how sweet it is". *Food Technol* 42: 11, 98.

ISO. 1994. *Sensory-Analysis—Identification, Selection of Descriptors for Establishing a Sensory Profile by a Multidimensional Approach*, International Organization for Standardization, ISO 11035.

Jack, F. R., J. R. Piggott, and A. Paterson (1994). Analysis of textural changes in hard cheese during mastication by progressive profiling. *J Food Sci* 59:539–43.

Jeltema, M. A. and E. W. Southwick (1986). Evaluation and application of odor profiling. *J Sens Stud* 1(2): 123–36.

Johnsen, P. B., G. V. Civille, J. R. Vercellotti, T. H. Sanders, and C. A. Dus (1988). Development of a lexicon for description of peanut flavor. *J Sens Stu* 3(1): 9–18.

Keane, P. (1992). The flavor profile. In R. C. Hootman (ed.), *Descriptive Analysis Testing for Sensory Evaluation, ASTM Manual 13* (pp. 5–14). Philadelphia, PA: ASTM International.

Kuesten, C., J. Bi, and Y. Feng (2013). Exploring taffy product consumption experiences using a multi-attribute time-intensity (MATI) method. *Food Qual Prefer* 30(2): 260–73.

Laguna, L., P. Varela, A. Salvador, and S. Fiszman (2013). A new sensory tool to analyse the oral trajectory of biscuits with different fat and fibre contents. *Food Res Int* 51: 544–53.

Larson-Powers, N. and R. M. Pangborn (1978). Paired comparison and time–intensity measurement of the sensory properties of beverages and gelatins containing sucrose or synthetic sweeteners. *J Food Sci* 43: 41–6.

Lawless, L. J. R. and G. V. Civille (2013). Developing lexicons: A review. *J Sens Stud* 28(4): 270–81.

Lawless, H. T. and H. Heymann (1998). *Sensory Evaluation of Food. Principles and Practices*. New York: Chapman & Hall.

Leach, E. J. and A. C. Noble (1986). Comparison of bitterness of caffeine and quinine by a time–intensity procedure. *Chem Sens* 11(3): 339–45.

Lee, W. E. III and R.M. Pangborn (1986). Time–intensity: The temporal aspects of sensory perception. *Food Technology* 40(11): 71–82.

Marshall, R. J. and S. P. Kirby (1988). Sensory measurement of food texture by free-choice profiling. *J Sens Stud* 3: 63–80.

Meilgaard, M. C. and J. E. Muller (1987). Progress in descriptive analysis of beer and brewing products. *Tech Quart Master Brew Assoc Am* 24(3):79–85.

Methven, L., K. Rahelu, N. Economou, L. Kinneavy, L. Ladbrooke-Davis, O. B. Kennedy, D. S. Mottrama, and M. A. Gosney (2010). The effect of consumption volume on profile and liking of oral nutritional supplements of varied sweetness: Sequential profiling and boredom tests. *Food Qual Prefer* 21(8): 948–55.

Moskowitz, H. R. (1975). Application of sensory assessment to food evaluation. II. Methods of ratio scaling. *Lebensmittel-Wissenschaft & Technologie* 8(6): 249.

Moskowitz, H. R. (1978). Magnitude estimation: Notes on how, what, where and why to use it. *J Food Qual* 1: 195.

Moskowitz, H. R. (1979). Correlating sensory and instrumental measures in food texture. *Cereal Food World* 22: 223.

Muñoz, A. M., G. V. Civille, and B. T. Carr (1992). *Sensory Evaluation in Quality Control*. New York: Van Nostrand Reinhold.

Nestrud, M. A. and H. T. Lawless (2010). Perceptual mapping of apples and cheeses using projective mapping and sorting. *J Sens Stud* 25: 390–405.

Ng, M., J. B. Lawlor, S. Chandra, C. Chaya, L. Hewson, and J. Hort (2012). Using quantitative descriptive analysis and temporal dominance of sensations analysis as complementary methods for profiling commercial blackcurrant squashes. *Food Qual Prefer* 25: 121–34.

Oreskovich, D. C., B. P. Klein, and J. W. Sutherland (1991). Procrustes analysis and its applications to free-choice and other sensory profiling. In H. T. Lawless and B. P. Klein (eds.), *Sensory Science Theory and Application in Foods* (pp. 353–93). New York: Marcel Dekker.

Overbosch, P. (1986). A theoretical model for perceived intensity in human taste and smell as a function of time. *Chem Sens* 11(3): 315–29.

Overbosch, P., C. J. van den Enden, and B. M. Keur (1986). An improved method for measuring perceived intensity/time relationships in human taste and smell. *Chem Sen* 11(3): 331–8.

Pagès, J. and F. Husson (2001). Inter-laboratory comparison of sensory profiles: Methodology and results. *Food Qual Prefer* 12: 297–309.

Pangborn, R. M., M. J. Lewis, and J. F. Yamashita (1983). Comparison of time–intensity with category scaling of bitterness of iso-a-acids in model systems and in beer. *J I Brew* 89: 349–55.

Pecore, S., C. Findlay, and S. Kirkmeyer (2013a). *Temporal Methods Seminar*, Indianapolis: ASTM.

Pecore, S., C. Rathjen-Nowak, and T. Tamminen (2013b). Holistic assessment of the eating experience: Going beyond traditional descriptive flavor analysis. Paper presented at the biennial meeting for the Society of Sensory Professionals, October 27–29, Napa, CA.

Pecore, S., N. Stoer, S. Hooge, N. Holschuh, F. Hulting, and F. Case (2006). Degree of difference testing: A new approach incorporating control lot variability. *Food Qual Prefer* 17(7–8): 552–5.

Pineau, N., S. Cordelle, and P. Schlich (2003). Temporal dominance of sensations: A new technique to record several sensory attributes simultaneously over time. In *5th Pangborn Symposium*, July 20–24, p. 121.

Pineau, N., P. Schlich, S. Cordelle, C. Mathonnière, S. Issanchou, A. Imbert, M. Rogeaux, P. Etiévant, and E. Köster (2009). Temporal dominance of sensations: Construction of the TDS curves and comparison with time–intensity. *Food Qual Prefer* 20: 450–455.

Pineau, N., A. G. de Bouillé, M. Lepage, F. Lenfant, P. Schlich, N. Martin, and A. Rytz (2012). Temporal dominance of sensations: What is a good attribute list? *Food Qual Prefer* 26:159–65.

Risvik, E., J. A. McEwan, J. S. Colwill, R. Rogers, and D. H. Lyon (1994). Projective mapping: A tool for sensory analysis and consumer research. *Food Qual Prefer* 5:263–9.

Risvik, E., J. A. McEwan, and M. Rødbotten (1997). Evaluation of sensory profiling and projective mapping data. *Food Qual Prefer* 8:63–71.

Ross, C. F., K. M. Weller, and J. R. Allredge (2012). Impact of serving temperature on sensory properties of red wine as evaluated using projective mapping by a trained panel. *J Sens Stud* 27:463–70.

Le Révérend, F. M., C. Hidrio, A. Fernandes, and V. Aubry (2008). Comparison between temporal dominance of sensations and time intensity results. *Food Qual Prefer* 19: 174–8.

Schmitt, D. J., L. J. Thompson, D. M. Malek, and J. H. Munroe (1984). An improved method for evaluating time–intensity data. *J Food Sci* 49: 539–42.

Schwartz, N. (1975). Method to skin care products. *J Text Stud* 6: 33.

Shamil, S., G. G. Birch, A. A. S. F. Jackson, and S. Meek (1988). Use of intensity–time studies as an aid to interpreting sweet taste chemoreception. *Chem Sens* 13(4): 597–605.

Sinesio, F., E. Moneta, and M. Esti (2005). The dynamic sensory evaluation of bitterness and pungency in virgin olive oil. *Food Qual Prefer* 16: 557–64.

Skinner, E. Z. (1988). The texture profile method. In H. R. Moskowitz (ed.), *Applied Sensory Analysis of Foods* (pp. 89–107). Boca Raton, FL: CRC Press.

Sokolowsky, M. and U. Fischer (2012). Evaluation of bitterness in white wine applying descriptive analysis, time-intensity analysis, and temporal dominance of sensations analysis *Analytica Chimica Acta* 732: 46–52.

Steinberg, D. D. (1967). The word sort: An instrument for semantic analysis. *Psychon Sci* 8:541–2.

Stone, H. and J. L. Sidel (1992). *Sensory Evaluation Practices*, 2nd Ed. Orlando, FL: Academic Press.

Stone, H., J. Sidel, S. Oliver, A. Woolsey, and R. C. Singleton (1974). Sensory evaluation by quantitative descriptive analysis. *Food Technol* 28(11): 24–34.

Szczesniak, A. S. (1963). Classification of textural characteristics. *J Food Sci* 28(4): 397.

Szczesniak, A. S., M. A. Brandt, and H. H. Friedman (1963). Development of standard rating scales for mechanical parameters of texture and correlation between the objective and the sensory methods of texture evaluation. *J Food Sci* 28(4): 397–403.

Szczesniak, A. S., B. S. Loew, and E. Z. Skinner (1975). Consumer texture profile technique. *J Food Sci* 40: 1243.

Tarea, S., G. Cuvelier, and J. M. Sieffermann (2007). Sensory evaluation of the texture of 49 commercial apple and pear purees. *J Food Qual* 30:1121–31.

Torri, L., C. Dinnella, A. Recchia, T. Naes, H. Tuorila, and E. Monteleone (2013). Projective mapping for interpreting wine aroma differences as perceived by naïve and experienced assessors. *Food Qual Prefer* 29: 6–15.

Williams, A. A. and G. M. Arnold (1984). A new approach to sensory analysis of foods and beverages. In J. Adda (ed.), *Progress in Flavour Research, Proceedings of the 4th Weurman Flavour Research Symposium* (pp. 35–50). Amsterdam: Elsevier.

Wilkinson, C., G. B. Dijksterhuis, and M. Minekus. 2000. From food structure to texture. *Trend Food Sci Technol* 11:442–50.

Zimoch, J. and C. J. Findlay (1998). Effective discrimination of meat tenderness using dual attribute time intensity. *J Food Sci* 63:940–4.

12

Spectrum™ Descriptive Analysis Method

12.1 DESIGNING A DESCRIPTIVE METHOD

The name *Spectrum* covers a method designed by Civille in collaboration with a number of companies that were looking for a way to obtain reproducible sensory descriptive analysis of their products (Muñoz and Civille, 1992, 1998). Civille created the Spectrum method in the 1970s based on her experiences with the flavor profile and texture profile methods (detailed in Chapter 11) at General Foods. The method was first formally presented at the Institute for Food Technologists Sensory Course in 1979 and was officially named the Spectrum Descriptive Analysis Method in 1986 when Sensory Spectrum was incorporated. Deviations from the texture and flavor profile methods to the Spectrum method include a more discriminating scale (over 150 points), the utilization of statistics, and the expansion of descriptive analysis beyond foods. In addition to personal care and home care products, the Spectrum method can be used in unconventional applications such as the odor and/or sound of car or airplane cabin interiors or quick-serve restaurants. The Spectrum method differs from methods such as quantitative descriptive analysis (QDA) because references are provided for both attributes (e.g., distilled lemon in the evaluation of lemon-lime soda) and intensity (e.g., Bitter 2, Bitter 5, Bitter 10, Bitter 15).

The philosophy of the Spectrum method is pragmatic; it provides guidelines for the design of a descriptive procedure for a given product category. The descriptive analysis data are then used to guide product development to a better understanding of a product's sensory properties. Recommendations based on descriptive analysis aid the product developer either to make the same product (even if ingredients or processes shift) or create new and/or improved products.

Some variation in the Spectrum Method includes (1) that panelists may be selected and trained to evaluate only one product or a variety of products; (2) that products may be described in terms of only appearance, aroma, flavor, texture, or sound characteristics or all of these; and (3) that panelists may be trained to evaluate one or all of these modalities. Spectrum is a "custom design" approach to panel development, selection, training, and maintenance.

The principal tools used in the Spectrum method are the reference lists for attributes and for intensities contained in Appendices 12.1–12.3, together with the scaling procedures and methods of panel training described in Chapters 5 and 10. The aim is to choose the most practical approach to selecting the lexicon and scale given the product category in question, the overall sensory program, the specific project objective(s), and the desired level of statistical treatment of the data.

Courses teaching the basic elements of Spectrum are available and include a detailed manual. Examples of its application are given in Lawless et al. (2012), Civille et al. (2010), and Johnsen et al. (1988).

12.2 MYTHS ABOUT THE SPECTRUM DESCRIPTIVE ANALYSIS METHOD

Throughout the years, false rumors and murmurings concerning the Spectrum descriptive analysis method have challenged several aspects of this highly scientific approach to descriptive panel testing. At the June 2002 IFT meeting, Sensory Spectrum presented a paper debunking the myths surrounding the Spectrum method.

12.2.1 Myth 1: All Descriptive Methods Are the Same

The truth is that all descriptive methods measure sensory attributes and their intensities. The Spectrum method differs from other descriptive methods in that it yields a more technical profile. Other methods differ in the selection and training of panelists, scale type, and product focus. For more information on the comparison of descriptive methodology, see the *ASTM Manual on Descriptive Analysis Testing for Sensory Evaluation* (edited by Robert C. Hootman).

12.2.2 Myth 2: Concept Development Is Unnecessary in Training a Spectrum Panel

Concept development for attributes is critical to lexicon stability. Lexicons are based on common terminology agreed on by panelists. Clarifying the concept, through use of references and examples, stabilizes the communication among the panelists. Creating a "complex" concept allows panelists to account for parts of the whole that in turn allow the product developers to understand what attributes make up the whole concept. Examples of complexes are listed below:

Total Corn Complex	Total Amount of Residue	Total Amount of Color
Raw corn	Oily	Red
Cooked corn	Waxy	Yellow
Toasted corn	Greasy	Blue
Masa	Silicone	Green

12.2.3 Myth 3: All Spectrum Training and Panel Leaders Are the Same; Anyone Can Do It

Although knowledge and familiarity are important, to be an effective trainer one must have both technical and group dynamic skills. To maximize the panel's learning, the trainer provides authority (clarifying technical issues) and structure (providing a framework for all panelists to learn). To develop the panel as an independent performing team, the trainer encourages growth and builds the panel's confidence. The success of the panel depends on it.

12.2.4 Myth 4: Consumer Terms Are Better than Technical Terms

It is not a question as to which are better—consumer terms or technical terms. The project objective dictates the type of terminology required. Consumer terms reflect the language of the user population (creamy, refreshing, soft). Technical terms provide direct feedback to product development (vanillin or vanilla). Technical terms can be directly related to the input of ingredients and process or the underlying physics or chemistry of the product.

12.2.5 Myth 5: Spectrum Panelists Are Forced to Use Canned Lexicons

Spectrum panelists discover the terms from within the samples. The process by which Spectrum panels develop lexicons is

1. Panelists experience the attributes (taste, touch, smell, etc.) in an array of products.
2. Panelists develop terms, interpret the experience, and record a draft lexicon.
3. Panelists are exposed to attribute references for clarification.
4. Panelists refine the precise terminology.
5. Panelists validate the complete lexicon by comparing a pair of different samples.

12.2.6 Myth 6: Spectrum Panelists Are Coerced into Intensity Calibration

Many descriptive methods practice qualitative calibration, in which panelists are provided with a common reference sample for an attribute (e.g., cooked oatmeal for "cooked oats"). The intensity calibration used in the Spectrum method is merely an extension of qualitative calibration. Providing intensity references increases panel reproducibility and communication. People look for boundaries in making decisions about amount ("Compared to what?"). The Spectrum method provides these boundaries by defining the limits of the sensory experience. Providing a series of levels for different stimuli encourages panelists to be consistent from time to time and across panelists despite potential differences in sensitivity. In particular, basic tastes and trigeminal sensations (e.g., bitter, heat/burn) can be highly variable from person to person; however, when all panelists agree that the reference is a 2.0 and rate the products accordingly, the effect of individual variations on final results is mitigated. In addition, using intensity references allows for a universal comparison across products and product categories.

12.2.7 Myth 7: The Universal Scale Cannot Show Small Differences

The number of things measured by one scale does not decrease its sensitivity. A ruler can measure the length of a multitude of objects. The length of the scale does not decrease sensitivity in a range of the scale as long as there are several points of discrimination. In cases where low discrimination is observed, the samples may be truly similar, or panelists may not be using all points of the scale (i.e., they are not using each tenth on the scale). In order to assist panelists in increasing discrimination, additional references representing additional scale points may be used. The benefit of being able to discuss differences across samples, across attributes, and across product categories makes the extra work needed to implement a scale worth it. With training and practice, statistically significant differences as low as 0.3 can be observed (Miller and Chambers, 2013).

12.2.8 Myth 8: Published References and Terms Are the Equivalent of a Training Manual

Panel success is a result of a skilled approach to concept development, the evaluation process, and confidence building. Terms and scales alone do not teach the process. The trainer provides understanding of the basic principles, builds on the basics to deepen understanding, fosters concept development, and encourages growth through practice and feedback. As in learning anything, the *coach matters*. In a learning situation, the book (or manual) is not nearly as critical to the learning as the teacher or coach.

12.2.9 Myth 9: Product Users Make the Best Panelists and Hedonics Influence Panel Ratings

A trained panelist does not need to use or like the product to be able to describe the product; liking does not equal knowing. The panelists' discrimination and description skills stem from their expertise (through training) and experience (through practice). Therefore, familiarity with a product does not influence the panelists' descriptive analysis skills.

12.2.10 Myth 10: Panelists Cannot Be Trained for an Array of Products

Panelists who can evaluate the appearance, fragrance, flavor, or texture of one product category are likely to be able to evaluate the appearance, fragrance, flavor, or texture of another product category, because the skills necessary to detect and describe attributes and intensities in one product category can readily be transferred to other categories.

12.2.11 Myth 11: Training for the Spectrum Method Is Too Time-Intensive

The length of a Spectrum training program is not predetermined; in fact, the length of training is determined by the scope of the panel. For panels with reduced scopes, the length of training could be significantly reduced. For example, a panel that only evaluates dairy products does not necessarily need to learn the attributes for grains or carbonated soft drinks. Furthermore, a panel that evaluates liquids only does not need to learn the texture terms for solids. Foundational skills such as scale usage and proper testing procedures should be emphasized and taught across all Spectrum panels.

12.2.12 Myth 12: The Spectrum Method Is Consensus Only

The Spectrum Method may be used with individual or consensus data collection; in fact, most users collect individual panelist data. Using the whole panel as an instrument or using individual panelists as instruments has advantages and disadvantages. Consensus data offers the flexibility to have critical discussions about attributes and intensity perceptions. These discussions lead to more precise profiles and allow each panelist to sharpen and enhance his/her tasting skills. Panelists have a chance to voice confusion over something they perceive and to utilize the skill set of the entire group to assign the appropriate term(s). When executed proficiently, by a highly skilled panel leader, consensus profiling is a panel-empowering method that delivers a highly nuanced product profile. On the other hand, when known variability exists (panelist or product), individual data typically capture the variability more effectively. Regardless of which method of profiling is being utilized, the skill of the panel leader cannot be overemphasized. It is critical for each method that an expert panel leader conducts the panel and instructs the panelists according to the chosen data collection approach.

12.2.13 Myth 13: Consensus Profiling Prevents Statistical Analysis of Panel Data

Consensus profiling does not prevent the use of statistics. In fact, all that is required to perform standard analysis of variance is two or more replications. For replicated consensus profiles, the statistical analysis of variance model contains "product" and "replication" and does not include "judge," which can be used when individual data is collected.

12.2.14 Myth 14: Difficult-to-Find References Prevent Universality of the Spectrum Scale

The status of reference availability does not necessarily affect the application of the Spectrum method. Reference substitutions can be made with the collaboration of the panel leader and panelists. In fact, such substitutions are common when the method is applied in a new geographical area for the first time. For example, when reference products on the semisolid texture scales (Appendix 12.2) were not readily available in Europe, recipes for meringues, custards, puddings, and butter and oil mixtures were developed to use local ingredients and generated recipes functioned as intensity references on the firmness and denseness scales.

12.3 TERMINOLOGY AND LEXICON DEVELOPMENT

The choice of terms may be broad or narrow according to the panel's objective—only aroma characteristics, or all sensory modalities. However, the method requires that all terminology is developed and described by a panel that has been exposed to the underlying technical principles of each modality to be described. For example, a panel describing color must understand color intensity, hue, and chroma. A panel involved in oral sensation, skinfeel, and/or fabric texture needs to understand what the tactile effects of rheology and mechanical characteristics are and how these in turn are affected by moisture level and particle size. The chemical senses pose an even greater challenge in requiring panelists to demonstrate a valid response to changes in ingredients and processing. Words such as vanilla, cocoa, and distilled orange oil require separate terms and references. If the panel hopes to attain the status of "expert panel" in a given field, it must demonstrate that it can use a concrete list of descriptors based on an understanding of the underlying technical differences among the attributes of a product.

Panelists begin to develop their list of best descriptors or their "lexicon" by completing five steps, which rest on three assumptions (Lawless and Civille, 2013). The lexicon development process is as follows.

Three basic assumptions.

1. Panelists are highly trained and suited to the task. For instruction on how to select and train panelists, see Chapter 10.
2. The product space is represented by a frame of reference. For expansive categories, a wide variety of products (50 or more) may be screened by two to four of the best-performing panelists to reduce the number of samples presented to the entire panel. This process ensures that the sensory space is represented while still maintaining cost efficiency. The screening approach is also utilized in flavor/fragrance targeting, which may be employed when a company seeks to create a product that represents an established brand. For example, grocery manufacturers may want to capitalize on a quick-serve restaurant's popularity and associated iconic flavor profiles. Screening may give the product developers the frame of reference for the brand; thus, they understand the flavor combinations for which to aim.
3. The developed protocol sets the appropriate test conditions to ensure that the data are reproducible. Controls for sample procurement, preparation, and test execution are established.

Five-step process.

1. First, panelists evaluate a broad array of products (commercial brands, competitors, pilot plant runs, etc.) that define the product category.
2. After some initial experience with the category, each panelist produces a list of terms to describe the sample set. The panel, in collaboration with the panel leader, determines the best terms that describe the set. The terms are then organized according to related groups. For example, if bread is being evaluated, toasted white wheat and cooked white wheat may be grouped together under grain complex.
3. Qualitative references that represent the generated terms are collected and presented to the panel. Additional terms and references may be taken from the literature (e.g., from published flavor lexicons) (Johnsen et al., 1988; Civille and Lyon, 1996). Panel leaders need to exercise caution in using published lexicons, since they may or may not be based on the findings of a strong panel.
4. Examples are collected and shown to promote further understanding of the attributes and how they relate to each other. An "example" is distinct from a reference

because examples are less singular. References are generally more singular, or the attribute of interest is prominent. For instance, ale could be an *example* for caramelized, but it is not a *reference* for it because ale also has bitterness, hop flavor, and so on.
5. The terms are then compiled or organized into a list that is comprehensive yet not overlapping.

Sensory scientists often look through the literature to find existing lexicons that might work for a given product category (for review, see Lawless and Civille, 2013). It is possible to use existing terms, but it is critical that the sensory scientist and the panel review a frame of reference of products in the chosen category to determine the words that actually describe the product set. Most of the lexicons described in this text started with products, and then the lexicons were developed from scratch based on the sample set. An example of the adaptation of existing underlying terms to a specific product category is the work on noodles by Janto et al. (1998). Several standard terms apply to noodles, but the vast Asian noodle frame of reference called for additional terms such as "starch between teeth" and "slipperiness between lips."

Some lexicons that represent whole categories are presented as wheels in a tiered structure. Typically, wheels organize attributes according to their general terms, followed by attributes that fall within the general term, and finally the most nuanced attributes. Since wheels offer a "snapshot" of a category's sensory world, they can be used as training tools for new descriptive panelists or in presentations in which sensory analysts must explain descriptive analysis to their business partners. Currently, published wheels include those for spirits, wine, beer, spices, and tea among others (Stapleton and Seltsam, 2010; Clapperton et al., 1976; Meilgaard et al., 1979; Noble et al., 1984, 1987; Lawless et al., 2012; Koch et al., 2012; Lawless and Civille, 2013). The spirits wheel (Figure 12.1) is as shown.

12.4 INTENSITY

The key properties of the Spectrum scale are universality and the high number of points of discrimination along the scale. Universality refers to the scale's inclusion of the full sensory "world" of products. In other words, the same intensity scale is applied to all product categories.

If product differences require a large number of points of discrimination to clearly define intensity differences both within and between attributes, the panel leader may choose to use (1) a 15-point scale measured in tenths, which allows for 151 points of discrimination; or (2) a category scale with 30 points or more. The most common application of the Spectrum scale is the use of 0–15 points (measured in tenths, yielding 151 points of discrimination) for most foods and 0–10 (measured in tenths, yielding 101 points of discrimination) for consumer products. The panelists write (or enter) the actual number for the intensity of each attribute.

The Spectrum method is based on extensive use of intensity reference points, which may be chosen according to the guidelines in Appendix 12.2. These are derived from the collective data of several panels over several replicates. The most intense reference signifies the uppermost part of the scale, and points along the scale are strategically

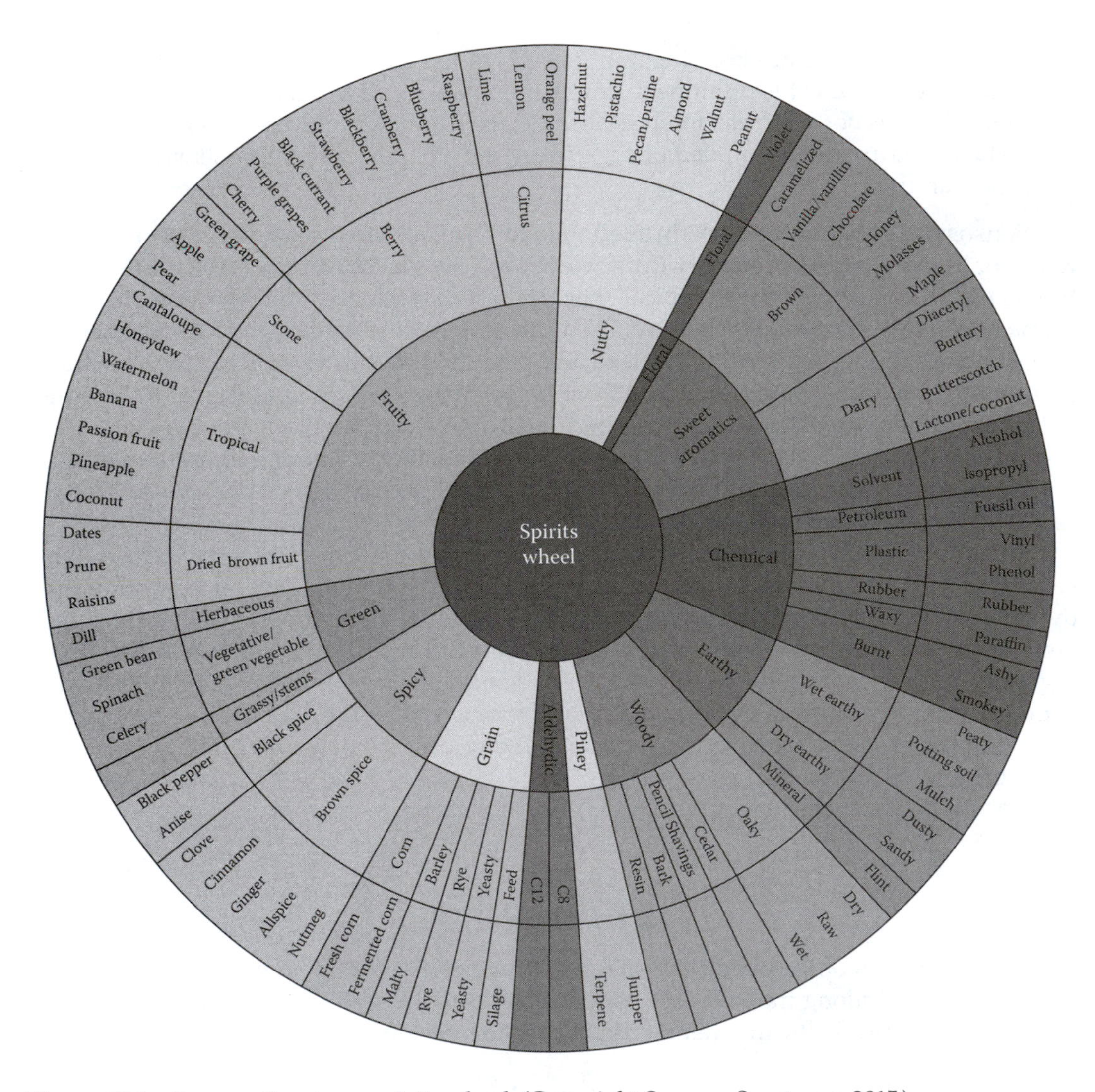

Figure 12.1 Sensory Spectrum spirits wheel. (Copyright Sensory Spectrum, 2015.)

chosen to depict commensurable points. References help teach the panel to assign quantitative measurements to the attributes generated during lexicon development while further refining the panel's qualitative frame of reference. Panelists are trained to rate product attributes proportionally. As such, they could choose to rate attribute intensities beyond the uppermost part of the scale as may be the case most often for ingredients and condiments. Examples include the sourness of pure lemon juice (30) and the sweetness of pure corn syrup (24). The scale must have at least two and preferably from three to five reference points distributed across the normal range for food (0–15). A set of well-chosen reference points greatly reduces panel variability, allowing for a comparison of data across time and products.

Such data also allow more precise correlation with stimulus changes (stimulus/response curve) and with instrumental data (sensory/instrumental correlations). For example, since Salt 5 represents the same intensity as Sweet 5 both today and last year, a product developer can compare data collected today with historical data. Thus, the data collected with a Spectrum panel parallels other laboratory instruments in the sense that the results are consistent over time. Fundamentally, this scaling approach is differentiated from other descriptive methods because products are rated according to the universal scale and not relative to each other. When the Spectrum scale is applied in practice, comparisons can be made across product categories, within a product category, or between attributes for given products (Table 12.1). The reference points on the universal scale serve as a cross validation to ensure that rated intensities are consistent across product attributes (Appendices 12.2 and 12.3) and even product categories and industries. These potential intercomparisons are the primary advantages of the Spectrum method (Muñoz and Civille, 1998). Other advantages include the ability of one panel to evaluate a wide range of products. For example, a food and beverage descriptive panel trained in the Spectrum method could evaluate wine, chips, bananas, and candy. When a company's needs are this varied, using a Spectrum method trained panel is resource efficient. The alternative to the Spectrum method would necessitate a product-specific panel for each of the categories (one for wine, one for chips, etc.). Furthermore, the universal nature of the Spectrum scale permits panelists to supersede the attribute intensities of the category's current products, which allows the potential development of new prototypes that stretch the category's sensory properties.

12.5 COMBINING THE SPECTRUM DESCRIPTIVE ANALYSIS METHOD WITH OTHER MEASURES

In sensory tests, the tools of the Spectrum method can be combined with difference-from-control or degree-of-difference (DFC or DOD) tests, quality scoring, consumer testing, and other measures. The basic philosophy of the Spectrum method, as mentioned, is to train the panel to fully define each and all of a product's attributes, to rate the intensity of each, and to include other relevant characterizing aspects such as changes over time, differences in the order of appearance of attributes, and integrated total aroma and/or flavor impact. Correlating the Spectrum data with other data (such as consumer responses) uncovers additional information to provide direction to the product developer.

12.5.1 Using the Spectrum Method Simultaneously with Other Methods

When combining descriptive analysis and DOD testing, the DOD scores indicate *how* different two samples are, and the accompanying sensory profiles indicate *where* the differences are. For instance, when two nutritional beverage products are compared, a DOD score of 5 on a 10-point scale indicates a moderate difference. The descriptive profiles may show one product having a note of degraded protein at intensity 3.0 and chocolate essence at 2.0, while the other sample has no degraded protein and chocolate essence at 4.0. Thus, the descriptive information complements the DOD score and vice versa. Additionally,

Table 12.1 Universal Scale References for Aromatics and Basic Tastes

Intensity	Aromatics	Sweet	Salt	Sour	Bitter
2.0	Aromatics in vegetable oil	2.0 % Sucrose solution in water	0.2 % NaCl solution in water	0.05 % Citric acid solution in water	0.05 % Caffeine solution in water
5	Cooked apple note in Motts Natural Applesauce	5.0 % Sucrose solution in water	0.35 % NaCl solution in water	0.10 % Citric acid solution in water	0.08 % Caffeine solution in water
7.5	Minute Maid Frozen Orange Juice from concentrate				
10	Grape note in Welch's concord grape juice	10.0 % Sucrose solution in water	0.55 % NaCl solution in water	0.15 % Citric acid solution in water	0.15 % Caffeine solution in water
12	Cinnamon note in Big Red gum				
15		16.0 % Sucrose solution in water	0.7 % NaCl solution in water	0.20 % Citric acid solution in water	0.20 % Caffeine solution in water

DOD scaling with a Spectrum panel can achieve results aligned with more extensive consumer testing while managing β risk. In a study using two different beverage systems, Spectrum-trained panel DOD scores were highly correlated with the number of correct responses in a consumer triangle test (Koelliker et al., 2014).

A parallel use of trained panelists can be in development of quality ratings. The goal of assigning a quality rating is to quantify product characteristics in the context of a high quality target. A high quality score indicates that the attributes of the product or prototype are within their respective target ranges. The process is based on the classification of positive and negative attributes and the intensities of those attributes in the products being tested. Recommendations for how to increase positive attributes and eliminate negative attributes are generally included with the quality rating. Using trained panelists for this work is beneficial due to their experience with the product category and their precise use of descriptive terms.

12.5.2 Combining the Spectrum Method with Other Sources of Sensory Data

Descriptive data can be combined with both quantitative and qualitative consumer data. The combination of descriptive analysis, quantitative consumer data, and statistical analysis has the potential to identify those attributes that are key positive drivers and those that are key negative drivers of consumer liking or acceptance; thus, the product developer has clear direction to improve the product (Chapter 18). Qualitative consumer methods such as sequence mapping (i.e., the tracking of the consumer's interactions with the product from the store through disposal) can be coupled with descriptive analysis to reveal product–consumer interaction sequences. In both of these cases, descriptive analysis is not acting as a standalone method; rather, the synergistic effects of the combination of methods yields deeper information (Chapter 18).

The creative and diligent sensory analyst can construct the optimal descriptive technique by selecting from the spectrum of terms, scaling techniques, and other optional components that are available at the start of each panel development.

12.6 SPECTRUM DESCRIPTIVE PROCEDURES FOR QUALITY ASSURANCE, SHELF-LIFE STUDIES AND SO ON

The reproducibility and consistency of the Spectrum descriptive analysis method makes it ideal for quality control and shelf-life applications. Since panelists are consistent across time, analysts can make judgments about quality degradation over time. Quality control is explored in detail in Chapter 17.

REFERENCES

ASTM Stock #DS72. 2011. *Lexicon for Sensory Evaluation: Aroma, Flavor, Texture, and Appearance*. ASTM International West Conshohocken, PA. www.astm.org.

Civille, G. V., and B. G. Lyon (1996). *Aroma and Flavor Lexicon for Sensory Evaluation. Terms, Definition, References and Examples*. ASTM Data Series Publication DS 66. West Conshohocken, PA: ASTM International.

Civille, G. V., K. Lapsley, G. Huang., S. Yada, and J. Seltsam (2010). Development of an almond lexicon to assess the sensory properties of almond varieties. *J Sensory Studies* 25: 146–62.

Clapperton, J. F., C. E. Dalgliesh, and M. C. Meilgaard (1976). Progress towards an international system of beer flavor terminology. *J Inst Brew* 82: 7–13.

Janto, M., S. Pipatsattayanuwong, M. W. Kruk, G. Hou, and M. R. McDaniel (1998). Developing noodles from US wheat varieties for the Far East market: Sensory perspective. *Food Qual Pref* 9(6): 403–12.

Johnsen, P. B., G. V. Civille, J. R. Vercellotti, T. H. Sanders, and C. A. Dus (1988). Development of a lexicon for the description of peanut flavor. *J Sensory Stud* 3(1): 9.

Koch, I. S., M. Muller, E. Joubert, M. Van Der Rijst, and T. Næs (2012). Sensory characterization of rooibos tea and the development of a rooibos sensory wheel and lexicon. *Food Res Int* 46: 217–28.

Koellicker, Y., L. J. R. Lawless, A. N. R. Krogmann, A. Sobel, and G. V. Civille (2014). Comparing a degree of difference scale to triangle testing. In *Presentation at Sensometrics*, Chicago, IL, July 29–August 1.

Lawless, L. J. R. and G. V. Civille (2013). Developing lexicons: A review. *J Sensory Stud* 28: 270–81.

Lawless, L. J. R., A. Hottenstein, and J. Ellingsworth (2012). The McCormick spice wheel: A systematic and visual approach to sensory lexicon development. *J Sens Stud* 27: 37–47.

Meilgaard, M. C., C. E. Dalgliesh, and J. F. Clapperton (1979). Beer flavor terminology. *J Inst Brew* 85: 38–42.

Miller, A. and D. H. Chambers (2013). Descriptive analysis of flavor characteristics for black walnut cultivars. *J Sens Stud* 78: S887–S893.

Muñoz, A. M. and G. V. Civille (1992). The Spectrum descriptive analysis method. In *ASTM Manual Series MNL 13, Manual on Descriptive Analysis Testing for Sensory Evaluation* (pp. 22–34), ed. R. C. Hootman. West Conshohocken, PA: American Society for Testing and Materials.

Muñoz, A. M. and G. V. Civille (1992). The Spectrum descriptive analysis method. In R. C. Hootman (ed.), *ASTM Manual Series MNL 13, Manual on Descriptive Analysis Testing* (pp. 22–34). West Conshohocken, PA: ASTM International.

Muñoz, A. M. and G. V. Civille (1998). Universal, product and attribute scaling and the development of common lexicons in descriptive analysis. *J Sens Stud* 13(1): 57–75.

Muñoz, A. M., G. V. Civille., and B. T. Carr (1992). *Sensory Evaluation in Quality Control*. New York: Chapman & Hall.

Noble, A. C., R. A. Arnold, B. M. Masuda, S. D. Pecore, J. O. Schmidt, and P. M. Stern (1984). Progress towards a standardized system of wine aroma terminology. *Am J Enol Vitic* 35: 107–9.

Noble, A. C., R. A. Arnold, J. Buechsenstein, E. J. Leach, J. O. Schmidt, and P. M. Stern (1987). Modification of a standardized system of wine aroma terminology. *Am J Enol Vitic* 38: 143–6.

Stapleton, L., and J. Seltsam (2010). Drinking from a pool of words: Use of a master lexicon for evaluation of spirits categories. Presented at the Society of Sensory Professionals Conference, October 27–29, 2010, Napa, CA.

APPENDIX 12.1 SPECTRUM TERMINOLOGY FOR DESCRIPTIVE ANALYSIS

The following lists of terms for appearance, flavor, and texture can be used by panels suitably trained to define the qualitative aspects of a sample.

When required, each of the terms can be quantified using a scale chosen from Chapter 5. Each scale must have at least two, and preferably three to five, chosen reference points, for example, from Appendix 12.2.

A simple scale can have general anchors:

None -------------------------------- Strong

or a scale can be anchored using bipolar words (opposite):

Smooth ---------------------------------Lumpy
Soft -- Hard

Attributes perceived via the chemical senses in general use a unipolar intensity scale (None–Strong), while for appearance and texture attributes, a bipolar scale is best, as shown below.

A. Terms Used to Describe Appearance

1. Color

a. Color hue — The actual color name or hue, such as red, blue, and so on. The description can be expressed in the form of a scale range if the product covers more than one hue:
[Red -- Orange]

b. Intensity — The intensity or strength of the color from light to dark:
[Light -- Dark]

c. Chroma — The chroma (or purity) of the color, ranging from dull, muddied to pure, bright color. Fire-engine red is a brighter color than burgundy red:
[Dull--- Bright]

d. Evenness — The evenness of distribution of the color, not blotchy:
[Uneven/blotchy ---Even]

2. Consistency/Texture

a. Visual viscosity — The visual thickness of the liquid:
[Thin -- Thick]

b. Roughness — The amount of irregularity, protrusions, grains, or bumps that can be seen on the surface of the product. Smoothness is the absence of surface particles:

[Smooth --- Rough]
Graininess is caused by small surface particles:
[Smooth --- Grainy]
Bumpiness is caused by large particles:
[Smooth --- Bumpy]

c. Particle interaction — The amount of stickiness among particles or the amount of agglomerations of small particles.

(Stickiness): [Not sticky---Sticky]

(Clumpiness): [Loose particles ---Clumps]

3. Size/Shape

a. Size — The relative size of the pieces or particles in the sample:
[Small -- Large]
[Thin ---Thick]

b. Shape — Description of the predominant shape of particles: flat, round, spherical, square, and so on.
[No scale]

4. Surface Shine

Amount of light reflected from the product's surface:
[Dull---Shiny]

B. Terms Used to Describe Flavor (General and Baked Goods)

The full list of fragrance and flavor descriptors is too unwieldy to reproduce here; the list of aromatics alone contains over a thousand words. In the following, aromatics for baked goods are shown as an example.

Flavor is the combined effects of the

- Aromatics
- Tastes
- Chemical feelings

stimulated by a substance in the mouth. For baked goods, it is convenient to subdivide the aromatics into

- Grainy aromatics
- Grain-related terms
- Dairy terms
- Other processing characteristics
- Sweet aromatics
- Added flavors/aromatics
- Aromatics from shortening
- Other aromatics

Example: Flavor Terminology of Baked Goods

1. Aromatics (of Baked Goods)

a. Grainy aromatics

Those aromatics or volatiles that are derived from various grains; the term *cereal* can be used as an alternative, but it implies finished and/or toasted character and is, therefore, less useful than *grainy*. Grainy: the general term to describe the aromatics of grains that cannot be tied to a specific grain by name.

Terms pertaining to a specific grain: corn, wheat, oat, rice, soy, rye.

Grain character modified or characterized by a processing note, or lack thereof:

Raw corn	Cooked corn	Toasted corn
Raw wheat	Cooked wheat	Toasted wheat
Raw oat	Cooked oat	Toasted oat
Raw rice	Cooked rice	Toasted rice
Raw soy	Cooked soy	Toasted soy
Raw rye	Cooked rye	Toasted rye

Definitions of processed grain terms:

Raw (name) flour: the aromatics perceived in a particular grain that has not been heat treated.

Cooked (name) flour: the aromatics of a grain that has been gently heated or boiled; Cream of Wheat has cooked wheat flavor; oatmeal has cooked oat flavor.

Baked/toasted (name) flour: the aromatics of a grain that has been sufficiently heated to caramelize some of the starches and sugars.

b. Grain-related terms

Green: the aromatic associated with unprocessed vegetation, such as fruits and grains; this term is related to raw, but has the additional character of hexenals, leaves, and grass.

Hay-like/grassy: grainy aromatic with some green character of freshly mowed grass, air-dried grain, or vegetation.

Malty: the aromatics of toasted malt.

c. Dairy terms

Those volatiles related to milk, butter, cheese, and other cultured dairy products. This group includes the following terms:

Dairy: as above.

Milky: more specific than dairy, the flavor of regular or cooked cow's milk.

Buttery: the flavor of high-fat fresh cream or fresh butter; not rancid, butyric, or diacetyl-like.

Cheesy: the flavor of milk products treated with rennet, which hydrolyzes the fat, giving it a butyric or isovaleric acid character.

d. Other processing — Caramelized: a general term used to describe starches characteristics and sugars characteristics that have been browned; used alone when the starch or sugar (e.g., toasted corn) cannot be named.

Burnt: related to overheating, overtoasting, or scorching the starches or sugars in a product.

e. Added flavors/aromatics — The following terms relate to specific ingredients that may be added aromatics to baked goods to impart specific character notes; in each case, references for the term are needed:

Nutty: peanut, almond, pecan, and so on.

Chocolate: milk chocolate, cocoa, chocolate-like.

Spices: cinnamon, clove, nutmeg, and so on.

Yeasty: natural yeast (not chemical leavening).

f. Aromatics from shortening — The aromatics associated with oil or fat-based shortening agents used as shortening in baked goods:

Buttery: see dairy.

Oil flavor: the aromatics associated with vegetable oils, not to be confused with an oily film on the mouth surfaces, which is a texture characteristic.

Lard flavor: the aromatics associated with rendered pork fat.

Tallowy: the aromatics associated with rendered beef fat.

g. Other aromatics — The aromatics that are not usually part of the normal product profile and/or do not result from the normal ingredients or processing of the product:

Vitamin: aromatics resulting from the addition of vitamins to the product.

Cardboard flavor: aromatics associated with the odor of cardboard box packaging, which could be contributed by the packaging or by other sources, such as staling flours.

Rancid: aromatics associated with oxidized oils, often also described as painty or fishy.

Mercaptan: aromatics associated with the mercaptan class of sulfur compounds. Other terms that panelists may use to describe odors arising from sulfur compounds are skunky, sulfitic, rubbery.

(End of section referring to baked goods only.)

2. Basic Tastes

a. Sweet — The taste stimulated by sucrose and other sugars, such as fructose, glucose, and so on, and by other sweet substances such as Rebaudioside A, saccharin, aspartame, and Acesulfam K.

b. Sour — The taste stimulated by acids, such as citric, malic, phosphoric, and so on.

c. Salty — The taste stimulated by sodium salts, such as sodium chloride and sodium glutamate, and in part by other salts, such as potassium chloride.

d. Bitter — The taste stimulated by substances such as quinine, caffeine, and hop bitters.

3. Chemical Feeling Factors

Those characteristics that are the response of tactile nerves to chemical stimuli.

a. Astringency — The shrinking or puckering of the tongue surface caused by substances such as tannins or alum.

b. Heat — The burning sensation in the mouth caused by certain substances such as capsaicin from red or piperine from black peppers; mild heat or warmth is caused by some brown spices.

c. Cooling — The cool sensation in the mouth or nose produced by substances such as menthol and mints.

C. Terms Used to Describe Semisolid Oral Texture

These terms are those specifically added for semisolid texture. Solid oral texture terms also may be used when applicable to any product or sample. Each set of texture terms includes the procedure for manipulation of the sample.

1. First Compression

Place 1/2 tsp. of sample on tongue; compress between tongue and palate.

a. Slipperiness — Ease to slide tongue over product:
[Drag--Slip]

b. Firmness — The force required to compress between tongue and palate:
[Soft -- Firm]

c. Cohesiveness — The amount the sample deforms/strings rather than shears/cuts:
[Shears/short--Deforms/cohesive]

d. Denseness — Compactness of the cross section:
[Airy ---Dense/compact]

2. Manipulation

Compress sample several more times (3–8 times).

a. Particle amount — The relative number/amount of particles in the mouth:
[None -- Many]

b. Particle size — The size of the particle in the mass:
[Small -- Large]

3. After Feel

Swallow or expectorate.

a. Mouthcoating	The amount of film left on the mouth surfaces: [None --- High]

Example: Semisolid Texture Terminology—Oral Texture of Peanut Butter

1. Surface	Hold 1/2 tsp. on spoon; feel surface with lips. Moistness—amount of wetness or oiliness (or both) on surface: [Dry --- Oily/moist] Stickiness—amount of product adhering to lips: [None --- Very Sticky] Roughness—overall amount of small and large particles in the surface: [Smooth --- Rough]
2. First compression	Place 1/2 tsp. of peanut butter in mouth and compress between tongue and palate. Semisolid slipperiness—ease to slide tongue over product: [Drag --- Slip] Semisolid firmness—force to compress sample: [Soft --- Firm] Semisolid cohesiveness—amount sample deforms/strings rather than shears/cuts: [Shears/short -- Deforms/cohesive] Adhesiveness (palate)—amount of force to remove sample from roof of mouth using tongue: [No force --- High force] Stickiness—amount of product that adheres to oral surfaces: [Not sticky --- Very sticky]
3. Breakdown/ manipulation	Manipulate between tongue and palate until phase change. Evaluate for: Moisture absorption—amount of saliva absorbed by the sample: [No Absorption -- High Absorption] Semisolid cohesiveness of mass— Amount of sample that deforms rather than crumbles, cracks, or breaks: [Crumbles --- Deforms] Stickiness of mass—amount of mass that adheres to oral surfaces: [None -- Very sticky]

4. Residual — Feel mouth surface and teeth with tongue after product is swallowed or expectorated. Evaluate for:
Mouthcoating—amount of film (total) left on mouth surface:
[None -- High]
Oily film—amount of oily residue on oral surface:
[None -- High]
Toothstick—amount of product adhering to the teeth:
[None --High]

D. Terms Used to Describe Solid Oral Texture

Each set of texture terms includes the procedure for manipulation of the sample.

1. Surface Texture

Feel surface of sample with lips.

a. Surface roughness — The overall roughness of the surface of the sample:
[Smooth --- Rough]
Macroroughness—the degree of roughness attributed to bumps/lumps in surface:
[Smooth --- Rough]
Microroughness—the degree of roughness attributed to small particles in surface:
[Smooth ---Rough]

b. Loose particles — Amount of loose particles free of the surface:
[None --Many]

c. Surface moisture — The amount of wetness or oiliness (or both) on surface:
[None --- High (wet/oily/moist)]

2. Partial Compression

Compress partially (specify with tongue, incisors, or molars) without breaking, and release.

a. Springiness (rubberiness) — Degree to which sample returns to original shape after a certain time period:
[No recovery -- Very springy]

3. First Bite (with Incisors)/First Chew (with Molars)

Bite through a predetermined size sample with incisors or molars as appropriate.

a. Hardness — Force required to compress through the product:
[Low force ("soft") ---------------------High force ("hard")]

b. Cohesiveness — Amount of sample that deforms rather than crumbles, cracks, or breaks:
[Crumbles--Deforms]

c. Fracturability	The force with which the sample breaks: [Crumbles--Fractures]
d. Uniformity of bite	Evenness of force throughout bite: [Not uniform force --------------------- Uniform force]
e. Moisture release/ juiciness	Amount of wetness/juiciness (oil, water) perceived in the mouth: [None --------------------- Very juicy]
f. Amount of particles	Amount of particles resulting from bite or detected in center of sample: [None --- Many]
g. Denseness	Compactness of cross section: [Airy -- Dense]
h. Crispness	The force (noise) with which the sample breaks or fractures, characterized by many, small breaks: [Not crisp ---Very crisp]
i. Crunchiness	The force (noise) with which the sample breaks or fractures, characterized by few, large breaks: [Not crunchy--- Very crunchy]

4. Chewdown

Chew sample with molars until phase change (to bolus):

a. Moisture absorption	Amount of saliva absorbed by product: [No absorption --High absorption]
b. Cohesiveness of mass	Degree to which sample holds together in a mass: [None --Tight mass]
Moistness of mass	The amount of wetness/oiliness (or both) on the surface of the mass at phase change: [Dry ---Wet/Oily/Moist]
Roughness of mass	The amount of roughness on the surface of the mass (can further characterize as gritty, grainy, coarse, lumpy, etc.): [None --High]
c. Adhesiveness to palate	Amount of force to remove sample from roof of mouth using tongue: [No force --- High force]
d. Flinty/glassy	The amount of sharp abrasive pieces in the mass: [None --Very many pieces]
e. Toothstick	Amount of product adhering to the teeth: [None -- Strong Adhesion]
f. Rate of melt	The rate of which the product melts during chew and manipulation: [Slow --- Fast]

5. Residual

Swallow or expectorate sample.

a. Loose particles	Amount of particles left in mouth: [None -- Many]
b. Oily film	Amount and degree of oil felt by the tongue when moved over the surfaces of the mouth: [None -- High]
c. Sticky mouthcoating	Stickiness/tackiness of coating when tapping tongue on roof of mouth: [Not sticky -- Very sticky]
d. Toothpack	Amount of product left in the crevices of teeth: [None -- Highly packed]

Example: Solid Texture Terminology of Oral Texture of Cookies

1. Surface	Place cookie on surface of lips and evaluate for: Roughness—overall amount of small and large particles in the surface: [Smooth -- Rough] Loose particles—amount of loose particles on surface: [None -- Many] Surface moisture—amount of wetness or oiliness (or both) on surface: [None -- High (wet/oily/moist)]
2. First bite	Place one third of cookie between incisors, bite down, and evaluate for: Fracturability—force with which the sample breaks: [Crumbly -- Fractures] Hardness—force required to compress through sample: [Soft -- Hard] Particle size—size of crumb pieces: [Small -- Large]
3. First chew	Place one third of cookie between molars, bite through, and evaluate for: Denseness—compactness of cross section: [Airy -- Dense] Uniformity of chew—evenness of force throughout chew: [Uneven -- Even]
4. Chewdown	Place one third of cookie between molars, chew to bolus, and evaluate for Moisture absorption—amount of saliva absorbed by the sample: [No absorption -- High absorption]

	Cohesiveness of mass—degree to which mass holds together:
	[No mass -- Tight mass]
	Toothpack—amount of sample left in the crevices of teeth:
	[None -- Highly packed]
	Grittiness between teeth—amount of small, hard particles between teeth during chew:
	[None -- High]
5. Residual	Swallow sample and evaluate residue in mouth:
	Oily—amount of oily residue on oral surface:
	[None --High]
	Particles—amount of particles left in mouth:
	[None -- Many]
	Chalky—degree to which mouth feels chalky:
	[Not chalky-- Very chalky]

E. Terms Used to Describe Skinfeel of Lotions and Creams

1. Appearance

In a petri dish, dispense the product in a spiral shape. Using a nickel-size circle, fill from edge to center.

a. Integrity of shape	Degree to which product holds its shape: [Flattens--Retains shape]
b. Integrity of shape	Degree to which product holds its shape after 10 s [Flattens--- Retains shape]
	Tap finger together lightly three times.
c. Gloss	The amount of reflected light from product: [Dull/flat-- Shiny/glossy]

2. Pick Up

Using automatic pipette, deliver 0.1 cc of product to tip of thumb or index finger. Compress product slowly between finger and thumb one time.

a. Firmness	Force required to fully compress product between thumb and index finger: [No force -- High force]
b. Stickiness	Force required to separate fingertips: [Not sticky-- Very sticky]
c. Cohesiveness	Amount sample strings rather than breaks when fingers are separated: [No strings---High strings]
d. Amount of peaking	Degree to which product makes stiff peaks on fingertips: [No peaks/flat --Stiff peaks]

3. Rub Out

Using automatic pipette, deliver 0.05 cc of product to center of 2 in. circle on inner forearm. Gently spread product within the circle using index or middle finger, at a rate of two strokes per second.

After 3 rubs,

a. Wetness	Amount of water perceived while rubbing: [None -- High amount]
b. Spreadability	Ease of moving product over the skin: [Difficult/drag -- Easy/slip]

After 12 rubs,

c. Thickness	Amount of product felt between fingertip and skin: [Thin, almost no product ------------- Thick, lots of product]

After 15–20 rubs,

d. Oil	Amount of oil perceived in the product during rub out: [None -- Extreme]
e. Wax	Amount of wax perceived in the product during rub out: [None -- Extreme]
f. Grease	Amount of grease perceived in the product during rub out: [None -- Extreme]

Continue rubbing and evaluate for:

g. Absorbency	The number of rubs at which the product loses wet, moist feel and a resistance to continue is perceived (upper limit = 120 rubs).

4. Afterfeel (Immediate)

a. Gloss	Amount or degree of light reflected off skin: [Dull-- Shiny]
b. Sticky	Degree to which fingers adhere to product: [Not sticky -- Very sticky]
c. Slipperiness	Ease of moving fingers across skin: [Difficult/drag -- Easy/slip]
d. Amount of residue	Amount of product on skin: [None -- Large amount]
e. Type of residue	Oily, waxy, greasy, silicone, powdery, chalky.

F. Terms Used to Describe Handfeel of Fabric or Paper

1. Force to gather	The amount of force required to collect/gather the sample toward the palm of the hand: [Low force --- High force]
2. Force to compress	The amount of force required to compress the gathered sample into the palm: [Low force --High force]
3. Stiffness	The degree to which the sample feels pointed, ridged, and cracked; not pliable, round, curved: [Pliable/round -- Stiff]
4. Fullness	The amount of material/paper/fabric/sample felt in the hand during manipulation: [Low amount of sample/flimsy -- High amount of sample/body]
5. Compression resilience	The force with which the sample presses against cupped hands: [Creased/folded -- Original shape]
6. Depression depth	The amount that the sample depresses when downward force is applied: [No depression --------------------------------------- Full depression]
7. Depression resilience/ springiness	The rate at which the sample returns to its original position after resilience/depression is removed: [Slow --Fast/springy]
8. Tensile stretch	The degree to which the sample stretches from its original shape: [No stretch-- High stretch]
9. Tensile extension	The degree to which the sample returns to original shape after tensile force is removed (Note: This is a visual evaluation): [No return --Fully returned]
10. Hand friction	The force required to move the hand across the surface: [Slip/no drag --Drag]
11. Fabric friction	The force required to move the fabric over itself: [Slip/no drag -- Drag]
12. Roughness	The overall presence of gritty, grainy, or lumpy particles in the surface; lack of smoothness: [Smooth --- Rough]
13. Gritty	The amount of small, abrasive picky particles in the surface of the sample: [Smooth/not gritty --Gritty]
14. Lumpy	The amount of bumps, embossing, large fiber bundles in the sample: [Smooth/not lumpy --Lumpy]

15. Grainy — The amount of small, rounded particles in the sample:
    [Smooth/not grainy -- Grainy]
16. Fuzziness — The amount of pile, fiber, fuzz on the surface:
    [Bald--Fuzzy/nappy]
17. Thickness — The perceived distance between thumb and fingers:
    [Thin -- Thick]
18. Moistness — The amount of moistness on the surface and in the interior of the paper/fabric. Specify if the sample is oily versus wet (water) if such a difference is detectable:
    [Dry --- Wet]
19. Warmth — The difference in thermal character between paper/fabric and hand:
    [Cool -- Warm]
20. Noise intensity — The loudness of the noise:
    [Soft -- Loud]
21. Noise pitch — Sound frequency of the noise:
    [Low/bass -- High/treble]

G. Terms Used to Describe the Feel of Hair (Wet and Dry)

Wet Hair Evaluation Procedure

1. Preparation before Application
Measure length of hair swatch from the end of the card to the end of the hair. Record the measurement. Pull hair swatch taut and measure as above. Record measurement.

Evaluate for

a. Sheen — Amount of reflected light:
   [Dull-- Shiny]

Comb through swatch with rattail comb. At third stroke of combing, evaluate for:

b. Combability (top half of swatch) (dry) — Ease with which comb can be moved down hair shafts without resistance or hair tangling:
   [Difficult --Easy]
c. Combability (bottom half of swatch) (dry) — Ease with which comb can be moved down hair shafts without resistance or hair tangling:
   [Difficult -- Easy]

d. "Fly away" hair	The tendency of the individual hairs to repel each other during combing after three strokes of combing down hair shafts: [None -- Much]

2. Application of Lotion

Dip hair swatch into cup of room temperature (72 °F) tap water. Thoroughly wet hair swatch. Squeeze out excess water. Pipet 0.125 cc of hair lotion onto edge of palm of hand. Using opposite index and middle fingers, rub onto edge of palm two to three times to distribute lotion. Pick up hair swatch by the card. Using long, even strokes from the top to bottom, apply lotion to hair swatch, turning card after each stroke and rubbing ends of swatch with index and middle fingers. Evaluate for:

a. Ease of distribution	Ease of rubbing product over hair: [Difficult --Easy]
b. Amount of residue	The amount of residue left on the surface of the hands: (Untreated skin = 0) [None --Extreme]
c. Type of residue	Oily, waxy, greasy, silicone

3. Evaluation

Clean hands with water before proceeding. Comb through hair swatch with a rattail comb one time and evaluate for:

a. Ease of detangling	Ease to comb through hair: [Very tangled, hard to comb-----Not tangled, easy to comb]
At the third stroke of combing evaluate for:	
b. Combability (top half of swatch) (wet)	Ease with which comb can be moved down hair shafts without resistance or hair tangling: [Difficult ------------------------------------- Easy]
c. Combability (bottom half of swatch) (wet)	Ease with which comb can be moved down hair shafts without resistance or hair tangling: [Difficult ------------------------------------Easy]
d. Stringiness (visual)	The sticking of individual hairs together in clumps: [Unclumped------------------------------------ Clumped]
e. Wetness (tactile)	The amount of perceived moisture: [Dry -- Wet]
f. Coldness (tactile)	Thermal sensation of lack of heat: [Hot------------------------------------ Cold]

g. Slipperiness (tactile)	Lack of drag or resistance as moving hairs along between fingers: [Drags------------------------------------Slips]
h. Roughness (tactile)	A rough, brittle texture of hair shafts: [Smooth -------------------------------------Rough]
i. Coatedness (tactile)	The amount of residue left on the hair shaft: [None, uncoated------------------------------------ Very coated]
j. Stickiness of hair to skin (tactile)	The tendency of the hair to stick to the fingers: [Not sticky------------------------------------ Very sticky]

4. Evaluation after Drying

Let hair swatch dry for 30 min lying on clean paper towels, checking swatch at 5 min intervals, and evaluating earlier if dried. At the third stroke of combing evaluate for:

a. Combability (top half of swatch) (dry)	Ease with which comb can be moved down hair shafts without resistance or hair tangling: [Difficult ------------------------------------ Easy]
b. Combability (bottom half of swatch) (dry)	Ease with which comb can be moved down hair shafts without resistance or hair tangling: [Difficult -------------------------------------Easy]
c. "Fly away" hair	The tendency of the individual hairs to repel each other during combing after three strokes of combing down hair shafts: [None ------------------------------------Much]
d. Stringiness (visual)	The sticking of individual hairs together in clumps: [Unclumped------------------------------------ Clumped]
e. Sheen	Amount of reflected light: [Dull------------------------------------ Shiny]
f. Roughness (tactile)	A rough, brittle texture of hair shafts: [Smooth -------------------------------------Rough]
g. Coatedness (tactile)	The amount of residue left on the hair shaft: [None, uncoated------------------------------------ Very coated]

Dry Hair Evaluation Procedure

1. Preparation before Application

Measure length of hair swatch from the end of the card to the end of the hair. Record the measurement. Pull hair swatch taut and measure as above. Record measurement. Visually evaluate hair for

a. Sheen — Amount of reflected light:
[Dull---------------------------------------Shiny]

Comb through hair with rattail comb. At third stroke of combing, evaluate for:

b. Combability (top half of swatch) (dry) — Ease with which comb can be moved down hair shafts without resistance or hair tangling:
[Difficult ------------------------------------ Easy]

c. Combability (bottom half of swatch) (dry) — Ease with which comb can be moved down hair shafts without resistance or hair tangling:
[Difficult ------------------------------------ Easy]

d. "Fly away" hair — The tendency of the individual hairs to repel each other during combing after three strokes of combing down hair shafts:
[None ------------------------------------ Much]

2. Application of Lotion

Pipet 0.125 cc of hair lotion onto edge of palm of hand. Using opposite index and middle fingers, rub onto edge of palm two to three times to distribute lotion. Pick up hair swatch by the card. Using long, even strokes, from the top to bottom, apply lotion to hair swatch, turning card after each stroke, rubbing ends of swatch with index and middle fingers. Evaluate for:

a. Ease of distribution — Ease of rubbing product over hair:
[Difficult ------------------------------------ Easy]

b. Amount of residue — The amount of residue left on the surface of the hands: (Untreated skin = 0)
[None ------------------------------------ Extreme]

c. Type of residue — Oily, waxy, greasy, silicone

3. Evaluation

Clean hands with water before proceeding. Comb through hair swatch with a rattail comb. At the third stroke of combing evaluate for:

a. Combability (top half of swatch) (wet) — Ease with which comb can be moved down hair shafts without resistance or hair tangling:
[Difficult -- Easy]

b. Combability (bottom half of swatch) (wet)	Ease with which comb can be moved down hair shafts without resistance or hair tangling: [Difficult ---------------------------------- Easy]
c. Stringiness (visual)	The sticking of individual hairs together in clumps: [Unclumped ---------------------------------- Clumped]
d. Wetness (tactile)	The amount of perceived moisture: [Dry ---------------------------------- Wet]
e. Coldness (tactile)	Thermal sensation of lack of heat: [Hot-- Cold]
f. Slipperiness (tactile)	Lack of drag or resistance as moving hairs along between fingers: [Drags---------------------------------- Slips]
g. Roughness (tactile)	A rough, brittle texture of hair shafts: [Smooth ---------------------------------- Rough]
h. Coatedness (tactile)	The amount of residue left on the hair shaft: [None, uncoated---------------------------------- Very coated]
i. Stickiness of hair to skin (tactile)	The tendency of the hair to stick to the fingers: [Not sticky ---------------------------------- Very sticky]

4. Evaluation after Drying

Let hair swatch dry for 30 min lying on clean paper towels, checking swatch at 5 min intervals and evaluating earlier if dried. Record drying time. Measure and record length of hair swatch from the end of the card to the end of the hair. Pull hair swatch taut and measure as above. Record measurement. Comb through hair swatch with rattail comb. At the third stroke of combing evaluate for:

a. Combability (dry) (top half of swatch)	Ease with which comb can be moved down hair shafts without resistance or hair tangling: [Difficult ---------------------------------- Easy]
b. Combability (dry)	Ease with which comb can be moved down hair shafts without bottom half of swatch resistance or hair tangling: [Difficult ---------------------------------- Easy]
c. "Fly away" hair	The tendency of the individual hairs to repel each other during combing after three strokes of combing down hair shafts: [None ---------------------------------- Much]

d. Stringiness (visual)	The sticking of individual hairs together in clumps: [Unclumped------------------------------------Clumped]
e. Sheen	Amount of reflected light: [Dull------------------------------------Shiny]
f. Roughness (tactile)	A rough, brittle texture of hair shafts: [Smooth ------------------------------------ Rough]
g. Coatedness (tactile)	The amount of residue left on the hair shaft: [None, uncoated------------------------------------Very coated]

H. Terms Used to Describe the Lather and Skinfeel of Bar Soap

Full Arm Test

1. Preparation for Skinfeel Test

Instruct panelists to refrain from using any type of moisturizing cleanser on evaluation days (to include bar soap, cleansing cream, bodywash, and cleansing oil) and from applying lotions, creams, or other moisturizers to their arms. Panelists may rinse arms with water and pat dry.

Limit panelists to evaluation of no more than two samples per day (one sample per site, beginning with the left arm). For the second soap sample, repeat the washing procedure on the right arm evaluation site. Wash each site once only.

2. Baseline Evaluation of Site

Visually evaluate skin for

a. Gloss	The amount or degree of light reflected off skin: [Dull--Shiny]
b. Visual dryness	The degree to which the skin looks dry (ashy/flaky): [None --Very dry]

Stroke cleansed fingers lightly across skin and evaluate for

c. Slipperiness	Ease of moving fingers across the skin: [Drag-- Slip]
d. Amount of residue	The amount of residue left on the surface of the skin: [None --Extreme]
e. Type of residue	Indicate the type of residue: Soap film, oily, waxy, greasy, powder
f. Dryness/ roughness	The degree to which the skin feels rough: [Smooth -- Rough]
g. Moistness	The degree to which the skin feels moist: [Dry -- Moist]

h. Tautness — The degree to which the skin feels taut or tight:
[Loose/pliable --Very tight]

Using edge of fingernail, scratch a line through the test site. Visually evaluate for:

i. Whiteness — The degree to which the scratch appears white:
[None -- Very white]

3. Evaluation of Lather and Skinfeel

Application and washing procedure: Apply wet soap bar to wet evaluation site. Apply with up–down motion (1 up–down lap = ½ s).

a. Amount of lather observed during application:
At 10, 20, 30 laps — [None -- Extreme]

At 30 laps continue with

b. Thickness of lather — Amount of product felt between fingertips and skin:
[Thin -- Thick]

c. Bubble size variation — The variation seen within the bubble size (visual):
[Homogeneous -- Heterogeneous]

d. Bubble size — The size of the soap bubbles in the lather (visual):
[Small -- Large]

Rinsing procedure: Rinse site by placing arm directly under warm running water. Use free hand to stroke gently with up–down lap over the site. Rinse for 15 laps. (1 lap = 1 s). Also rinse evaluation fingers.

Evaluation before Drying

a. Rinsability — The degree to which the sample rinses off (visual):
[None -- All]

Gently stroke upward on skin site with a clean finger and evaluate for:

b. Slipperiness — Ease of moving fingers across the skin:
[Drag --Slip]

c. Amount of residue — The amount of residue left on the surface of the skin:
[None --Extreme]

d. Type of residue — Indicate the type of residue:
soap film, oily, waxy, greasy, powder

Evaluation after drying: Dry the site by covering it with a paper towel and patting dry 3 times along the site. Also thoroughly dry evaluation finger. Visually evaluate skin for

a. Gloss	Visual: amount of light reflected on the surface of the skin: [Dull-- Shiny/glossy]
b. Visual dryness	The degree to which the skin looks dry (ashy/flaky): [None -- Very dry]

Tap dry, cleansed finger over treated skin. Gently stroke skin site with clean finger and evaluate for:

c. Stickiness	The degree to which fingers stick to residual product on the skin: [Not sticky--Very sticky]
d. Slipperiness	Ease of moving fingers across the skin: [Drag--Slip]
e. Amount of residue	The amount of residue left on the surface of the skin: [None -- Extreme]
f. Type of residue	Indicate the type of residue: Soap film, oily, waxy, greasy, powder.
g. Dryness/ roughness	The degree to which the skin feels dry/rough: [Smooth -- Dry/rough]
h. Moistness	The degree to which the skin feels moist, wet: [Dry -- Moist]
i. Tautness	The degree to which the skin feels taut or tight: [Loose/pliable -- Very taut]

Using the edge of the fingernail, scratch through test site and evaluate for:

j. Whiteness	The degree to which the scratch appears white: [None -- Very white]

I. Terms Used to Describe the Skinfeel of Antiperspirants

Roll-On/Solids/Gels

1. Preparation of Skin

Evaluation site (crook of arm) is washed with nonabrasive, nondeodorant soap more than 10 min prior to the start of the evaluation. No lotion is applied to arms before the evaluation. Panelists should wear short-sleeved or sleeveless shirts to panel to avoid rubbing any product off onto clothing during evaluation.

A 6 in. × 2 in. rectangle is marked on the crook of the arm so the fold bisects the rectangle. Divide the rectangle into four 1.5 in. sections to allow for undisturbed area for each timed evaluation point.

2. Baseline Evaluation
Prior to application, instruct panelists to evaluate untreated sites for baseline references. Visually evaluate skin for

a. Gloss	The amount or degree of light reflected off skin: [Dull--- Shiny]
b. Visual dryness	The degree to which the skin looks dry (ashy/flaky): [None --Very dry]

Stroke cleansed fingers lightly across skin and evaluate for:

c. Slipperiness	Ease of moving fingers across the skin: [Drag--Slip]
d. Amount of residue	The amount of residue left on the surface of the skin: [None --Extreme]
e. Type of residue	Indicate the type of residue: Soap film, oily, waxy, greasy, powder
f. Dryness/ roughness	The degree to which the skin feels rough: [Smooth --Rough]
g. Moistness	The degree to which the skin feels moist: [Dry ---Moist]
h. Tautness	The degree to which the skin feels taut or tight: [Loose/pliable -- Very tight]

Using edge of fingernail, scratch a line through the test site. Visually evaluate for:

i. Whiteness	The degree to which the scratch appears white: [None --Very white]

3. Application of Antiperspirant
Follow a statistical design and adhere to the right–left arm balance.

Roll-on, Direct Application*:* Sample is primed prior to application. Arm is held out at right angle to the body with the palm of hand perpendicular to the floor. The panel leader applies the product by stroking in a zigzag pattern to evenly cover the surface area of the rectangle. Count the number of strokes to achieve full coverage. Panel leader will weigh the sample before and after and panelists will record the total weight on the ballot.

Solids/gels*:* The panel leader applies the product by stroking up the arm once through the 6 in. × 2 in. rectangle (force to apply), then back down and up the arm three times (ease to spread), using a consistent pressure to get the product on the arm. A tare weight is taken after the fourth stroke. The weight must fall between the 0.20 and 0.25 g range. The panel leader continues with additional strokes until weight is reached and then records the weight and the number of strokes.

During application for roll-ons: Using the first two pads of just the index finger, run product across the entire site using a gentle oval motion—stroke at a rate of two strokes per second (This is done to ensure that the entire site is covered evenly.)

After three rubs, evaluate for:

a. Wetness	The amount of water perceived on the skin: [None -- High amount]
b. Spreadability	The ease with which the product is spread on the skin: [Difficult -- Easy]

4. Immediate Evaluation

Immediately after application, evaluate for:

a. Coolness	The degree to which the sample feels "cool" on the skin (somesthetic): [Not at all cool -- Very cool]
b. Gloss	The amount of reflected light from the skin: [Not at all shiny --Very shiny]
c. Whitening	The degree to which the skin turns white: [None -- Very white]
d. Amount of residue (visual)	The amount of product visually perceived on the skin: [None --Large amount]
e. Tautness	The degree to which the skin feels taut or tight: [Loose/pliable --Very tight]

Fold arm to make contact. Hold for 5 s. Unfold arm and evaluate for:

f. Stickiness (fold)	Degree to which arm sticks to itself: [Not at all --Very sticky]

Stroke finger lightly across skin on one section of rectangle and evaluate for:

g. Wetness	The amount of water perceived on the skin: [None --High amount]
h. Slipperiness	Ease of moving fingers across the skin: [Drag -- Slip]
i. Amount of residue	The amount of residue perceived on skin (tactile). Evaluate by stroking finger across site: [None --Extreme]
j. Oil	The amount of oil perceived on skin: [None --Extreme]
k. Wax	The amount of wax perceived on skin: [None -- Extreme]
l. Grease	The amount of grease perceived on skin: [None --Extreme]

m. Powder/chalk/grit	The amount of powder, chalk, and/or grit perceived on skin: [None --Extreme]
n. Silicone	The amount of silicone perceived on skin: [None -- Occluded]

5. Evaluation after 5, 10 or 15, and 30 Min. Evaluate for:

a. Occlusion	The degree to which the sample occludes or blocks the air passage to the skin: [None -- Occluded]
b. Whitening	The degree to which the skin turns white: [None --Large amount]
c. Amount of residue (visual)	The amount of product visually perceived on skin: [None -- Large amount]
d. Tautness	The degree to which the skin feels taut or tight: [Loose/pliable -- Very tight]

Fold arm to make contact. Hold 5 sec. Unfold arm and

e. Stickiness	The degree to which arm sticks to itself: [Not at all sticky-- Very sticky]

Stroke fingers lightly across skin on one section of rectangle and

f. Wetness	The amount of water perceived on the skin: [None -- High amount]
g. Slipperiness	Ease of moving fingers across the skin: [Drag-- Slip]
h. Amount of residue	The amount of residue perceived on skin (tactile). Evaluate by stroking finger across site: [None --Extreme]
i. Oil	The amount of oil perceived on skin: [None --Extreme]
j. Wax	The amount of wax perceived on skin: [None -- Extreme]
k. Grease	The amount of grease perceived on skin: [None --Extreme]
l. Powder/chalk/grit	The amount of powder, chalk, and/or grit perceived on skin: [None --Extreme]
m. Silicone	The amount of silicone perceived on skin: [None --Extreme]

6. Evaluation after 30 Min

Place a swatch of black fabric over test site. Fold arm so fingertips touch the shoulder. With arm still folded, pull fabric from crook of arm. Compare residue on swatch to visual scale provided.

a. Rub-off whitening — The amount of residue on the dark fabric:
[None -- Large amount]

APPENDIX 12.2 SPECTRUM INTENSITY SCALES FOR DESCRIPTIVE ANALYSIS

The scales below (all of which run from 0 to 15) contain intensity values for aromatics (A) and for tastes (B) that were derived from repeated tests with trained panels at Hill Top Research, Inc., Cincinnati, Ohio, and with trained panels at Sensory Spectrum; and also for various texture characteristics (C and D) that were obtained from repeated tests at Hill Top Research or at Sensory Spectrum, or that were developed at Bestfoods Technical Center, Somerset, New Jersey.

New panels can be oriented to the use of the 0–15 scale by presentation of the basic tastes using concentrations of caffeine, citric acid, NaCl, and sucrose, which are listed under Section B. If a panel is developing a descriptive system for an orange drink product, the panel leader can present three "orange" references:

1. Orange drink Hi-C labeled "Orange Complex 3.0"
2. Reconstituted Minute Maid concentrate labeled "Orange Complex 7.5 and Orange Peel 2.0"
3. Tang labeled "Orange Complex 9.5 and Orange Peel 4.0"

At each taste test of any given product, labeled reference samples related to its aromatic complex can be presented so as to standardize the panel's scores and keep panel members from drifting.

A Intensity Scale Values (0–15) for Some Common Aromatics

Term	Reference	Scale Value
Baked white wheat	Ritz crackers (Nabisco)	6.5
Caramelized sugar	Tortilla chips (Frito Lay)	2
	Ketchup (Heinz)	3
	Bugles (General Mills)	4
	Bordeaux cookies (Pepperidge Farm)	4.5
Celery	V-8 vegetable juice (Campbell)	5
Cheese	American cheese, slices (Kraft Singles)	5
Cinnamon	Big Red gum (Wrigley)	12.5

(Continued)

A Intensity Scale Values (0–15) for Some Common Aromatics (Continued)

Term	Reference	Scale Value
Coffee impact	Coffee (Maxwell House)	3
Cooked apple	Applesauce, natural (Mott's)	5
Cooked orange	Frozen orange concentrate (Minute Maid)—reconstituted	5
Cooked white wheat	Pound cake (Sara Lee)	2
Egg	Mayonnaise (Hellmann's)	5
Grain complex	Wheatina cereal	9
	Triscuit (Nabisco)	8
	Ritz cracker (Nabisco)	6.5
	Cream of Wheat (Nabisco)	4.5
	Penne, whole grain pasta (Barilla) cooked, chilled	4
	Penne white pasta (Barilla) cooked, chilled	3
Grape	Kool-Aid, grape	5
	Grape juice (Welch's Concord)	10
Lemon	Lemonade (Country Time)	5
Milky complex	American cheese, slices (Kraft Singles)	3
	Powdered milk (Carnation)	4
	Whole milk	5
Mint	Doublemint gum (Wrigley)	11
Oil	Potato chips (Pringles)	1
	Soybean oil (Crisco Vegetable Oil)	2
	Heated oil (Crisco Vegetable Oil)	4
Orange complex	Orange drink (Hi-C)	3
	Frozen orange concentrate (Minute Maid)—reconstituted	7
	Orange concentrate—reconstituted (Tang)	9.5
Orange peel	Soda (Orange Crush)	2
	Frozen orange concentrate (Minute Maid)—reconstituted	2
	Orange concentrate—reconstituted (Tang)	4
Peanut	Medium roasted regular cocktail peanuts (Planters)	7
Potato	Potato chips (Pringles)	4.5
Vanillin	Powdered doughnut (Hostess)	2
	Honey bun (Little Debbie)	2.5

B Scales Values (0–15) for the Four Basic Tastes

	Sweet	Salt	Sour	Bitter
American cheese, slices (Kraft)		7	5	
Applesauce, natural (Mott's)	5		4	
Applesauce, regular (Mott's)	8.5		2.5	
Big Red gum (Wrigley)	11.5			
Bordeaux cookies (Pepperidge Farm)	11.5			
Basic taste blends				
5% Sucrose/0.1% Citric acid	6		7	
5% Sucrose/0.55% NaCl	7	9		
0.1% Citric acid/0.55% NaCl		11	6	
5% Sucrose/0.1% Citric Acid/0.35% NaCl	5	5	3.5	
5% Sucrose/0.1% Citric acid/0.55% NaCl	4	11	6	
Caffeine, solution in water				
0.05%				2
0.08%				5
0.15%				10
0.20%				15
Celery seed				9
Chocolate bar (Hershey's)	10		5	4
Citric acid, solution in water				
0.05%			2	
0.10%			5	
0.15%			10	
0.20%			15	
Coca-Cola Classic	9			
Endive, raw				7
Fruit punch (Hawaiian)	10		3	
Grape juice (Welch's Concord)	6		7	2
Grape Kool-Aid	10		1	
Kosher dill pickle (Vlasic)		12	10	
Lemon juice (ReaLemon)			15	
Lemonade (Country Time)	7		5.5	
Mayonnaise (Hellmann's)		12	4.5	
NaCl, solution in water				
0.2%		2		
0.35%		5		
0.55%		10		
0.7%		15		
Soda (Orange Crush)	10		2	
Frozen orange concentrate (Minute Maid)—reconstituted	7.5		3.5	
Orange concentrate—reconstituted (Tang)	9			

B Scales Values (0–15) for the Four Basic Tastes (Continued)

	Sweet	Salt	Sour	Bitter
Potato chips (Lay's)	3.8	13.5		
Potato chips (Pringles)	6	13		
Snack cracker (Ritz)	4	8		
Soda cracker (Premium)		5		
Spaghetti sauce (Ragu)	8	12		
Sucrose, solution in water				
2.0%	2			
5.0%	5			
10.0%	10			
16.0%	15			
Sweet pickle (Gherkin, Vlasic)	8.5		8	
Orange concentrate—reconstituted (Tang)	11.5		5	
V-8 vegetable juice (Campbell)		8		
Wheatina cereal		6		2.5
Whole grain wheat cracker (Triscuit)		9.5		

C Intensity Scale Values (0–15) for Semisolid Oral Texture Attributes

Scale Value	Reference	Brand/Type/Manufacturer	Sample Size
1. Slipperiness			
2.0	Classic hummus	Sabra	1 oz.
3.5	Baby food—peas	Beechnut Stage 2	1 oz.
7.5	Chocolate pudding, instant, made with whole milk	Jello	1 oz.
12.0	Sour cream, full fat	Breakstone	1 oz.
2. Firmness			
3.0	Aerosol whipped cream	Reddi-Wip (Con Agra)	1 oz.
5.0	Miracle Whip	Kraft Foods	1 oz.
9.0	Cheez Whiz	Kraft Foods	1 oz.
11.0	Peanut butter	Hormel/Skippy	1 oz
13.0	Cream cheese	Kraft/Philadelphia	1 oz.
3. Cohesiveness			
2.0	Instant gelatin dessert	Jello, Kraft Foods	½ in. cube
6.5	Instant vanilla pudding	Jello, Kraft Foods made with whole milk	1 oz.
10	Baby food—bananas	Gerber, Stage 1	1 oz.
14.0	Tapioca pudding	SnackPack shelf stable (Con Agra)	1 oz.

(Continued)

C Intensity Scale Values (0–15) for Semisolid Oral Texture Attributes (Continued)

Scale Value	Reference	Brand/Type/Manufacturer	Sample Size
4. Denseness			
2.5	Aerosol whipped cream	Reddi-wip	1 oz.
5.0	Mousse mix	Oetker Milk Chocolate Mousse Mix made with whole milk	1 oz.
7.0	Marshmallow Fluff	Fluff-Durkee-Mower	1 oz.
9.0	Whipped cream cheese	Kraft Foods/Philadelphia	1 oz.
15.0	Block cream cheese	Kraft Foods/Philadelphia	½ in. cube
5. Amount of Particles			
2.5	1 c. refrigerated chocolate pudding + ½ t. yellow corn meal	Kozy Shack or equivalent, Quaker yellow corn meal or equivalent	1 oz.
6.0	1 c. refrigerated chocolate pudding + 1 t. yellow corn meal	Kozy Shack or equivalent, Quaker yellow corn meal or equivalent	1 oz.
8.0	1 c. refrigerated chocolate pudding + 1½ t. yellow corn meal	Kozy Shack or equivalent, Quaker yellow corn meal or equivalent	1 oz.
10.0	1 c. refrigerated chocolate pudding + 2 t. yellow corn meal	Kozy Shack or equivalent, Quaker yellow corn meal or equivalent	1 oz.
6. Particle size			
4.0	Small pearl tapioca*		2 oz.
8.0	Boba Tea tapioca	Pearl Milk Tea Tapioca or Bubble Tea	2 oz.
15.0	Large tapioca balls*		2 oz.
7. Amount of Mouthcoating Film of Liquids			
3.0	Whole milk		1 oz.
6.0	Maple-type syrup	Log Cabin Original Syrup	1 oz.
7.5	Heavy cream		1 oz.

* Contact Sensory Spectrum for details on sample preparation.

D Intensity Scale Values (0–15) for Solid Oral Texture Attributes

Scale Value	Reference	Brand/Type/Manufacturer	Sample Size
1. Standard Surface Roughness Scale[a]			
0.0	Gelatin dessert	Jello	2 tbsp
5.0	Orange peel	Peel from fresh orange	½ in. piece
8.0	Potato chips	Pringles	5 pieces
12.0	Hard granola bar	Nature Valley Oats n Honey	½ bar
15.0	Rye wafer	Finn Crisp/Vaasan OY	½ in. sq.

D Intensity Scale Values (0–15) for Solid Oral Texture Attributes (Continued)

Scale Value	Reference	Brand/Type/Manufacturer	Sample Size
Technique:	Place sample against lips; feel the surface to be evaluated with the lips and tongue.		
Definition:	The overall amount of small and large particles in the surface. [Smooth -- Rough]		
2. Standard Surface Moisture Scale			
0.0	Unsalted premium cracker	Nabisco	1 cracker
3.0	Carrots	Uncooked, fresh, unpeeled	½ in. slice
7.5	Apples	Red Delicious, uncooked, fresh, unpeeled	½ in. slice
10.0	Ham	Oscar Mayer Chopped Ham	½ in. piece
15.0	Water	Filtered, room temp.	½ tbsp
Technique:	Hold the sample against lips; evaluate the amount of moisture on sample.		
Definition:	The amount of moisture (oil or water) on the surface. [Dry -- Wet]		
3. Standard Stickiness to Lips Scale			
1.0	Cherry tomato	Uncooked, fresh, unpeeled	½ in. slice
4.0	Nougat center (Remove chocolate first)	Three Musketeers/Mars	½ in. cube
7.5	Pretzel rod	Bachman	1 piece
15.0	Rice Krispies	Kellogg's	1 tsp
Technique:	Moisten lips by running tongue over lips. Hold sample near mouth; compress sample lightly between lips and release.		
Definition:	The degree to which the sample adheres to the lips. [None -- High adhesion]		
4. Standard Springiness Scale			
0.0	Cream cheese	Kraft Foods/Philadelphia	½ in. cube
5.0	Frankfurter	Cooked 5 min/Hebrew National Beef	½ in. slice
9.0	Marshmallow	Miniature marshmallow/Kraft Foods	3 pieces
15.0	Gelatin dessert	Jello, Knox (see Note)	½ in. cube
Technique:	Place sample between molars; compress partially without breaking the sample structure; release.		
Definition:	(1) The degree to which sample returns to original shape or (2) The rate with which sample returns to original shape. [No recovery -- Very springy]		

Note: One package Jello and one package Knox gelatin are dissolved in 1½ cups hot water and refrigerated for 24 h.

[a] The roughness scale measures the amount of irregular particles in the surface. These may be small (chalky, powdery), medium (grainy), or large (bumpy).

D Intensity Scale Values (0–15) for Solid Oral Texture Attributes (Continued)

Scale Value	Reference	Brand/Type/Manufacturer	Sample Size
5. Standard Hardness Scale			
1.0	Cream cheese	Kraft Foods/Philadelphia block cream cheese	½ in. cube
4.5	Cheese	Yellow American pasteurized process-deli/Land O'Lakes	½ in. cube
7.0	Frankfurter	Large, cooked 5 min/Hebrew National Beef	½ in. slice
9.0	Peanuts	Cocktail type in vacuum tin/Planters	1 nut, whole
11.0	Almonds	Shelled/Planters or Blue Diamond	1 nut
14.5	Hard candy	Life Savers	3 pieces, one color
Technique:	For solids, place food between the molars and bite down evenly, evaluating the force required to compress the food. For semisolids, measure hardness by compressing the food against palate with tongue once. When possible, the height for hardness standards is ½ in.		
Definition:	The force to attain a given deformation, such as • Force to compress between molars, as above • Force to compress between tongue and palate • Force to bite through with incisors [Low force ("Soft") --High force ("Hard")]		
6. Standard Cohesiveness Scale			
1.0	Corn muffin	Jiffy Standard Cornbread Recipe	½ in. cube
4.5	Cheese	Yellow American pasteurized process-deli/Land O'Lakes	½ in. cube
8.0	Pretzel	Soft pretzel—any frozen, baked	½ in. piece
10.0	Dried fruit	Sun-dried seedless raisins/Sun-Maid	1 tsp
12.5	Candy chews	Starburst/Mars	1 piece
15.0	Chewing gum	Freedent/Wrigley	1 stick
Technique:	Place sample between molars; compress fully (can be done with incisors).		
Definition:	The amount to which sample deforms rather than crumbles, cracks, or breaks. [Crumbles--Deforms]		
7. Standard Fracturability Scale			
1.0	Corn muffin	Jiffy Mix Cornbread Recipe (1/4 c. milk, 350°F for 30 min)	½ in. cube
2.5	Egg Jumbos	Stella D'oro/Synder's Lance	½ in. cube
4.2	Graham crackers	Original Graham Crackers/Nabisco	½ in. cube
6.7	Melba toast	Plain, rectangular/Devonsheer, Melba Co.	½ in. sq.
8.0	Ginger snaps	Nabisco	½ in. sq.

D Intensity Scale Values (0–15) for Solid Oral Texture Attributes (Continued)

Scale Value	Reference	Brand/Type/Manufacturer	Sample Size
10.0	Rye wafers	Finn Crisp/Vaasan OY	½ in. sq. (2 pcs. stacked)
11.5	Pita chips	Stacy's	½ in. sq. (2 pcs. stacked)
13.0	Peanut brittle	Brand available	½ in. sq. candy part
14.5	Hard candy	Life Savers	1 piece
Technique:	Place sample between molars and bite down evenly until the food crumbles, cracks, or shatters.		
Definition:	The force with which the sample breaks. [Crumbles-- Fractures]		
8. Standard Viscosity Scale			
1.0	Water	Bottled Mountain Spring (room temp.)	1 tsp
2.0	Light cream	Brand available, not ultrapasteurized (refrigerated)	1 tsp
4.0	Heavy cream	Brand available, not ultrapasteurized (refrigerated)	1 tsp
6.0	Maple syrup	Pure maple syrup, grade A (room temp.)	1 tsp
9.0	Chocolate syrup	Hershey's (room temp.)	1 tsp
12.0	Mixture: 1 cup sweetened condensed milk +2 tbsp heavy cream	Magnolia or Eagle Brand/Eagle Family Foods (refrigerated)	1 tsp
14.5	Sweetened condensed milk	Eagle Brand/Eagle Family Foods (room temp.)	1 tsp
Technique:	(1) Place 1 tsp of product close to lips; draw air in gently to induce flow of liquid; measure the force required. (2) Once product is in mouth, allow to flow across tongue by moving tongue slowly to roof of mouth; measure rate of flow (the force here is gravity).		
Definition:	The rate of flow per unit force: • The force to draw between lips from spoon • The rate of flow across tongue [Not viscous-- Very viscous]		
9. Standard Denseness Scale			
0.5	Cool Whip	Kraft Foods	2 tbsp
2.5	Marshmallow Fluff	Fluff/Durkee-Mower	2 tbsp
4.0	Nougat center (Remove chocolate first)	Three Musketeers/Mars	½ in. cube
6.0	Malted milk balls	Whopper/The Hershey Company	5 pieces

D Intensity Scale Values (0–15) for Solid Oral Texture Attributes (Continued)

Scale Value	Reference	Brand/Type/Manufacturer	Sample Size
			(Continued)
9.5	Frankfurter	Cooked 5 min, Oscar Mayer Beef	5, ½ in. slices
15.0	Fruit jellies	Chuckles/Ferrara Candy Co.	3 pieces
Technique:	Place sample between molars and compress.		
Definition:	The compactness of the sample cross section.		
	[Light/airy -- Dense/compact]		
10. Standard Crispness Scale			
2.0	Granola bar	Quaker Low Fat Chewy Chunk	1/3 bar
5.0	Club cracker	Keebler	½ cracker
6.5	Graham cracker	Honey Maid/Nabisco	1 in. sq.
7.0	Oat cereal	Cheerios /Toasted Whole Grain	1 oz.
14.0	Corn flakes	Kellogg's	1 oz.
17.0	Melba toast	Devonsheer	½ cracker
Technique:	Place sample between molar teeth and bite down evenly until the food breaks, crumbles, cracks, or shatters.		
Definition:	The force (noise) with which a product breaks or fractures, characterized by many, small breaks.		
	[Not crisp -- Very crisp]		
11. Standard Moisture Release/Juiciness Scale			
1.0	Banana	Banana	½ in. slice
2.0	Carrot	Raw carrot	½ in. slice
4.0	Mushroom	Raw mushroom	½ in. slice
7.0	Snap bean	Raw snap bean	5 pieces
8.0	Cucumber	Raw cucumber	½ in. slice
10.0	Apple	Red Delicious apple	½ in. wedge
12.0	Honeydew melon	Honeydew melon	½ in. cubes
15.0	Orange	Florida juice orange	½ in. wedge
15.0	Watermelon	Watermelon	½ in. cube (no seeds)
Technique:	Chew sample with the molar teeth for up to five chews.		
Definition:	The amount of juice/moisture (oil, water) perceived in the mouth.		
	[None --Very juicy]		
12. Standard Flinty/Glassy Scale			
2.0	Bugles corn snacks	General Mills	1 oz.
8.0	Frosted flakes	Kellogg's	1 oz.
14.0	Dehydrated potatoes	Betty Crocker Au Gratin Potatoes	1 oz.
Technique:	Chew sample three times and using the tongue to measure the degree of pointiness of pieces and amount of pointy shards present.		
Definition:	The degree to which the sample breaks into pointy shards and the amount present after three chews.		

D Intensity Scale Values (0–15) for Solid Oral Texture Attributes (Continued)

Scale Value	Reference	Brand/Type/Manufacturer	Sample Size
	[None -- Very many pieces]		
13. Standard Moisture Absorption Scale			
0.0	Licorice	Shoestring	1 piece
3.5	Licorice, red	Twizzlers/Hershey's	1 piece
9.0	Popcorn	Bagged/Bachman Air Popped Popcorn	2 tbsp
10.0	Potato chips	Wise	2 tbsp
13.0	Cake	Pound cake, frozen type/Sara Lee	1 slice
15.0	Saltines	Unsalted top premium cracker/Nabisco	1 cracker
Technique:	Chew sample with molars until phase change.		
Definition:	The amount of saliva absorbed by sample during chewdown.		
	[No absorption -- High absorption]		
14. Standard Cohesiveness of Mass Scale			
0.0	Licorice	Shoestring	1 piece
2.0	Carrots	Uncooked, fresh, unpeeled	½ in. slice
4.0	Mushroom	Uncooked, fresh	½ in. slice
7.5	Frankfurter	Cooked 5 min/Hebrew National Beef	½ in. slice
10.0	Cheese, yellow	American pasteurized process- deli/ Land O'Lakes	½ in. cube
14.0	Fig Newton	Nabisco	
Technique:	Chew sample with molars until phase change.		
Definition:	The degree to which chewed sample holds together in a mass.		
	[None/no mass--Compact/tight mass]		
15. Standard Toothstick Scale			
2.0	Mushrooms	Uncooked, fresh, unpeeled	½ in. slice
7.5	Graham cracker	Original Graham Crackers/Nabisco	½ in. sq.
10.0	Cheese	Yellow American pasteurized process-deli/Land O'Lakes	½ in. cube
15.0	Candy	Jujyfruits/Ferrara Candy Company	3 pieces
Technique:	Place sample in mouth, chew fully, and swallow or expectorate. Feel tooth surfaces with tongue.		
Definition:	The amount of product adhering to the teeth after mastication of the product.		
	[None --Large Amount]		

E Intensity Scale Values (0–100) for Skinfeel Texture Attributes

Scale Value	Product	Manufacturer
1. Integrity of Shape (Immediate)		
7	Baby oil	Johnson & Johnson
40	Keri Lotion, Original	Novartis
85	Vaseline Intensive Care	Unilever
92	Neutrogena Norwegian Hand Cream	Johnson & Johnson
2. Integrity of Shape (After 10 sec)		
3	Baby oil	Johnson & Johnson
30	Keri Lotion, Original	Novartis
80	Vaseline Intensive Care	Unilever
92	Neutrogena Norwegian Hand Cream	Johnson & Johnson
3. Gloss		
5	Gillette Foamy Reg. Shave Cream	Procter & Gamble
36	Noxzema	Unilever
70	Neutrogena Norwegian Hand Cream	Johnson & Johnson
78	Vaseline Intensive Care	Unilever
98	Baby oil	Johnson & Johnson
4. Firmness		
0	Baby oil	Johnson & Johnson
30	Olay Active Hydrating Beauty Fluid Lotion	Procter & Gamble
55	Ponds Cold Cream, Original	Unilever
84	Petrolatum	Generic
98	Lanolin HPA	Lansinoh
5. Stickiness		
1	Baby oil	Johnson & Johnson
26	Vaseline Intensive Care	Unilever
84	Petrolatum	Generic
99	Lanolin HPA	Lansinoh
6. Cohesiveness		
0	Baby oil	Johnson & Johnson
10	Vaseline Intensive Care	Unilever
60	Dove Cream Oil Intensive Body Lotion	Unilever
82	Petrolatum	Generic
7. Peaking		
0	Baby oil	Johnson & Johnson
23	Keri Lotion, Original	Novartis
36	Vaseline Intensive Care	Unilever
77	Zinc oxide	Generic
96	Petrolatum	Generic

E Intensity Scale Values (0–100) for Skinfeel Texture Attributes

Scale Value	Product	Manufacturer
8. Wetness		
0	Talc powder	Johnson & Johnson
22	Petrolatum	Generic
35	Baby oil	Johnson & Johnson
60	Vaseline Intensive Care	Unilever
70	Aloe Vera gel	Fruit of the Earth
100	Water	—
9. Spreadability		
29	Petrolatum	Generic
66	Vaseline Intensive Care	Unilever
97	Baby oil	Johnson & Johnson
10. Thickness		
5	Isopropyl alcohol	Generic
30	Vaseline Intensive Care	Unilever
65	Petrolatum	Generic
86	Neutrogena Norwegian Hand Cream	Johnson & Johnson
11. Amount of Residue		
0	Untreated skin	—
15	Vaseline Intensive Care	Unilever
65	Petrolatum	Generic
85	Baby Oil	Johnson & Johnson

F Intensity Scale Values (0–15) for Fabric-Feel Attributes

Scale Value	Fabric Type	Testfabrics ID#[a]
1. Stiffness		
1.3	Polyester/cotton 50/50 single knit tubular	7421
4.7	Mercerized cotton print cloth	400M
8.5	Mercerized combed cotton poplin	407
14.0	Cotton organdy	447
2. Force to Gather		
1.5	Polyester cotton 50/50 single knit tubular	7421
3.5	Cotton cloth greige	400R
7.0	Bleached cotton terry cloth	420BR
14.5	#10 Cotton duck greige	426
3. Force to Compress		
1.5	Polyester/cotton 50/50 single knit tubular	7421
3.4	Cotton cloth greige	400R

F Intensity Scale Values (0–15) for Fabric-Feel Attributes (Continued)

Scale Value	Fabric Type	Testfabrics ID#[a]
		(*Continued*)
8.0	Bleached cotton terry cloth	420BR
14.5	#10 Cotton duck greige	426
4. Depression Depth		
0.7	Cotton print cloth	400
1.8	Desized, bleached cotton duck	464
6.4	Texturized polyester interlock knit fabric	730
13.6	Bleached cotton terry cloth	420BR
5. Springiness		
0.7	Cotton print cloth	400
1.8	Desized, bleached cotton duck	464
6.2	Texturized polyester interlock knit fabric	730
10.5	Bleached cotton terry cloth	420BR
13.5	Texturized polyester double knit jersey	720
6. Fullness/Body		
1.6	Combed cotton batiste	435
4.0	Cotton sheeting	493
7.8	Cotton single knit	473
13.3	Cotton fleece	484
7. Tensile Stretch		
0.5	#8 Cotton duck greige	474
2.6	Spun viscose challis	266W
13.0	Texturized polyester double knit jersey	720
15.0	Texturized polyester interlock knit fabric	730
8. Compression Resilience: Intensity		
0.9	Polyester/cotton 50/50 single knit fabric	7421
3.8	Cotton cloth greige	400R
9.5	Acetate satin bright ward, delustered filling	105B
14.0	#10 Cotton duck greige	426
9. Compression Resilience: Rate		
1.0	Polyester/cotton 50/50 single knit tubular	7421
7.0	Filament nylon 6.6 semidull taffeta	306A
14.0	Dacron	738
10. Thickness		
1.3	Filament nylon 6.6 semidull taffeta	306A
3.3	Cotton print cloth	400
7.7	Cotton sheeting	493
13.0	#10 Cotton duck greige	426

F Intensity Scale Values (0–15) for Fabric-Feel Attributes (Continued)

Scale Value	Fabric Type	Testfabrics ID#[a]
11. Fabric-to-Fabric Friction		
1.7	Filament nylon 6.6 semidull taffeta	306A
5.0	Dacron	738
10.0	Acetate satin bright ward, delustered filling	105B
15.0	Cotton fleece	484
12. Fuzzy		
0.7	Dacron	738
3.6	Cotton crinkle gauze	472
7.0	Cotton T-shirt, tubular	437W
13.6	Cotton fleece	484
13. Hand Friction		
1.4	Filament nylon 6.6 semidull taffeta	306A
3.5	Bleached, mercerized combed broadcloth	419
7.2	Cotton print cloth	400
10.0	Cotton flannel	425
15.0	Bleached cotton terry cloth	420BR
14. Noise intensity		
1.6	Cotton flannel	425
2.7	Cotton crinkle gauze	472
6.3	Cotton organdy	447
14.5	Dacron 56 taffeta	738
15. Noise Pitch		
1.5	Cotton flannel	425
2.5	Cotton crinkle gauze	472
7.2	Cotton organdy	447
14.5	Dacron 56 taffeta	738
16. Gritty		
0.5	Polyester/cotton 50/50 single knit tubular	7421
6.0	Cotton cloth, greige	400R
10.0	Cotton print cloth	400
11.5	Cotton organdy	447
17. Grainy		
2.1	Mercerized combed cotton poplin	407
4.9	Carded cotton sateen bleached	428
9.5	Cotton tablecloth fabric	455-54
13.6	#8 Cotton duck greige	474

[a] Testfabrics identification numbers are the product numbers of Testfabrics Inc., P.O. Box 26, West Pittston, PA 18643, www.testfabrics.com

APPENDIX 12.3 STREAMLINED APPROACH TO SPECTRUM REFERENCES

Central to the Spectrum method is the use of intensity references. During training, Spectrum panelists are typically oriented to dozens of intensity references for flavor and texture. However, time or budget constraints often lead companies to seek ways to reduce the volume of references used, or panel leaders desire a smaller set of references for daily panel use. Also, many panels are more comfortable using references within product categories they typically evaluate. To address this, Sensory Spectrum has developed a streamlined approach to Spectrum references using a dozen foods commonly available in the United States. These 12 products can serve as the building blocks for flavor and texture intensity references and provide a panel leader with readily available reference products requiring minimal preparation. Since the Spectrum method does not depend on the availability of a single reference product, references that best meet the needs of the panel should be selected. The list below is intended to represent the average attribute intensity values for those products. Slight product variability and individual differences in panelist sensitivity (ex. bitterness) are expected. A panel leader might select 3–7 products to use regularly, providing panelists with the in-context references they tend to crave. In addition, Sensory Spectrum advocates the selection of panel specific internal control products for which complete profiles are developed then presented and reviewed at each panel session as a tool to standardize the panel's scores and minimize intensity drift. This approach is commonly used in Spectrum skinfeel panels.

1. Pepperidge Farm Bordeaux Cookies

AROMATICS	
Total impact	7.0
Grain complex	3.0
Toasted grain	3.0
Dairy complex	1.5
Butter/milk fat	1.5
Sweet aromatics	5.0
Vanilla/vanillin	1.0
Caramelized	4.5
Basic Tastes	
Sweet	12.5
Salt	4.0
TEXTURE	
Surface	
Microroughness	10.0
Macroroughness	3.0
Loose particles	5.0
Oily lips	2.0
Chewdown	
Hardness	8.0

Crispness	8.0
Denseness	7.0
Moisture absorption	10.0
Cohesiveness of mass	2.0
Roughness of mass	9.0 Gritty/coarse
Moistness of mass	10.0
Persistence of crisp	8.0
Dissolvability	7.0
Residual	
Toothpack	5.0
Loose particles	3.5

2. Sara Lee All Butter Pound Cake

Thawed in refrigerator, edges/crust removed, room temperature

AROMATICS	
Total impact	6.8
White wheat complex	2.5
Cooked white wheat	2.5
Eggy	1.5
Sweet aromatics	4.0
Caramelized	1.5
Vanilla	2.5
Dairy complex	2.0
Butterfat	2.0
Basic Tastes	
Sweet	10.5
Salt	3.5
TEXTURE	
Surface (crumb)	
Microroughness	4.5
Loose particles	5.0
Oily lips	2.5
Partial Compression	
Springiness	9.0
First Bite	
Hardness	3.5
Uniformity of bite	14.0
First Chew	
Denseness	7.0
Cohesiveness	5.5
Chewdown (10 chews)	
Moisture absorption	13.0

Cohesiveness of mass	8.5
Moistness of mass	10.0
Roughness of mass	4.0 grainy
Adhesiveness to palate	2.0
Residual	
Loose particles	3.5
Mouthcoating	5.0
Oily/greasy	2.5
Chalky	2.5
Toothstick	1.0

3. Skippy Creamy Peanut Butter

AROMATICS	
Total impact	8.0
Peanut complex	6.0
Roasted peanut	4.0
Woody/hulls/skins	2.0
Basic Tastes	
Sweet	7.5
Sour	1.0
Salt	9.0
Bitter	2.0
Chemical Feeling Factors	
Astringent	2.0
TEXTURE	
Semisolid First Compression	
Firmness	11.0
Cohesiveness	11.0
Denseness	15.0
Chewdown	
Grit between teeth	1.0
Adhesiveness to palate	12.0
Mixes with saliva	7.0
Cohesiveness of mass	7.0
Roughness of mass	1.0
Residual	
Film	10.0 Oily/particulate

4. Hellmann's Mayonnaise
Room temperature

AROMATICS	
Total impact	7.0
Eggy	5.5
Mustard	4.0
Vinegar	4.0
Lemon	2.0
Vegetable oil	2.0
Onion	1.0
Basic Tastes	
Sweet	4.0
Sour	4.5
Salt	12.0
Chemical Feeling Factors	
Burn	1.5
Astringent	3.5
TEXTURE	
Surface	
Slipperiness	12.5
Firmness	3.5
Denseness	9.0
Cohesiveness	7.0
Mixes with saliva	10.5
Cohesiveness of mass	6.0
Adhesiveness of mass	7.0
Residual	
Oily film	5.0

5. Land O'Lakes American Cheese (from Deli or Local Grocer)
½″ cubes; Note: If sourced from other merchandiser, please contact us for alterations to various texture scales.

AROMATICS	
Total impact	7.0
Dairy complex	6.0
Cooked milky	3.0
Butter fat	1.0
Soured/cheesy/butyric	3.0
Nutty	2.0
Basic Tastes	
Sweet	3.5

Sour	4.5
Salt	14.0
Chemical Feeling Factors	
Astringent	2.5
TEXTURE	
Surface	
Roughness	0.0
Moistness	2.5
Partial Compression	
Springiness	0.8
First Bite/Chew	
Hardness	4.5
Denseness	15.0
Cohesiveness	4.5
Chewdown	
Mixes with saliva	5.5
Cohesiveness of mass	10.0
Moistness of mass	9.0
Adhesiveness to palate	4.0
Roughness of mass	3.0 Lumpy
Toothstick	9.0
Residual	
Toothpack	2.0
Mouthcoating	2.5
Dairy film	2.0

6. Lay's Classic Potato Chips

AROMATICS	
Total impact	6.9
Potato complex	4.8
Cooked potato	3.1
Browned/toasted/caram. Potato	1.5
Earthy/skins	*0, but 1.0 if present on individual chips
Heated oil	2.2
Basic Tastes	
Salt	13.5
Sweet	3.8
Sour	1.1
Bitter	0.8
TEXTURE	
Surface	

Microroughness	7.0
Macroroughness	2.5
Oily lips	6.3
Manual oiliness	5.7
Manual particles	5.2
First Chew	
Hardness	5.4
Denseness	5.8
Fracturability	5.7
Chewdown	
Persistence of crisp/crunch	10.0
Moisture absorption	10.1
Moistness of mass	8.4
Cohesiveness of mass	6.8
Roughness of mass	6.0
Grainy	5.0
Lumpy	1.0
Toothstick	2.3
Residual	
Toothpack/toothstick	4.8
Oily film/residue	3.0
Loose particles	2.4

7. Minute Maid Orange Juice: Frozen Concentrate Reconstituted

AROMATICS	
Total impact	7.5
Orange complex	7.5
Raw	1.0
Cooked	5.0
Expressed orange oil	2.0
Other citrus	1.0
Basic Tastes	
Sweet	8.0
Sour	3.8
Bitter	1.5
Chemical Feeling Factors	
Astringent	2.5
Burn	2.0
TEXTURE	
Viscosity	2.0
Particulates	1.5
Mixes with saliva	12.0

8. Oscar Meyer Beef Hot Dogs

AROMATICS	
Total impact	8.0
Cured meat complex	4.0
Beef/pork	4.0
Smoke	2.5
Spice complex	2.0
Brown spice complex	2.0
Garlic	2.5
Basic Tastes	
Salt	16.0
Sweet	4.0
Sour	2.0
TEXTURE	
Surface (skin)	
Moisture	3.5
Roughness	2.0
Oiliness	3.0
First Chew (with molars)	
Springiness	6.0
First Bite/Chew	
Firmness	7.0
Cohesiveness	6.0
Denseness	9.5
Juiciness	8.0
Chewdown (10 chews)	
Cohesiveness of mass	5.5
Moistness of mass	9.0
Skin awareness	5.0
Roughness of mass	6.0
Residual	
Oily/greasy	4.0
Loose particles	3.0

9. DeCecco Spaghetti (12 min cook)

AROMATICS	
Total impact	4.5
White wheat complex	4.5
Raw white wheat	0.5
Cooked white wheat	4.0

Basic Tastes	
Sweet	1.5
Salt	1.0
TEXTURE	
Surface	
Wetness	1.0
Microroughness	1.0
Stickiness	7.0
First Chew	
Hardness	6.0
Denseness	15.0
Cohesiveness	7.0
Toothpull	3.0
Chewdown	
Mixes with saliva	3.0
Cohesiveness of mass	2.0
Geometrical of mass	Beady/grainy
Residual	
Loose particles	2.0
Chalky film	2.0
Toothstick	2.0

10. Heinz Tomato Ketchup

AROMATICS	
Total impact	9.0
Tomato complex	5.0
Cooked tomato	5.0
Vinegar	3.8
Green herb complex	3.0
Celery	3.0
Brown spice complex	5.5
Clove	5.0
Black pepper	3.0
Sweet aromatics	3.8
Caramelized	3.8
Cooked onion	2.0

Basic Tastes	
Sweet	9.5
Sour	5.8
Salt	11.5
Bitter	1.5
Chemical Feeling Factors	
Astringent	5.0
Burn	2.0
TEXTURE	
Surface	
Slipperiness	8.0
First Manipulation	
Firmness	2.0
Cohesiveness	4.5
Manipulation (5)	
Mixes with saliva	14.0
Cohesiveness of mass	2.5
Adhesiveness of mass	2.5

11. Häagen-Dazs Vanilla Ice Cream

AROMATICS	
Total impact	8.0
Dairy complex	4.0
Cooked	2.0
Butter fat	2.0
Eggy (cooked)	2.0
Vanilla impression	4.8
Vanillin	1.0
Bourbon/alcohol	2.0
Dried fruit	2.5
Basic Tastes	
Salt	2.0
Sweet	12.0
Sour	1.5
Bitter	0.0

TEXTURE	
First Compression	
Semisolid firmness	8.5
Semisolid denseness	14.0
Slipperiness	13.0
Manipulation (5)	
Mixes with saliva	13.0
Viscosity of liquid	2.0
Manipulations to melt	5.0
Residual	
Fatty/oily film	2.0
Dairy film	2.0

12. Yoplait Original Strawberry Yogurt

AROMATICS	
Total impact	7.5
Dairy complex	3.5
Cultured dairy	2.5
Cooked dairy	1.0
Strawberry complex	5.0
Cooked strawberry	2.0
Ethyl maltol	3.0
Basic Tastes	
Sweet	11.0
Sour	3.0
Chemical Feeling Factors	
Astringent	4.0
TEXTURE	
Compression	
Semisolid firmness	4.0
Semisolid cohesiveness	6.0
Semisolid denseness	15.0
Manipulation (5)	
Mixes with saliva	8.5
Particulate	4.0
Chalky	3.0
Lumpy	1.0
Residual	
Fatty/oily film	1.0
Chalky film	3.0

APPENDIX 12.4 SPECTRUM DESCRIPTIVE ANALYSIS: PRODUCT LEXICONS

A. White Bread Flavor

1. Aromatics
   - Grain Complex
     - Raw white wheat (dough)
     - Cooked white wheat
     - Toasted
     - Cornstarch
     - Whole grain
   - Yeasty/fermented
   - Dairy complex
     - Milk, cooked milk
     - Buttery, brown butter
   - Eggy
   - Sweet Aromatic Complex:
     - Caramelized/honey/malty/fruity
   - Mineral: inorganic, stones, cement, metallic
   - Baking soda
   - Vegetable oil

Other Aromatics: Mushroom, carrot, earthy, fermented, acetic, plastic, cardboard, chemical leavening

2. Basic Tastes
   - Salty
   - Sweet
   - Sour
   - Bitter

3. Chemical Feeling Factors
   - Metallic
   - Astringent/drying
   - Phosphate
   - Baking soda feel

B. White Bread Texture

1. Surface
   - Crumb texture
     - Roughness
     - Loose particles
     - Moistness
   - Crust Texture
     - Roughness
     - Loose particles
     - Moistness

2. First Chew
   - Crumb denseness
   - Crumb cohesiveness
   - Crumb firmness
   - Crust hardness
   - Crust denseness
   - Crust cohesiveness

3. Partial Compression
   - Crumb springiness

4. Chew down
   - Moisture absorption
   - Moistness of mass
   - Adhesive to palate
   - Cohesiveness of mass
   - Lumpy
   - Grainy

5. Residual
   - Loose particles
   - Toothstick
   - Toothpack
   - Tacky film

C. Toothpaste Flavor

1. Before Expectoration Aromatics
   - Mint complex
     - Peppermint/menthol
     - Spearmint
     - Wintergreen
   - Base/chalky
   - Bicarbonate

   - Anise
   - Fruity
   - Brown Spice
   - Citrus
   - Soapy

2. After Rinsing Aromatics
   - Minty
   - Fruity
   - Brown spice
   - Anise

3. Basic Tastes
   - Sweet
   - Bitter
   - Salty

4. Chemical Feeling Factors
   - Burn
   - Bicarbonate feel
   - Cool
   - Astringent
   - Metallic

D. Toothpaste Texture

1. Brush on front teeth 10 times
   - Firmness
   - Sticky
   - Number of brushes to foam
   - Ease to disperse
   - Denseness of foam
2. Expectorate
   - Chalky
   - Gritty
   - Slickness of teeth

3. 20 Brushes (back teeth)
   - Grittiness between teeth
   - Amount of foam
   - Slipperiness of foam

4. Rinse
   - Slickness of teeth

E. Potato Chip Flavor

1. Aromatics
   - Potato complex
     - Raw potato/green
     - Cooked potato
     - Browned
     - Dehydrated
   - Earthy/potato skins

2. Basic Tastes
   - Salty
   - Sweet
   - Sour
   - Bitter

Sweet potato

Oil complex
- Heated vegetable oil
- Overheated/abused oil

Sweet caramelized
Cardboard
Painty
Spice

3. Chemical Feeling Factors
   - Tongue burn
   - Astringent

F. Potato Chip Texture

1. Surface
   - Oiliness
   - Roughness, macro
   - Roughness, micro
   - Loose Crumbs

2. First Bite/First Chew
   - Hardness
   - Crispness
   - Denseness

Particles after 4–5 chews

3. Chewdown
   - Moisture absorption
   - # Chews to bolus
   - Persistence of crisp
   - Abrasiveness of mass
   - Moistness of mass
   - Cohesiveness of mass

4. Residual Toothpack
   - Chalky mouthcoat
   - Oily film

G. Mayonnaise Flavor

1. Aromatics
   - Vinegar (type)
   - Cooked egg/eggy
   - Dairy milky/cheesy/butter
   - Mustard (type)
   - Onion/garlic
   - Lemon/citrus
   - Pepper (black/white)
   - Lemon juice
   - Fruity (grape/apple)
   - Brown spice (clove)
   - Paprika
   - Vegetable oil (aromatic)
   - Other aromatics: Cardboard (stale oil), starch, paper, nutty/woody, sulfur, painty (rancid oil), caramelized, fishy

2. Basic Tastes
   - Salty
   - Sweet
   - Sour
   - Bitter

3. Chemical Feeling Factors
   - Astringent
   - Tongue burn/heat
   - Prickly/pungent

H. Mayonnaise Texture

1. Surface Compression
   Slipperiness

2. First Compression
   Firmness
   Cohesiveness
   Stickiness to palate

3. Manipulation
   Cohesiveness of mass
   Lumpy mass
   Adhesive mass
   Rate of breakdown

4. Residual
   Oily film
   Sticky/tacky film
   Chalky film

I. Corn Chip Flavor

1. Aromatics
   Corn Complex
      Raw corn
      Cooked corn
      Toasted/browned corn
      Masa/fermented
   Caramelized
   Oil Complex
      Heated oil
      Heated corn oil
      Hydrogenated
   Other grain (type)
   Burnt
   Earthy/green husks

2. Basic Tastes
   Salty
   Sweet
   Sour
   Bitter

3. Chemical Feeling Factors
   Astringent
   Burn

J. Corn Chip Texture

1. Surface
   Roughness, macro
   Roughness, micro
   Manual oiliness
   Oiliness on lips
   Loose particles

2. First bite/first chew
   Hardness
   Crispness/crunchiness
   Denseness
   Amount of particles

3. Chewdown
   Moisture absorption
   # Chews to bolus
   Moistness of mass
   Persistence of crunch/crisp
   Cohesiveness of mass
   Graininess of mass

4. Residual
   Toothpack
   Grainy particles
   Chalky mouthfeel
   Oily/greasy mouthfeel

K. Cheese Flavor

1. Aromatics
   - Dairy complex
     - Milky
     - Cooked milk/caramelized
     - Dairy fat
     - Non-Fat Dry Milk (NFDM)
     - Cheesy
     - Cultured dairy complex
     - Sour cream/buttermilk
     - Cottage cheese
     - Yogurt
   - Cheese acids complex
     - Butyric
     - Propionic
     - Isovaleric
     - Caproic/caprylic
   - Smoky
   - Nutty/woody
   - Fruity
     - Tropical
   - Herb complex
   - Musty/moldy
   - Degraded protein/casein/animal
   - Para-cresol/barnyard
   - Plastic/vinyl
   - Cardboard
2. Basic Tastes
   - Sweet
   - sour
   - Salty
   - Bitter
3. Chemical Feeling Factors
   - Astringent
   - Bite/sharp
   - Burn/heat
   - Nasal pungency

L. Cheese Texture

1. Surface
   - Rough macro-bumpy
   - Rough micro-grainy/gritty or chalky
   - Wetness
   - Oily/fatty
   - Loose particles
2. First Bite/First Chew
   - Firmness
   - Hardness
   - Denseness
   - Cohesiveness
   - Toothstick
   - Number of pieces
4. Chewdown
   - Mixes with saliva
   - Rate of melt
   - Cohesiveness of mass
   - Moistness of mass
   - Adhesiveness of mass
   - Lumpiness of mass
   - Grainy mass
   - Toothstick
5. Residual
   - Toothstick
   - Mouthcoat
   - Oily film

3. Partial Compression
   - Springiness
   - Particles left

Chalky film
Tacky
Dairy film
Sticky film

M. Caramel/Confections Flavor

1. Aromatics
   - Caramelized sugar
   - Dairy complex
     - Baked butter
     - Cooked milk
   - Sweet Aromatics
     - Vanilla
     - Vanillin
   - Diacetyl
   - Scorched
   - Yeasty (dough)

Other aromatics: cellophane, phenol, cardboard, painty

2. Basic Tastes
   - Sweet
   - Sour
   - Salty

3. Chemical Feeling Factors

Tongue burn

N. Caramel Texture

1. Surface
   - Lipstick
   - Moistness
   - Roughness

2. First Bite/First Chew
   - Hardness
   - Denseness
   - Cohesiveness
   - Toothstick

3. Chewdown
   - # of chews to bolus
   - Mixes with saliva
   - Cohesiveness of mass
   - Moistness of mass
   - Roughness of mass
   - Toothpull
   - Adhesiveness to palate
   - # of chews to Swallow

4. Residual
   - Oily/greasy film
   - Tacky film
   - Toothstick

O. Chocolate Chip Cookie Flavor

1. Aromatics
   - White wheat complex
     - Raw white wheat
     - Cooked white wheat
     - Toasted/browned
   - Chocolate/cocoa complex
     - Chocolate
     - Cocoa
   - Dairy complex
     - NFDM
     - Baked butter
     - Cooked milk
   - Sweet aromatics complex
     - Brown sugar/molasses
     - Vanilla, vanillin
     - Caramelized
     - Coconut
   - Nutty
   - Fruity
   - Baked egg
   - Shortening (heated oil, hydrogenated vegetable fat)
   - Baking soda
   - Cardboard

2. Basic Tastes
   - Sweet
   - Salty
   - Bitter

3. Chemical Feeling Factors
   - Burn

P. Chocolate Chip Cookie Texture

1. Surface
   - Roughness, micro
   - Roughness, macro
   - Loose crumbs/particles
   - Oiliness
   - Surface moisture

2. First Bite/First Chew
   - Firmness/hardness
   - Crispness
   - Denseness
   - Cohesiveness
   - Crumbly

3. Chewdown
   - # chews to bolus
   - Moisture absorption
   - Cohesiveness of mass
   - Moistness of mass
   - Awareness of chips
   - Roughness of mass
   - Persistence of crisp

4. Residual
   - Toothpack
   - Toothstick
   - Oily/greasy film
   - Grainy particles
   - Loose particles
   - Mouthcoating

Q. Spaghetti Sauce Flavor

1. Aromatics
   - Tomato complex
     - Raw
     - Cooked
   - Tomato character
     - Seedy/skin
     - Fruity
     - Fermented/soured
     - Viney
     - Skunky
   - Caramelized
   - Vegetable complex
     - Bell pepper, mushroom, other
   - Onion/garlic
   - Green herb complex
     - Oregano, basil, thyme
   - Black pepper
   - Cheese/Italian
   - Other
     - Fish, meat, metallic
2. Basic Tastes
   - Salty
   - Sweet
   - Sour
   - Bitter
3. Chemical Feeling Factors
   - Astringent
   - Heat
   - Bite

R. Spaghetti Sauce Texture

1. Surface
   - Wetness
   - Oiliness
   - Particulate
2. First Compression
   - Viscosity/Thickness
   - Cohesiveness
   - Pulpy matrix/base
     - Amount
     - Size
   - Amount of large particles
   - Amount of small particles
3. Manipulation
   - Amount of particles/chunks
     - Largest size
     - Smallest size
   - Chew particles
     - Hardness
     - Crispness
     - Fibrousness (vegetables and herbs)

Manipulate 5 times
   - Mixes with saliva
   - Amount of particles

4. Residual
   - Oily mouthcoat
   - Loose particles

S. Facial Wipes Handfeel Texture

1. Surface
   - Amount of Product
   - Gritty
   - Grainy
   - Lumpy
   - Fuzzy
   - Slipperiness
   - Thickness
2. Manipulation
   - Force to Gather
   - Stiffness
   - Fullness/Body

T. Facial Wipes Skinfeel Appearance and Texture

1. In Use
   - Amount of lather (visual)
   - Bubble size (visual)
   - Bubble size variation (visual)
   - Thickness of lather
2. Before Drying
   - Rinsability
   - Stickiness
   - Slipperiness
   - Amount of residue
   - Type of residue
3. After Feel
   - Cool
   - Gloss (visual)
   - Facial lines/creases (visual)
   - Stickiness
   - Slipperiness
   - Amount of residue
   - Type of residue
   - Skin roughness
   - Moistness
   - Tautness

U. Mascara Evaluation

1. Baseline and wear
   - Lash visibility
   - Color intensity base/tips
   - Length
   - Thickness
   - Density
   - Degree of lash curl
   - Gloss
   - Tangling
   - Separation
2. Application
   - Ease of application (strokes)
3. Wear (multiple time points)
   - Lash wetness
   - Top/bottom lash stickiness
   - Transfer
   - Color intensity base
   - Color intensity tips
   - Length
   - Thickness
   - Density
   - Degree of lash curl
   - Gloss
   - Tangling
   - Separation
   - Clumping
   - Spiking
   - Fibers
   - Beading
   - Flaking
   - Smudging

V. Fragrance

Floral
White flower
Rose
Muguet
Violet
Floral/other
Citrus
Aldehydic
Fruity
Stone Fruit
Berry
Melon
Tropical
Fruit other
Fougere
Pine
Spice
Black
Brown
Sweet
Amber
Caramelized
Vanillin
Powdery
Camphor
Herbaceous
Woody
Resinous
Green
Moss/chypre
Ozonic/marine
Animal

APPENDIX 12.5 SPECTRUM DESCRIPTIVE ANALYSIS: EXAMPLES OF FULL PRODUCT DESCRIPTIONS

A. White Bread

	Standard	Premium
1. Appearance	Golden brown	Golden brown
Color of crust	10	12
Evenness color of crust	12	12
Color of crumb	Yellow	Yellow
Chroma of crumb	10	9
Cell size	7	11
Cell uniformity	12	8
Uniformity of shape	12	9
Thickness	10	7
Distinctiveness of cap	2	7
2. Flavor		
2.1 Aromatics		
Grain complex		
Raw	5.5	7
Cooked	2	0
Browned	1	2.5
Bran	0	0
Dairy/buttery	0	3.5
Soured (milky, cheese, grain)	2.5	0
Caramelized	0	3
Yeasty/fermented	2	4
Plastic	1	0
Chemical leavening	4	0
Baking soda	0	0
2.2 Basic Tastes		
Sweet	2.5	5
Salty	8	7
Sour	3	2
Bitter	1.5	0
2.3 Chemical Feeling Factors		
Metallic	1.5	0
Astringent	3	1.5
Baking soda feel	0	0
3. Texture		
3.1 Surface		
Roughness of crumb	6	5
Initial moistness	6.5	9

(Continued)

A. White Bread (Continued)

	Standard	Premium
3.2 First Chew		
Crust firmness	5	3.5
Crust cohesiveness	7	2
Firmness of crumb	3	3.5
Denseness of crumb	3	8
Cohesiveness of crumb	10	6.5
Uniformity of chew	6.5	12
3.3 Chewdown (10 chews)		
Moisture absorption	12	14
Cohesiveness of mass	10	11
Moistness of mass	8	12
Roughness of mass	6	4
Lumpy	5	1.5
Grainy	1	3
Adhesiveness to palate	6	4
Stickiness to teeth	4	2
3.4 Residual		
Loose particles	3	1
Tacky film	2	0

B. Toothpaste

	Standard Mint Paste	Mint Gel
1. Appearance		
Extruded	5	6
Cohesive	9	20
Shape	9	8
Gloss	6.5	15
Particulate	0	0
Opacity	15	2
Color intensity	3.5	9
Chroma	10	12
2. Flavor		
2.1 First Foam		
Mint complex	11	6
Peppermint/menthol	0	6
Spearmint	0	0
Wintergreen	11	0
Brown spice complex	3.5	0

(*Continued*)

B. Toothpaste (Continued)

	Standard Mint Paste	Mint Gel
Cinnamon	1	0
Clove	2	0
Anise	0	3.5
Floral	0	2
Base/chalky	3.5	3
Soapy	1.5	2.5
Sweet	9	9
Salty	2	0
Bitter	3	5
Sour	0	0
2.2 Expectorate		
Minty	7	1.5
Brown spice	1	0
Floral	0	2
Burn	2	4
Cool	9	14
Astringency	4	7
Base	1.5	3
2.3 Rinse		
Brown spice	1.5	0
Fruity	0	0
Minty	3.5	1.5
Base	1.5	2
Salty	0	0
Sweet	4	4
Burn	1.5	2.5
Cool	8	11
Bitter	1.5	4
Soapy	0	1
2.4 Five Minutes		
Fruity	0	0
Minty	3	1
Soapy	1.5	1
Cool	7	6
Bitter	2	5
Brown spice	0	0
Anise	0	3
3. Texture		
3.1 Brush on front teeth 10×		
Firmness	4.5	6

B. Toothpaste (Continued)

	Standard Mint Paste	Mint Gel
Sticky	8	9
3.2 First Foam		
Amount of foam	8	7
Slipperiness of foam	7	4
Denseness of foam	11	9.5
3.3 Expectorate		
Chalky	4.5	7
Slickness of teeth	5	3.5

C. Peanut Butter

	Local Brand	National Brand
1. Appearance		
Color intensity	7.0	7.5
Chroma	5.4	6.0
Gloss	5.2	5.1
Visible particles	2.5	2.0
2. Flavor		
2.1 Aromatics		
Roasted peanut	3.0	6.1
Raw/beany	2.3	1.3
Over roasted	0.6	3.0
Sweet aromatic	3.1	4.5
Woody/hull/skins	4.4	1.6
Fermented fruit	0	0
Phenol	0	0
Cardboard	0.4	0
Burnt	0	0
Musty	0.3	0
Green	0.1	0
Painty	0.1	0
Soy	1.0	0
2.2 Basic Tastes		
Salt	11.9	9.1
Sweet	9.2	7.4
Sour	1.9	1.1
Bitter	3.1	1.6
2.3 Chemical Feeling Factors		
Astringent	2.5	2.0

(Continued)

C. Peanut Butter (Continued)

	Local Brand	National Brand
3. Texture		
3.1 Surface		
Surface roughness	2.5	1.3
3.2 First Compression		
Firmness	7.0	5.7
Cohesiveness	6.9	7.0
Denseness	15	15
Adhesive	11.4	9.8
3.3 Manipulation		
Mixes with saliva	8.4	9.9
Adhesiveness of mass	4.9	2.6
Cohesiveness of mass	5.4	4.1
Roughness of mass	1.8	1.0
4. Residual		
Loose particles	0.1	0
Oily film	1.6	1.5
Chalky film	1.7	1.1

D. Mayonnaise

	National Brand Mayonnaise	National Brand Dressing
1. Appearance		
Color	Cream/yellow	White
Color intensity	2	1
Chroma	12	10
Shine	10	12.5
Lumpiness	9	4
Bubbles	5	2
2. Flavor		
2.1 Aromatics		
Eggy	6.8	1.5
Mustard	4.5	3.5
Vinegar	4.5	9
Lemon	3.5	1
Oil	1.5	0
Starchy	0	1.5
Onion	1.5	0
Clove	0	4.8
2.2 Basic Tastes		
Salty	8	7

D. Mayonnaise (Continued)

	National Brand Mayonnaise	National Brand Dressing
Sour	3	8
Sweet	3	8
2.3 Chemical Feeling Factors		
Burn	2	3
Pungent	2	3
Astringent	3.5	6
3. Texture		
3.1 Surface		
Adhesiveness to lips	6	10
3.2 First Compression		
Firmness	8.5	9
Denseness	11	12.5
Cohesiveness	6	10
3.3 Manipulation		
Cohesiveness of mass	7	8.5
Adhesiveness of mass	7	5
Mixes with saliva	11.5	8
3.4 Residual		
Oily film	4	1.5
Tackiness	0	0
Chalkiness	0	1

E. Marinara Sauce

	Shelf-Stable (Jar)	Fresh–Refrigerated
1. Appearance		
Color	Red/orange	Red/orange
Color intensity	11	13
Chroma	12	8
Shine	7.5	7.5
Total particles		
Micro particles	10	8
Macro particles	5	12
2. Flavor		
2.1 Aromatics		
Tomato complex	8	7
Raw	1.5	5
Cooked	6.8	3

(Continued)

E. Marinara Sauce (Continued)

	Shelf-Stable (Jar)	Fresh–Refrigerated
Tomato character (Seedy/ skin, red fruity, viney, skunky)	8	7
Fermented/soured	0	0
Caramelized	4	2
Vegetable complex		
Bell pepper, mushroom, other	2	4
Onion/garlic	5	6.5
Green herbs complex		
Oregano, basil, thyme	5	7.8
Black pepper	1.5	4
Cheese/Italian	3.5	1
2.2 Basic Tastes		
Sweet	7	5.5
Sour	2.5	2
Salty	9	7
2.3 Chemical Feeling Factors		
Astringent	4	4.5
Heat	1.5	4
3. Texture		
3.1 First Compression		
Cohesiveness	3	1
Pulpy matrix/base	5.5	9.5
3.2 Manipulation		
Amount of particles/chunks	4	10
Largest size	3	8
Smallest size	1	2.5
3.3 Chew Particles		
Hardness	3	5.5
Crispness	2	6
Fibrousness (vegetables & herbs)	4	5
3.4 Manipulate Five Times		
Mixes with saliva	11	12
4. Residual		
Oily mouthcoat	2	4
Loose particles	1	4

APPENDIX 12.6 SPECTRUM DESCRIPTIVE ANALYSIS TRAINING EXERCISES

A. Basic Taste Combinations Exercise

1. Scope

This exercise serves as a basic panel calibration tool. A product's flavor often includes a combination of two or three taste modalities, and the blends of salt, sweet, and sour provide the panel with an opportunity to develop the skill of rating taste intensities without the distraction of aromatics.

2. Test Design

Trainees begin by familiarizing themselves with the reference set, consisting of six cups with single component solutions. The cups carry labels such as Sweet 5, Salt 10, and so on, where 5 = weak, 10 = medium, and 15 = very strong. The reference set remains available for the duration of the exercise.

The evaluation set consists of equal proportion blends of two or three of the reference solutions. The panel leader can prepare some or all of the blends in the evaluation set. The panel leader hands out one blend at a time, and the trainees record their impressions using the score sheet below.

At the end of the exercise, the sheet marked "average results" is made available. The panel leader should expect the panel means to fall within one point of these averages.

3. Materials

Assume 15 participants and 10 mL serving size: Prepare 1 L of each reference solution, which requires 150 g white sugar, 8.5 g salt, and 3 g citric acid. Serving items needed are

300 plain plastic serving cups, 2 oz size
15 individual serving trays
15 large opaque cups with lid (spit cups), for example, 16 oz size
15 water rinse cups, 6 oz size
6 water serving pitchers
1 packet napkins
60 tasting spoons (white plastic) if anyone requires those

4. Reference Set

Label	Content
Salt—5	0.35% NaCl
Salt—10	0.55% NaCl
Sweet—5	5% Sucrose
Sweet—10	10% Sucrose
Sour—5	0.1% Citric acid
Sour—15	0.2% Citric acid

Prepare solutions using water free of off flavors. Solutions may be prepared 24–36 h prior to use. Refrigerate prepared samples. On day of evaluation, allow to warm to 70 °F and serve 10 mL per participant.

5. Evaluation Set

Contents	Code
5% Sucrose/0.1% Citric acid	232
5% Sucrose/0.2% Citric acid	715
10% Sucrose/0.1% Citric acid	115
5% Sucrose/0.35% NaCl	874
5% Sucrose/0.55% NaCl	903
10% Sucrose/0.35% NaCl	266
0.1% Citric acid/0.35% NaCl	379
0.2% Citric acid/0.35% NaCl	438
0.1% Citric acid/0.55% NaCl	541
5% Sucrose/0.1% Citric acid/0.35% NaCl	627
10% Sucrose/0.2% Citric acid/0.55% NaCl	043
10% Sucrose/0.1% Citric acid/0.35% NaCl	210
5% Sucrose/0.2% Citric acid/0.35% NaCl	614
5% Sucrose/0.1% Citric acid/0.55% NaCl	337

Basic Taste Combinations Exercise: Composition of Evaluation Set

Code	% Sucrose	% Citric Acid	% NaCl
232	5	0.10	
715	5	0.20	
115	10	0.10	
874	5		0.35
903	5		0.55
266	10		0.35
379		0.10	0.35
438		0.20	0.35
541		0.10	0.55
627	5	0.10	0.35
043	10	0.20	0.55
210	10	0.10	0.35
614	5	0.20	0.35
337	5	0.10	0.55

Basic Taste Combinations Exercise: Average Results

Sample	Sweet	Sour	Salty
232	6	7	
715	4	8.5	
115	9.5	4	
874	6		6
903	7		9
266	11		7
379		9	9
438		10	6.5
541		6	11
627	5	3.5	5
043	8	8	9
210	9	4	6
614	3	9	8
337	4	6	11

B. Cookie Variation Exercise

1. Scope

This exercise teaches the Spectrum lexicon (list of terms) for baked cookies by exposing the trainees to a set of samples of increasing complexity, adding one ingredient at a time. Many products that are combinations of ingredients can be handled in this manner, by constructing the flavor complex one or two terms at a time.

2. Test Design

Trainees begin by evaluating cookie 1, baked from flour and water. They are asked to suggest terms to describe this sample. Together, the panel leader and the trainees discuss the terms, for example, cooked wheat/pasta-like/cream of wheat/breadcrumb, and doughy/raw/raw wheat/raw flour. Then, they select a single descriptor to represent each set of linked terms, for example, cooked wheat and raw wheat. Trainees record the results on the score sheet marked "vocabulary construction."

The panel leader hands out cookie 2, baked from flour, water, and butter, and trainees suggest terms for the added aromatics. Again, the group selects a single descriptor to cover each sequence of linked (overlapping) terms.

Once the lexicon is developed, it can be validated by comparing any two of the reference samples and determining whether the lexicon works to discriminate and describe the samples appropriately.

The score sheet marked "possible full vocabulary" can then be used to describe any pair of the samples, using a scale of 0 = absent, 5 = weak, 10 = medium, and 15 = very strong for the intensity of each attribute.

3. Reference Set

1. Flour, water
2. Flour, water, butter
3. Flour, water, margarine
4. Flour, water, shortening
5. Flour, water, shortening, salt
6. Flour, water, shortening, baking soda
7. Flour, water, sugar
8. Flour, water, brown sugar
9. Flour, water, butter, sugar
10. Flour, water, margarine, sugar
11. Flour, water, shortening, sugar
12. Flour, water, sugar, egg, margarine
13. Flour, water, sugar, egg, margarine, vanilla extract
14. Flour, water, sugar, egg, margarine, almond extract

4. Cookie Recipes

Prepare each recipe as shown in the table using the following information:

1. All recipes will serve up to 20–25 participants. Please see shopping list for proper ingredients to purchase. Prepare each recipe as you would any standard cookie dough (i.e., cream fat (if any) first; add sugar, flavorings, and egg (if any) and then add dry ingredients and liquid alternately. Mix just until all ingredients are combined. Use a standard mixer or food processor for mixing dough.
2. Spread evenly into a 9 × 13 in. rectangular baking pan that has been lined with coded parchment paper. Be sure to mark the appropriate cookie number on the parchment paper.
3. Note that some dough will be very thick. Use a roller to spread dough evenly in pan. Rollers can be wallpaper rollers from the hardware store or small pastry rollers from a cooking specialty shop or catalogue. Plastic wrap may be placed on top of the cookie dough while rolling to prevent sticking. Remove plastic before cutting and baking.
4. Precut cookies into 32 squares before baking (8 pieces × 4 pieces).
5. Bake in a preheated oven at 350°F for 35–40 min until slightly browned. Oven temperatures and time may vary for your oven.
6. All cookies should be very similar in color when baked—except for #8 (darker).
7. Cut cooled baked cookies on a cutting board. Remove dried edges and cut each batch into approximately 20–32 usable cookies. Use the best samples.
8. Cookies may be prepared up to 36 h in advance, wrapped in waxed paper or parchment paper (completely) and then wrapped in aluminum foil and stored at room temperature. Do not allow plastic wrap or foil to contact cookies as flavor transfer may occur. Label the foil.
9. Cookies may also be frozen for up to 4 weeks prior to the exercise, wrapped as indicated above. Bring cookies out of freezer to thaw 12–24 h before the exercise.
10. Serve one cookie per participant.

Serving suggestions:

Obtain one 9 × 11 inch Styrofoam tray (Standard size 12S) for each participant. Mark the tray into 15 sections (3 × 5) with a wax or grease pencil. Mark sections 1 thru 14. Place corresponding cookie in numbered section.
Alternately, you may label cupcake liners and place cookies into the liners. Serve liners on a tray.

1	2	3	4
2 1/2 cups flour 1 cup water	2 1/2 cups flour 1/4 cup water 1/2 cup +2 tablespoons butter	2 1/2 cups flour 1/4 cup water 1/2 cup +2 tablespoons margarine	2 1/2 cups flour 1/4 cup water 1/2 cup +2 tablespoons shortening
5	**6**	**7**	**8**
2 1/2 cups flour 1/4 cup water 1/2 cup + 2 tablespoons shortening 1 teaspoon salt	2 1/2 cups flour 1/4 cup water 1/2 cup + 2 tablespoons shortening 1/3 teaspoon baking soda	2 1/2 cups flour 3/4 cup water 1 cup white granulated sugar	2 1/2 cups flour 3/4 cup water 1 cup brown sugar
9	**10**	**11**	**12**
2 1/2 cups flour 1/4 cup water 1/2 cup + 2 tablespoons butter 1 cup white granulated sugar	2 1/2 cups flour 1/4 cup water 1/2 cup + 2 tablespoons margarine 1 cup white granulated sugar	2 1/2 cups flour 1/4 cup water 1/2 cup + 2 tablespoons shortening 1 cup white granulated sugar	2 1/2 cups flour 1/4 cup water 1 cup white granulated sugar 1/2 cup +2 tablespoons margarine
13	**14**		
2 1/2 cups flour 1/4 cup water 1 cup white granulated sugar 1 egg 1/2 cup +2 tablespoons margarine 1 teaspoon pure vanilla extract	2 1/2 cups flour 1/4 cup water 1 cup granulated white sugar 1 egg 1/2 cup +2 tablespoons margarine 1 teaspoon almond extract		

5. Materials at Each Participant's Station

Opaque cup with lid (spit cup)
Translucent water rinse cup

Rinse water
Napkin
Cupcake paper liners coded: 1–14
or marked Styrofoam tray
Rinse water serving pitchers

6. Groceries and Paper Products

Purchase the total amount to serve the appropriate amount of each sample to each participant.

All-purpose flour
Butter, unsalted
Margarine, store brand
Shortening
White granulated sugar
Dark brown sugar
Eggs
Baking soda
Salt
Pure vanilla extract
Pure almond extract

Styrofoam trays (12S) and wax pencil or cupcake paper cups (14 per participant)
Individual serving trays (1 per participant)
Styrofoam (opaque) cups with lids (spit cups)
Water rinse cups
Napkins
Water serving pitchers

COOKIE VARIATION EXERCISE—VOCABULARY CONSTRUCTION

1. Flour, water	________________
2. Flour, water, butter	________________
3. Flour, water, margarine	________________
4. Flour, water, shortening	________________
5. Flour, water, shortening, salt	________________
6. Flour, water, shortening, baking soda	________________
7. Flour, water, sugar	________________
8. Flour, water, brown sugar	________________
9. Flour, water, butter, sugar	________________
10. Flour, water, margarine, sugar	________________

11. Flour, water, shortening, sugar	____________
12. Flour, water, sugar, egg, margarine	____________
13. Flour, water, sugar, egg, margarine, vanilla extract	____________
14. Flour, water, sugar, egg, margarine, almond extract	____________

COOKIE VARIATION EXERCISE—EXAMPLE OF RESULTS

1. Flour, water	raw wheat/dough/raw flour
	cooked wheat/paste/cream of wheat/ breadcrumb
2. Flour, water, butter	as #1 plus: butter/baked butter/ browned/butter
	toasted wheat
3. Flour, water, margarine	as #1 plus: heated vegetable oil; toasted wheat
4. Flour, water, shortening	as #1 plus: heated vegetable fat/Crisco
	toasted wheat/pie crust
5. Flour, water, shortening, salt	as #4 plus: salty
6. Flour, water, shortening, baking soda	as #5 plus: baked soda aromatic, salty
	baking soda feeling factor
7. Flour, water, sugar	as #1 plus caramelized, sweet
	toasted wheat
8. Flour, water, brown sugar	as #7 plus molasses
9. Flour, water, butter, sugar	as #2 plus sweet, caramelized
10. Flour, water, margarine, sugar	as #3 plus sweet, caramelized
11. Flour, water, shortening, sugar	as #4 plus sweet, caramelized
12. Flour, water, sugar, egg, margarine	as #11 plus baked eggy
13. Flour, water, sugar, egg, margarine, vanilla	as #12 plus: vanilla/vanillin/cake
14. Flour, water, sugar, egg, margarine, almond extract	as #12 plus cherry/almond

COOKIE VARIATION EXERCISE—POSSIBLE FULL VOCABULARY

CHARACTERISTICS	#379	#811
White wheat complex	______	______
Raw	______	______
Cooked	______	______
Toasted	______	______
Eggy	______	______
Shortening complex	______	______
Butter, baked	______	______
Heated vegetable oil	______	______
Sweet aromatics	______	______
Caramelized	______	______
Vanilla/vanillin	______	______
Almond/cherry	______	______
Molasses	______	______
Other aromatics (baking soda, etc.)	______	______
Sweet	______	______
Salty	______	______
Baking soda feel	______	______
______	______	______
______	______	______

13

Affective Tests

Consumer Tests and In-House Panel Acceptance Tests

13.1 PURPOSE AND APPLICATIONS

The primary purpose of affective tests is to assess the personal response (preference or acceptance) of current or potential customers to a product, a product idea, or specific product characteristics.

Affective tests are used mainly by producers of consumer goods but also by service providers such as hospitals, banks, and the armed forces, where many tests were first developed (Chapter 1, Section 1.2). Every year, the use of consumer tests becomes more common. They have proven highly effective as a tool used to design products and services that will sell in larger quantities or command a higher price. Prosperous companies tend to excel in consumer-testing knowledge and, consequently, in knowledge about their consumers.

This chapter establishes rough guidelines for the design of consumer tests and in-house affective tests. More detailed discussions are given by Amerine et al. (1965); Schaefer (1979); Moskowitz (1983); Civille et al. (1987); Wu and Gelinas (1989, 1992); Stone and Sidel (2004); Resurreccion (1998); and Lawless and Heymann (2010). One question that divides these authors is the use of in-house panels for acceptance testing. This chapter adopts the opinion that the appropriate choice of a panel is dependent on the product category being tested: Baron Rothschild may not rely on consumer tests for his wines, but ConAgra and Kraft Foods need them. For the average company's products, the amount of testing generated by intended and unavoidable variations in process and raw materials far exceeds the capacity of consumer panels, so in-house panels may be considered as an appropriate tool for most jobs and are calibrated against consumer tests as often as possible.

Results from consumer tests are more widely used than ever before. With the constantly changing marketplace offering more variety, niche products, and increasingly discriminating consumers, it is becoming more difficult to predict consumer preferences. This change in the market has increased the emphasis on the collection of consumer opinions. Consumer studies are expensive, with costs increasing an average of 5%–10% annually. The result is that there are many options available, and exploration into alternative approaches is an ongoing endeavor. Consumers are more connected than ever before, and "digital" research involving online and mobile testing represents an opportunity for in the moment data collection. This type of consumer testing involves different opportunities with some challenges and continues to evolve in the dynamic technological environment. Most people today have participated in some form of consumer test. Typically, a test involves 100–500 target consumers divided over three or four cities. A target consumer represents the population for whom the product is intended. The use of qualitative and quantitative testing including in-house panels, home use tests, focus groups, and online research is expanding. Researchers have a responsibility to ensure that the tests are appropriate and cost-effective. Appropriate use of consumer studies includes product screening prior to larger scale market research tests, assessing the viability of new products unbranded for ingredient substitutions or cost reductions on major brands, or gaining insights for product development. Inappropriate testing derives from poor testing systems (protocols), questionnaires, and consumer screeners; or the misuse of results based upon testing the wrong products, testing at the wrong time, testing with the wrong consumers, or using testing as a substitute for market research studies.

An example of a typical consumer study on carbonated beverages would require that it be conducted in two cities with recruitment of males 18–34 who had purchased a carbonated beverage at a convenience store within the previous 2 weeks. Potential respondents are screened by interview over the phone or in a shopping mall. Those selected and willing to participate are given a variety of beverages together with a scorecard requesting they rate their preference and state their reasons, along with requesting information on past buying habits and various demographic questions such as age, income, employment, ethnic background, and so on. Results are calculated in the form of preference scores overall and for various subgroups.

Consumer tests become more valuable when true insights are uncovered. Insights are the "ah ha's" that were previously not known. Insights vary and can be identified from both quantitative and qualitative research. For example, insights may include the way a child opens the peanut butter jar and spreads the peanut butter on a piece of bread, or the sensory signal of freshness that is identified when seeing the steam from a hot cup of coffee. Insights flow from new information that has not previously been heard or the way different sets of data are merged and mined.

One exciting change in recent years in the field of sensory consumer research involves working at the beginning of the product development process at the fuzzy front end. The advent of the fuzzy front end approach to creating new products has provided an opportunity to uncover and discover unarticulated consumer needs and product dynamics early in the development process. Working at the fuzzy front end is further discussed in Section 18.2.

Study designs need to be carefully tailored to the expected consumer group. The globalization of products often requires different study designs for different audiences. There has been a significant increase in global research from a quantitative and qualitative perspective. The American Society for Testing and Materials (ASTM) E18 has developed guidelines for consumer research across countries and cultures.

The most effective tests for preference or acceptance are based on carefully designed test protocols run among carefully selected subjects with representative products. The choice of test protocol and subject is based on the project objective. Nowhere in sensory evaluation is the definition of the project objective more critical than with consumer tests, which often cost from $10,000 to over $100,000. In-house affective tests are also expensive; the combined cost in salaries and overhead can run $400–$2000 for a 20 min test involving 20–40 people.

From a project perspective, the reasons for conducting consumer tests usually fall into one of the following categories:

- Product maintenance
- Product improvement/optimization
- Development of new products
- Assessment of market potential
- Product category review/benchmarking
- Support for advertising claims

13.1.1 Product Maintenance

In a typical food or personal care company, a large proportion of the product work carried out by research and development (R&D) and marketing departments deals with the maintenance of current products, their market shares, and sales volumes. R&D projects may involve cost reduction, substitution of ingredients, process and formulation changes, and packaging modifications, in each case without affecting the product characteristics and overall acceptance. Sensory evaluation tests used in such cases are often discrimination tests designed to assess differences or similarities among products. However, when a match is not possible, it is necessary to take one or more "near misses" out to the consumer to determine if these prototypes will at least achieve parity (in acceptance or preference) with the current product and, perhaps, with the competition.

Product maintenance is a key issue in quality control/quality assurance and shelf-life/storage projects. Initially, it is necessary to establish the "affective status" of the standard or control product with consumers. After this is done, internal tests can be used to measure the magnitude and type of change over time, condition, production site, raw material sources, and so on, with the aid of quality control (QC) or storage testing. The sensory differences detected by internal tests, large and small, may then be evaluated again by consumer testing to determine how large a difference is sufficient to reduce (or increase) the acceptance rating or percent preference compared to the control or standard specifications.

13.1.2 Product Improvement/Optimization

Because of the intense competition for shelf space, companies are constantly seeking to improve and optimize products so that they deliver what the consumer is seeking and thus fare better than the competition. A product improvement project generally seeks to "fix" or upgrade one or two key product attributes that consumers have indicated need improvement. A product optimization project typically attempts to manipulate a few ingredient or process variables so as to improve the desired attributes and the overall consumer acceptance. Both types of projects require the use of a good descriptive panel to (1) verify the initial consumer needs and (2) document the characteristics of the successful prototype. Examples of projects to improve product attributes are

- Increasing a key aroma and/or flavor attribute, such as lemon, peanut, coffee, chocolate, and so on

- Increasing an important texture attribute, such as crisp, moist, and so on, or reducing negative properties such as soggy or chalky
- Decreasing a perceived off note (e.g., crumbly dry texture, stale flavor or aroma, artificial rather than natural fruit flavor)
- Improving perceived performance characteristics, such as longer lasting fragrance, brighter shine, more moisturized skin, and so on

In product improvement, prototypes are made, tested by a descriptive or attribute panel to verify that the desired attribute differences are perceptible, and then tested with consumers to determine the degree of perceived product improvement and its effect on overall acceptance or preference scores.

For product optimization (IFT, 1979; Moskowitz, 1983; Carr, 1989; Gacula, 1993; Resurreccion 1998; Stone and Sidel, 2004) ingredients or process variables are manipulated; the key sensory attributes affected are identified by descriptive analysis, and consumer tests are conducted to determine if consumers perceive the change in attributes and if such modifications improve the overall ratings.

The study of attribute changes, together with consumer scores, enables the company to identify and understand those attributes and/or ingredients or process variables that drive overall acceptance in the market.

13.1.3 Development of New Products

During the typical new product development cycle, affective tests are needed at several critical junctures. For example, the development of a new product may require

- Focus groups to evaluate a concept or prototype
- Feasibility studies in which the test product is presented to consumers, allowing them to see and touch it
- A series of quantitative central location tests throughout product development to confirm that the product characteristics do offer the expected advantage over the competition
- Renewed comparisons during the reduction-to-practice stage to confirm that the desired characteristics survive into larger-scale production
- Central location and home use tests during the growth phase to determine the degree of success enjoyed by the competition as it tries to catch up

Depending on test results at each stage and the ability of R&D to reformulate or scale up at each step, the new product development cycle can take from a few months to a few years. This process requires the use of several types of affective tests designed to measure, for example, responses to the first concepts, chosen concepts versus prototypes, different prototypes, and competition versus prototypes. At any given time during the development process, the test objectives may resemble those of a product maintenance project, for example, a pilot plant scale-up or an optimization project, as described previously. Rapid prototype development is used when the time to market is short and there is an urgent need to make a product decision. This approach utilizes ongoing frequent contact with the target consumer population to collect immediate feedback. The feedback is provided to product development to make rapid changes, which are then submitted to the target audience for further feedback. Test methods utilized with this approach include small-scale central location tests (CLTs) or qualitative research.

13.1.4 Assessment of Market Potential

Typically, the assessment of market potential is a function of the marketing department, which, in turn, will consult with the sensory evaluation department about aspects of the questionnaire design (such as key attributes that describe differences among products), the method of testing, and data previously collected by sensory evaluation. Questions about intent to purchase, purchase price, current purchase habits, consumer food habits (Barker, 1982; Meiselman, 1984), and the effects of packaging, advertising, and convenience are critical for the acceptance of branded products. The sensory analyst's primary function is to guide research and development. Whether the sensory analyst should also include market-oriented questions in consumer testing is a function of the structure of the individual company, the ability of the marketing department to provide such data, and the ability of the sensory analyst to assume responsibility for assessing market conditions.

13.1.5 Category Review/Benchmarking

When a company wishes to study a product category for the purpose of understanding the position of its brand within the competitive set or for the purpose of identifying areas within a product category where opportunities may exist, a category review is recommended (Lawless and Heymann, 2010). Descriptive analysis of the broadest array of products and/or prototypes that defines or covers the category yields a category map. Using multivariate analysis techniques, the relative position of both the products and the attributes can be displayed in graphic form (see Chapter 15). This permits researchers to learn (1) how products and attributes cluster within the product/attribute space, (2) where the opportunities may be in that space for new products, and (3) which attributes best define which products. A detailed example of a category appraisal is that of frankfurters by Muñoz et al. (1996), in which consumer data and descriptive panel data are related statistically.

Additional testing of several of the same products with consumers can permit the projection of other vectors into the space. These other vectors may represent consumers' overall liking and/or consumers' integrated terms, such as "creamy", "rich", "fresh", or "soft". The identification of consumers segments based on patterns of liking for different products is also possible. Consumer liking within each segment is driven by specific key descriptive features of the product category being studied.

13.1.6 Support for Advertising Claims

Product and service claims made in print or on radio, television, or the Internet require valid data to support the claims. Sensory claims of parity ("tastes as good as the leading brand") or superiority ("cleans windows better than the leading brand") need to be based on consumer research and/or panel testing using subjects, products, and test designs that provide credible evidence of the claim. For specific information on the requirements and design considerations for this type of test, refer to ASTM E1958-12; see also chapter 9 of Gacula (1993). This topic is explored further in Section 18.10.

13.1.7 Uncovering Consumer Needs

Working on projects that result in the creation of new unique opportunities or improved products that are specifically designed to meet a consumer need is a developing area for sensory evaluation. Understanding the consumers' articulated and unarticulated needs, wants, wishes, and behaviors results in products designed to meet specific needs based on better target definitions for concept and product requirements, and in stronger business cases developed based on facts and specific consumer directed information. Techniques used to collect this information are often focused in the qualitative area using observational research, one-on-one interviews, point-of-purchase interviews, diaries, and home visits. Application of approaches uncovers how products within the category are "really" used and identifies key sensory properties, including "must have" and "nice to have" features as well as those that signal benefits or performance to consumers.

13.2 SUBJECTS/CONSUMERS IN AFFECTIVE TESTS

13.2.1 Sampling and Demographics

Whenever a sensory test is conducted, a group of subjects is selected as a sample of some larger population about which the sensory analyst hopes to draw some conclusion. In the case of discrimination tests (difference tests and descriptive tests), the sensory analyst chooses individuals with average or above-average abilities to detect differences. It is assumed that if these individuals cannot "see" a difference, the larger human population will be unable to see it. In the case of affective tests, however, it is not sufficient to merely select randomly from the vast human population. Consumer goods and services try to meet the needs of target populations, select markets, or carefully chosen segments of the population. Such criteria require that the sensory analyst first determine the population for whom the product (or service) is intended; for example, for a sweetened breakfast cereal, the target population may be children between the ages of 4 and 12; for a sushi and yogurt blend, the select market may be southern California; and for a high-priced jewelry item, article of clothing, or automobile, the segment of the general population may be young, 25–35-year-old, upwardly mobile professionals, both married and unmarried.

Consumer researchers who are faced with the task of balancing the need to identify and use a sample of consumers who represent the target population with the cost of having a very precise demographic model, use a screener. Proper screening requires thought and input, not only from the consumer researcher but also from the client, such as product development or market research. The information collected from the screening process helps determine similarities and differences among groups of people and the subsequent influence of those similarities and differences on product liking and purchase. An effective screener starts with a clear understanding of the research objective. Detailed screening criteria for qualitative and quantitative tests may differ; however, there is a series of broad questions that are typically asked, such as age, gender, occupation/profession, ethnicity, income, general usage, sensitivities, time availability, and willingness to participate. With widely used products such as cold cereals, soft drinks, beer, cookies, and facial tissues, research guidance consumer tests may require selection only of users or potential users of the product brand or category. The cost of stricter demographic criteria may be justified for the later stages of consumer research guidance or for marketing research tests. Among the demographics to be considered in selecting sample subjects are

13.2.1.1 User Group

Based on the rate of consumption or use of a product by different groups within the population, brand managers often classify users as light, moderate, or heavy users. These terms are highly dependent on the product type and its normal consumption (see Table 13.1). For products that are continually changing, such as electronics, the lead users will provide the most useful information for new product concepts. The lead-user segment recognizes a need well before the general population and attempts to fill that need. Another target group includes dissatisfied users who may use a product because of the lack of a better substitute in the market place. Market researchers may seek out this group for innovative ideation. For specialty products or new products with low incidence in the population, the cost of consumer testing radically increases because many people must be contacted before the appropriate sample of users can be found.

13.2.1.2 Age

The ages of 4–12 are best to test toys, sweets, and cereals; teenagers at 12–19 buy clothes, magazines, snacks, soft drinks, and entertainment. Young adults at 20–35 receive the most attention in consumer tests (1) because of population numbers, (2) because of higher consumption made possible by the absence of family costs, and (3) because lifelong habits and loyalties are formed in this age range. Above age 35, consumers buy houses and raise families; above age 65, they use healthcare tend to look for value in consumables with an eagle eye. If a product, such as a soft drink, has a broad age appeal, the subjects should be selected by age in proportion to their representation in the user population.

13.2.1.3 Gender

Although women still buy more consumer goods and clothes, and men buy more automobiles, alcohol, and entertainment, the differences in purchasing habits between the genders continue to diminish. Researchers should use very current figures on users by gender for products such as convenience foods, snacks, personal care products, and wine.

Table 13.1 Typical Frequency Use of Various Consumer Products

	Product			
User Classification	**Coffee**	**Peanut Butter, Air Freshener**	**Macaroni and Cheese**	**Rug Deodorizer**
Light	Up to 1 cup/day	1–4×/month	Once/2 months	1×/year
Moderate	2–5 cups/day	1–6×/week	1–4×/month	2–4×/year
Heavy	5 cups/day	1× or more/day	Over 2×/week	1×/month or more

13.2.1.4 Income

Meaningful groups for most items marketed to the general population per household and year are

- Under $30,000
- $30,000–$49,999
- $50,000–$79,999
- $80,000–$99,000
- Over $100,000

Different groups may be relevant at times, e.g., $500,000, $1,000,000, etc., for recruiting participants that own yachts over 50 ft.

13.2.1.5 Geographic Location

Because of the regional differences in preference for many products, for example, across the United States, it is often necessary to test products in more than one location and to avoid testing (or to use proportional testing) of products for the general population in areas with distinct local preferences, for example, New York, the deep South, or southern California. In addition, attention to urban, suburban, and rural representation can also influence test results.

Nationality, region, ethnicity, religion, education, and employment; and other factors such as marital status, number and ages of children in family, pet ownership, size of domicile, and so on may be important for the sampling of some products or services. The product researcher, brand manager, or sensory analyst must carefully consider all the parameters that define the target population before choosing the demographics of the sample for a given test.

Examples of step-by-step questionnaires used by marketing researchers to screen prospective respondents may be found in Meilgaard (1992) and Resurreccion (1998).

13.2.2 Source of Test Subjects

Consumer tests require sampling from the population that uses the product. There are three sources from which individuals are chosen to participate in studies: employees, local area residents who are recruited to join a database, and the general population.

13.2.2.1 Employees

The need to sample properly from the consuming population excludes, in principle, the use of employees and residents local to the company offices, technical center, or plants. However, because of the high cost and long turnaround time of consumer tests, companies see a real advantage in using employees or the local population for at least part of their affective testing.

In situations where the project objective is product maintenance employees and local residents do not represent a great risk as the test group. In a project oriented toward maintaining the "sensory integrity" of a current product, employees or local residents familiar with the characteristics of the product can render evaluations that are a good measure of the reaction of regular users. In this case, the employee or local resident judges the relative difference in acceptability or preference of a test sample vis-à-vis the well-known standard or control.

Employee acceptance tests can be a valuable resource when used correctly and when limited to maintenance situations. Because of their familiarity with the product and with testing, employees can handle more samples at one time and provide better discrimination, faster replies, and cheaper service. Employee acceptance tests can be carried out in a laboratory or in the style of a central location test, or the employees may take the product home.

However, for new product development, product optimization, or product improvement, employees or local residents should not be used to represent the consumer. The following are some examples of biases that may result from conducting affective tests with employees:

1. Employees tend to find reasons to prefer the products that they and their fellow employees helped to make or, if morale is bad, to find reasons to reject such products. It is therefore imperative that products be disguised. If this is not possible, a consumer panel must be used.
2. Employees may be unable to weigh desirable characteristics against undesirable ones in the same manner as a consumer. For example, employees may know that a recent change was made in the process to produce a paler color, and this will make them prefer the paler product and give too little weight to other characteristics. Again, in such a case, the color must be disguised, or if this is not possible, outside testing must be used.
3. Where a company makes separate products for different markets, outside tests will be distributed to the target population, but this cannot be done with employees. If required to test with employees, it is suggested that they be told that the product is destined for X market, but sometimes this cannot be done without violating the requirement that the test be blind. If so, outside testing must be used.

13.2.2.2 Local Area Residents

One approach to recruiting respondents for consumer testing is for a company to develop their own database of local area residents. This approach, although relatively cost-effective, requires internal support to develop, maintain, and recruit respondents to participate after it is established. Caution must be taken not to overuse the consumers in the database. It may be difficult to maintain confidentiality, especially if the test facilities are onsite. Determining when and at which stage of a project the consumer database can be used is important to insure that the information collected is appropriate to the project objective.

In summary, the test organizer must plan the test imaginatively and must be aware of all sources of bias. In addition, the validity of responses must be assured by frequent comparisons with consumer tests that use the broader consumer population on the same samples. In this way, the organizer and the employee panel members slowly develop a knowledge of what the market requires; this, subsequently, makes it easier to gauge the pitfalls and avoid them.

13.2.2.3 General Population

Testing from the general population typically captures the responses from the target user group by going into the field and recruiting consumers who meet specific predefined criteria. These respondents are most often selected from a database and contacted directly to participate, or they are recruited from a central location such as a shopping mall. The advantage of this approach is the ability to test with product users or potential users; the disadvantage is the added cost to recruit. It is important to include the appropriate screening criteria to eliminate professional evaluators.

13.3 CHOICE OF TEST LOCATION

The test location or test site has numerous effects on the results, not only because of its geographic location but also because the place in which the test is conducted defines several other aspects of product sampling and perceived sensory properties. It is possible to get different results from different test sites with a given set of samples and consumers. These differences occur as a result of differences in

- The length of time the products are used/tested
- Controlled preparation versus normal-use preparation of the product
- The perception of the product alone in a central location versus in conjunction with other foods or personal care items in the home
- The influence of family members on each other in the home
- The length and complexity of the questionnaire

For a more detailed discussion, see Resurreccion (1998).

13.3.1 Laboratory Tests

The advantages of laboratory tests are

- Product preparation and presentation can be carefully controlled.
- Employees can be contacted on short notice to participate.
- Color and other visual aspects that may not be fully under control in a prototype can be masked so that subjects can concentrate on the flavor or texture differences under investigation.

The disadvantages of laboratory tests are that

- The location suggests that the test products originate in the company or specific plant, which may influence biases and expectations because of previous experience. Experience with and knowledge of product(s) often results in increased sensitivities to differences. The reaction to perceived differences may not accurately reflect the target population.
- The lack of normal consumption (e.g., sip test rather than consumption of a full portion) may influence the detection or evaluation of positive or negative attributes.
- Standardized preparation procedures and product handling protocols might not necessarily mimic consumer behavior and experience at home.

13.3.2 Central Location Tests

Central location tests are usually conducted in an area where potential purchasers congregate or can be assembled. The organizer sets up a booth or rents a room at a fair, shopping mall, church, or test agency. A product used by schoolchildren may be tested in the school playground; a product for analytical chemists may be tested at a professional convention. Respondents are intercepted and screened in the open, and those selected for testing are led to a closed-off area. Subjects can also be prescreened by phone and invited to a test site prerecruited. This type of recruiting is more targeted and provides better control on balancing all the required demographics and product use history. Typically, 50–300 responses are collected per location. Products are prepared out of sight and served on uniform plates (cups, glasses) labeled with three-digit codes. The potential for distraction may be high, so instructions and questions should be clear and concise; examples of score sheets are provided in Appendix 13.3. In a variant of the procedure, products are dispensed openly from original packaging, and respondents are shown storyboards with examples of advertising and descriptions of how products will be positioned in the market.

The advantages of central location tests are that

- Respondents evaluate the product under conditions controlled by the organizer; any misunderstandings can be cleared up and a truer response obtained.
- The products are tested by the end users themselves; this assures the validity of the results.
- Conditions are favorable for a high percentage return of responses from a large sample population.
- Several products may be tested by one consumer during a test session, thus allowing for a considerable amount of information for the cost per consumer.

The main disadvantages of central location tests are that

- The product is being tested under conditions that are artificial in comparison to normal use at home or at parties, restaurants, and so on in terms of preparation, amount consumed, and length and time of use.
- The number of questions that can be asked may be limited versus testing in the home. This in turn limits the information obtainable from the data with regard to the preferences of different age groups, socioeconomic groups, and so on.

13.3.3 Home Use Tests

In most cases, home use tests (or home placement tests) represent the ultimate in consumer research. The product is tested under its normal conditions of use. The participants are selected to represent the target population. The entire family's opinion can be obtained, with the influence of one family member on another taken into account. In addition to the product itself, the home use test provides a check on the package to be used and the product preparation instructions, if applicable. Typical panel sizes are 75–300 per city in three or four cities. Often, two products are compared. The first is used for 4–7 days and its corresponding score sheet is completed, after which the second is supplied and rated. The two products should not be provided together because of the opportunities for using the wrong clues as the basis for evaluation or assigning responses to the wrong score sheet. Examples of score sheets are provided in Appendix 13.3.

The advantages of home use tests are (Moskowitz, 1983; Resurreccion, 1998) that

- The product is prepared and consumed under natural conditions of use.

- Information regarding preference between products will be based on stabilized (repeated) use rather than first impressions alone, as in a mall intercept test.
- Cumulative effect on the respondent from repeated use can provide information about the potentials for repeat sale.
- Statistical sampling plans can be fully utilized.
- Because more time is available for the completion of the score sheet, more information can be collected regarding the consumer's attitudes toward various characteristics of the product, including sensory attributes, packaging, price, and so on.

The disadvantages of the home use tests are that

- A home use test is time consuming, taking from 1 to 4 weeks to complete.
- It uses a much smaller set of respondents than a central location test; to reach many residences would be unnecessarily time consuming and expensive.
- The possibility of no response is greater; unless frequently reminded, respondents forget their tasks; haphazard responses may be given as the test draws to a close.
- A maximum of three samples can be compared; any larger number will upset the natural-use situation that was the impetus for choosing a home use test in the first place. Thus, multisample tests, such as optimization and category review, do not lend themselves to home use tests.
- The tolerance of the product for mistakes in preparation is tested. The resulting variability in preparation or use along with variability from the time of use and from other foods or products used with the test product, combine to produce a large variability across a relatively small sample of subjects.

13.4 AFFECTIVE METHODS: QUALITATIVE

13.4.1 Applications

Qualitative affective tests (e.g., interviews, focus groups, and observational research) are those that measure the subjective responses of a sample of consumers to the sensory properties of products by having those consumers talk about their feelings in an interview or small group setting or use the product at home or in a central location setting. Qualitative methods are used in the following situations:

- To uncover and understand consumer needs that are unexpressed; for example, "Why do people buy 4-wheel-drive cars to drive on asphalt?" Researchers that include anthropologists and ethnographers conduct open-ended interviews. See Section 18.2 for further information.
- To assess consumers' initial responses to a product concept and/or a product prototype. When product researchers need to determine if a concept has some general acceptance or, conversely, some obvious problems, a qualitative test can allow consumers to freely discuss the concept and/or a few early prototypes. The results, a summary, and a tape of such discussions permit the researcher to better understand the consumers' initial reactions to the concept or prototypes. Project direction can be adjusted at this point, in response to the information obtained.

- To learn consumer terminology to describe the sensory attributes of a concept, prototype, commercial product, or product category. In the design of a consumer questionnaire and advertising, it is critical to use consumer-oriented terms rather than those derived from marketing or product development. Qualitative tests permit consumers to discuss product attributes openly in their own words.
- To clarify and expand on consumers responses from quantitative research. Quantitative research is ideal for determining how consumers like a product or react to the sensory attributes. However, it does not always fully capture the nuances or the reasons behind the rating. More in-depth knowledge can be gained by asking consumers to remain after the quantitative portion for a one-on-one interview or by having them return at a later date for a focus group.
- To learn about consumer behavior regarding use of a particular product. When product researchers wish to determine how consumers use certain products (package directions) or how consumers respond to the use process (dental floss, feminine protection), qualitative tests probe the reasons for and practices of consumer behavior.

In the qualitative methods discussed below, a highly trained interviewer/moderator is required. Because of the high level of interaction between the interviewer/moderator and the consumers, the interviewer must learn group dynamics skills, probing techniques, techniques for appearing neutral, and summarizing and reporting skills.

13.4.2 Qualitative Screener Development

The best source of information for developing the screening criteria is the client. In addition to the broad areas or categories outlined in Section 13.2.1, a series of questions that probe usage habits, purchase criteria, allergies or sensitivities to the product, or ingredients and concept acceptance need to be asked. Because a small number of respondents participate in qualitative research, it is important to develop specific attitude or usage criteria to ensure respondents are representative of a diverse group of people who meet the critical criteria. A major component of qualitative screening addresses the consumer's willingness to contribute in a group discussion. The interviewer would ask the perspective participant if he/she is willing to openly voice his/her opinion in a group. Obviously, a candidate not willing to open up and share their feelings would not be a good choice for a qualitative discussion. A final selection criterion is the candidate's ability to express thoughts and feelings in an effective manner. An open-ended question asked at the end of the interview is typically used for this assessment. The question could be "If you could meet any one person in history, who would it be and why?" As a safeguard for a productive discussion and to be sure no one has sent a substitute, each chosen participant is asked a few additional questions upon arrival at the facility. This rescreening process should only require a few minutes to complete.

13.4.3 Types of Qualitative Affective Tests

13.4.3.1 Focus Groups

A small group of 8–12 consumers, selected on the basis of specific criteria (product usage, consumer demographics, etc.) meet for 1–2 h with the focus group moderator.

The moderator presents the subject of interest and facilitates the discussion using group dynamics techniques to uncover as much specific information from as many participants as possible, directed toward the focus of the session.

Typically, two or three such sessions, all directed toward the same project focus, are held to determine any overall trend of responses to the concept and/or prototypes. Notes are also made of unique responses apart from the overall trend. A summary of these responses, plus DVDs or tapes (audio or visual), are provided to the client researcher. Purists will say that $3 \times 12 = 36$ verdicts are too few to be representative of any consumer trend; in practice, however, if a trend emerges that makes sense, modifications are made based on this. The modifications may then be tested in subsequent groups or quantitative research.

The literature on marketing is a rich source of details on focus groups, for example, Krueger (1988), Casey and Krueger (1994), and Resurreccion (1998).

13.4.3.2 Focus Panels

In this variant of the focus group, the interviewer utilizes the same group of consumers two or three more times. The objective is to make some initial contact with the group, have some discussion on the topic, send the group home to use the product, and then have the group return to discuss its experiences. This approach is very effective when performing rapid prototype development. It allows consumers to participate in the development of a product and provide ongoing feedback and direction.

13.4.3.3 Mini Groups, Diads, Triads

Mini groups, diads, and triads are an alternative to focus groups of 8–12 consumers. Mini groups are usually comprised of four to six respondents, triads are three respondents, and diads are two respondents with one interviewer. This approach is often used when there is a need to go in depth on a particular discussion, if the subject being discussed is sensitive, or it is difficult to find respondents to meet the screening criteria. The format typically follows the same format as a focus group.

13.4.3.4 One-on-One Interviews

Qualitative affective tests in which consumers are individually interviewed in a one-on-one setting are appropriate in situations in which the researcher needs to understand and probe a great deal from each consumer or in which the topic is too sensitive for a focus group. These are often called in-depth interviews or IDIs. The interviewer conducts successive interviews with anywhere from 12 to 50 consumers, using a similar format with each but probing in response to each consumer's answers.

One unique variant of this method is to have a person use or prepare a product at a central interviewing site or in the consumer's home. Notes or a video are taken regarding the process, which is then discussed with the consumer for more information. Interviews with consumers regarding how they use a detergent or prepare a packaged dinner have yielded information about consumer behavior that was very different from what the company expected or what consumers said they did.

One-on-one interviews or observations of consumers can give researchers insights into unarticulated or underlying consumer needs, and this in turn can lead to innovative products or services that meet such needs.

13.5 AFFECTIVE METHODS: QUANTITATIVE

13.5.1 Applications

Quantitative affective tests are those that determine the responses of a large group of consumers (50 to several hundred) to a set of questions regarding choice, preference, liking, sensory attributes, and so on. Quantitative affective methods are applied in the following situations:

- To utilize survey tools to screen a larger number of possible prototypes into a manageable subset of the most-liked combinations.
- To determine overall preference or liking for a product or products by a sample of consumers who represent the population for whom the product is intended. Decisions about whether to use acceptance and/or preference questions are further discussed under each test method.
- To determine preference or liking for broad aspects of product sensory properties (aroma, flavor, appearance, and texture). Studying broad facets of product character can provide insight regarding the factors affecting overall preference or liking.
- To measure consumer responses to specific sensory attributes of a product. Use of intensity, hedonic, or "just right" scales can generate data that can then be related to the hedonic ratings discussed previously and to descriptive analysis data.

13.5.2 Design of Quantitative Affective Tests

13.5.2.1 Quantitative Screener Development

As with screening for qualitative discussion groups, understanding the research objective is crucial to identifying the population required for a quantitative study. Based on client input and what is known about the product category in general, screening criteria quotas can be established. For chocolate flavored milk, the segment of the population that should be targeted is boys (50%) and girls (50%) between the ages of 5 and 10 (50%) and 11 and 16 (50%). Quantitative studies can require several days to complete for many reasons, such as the total number of samples in the study versus the number that can be tested in one session. Therefore, during the screening interview, a candidate must agree to participate and be willing to come to the facility for more than one session to complete the study. To avoid no-shows halfway through the study, participants are told that to receive compensation they must complete all required sessions.

13.5.2.2 Questionnaire Design

In designing questionnaires for affective testing, the following guidelines are recommended:

1. Keep the length of the questionnaire in proportion to the amount of time the subject expects to be in the test situation. Subjects can be contracted to spend hours testing several products with extensive questionnaires. At the other extreme, a few questions may be enough information for some projects. Design the questionnaire to ask the minimum number of questions to achieve the project objective, then construct the test so that the respondents expect to be available for the appropriate time span.

2. Keep the questions clear and somewhat similar in style. Use the same type of scale—whether preference, hedonic, just about right, or intensity scale—within the same section of the questionnaire. Intensity and hedonic questions may be asked in the same questionnaire (see examples in Appendix 13.3), but they should be clearly distinguished. The questions and their responses should follow the same general pattern in each section of the questionnaire. For consistency and to insure accurate responses, the scales should be designed to go in the same direction, for example, [Too little······Too much], for each attribute, so that the subject does not have to stop and decode each question.
3. Direct the questions to address the primary differences between/among the products in the test. Attribute questions should relate to the attributes that are detectable in the products and which differentiate among them. This can be determined by previously conducted descriptive tests. Subjects will not give clear answers to questions about attributes they cannot perceive or differences they cannot detect.
4. Use only questions that are actionable. Do not ask questions to provide data for which there is no appropriate action. If one asks subjects to rate the attractiveness of a package and the answer comes back that the package is somewhat unattractive, does the researcher know what to "fix" or change to alter that rating?
5. Always provide spaces on a score sheet for open-ended questions. For example, ask the reason a subject responded the way he/she did to a preference or acceptance question, immediately following that question.
6. Place the overall question for preference or acceptance in the place on the score sheet that will elicit the most considered response. In many cases, the overall acceptance is of primary importance, and analysts rightly tend to place it first on the score sheet. However, in cases where a consumer is asked several specific questions about appearance and/or aroma before the actual consumption of the product, it is necessary to wait until those attributes are evaluated and rated before addressing the total acceptance or preference question. Appendix 13.3 provides two examples of acceptance questionnaires.

13.5.2.3 Protocol Design

Sensory tests are difficult enough to control in a laboratory setting (see Section 3.2). Outside the laboratory, in a central location or home use setting, the need for controls of test design, product handling, and subject/consumer selection is even greater. In developing and designing outside affective tests, the following guidelines are recommended:

13.5.2.3.1 Test Facility

In a central location test, the facility and test administrators must adhere to strict protocols regarding the size, flexibility, location, and environmental controls at each test site. The test should be conducted in locations that provide high access to the target population, and subjects should be able to reach the test site easily.

Based on the design of the study, consideration should be given to the ability of each facility to provide adequate space, privacy for each consumer/subject, proper environmental controls (lighting, noise control, odor control, etc.), space for product handling and preparation, and a sufficient number of administrators and interviewers.

13.5.2.3.2 Test Administrators

The administrators are required to be both trained and experienced in the specific type of test design developed by the sensory analyst. In addition to familiarity with the test design, test administrators must be given a detailed set of instructions for the handling of questionnaires, subjects, and samples for a specific study.

13.5.2.3.3 Test Subjects

Each test site requires careful selection of subjects based on demographic criteria that define the population of interest (see Section 13.2). Once selected, subjects are made aware of the location, the duration of the test, the type and number of products to be tested, and the type of payment. Consumers do not respond well to surprises regarding exactly what is expected of them.

13.5.2.3.4 Screen Samples

Prior to any affective test, samples must be screened to determine

- The exact sample source to be tested (bench, pilot plant, production, and code date)
- The storage conditions under which samples are to be held and shipped
- Packaging requirements for storage and shipping
- Shipping method (air, truck, refrigerated, etc.)
- Product sensory attributes using descriptive analysis for use in questionnaire design and in final data interpretation for the study

13.5.2.3.5 Sample Handling

As part of the test protocol that is sent to the test site, detailed and specific instructions regarding storage, handling, preparation, and presentation of samples are imperative for proper test execution.

Appendix 13.4 provides worksheets for the development of a protocol for an affective test and an example of a completed protocol.

13.5.3 Types of Quantitative Affective Tests

Affective tests can be classified into two main categories on the basis of the primary task of the test:

Task	Test and Type	Questions
Choice	Preference tests	Which combination of attributes do you prefer?
		Which sample do you prefer?
		Which sample do you like better?
Rating	Acceptance tests	What is your purchase intent for this product?
		How much do you like the product?
		How acceptable is the product?

In addition to these questions, which can be asked in several ways using various questionnaire forms, the test design often asks secondary questions about the reasons for the expressed preference or acceptance (see Appendix 13.3 for questionnaire examples and Section 13.5.4 on attribute diagnostics).

13.5.3.1 Preference Tests

The choice of preference or acceptance for a given affective test should be based on the project objective. If the project is specifically designed to pit one product directly against another in situations such as product improvement or parity with competition, then a preference test is indicated. The preference test forces a choice of one item over another or others. What it does not do is indicate whether any of the products are liked or disliked. Therefore, the researcher must have prior knowledge of the "affective status" of the current product or competitive product that he or she is testing against.

Preference tests can be classified as follows:

Test Type	No. of Samples	Preference
Paired preference	2	A choice of one sample over another (A–B)
Rank preference	3 or more	A relative order of preference of samples (A–B–C–D)
Multiple paired preference (all pairs)	3 or more	A series of paired samples with all samples paired with all others (A–B, A–C, A–D, B–C, B–D, C–D)
Multiple paired preferences (selected pairs)	3 or more	A series of paired samples with one or two select samples (e.g., control) paired with two or more others (not paired with each other) (A–C, A–D, A–E, B–C, B–D, B–E)

See Chapter 8 for a discussion of principles, procedures, and analysis of paired and multipaired tests.

Example 13.1: Paired Preference—Improved Peanut Butter

Problem/situation. In response to consumer requests for a product "with better flavor with more peanutty character," a product improvement project has yielded a prototype that was rated significantly more peanutty in an attribute difference test (such as discussed in Chapter 8, Sections 8.1–8.4). Marketing wishes to confirm that the prototype is indeed preferred to the current product, which is enjoying large volume sales.

Test objective. To determine whether the prototype is preferred over the current product.

Test design. This test is one sided as the prototype was developed to be more peanutty in response to consumer requests. A group of 100 subjects, prescreened as users of peanut butter, are selected and invited to a central location site where they receive the two samples in simultaneous presentation, half in the order A–B, the other half B–A. All samples are coded with three-digit random numbers. Subjects are encouraged to make a choice (see discussion of forced choice, Section 8.2). The score sheet is shown in Figure 13.1. The null hypothesis is H_0; the preference for the higher-peanut flavor prototype is $\leq 50\%$. The alternative hypothesis is H_a; the preference for the prototype is $>50\%$.

Screen samples. Samples used are those already subjected to the attribute difference test described earlier, in which a higher level of peanut flavor was confirmed.

Peanut Butter
Instructions 1. Taste the product on the left first and the product on the right second. Now that you've tasted both products, which one do you prefer? Please choose one: ☐ 463 ☐ 189
2. Please comment on the reasons for your choice: ________
Name ________ Date ________

Figure 13.1 Score sheet for paired preference test for Example 13.1: Improved peanut butter.

Conduct the test. The method described in Section 8.2.4, was used; 62 subjects preferred the prototype. It is concluded from Table 19.10 that a significant preference exists for the prototype over the current product.

Interpret results. The new product can be marketed in place of the current with a label stating: "More Peanut Flavor."

13.5.3.2 Acceptance Tests

When a product researcher needs to determine the "affective status" of a product, that is, how well it is liked by consumers, an acceptance test is the correct choice. The product is compared to a well-liked company product or that of a competitor, and a hedonic scale, such as those shown in Figure 13.2, is used to indicate degrees of unacceptable to acceptable, or dislike to like. The two lower scales, "KIDS" and "Snoopy," are commonly used with children of grade-school age.

From relative acceptance scores, one can infer preference; the sample with the higher score is preferred. The best (most discriminating, most actionable) results are obtained with scales that are balanced; that is, they have an equal number of positive and negative categories and have steps of equal size. The scales shown in Figure 13.3 are not as widely used because they are unbalanced, unevenly spaced, or both. The six-point "excellent" scale in Figure 13.3, for example, is heavily loaded with positive (good to excellent) categories and the space between "poor" and "fair" is clearly larger than that between "extremely good" and "excellent." The difference between the latter may be unclear to many people. Acceptance tests are, in fact, very similar to attribute difference tests (see Sections 8.1–8.4) except that the attribute here is acceptance or liking. Different types of scales such as category (as shown in Figure 13.2), line scales, or magnitude estimation scales can be used to measure the degree of liking for a product.

Example 13.2 Acceptance of Two Prototypes Relative to a Competitive Product—High-Fiber Breakfast Cereal

Problem/situation. A major cereal manufacturer has decided to enter the high-fiber cereal market and has prepared two prototypes. Another major cereal producer already has a

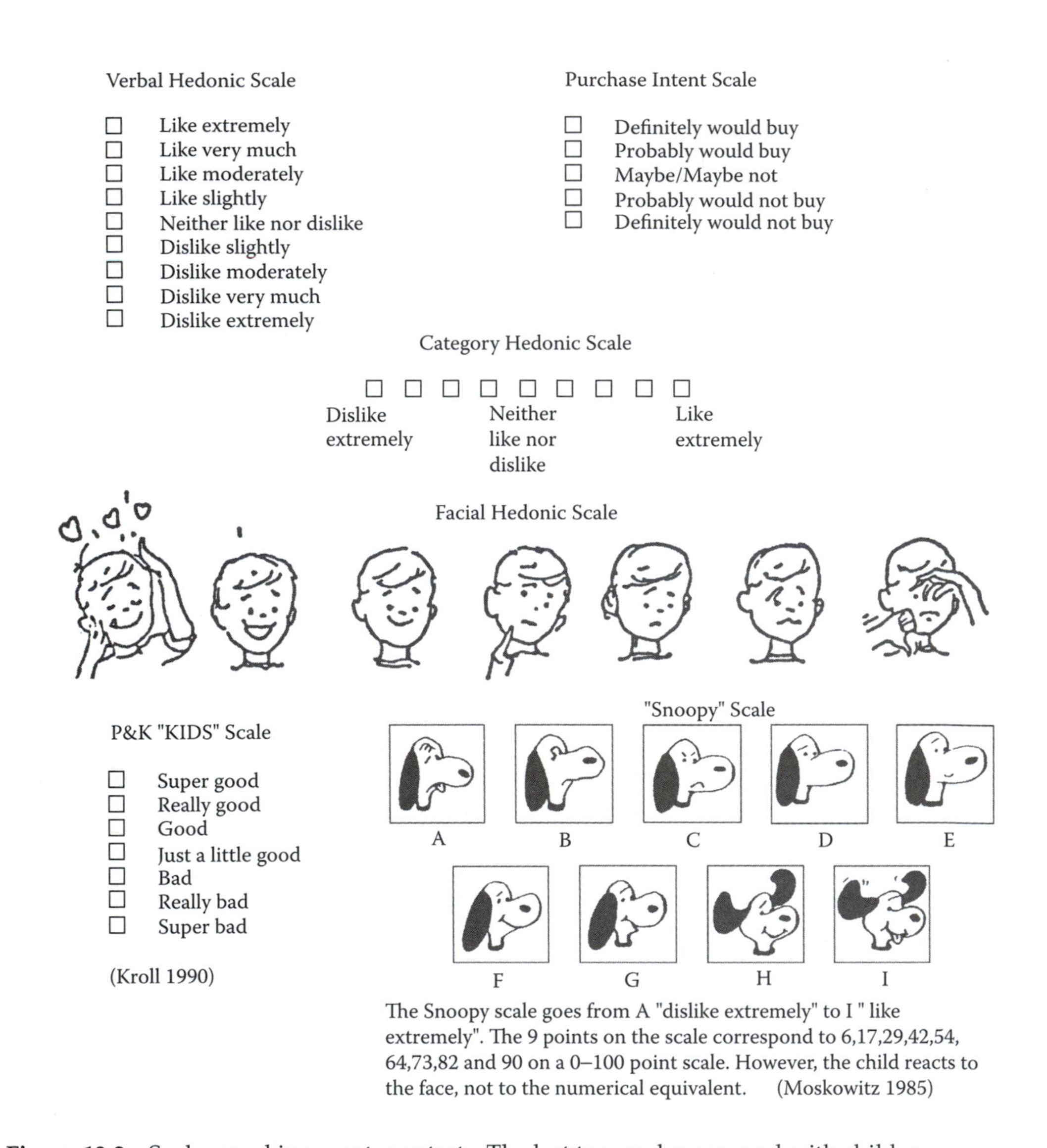

Figure 13.2 Scales used in acceptance tests. The last two scales are used with children.

brand on the market that continues to grow in market share and leads among the high-fiber brands. The researcher needs to obtain acceptability ratings for his two prototypes compared to the leading brand.

Project objective. To determine whether one or the other prototype enjoys sufficient acceptance to be test marketed against the leading brand.

Test objective. To measure the acceptability of the two prototypes and the market leader among users of high-fiber cereals.

Screen the samples. During a product review, several researchers, the brand marketing staff, and the sensory analyst taste the prototypes and the competitive cereal that are to be submitted to a home placement test.

Eight-point wonderful	Nine-point quartermaster (unbal.)	Six-point wonderful (unbalanced)
Think it's wonderful	Like extremely	Wonderful, think it's great
Like it very much	Like strongly	I like it very much
Like it quite a bit	Like very well	I like it some what
Like it slightly	Like fairly well	So-so, it's just fair
Dislike it slightly	Like moderately	I don't particularly like it
Dislike it quite a bit	Like slightly	I don't like it at all
Dislike it very much	Dislike slightly	
Think it's terrible	Dislike moderately	
	Dislike intensely	
Seven-point excellent	**Five-point (unbalanced)**	**Six-point excellent**
Excellent	Excellent	Excellent
Very good	Good	Extremely good
Good	Fair	Very good
Fair	Poor	Good
Poor	Terrible	Fair
Very poor		Poor
Terrible		

Figure 13.3 Examples of hedonic scales that are not clearly balanced nor evenly spaced.

Test design. Each prototype is paired with the competitor in a separate sequential evaluation in which each product is used for one week. The prototypes and the competitive product are each first evaluated in half of the test homes. Each of the 150 qualified subjects is asked to rate the products on the nine-point verbally anchored hedonic scale shown in Figure 13.2.

Conduct the test. One product (prototype or competition) is placed in the home of each prescreened subject for one week. After the questionnaire is completed and the first product is removed, the second product is given to the subject to use for the second week. The second questionnaire and remaining samples are collected at the end of the second week.

Analyze results. Separate paired *t*-tests (see Chapter 14) are conducted for each prototype versus the competition. The mean acceptability scores of the samples were as follows:

	Prototype	Competition	Difference
Prototype 1	6.6	7.0	–0.4
Prototype 2	7.0	6.9	+0.1

The average difference between prototype 1 and the competition was significantly different from zero; that is, the average acceptability of prototype 1 is significantly less than the competition. There was no significant difference between prototype 2 and the competition.

Interpret results. The project manager concludes that prototype 2 did as well as the competition, and the group recommends it as the company entry into the high-fiber cereal field.

13.5.4 Assessment of Individual Attributes (Attribute Diagnostics)

As part of a consumer test, researchers often endeavor to determine the reasons for any preference or rejection by asking additional questions about the sensory attributes (appearance, aroma/fragrance, sound, flavor, texture, and feel). Such questions can be classified into the following groups:

1. Affective responses to attributes:
   Preference: Which sample do you prefer for fragrance?
   Hedonic: How do you like the texture of this product?
   [Dislike extremely -------Like extremely]
2. Intensity response to attribute:
   Strength: How strong/intense is the crispness of this cracker?
   [None ------ Very strong]
3. Appropriateness of intensity:
   Just about right: Rate the sweetness of this cereal:
   [Not at all sweet enough ------ Much too sweet]

Figure 13.4 shows examples of attribute questions; others are discussed in Section 13.5.1. In the first example—a preference questionnaire with two samples—respondents are asked, for each attribute, which sample they prefer. In the second example—an "attribute diagnostics" questionnaire with a single sample—respondents rate each attribute on a scale from "like extremely" to "dislike extremely." Such questionnaires are considered less effective in determining the importance of each attribute because subjects often rate the attributes similar to the overall response, and the result is a series of attributes that have a "halo" of the general response. In addition, if one attribute does receive a negative rating, the researcher has no way of determining the direction of the dislike. If a product texture is disliked, is it "too hard" or "too soft"?—"too thick" or "too thin"?

The "just about right" scales shown in the third and fourth examples (see also Vickers, 1988; ASTM, 2009) allow the researcher to assess the intensity of an attribute relative to some mental criterion of the subjects. "Just right" scales cannot be analyzed by calculating the mean response, as the scale might be unbalanced or unevenly spaced, depending on the relative intensities and appropriateness of each attribute in the mind of the consumer. The following procedure is recommended:

1. Calculate the percentage of subjects who respond in each category of the attribute.

Example of Results for Attribute "Just about Right" Scales

% Response	5	15	40	25	15
Category	Much too little	Somewhat too little	Just about right	Somewhat too much	Much too much

2. Using a χ^2-test (see Chapter 14), compare the distribution of responses to that obtained by a successful brand.

A similar approach is to use an intensity scale (without midpoint) for each attribute (the fifth example). To assess the appropriateness of these attributes, the intensity values must

Example 1 Attribute Diagnostics: Examples of Attribute-by-Preference Questions

1. Which sample did you prefer overall? 467—— 813——
2. Which did you prefer for color? 467—— 813——
3. Which did you prefer for cola impact? 467—— 813——
4. Which did you prefer for citrus flavor? 467—— 813——
5. Which did you prefer for spicy flavor? 467—— 813——
6. Which did you prefer for sweetness? 467—— 813——

Example 2 Attribute Diagnostics Questionnaire with a Single Sample Using Hedonic Rating of Each Attribute

☐ Like extremely
☐ Like very much
☐ Like moderately
☐ Like slightly
☐ Neither like nor dislike
☐ Dislike slightly
☐ Dislike moderately
☐ Dislike very much
☐ Dislike extremely

Using the above scale rate the following:
[Scale could be repeated after each question]
How do you feel *overall* about this beverage?________
How do you feel about the color?________
How do you feel about the cola impact?________
How do you feel about the citrus flavor?________
How do you feel about the spice flavor?________
How do you feel about the sweetness?________
How do you feel about the body?________

Example 3 "Just about Right" Scales for Attributes (Stew)

Please indicate your opinion about the following characteristics:

Gravy color	☐ Too light	☐	☐ Just about right	☐	☐ Too dark
Amount of vegetables	☐ Too few	☐	☐ Just about right	☐	☐ Too many
Amount of beef flavor	☐ Too low	☐	☐ Just about right	☐	☐ Too high
Amount of saltiness	☐ Too low	☐	☐ Just about right	☐	☐ Too high
Spiciness	☐ Too low	☐	☐ Just about right	☐	☐ Too high
Thickness of gravy	☐ Too thin	☐	☐ Just about right	☐	☐ Too thick

Example 4 Attribute Diagnostics: Implied "Just about Right" Scales

1. Color ☐ ☐ ☐ ☐ ☐ ☐ ☐
Much too light — Much too dark
2. Cola flavor ☐ ☐ ☐ ☐ ☐ ☐ ☐
Much too weak — Much too strong
3. Citrus flavor ☐ ☐ ☐ ☐ ☐ ☐ ☐
Much too weak — Much too strong
4. Sweetness ☐ ☐ ☐ ☐ ☐ ☐ ☐
Not at all sweet enough — Much too sweet
5. Thickness ☐ ☐ ☐ ☐ ☐ ☐ ☐
Much too thin — Much too thick
6. Carbonation ☐ ☐ ☐ ☐ ☐ ☐ ☐
Not at all carbonated enough — Much too carbonated

Figure 13.4 Examples of scales used in attribute diagnostics tests.

Example 5 Attribute Diagnostics: Simple Intensity Scales

Please indicate the intensity of the following attributes of the sample of pasta:

Appearance

Attribute										
1. Color intensity	Light	☐	☐	☐	☐	☐	☐	☐	☐	☐ Dark
2. Surface smoothness	Rough	☐	☐	☐	☐	☐	☐	☐	☐	☐ Smooth
3. Broken pieces	None	☐	☐	☐	☐	☐	☐	☐	☐	☐ Many

Flavor

Attribute										
4. Cooked pasta flavor/taste	None	☐	☐	☐	☐	☐	☐	☐	☐	☐ Strong
5. Saltiness	None	☐	☐	☐	☐	☐	☐	☐	☐	☐ Strong
6. Eggy flavor/taste	None	☐	☐	☐	☐	☐	☐	☐	☐	☐ Strong
7. Fresh flavor/taste	None	☐	☐	☐	☐	☐	☐	☐	☐	☐ Strong

Texture

Attribute										
8. Stickiness	Not sticky	☐	☐	☐	☐	☐	☐	☐	☐	☐ Very sticky
9. Firmness	Very soft	☐	☐	☐	☐	☐	☐	☐	☐	☐ Very firm
10. Springiness	Very mushy	☐	☐	☐	☐	☐	☐	☐	☐	☐ Very springy
11. Starchy	None	☐	☐	☐	☐	☐	☐	☐	☐	☐ Very starchy

Figure 13.4 (Continued)

be related to overall acceptance or to acceptance for that attribute. The studies conducted by General Foods on the consumer texture profile method (Szczesniak et al. 1975) related consumer intensity ratings to their own ratings for an ideal; it showed high correlations between acceptance ratings and the degree to which various products approached the consumer's ideal.

13.5.5 Other Information

Attitudes and images of a brand or product category may change over time. Market researchers conduct frequent tracking studies called attitude and usage (A&U); attitude, awareness, and usage (AAU); and usage and attitude (U&A) on a regular basis to monitor consumer perceptions and behaviors. An A&U study, when periodically repeated, provides a means to capture how marketing activities are influencing the consumer's awareness of the brand or product. Awareness is tracked relative to the competition: how a brand image is changing, how usage patterns differ across target markets, and variables defining the target market, including demographics. These studies are designed to capture specific usage information, such as frequency of going to a fast food restaurant and menu items ordered, or specific purchase behavior for chocolate chip cookies. Information collected can be used as a basis for marketing planning, new product development, and competitive intelligence. Although studies are set up strictly to measure attitudes and usage, the questions used can be incorporated into a product study to determine if the sample is tracking in a similar manner to the typical target audience.

13.6 INTERNET RESEARCH

13.6.1 Introduction

"In Q2 2013, the tipping point occurred: smartphone sales exceeded feature phone sales for the first time. The global rise of hand held portable computing devices has enhanced researchers' ability to collect data in real time. Research study subjects can complete surveys, answer open ends, upload pictures, correspond with researchers, all from wherever they are." (Stapleton 2014). While this growing means of capturing timely information from consumers is attractive to many companies, it adds another layer of potential pitfalls that must be carefully considered and controlled in order to get reliable data.

13.6.2 Applications

Internet research, whether quantitative or qualitative in nature, is a powerful tool when used correctly. Traditional surveys and questionnaires (including liking and attribute intensity ratings) can be used online to yield quantitative data. Qualitative data can be collected via Internet-based focus groups or interviews, in which consumers may be invited to join a "chat" to discuss a particular product or product concept.

The Internet can be used in early stages of concept research to determine directions or strategies for early product development, or to screen through combinations of product features. In this type of front-end research, consumers are asked quantitative and open-ended questions regarding their thoughts, feelings, and impressions of new concepts, presented to them as text descriptions that may or may not be accompanied by graphics, audio, or video. The concepts that are most likely to optimize consumer utility (i.e. the satisfaction derived from consumption of goods) can thus be identified. Results from these Internet surveys can help direct the product development process, in which a vast number of possible products must be distilled into a handful of prototypes. Statistical tools such as conjoint analysis and graph theory can be of additional help.

Conjoint analysis is a technique frequently used with data from online surveys to aid in identifying conceptual features important to consumers (Moskowitz et al., 2001; much research on the use of conjoint analysis and the Internet has been conducted by Howard Moskowitz and Associates). The goal of conjoint analysis is to determine which attributes (or combinations of attributes) of a new product concept elicit the highest possible levels of consumer acceptance and satisfaction (Moskowitz et al. 2001). Conjoint analysis can be divided into two main categories, ratings-based and choice-based. Ratings-based conjoint analysis asks respondents to rate their acceptance (e.g. purchase intent) of a hypothetical product composed of a selected combination of attributes. In choice-based conjoint analysis, respondents choose their most preferred hypothetical product among a set of products composed of different combinations of attributes. Ratings of the hypothetical products are then analyzed (via regression analysis) to assess which attributes contribute most to consumer acceptance.

Similarly, graph theory is used to narrow a broad range of attributes into a subset of attribute combinations that would most likely elicit positive consumer responses. Respondents are shown all possible attribute pairs and are asked to indicate whether each pair is compatible. All attributes and their compatible pairs are then mapped onto a graph, which is used to predict which larger combinations of attributes would be acceptable to consumers. This approach reduces the number of permutations that are shown to respondents, which is advantageous in situations where the combinations generated by conjoint analysis are too vast for the practical restraints of survey research. Applications have included identifying the most suitable combination of salad ingredients and optimizing Meal-Ready-to-Eat (MREs) for military rations (Nestrud et al. 2011; Nestrud 2011), which show the feasibility of the approach in product and menu development applications.

13.6.3 Design of Internet Research

Internet research methods may be developed internally or externally. When designing Internet research internally, a number of research tools and software programs are available for guidance, including *Zoomerang* and *SurveyMonkey*, to name a few. These programs provide guidance in designing and deploying surveys via email or website posting, and in analyzing the data. Alternatively, an external research firm may be contracted for assistance. Numerous companies conduct Internet research–each with their own approach to recruiting, data delivery, and data analysis. The needs of the project and the sensory team will drive the selection of an appropriate Internet-based research method.

Proper communication with participating consumers is crucial when planning and conducting Internet research. Communication happens during initial contact, response, and follow-up, and optionally during a pre-notification period. Communicating with participants at all four of these time points will result in a higher response rate and higher participant satisfaction. Pre-notification emails have the added bonus of generating interest about the study and notifying participants to be on alert for the survey email. Follow-ups can be conducted via phone, email, direct mail, etc. and can remind a participant that a survey has been sent to their email. These communication modes can also be used to thank a participant for completing the survey.

The basic steps in the process of designing and implementing an online survey are outlined below, and the success of the research study is dependent upon all steps involved.

1. Define the survey objectives, including:
   - Specifying the population of interest (e.g., women age 35+ or cat owners)
   - Delineating the type of data to be collected (quantitative or qualitative)
   - Determining the desired precision of the results

2. Determine the sampling method
3. Create and test the survey instrument, including:
    - Choosing the response mode (mail, Web, or other)
    - Drafting the questions
    - Pretesting and revising the survey instrument
4. Contact respondents throughout the process by:
    - Pre-notifiying participants about the upcoming survey
    - Sending the survey via email or link
    - Reminding participants to complete the survey (if necessary)
    - Thanking participants after survey is completed
5. Collect, organize, and analyze data

13.6.4 Internet Research Considerations

13.6.4.1 Benefits and Pitfalls of Using the Internet for Research

Internet research (e.g. online, mobile) has increasingly become an attractive way to collect data, and has many advantages over traditional data collection, but it also has many potential pitfalls. This section will cover both the advantages and the potential pitfalls of Internet research and will describe how to account for them during study design.

Advantages:

- Control of question presentation
- Control of question completion (i.e. participant cannot continue survey without completing required questions)
- Use of other media in survey (e.g. images, video)
- More than one way for participants to record their data (e.g. text messaging or online via computer, phone, or tablet)
- No transcription errors
- Timestamp of when survey is completed
- Real time data collection
- Convenient for participants
- Broader audience and more diverse demographic profile
- Large study populations that would not be feasible with traditional consumer methods

As consumers increasingly spend more time on the Internet and hone their online skills, the list of reasons to capture data via this medium continues to grow. While the benefits of capturing data in this manner are many, there are inherent problems with Internet research that need to be recognized and accounted for when designing such studies and reviewing this type of data.

Pitfalls:

- Inconsistent computer skills among participants
- Time gap between product use and survey completion
- Participants' short online attention span may prevent use of long surveys
- Self-reported demographics and ethnographics may be inaccurate

- Varying Internet speeds and use of different operating systems may affect robustness/ ease-of-use of survey

Understanding how participants interface with the Internet on their computers and mobile devices is of utmost importance when designing Internet research. Researchers should consider questions such as:

- Is the participant Internet-savvy and used to accessing the Internet via their smartphone or tablet multiple times during the day?
- Can the participant upload a picture from their device onto the Internet?
- Is the participant capable of browsing the Internet for content at the same time they are taking your survey?

If these skills are required for the success of the study, it is important to ensure that they are addressed in screening questionnaires. Not taking the time to understand participants and their skill levels can lead to frustration on both ends and possibly less reliable data. If participants are asked to complete tasks that take more time than anticipated or result in undesired results (e.g. losing data, browser difficulties), the risk of non-completion or completion using erroneous data entered out of frustration increases. Poor data may be collected and/or the study may result in failure, which consequently may lead to poor decisions.

In the context of at-home product research using the Internet, understanding how participants interact with the product relative to where and when the survey will be completed is key to successful research design. Considering whether participants will be able to access the Internet while using the product or whether they will more likely have to enter their data once their product experience is complete, can inform the overall design of the survey and help to determine the type and format of survey questions. For instance, if there is a long time between the participant using the product and completing the survey (e.g. women's in-shower shaving cream), questions with pre-determined answer choices (e.g. dropdown lists, check all that apply (CATA), etc.) can help prompt the user's memory. However, in the moment "ah ha's" may be missed because no mechanism exists for the participant to record their revelation while in the shower. Including open-ended questions will allow for unprompted responses even if participants are no longer "in the moment." Alternately, if the product is used during a busy or distracting time (e.g. family meal, rush hour, etc.), keeping the survey short and direct will likely yield the most accurate and meaningful data without disengaging the participants.

When designing ethnographic and anthropologic studies, it is important to recognize that Internet-based research has a more narrow scope than field observation. Participants may self-edit, skip steps, and/or provide incomplete information when completing online surveys. Field observation permits more detailed observations of participants and can provide a context that may not be conveyed online. Understanding what nuances might be missing from data collected online can aid in realistic interpretation of the data (Stapleton 2014).

When utilizing "live" online platforms (e.g. text messaging, chat rooms, and live video) to obtain qualitative data, researchers need to consider how interaction with participants online will be different than in person. While true of all Internet-based data capture, live capture is especially fraught with the inability to read a participant's intended tone. Pauses or gaps in response may indicate that the participant needs to reflect before answering or alternatively, that a disconnection or distraction has occurred. Participants' typing/texting skills and the speed and quality of their Internet connections can also impact response times.

13.6.4.2 Platform

Online surveys should be simple and easy to use on a wide variety of platforms. Despite the rapidly growing Internet research market, many surveys fail to conform to all or even most operating systems. Just because a survey works on the system on which it was created, does not mean it will work for everyone. There are a multitude of different smartphones, tablets, computers, netbooks, etc. that utilize different operating systems and browsers, have varying speeds, and/or have a wide variety of viewing capabilities & screen sizes. Developing questions that are easy to read and answer across all devices will increase the probability that the survey collects meaningful data.

13.6.4.3 Recommendations and Checks & Balances

During the planning and development of Internet research, the following recommendations may prove useful:

- Before beginning, the researcher should ensure that the purpose and objectives truly lend themselves to Internet research.
- Thinking through possible challenges one's study may encounter and anticipating how these may impact the results will allow a more critical review of findings and lead to a more educated decision once data is collected.
- In ongoing, long-term Internet research, increasing the number of participants (in anticipation of attrition) should also be considered, at least until a clearer understanding of expectations is reached.
- Creating redundancy in the study by including other types of field research in addition to Internet research may provide a richer view of how consumers are engaging with the product and will indicate if online studies are mirroring more traditional results.

Case Study: Internet Research

Internet research intended to gather insights into consumers' perceptions is ideally used throughout the product development cycle, with added insights gained when used early in the process. Specific techniques/approaches can be implemented to decipher consumers' needs and preferences through an interactive process.

In this case study, the Consumer's Mind Internet research tool by Future Strategies was used for data collection and analysis. One hundred and sixteen female chocolate consumers, aged 20–65 years, were recruited via phone and Internet. Two phases of research were conducted: In phase one, the consumer responded to a series of questions on chocolate attributes such as level of sweetness or creamy smooth texture, their reasons for eating chocolate, and their selection criteria, including brand and price. For each of the questions, consumers were asked how important the attribute is and how satisfied they are with their current product. In phase two, the respondents were provided with six chocolate samples, three milk and three dark, to evaluate for the same attributes tested in phase one, in addition to liking on a nine-point hedonic scale. Phase one defines the critical aspects that drive a product's success by providing insights that help to understand underlying consumer needs. In the first-level comparative analysis, consumers were asked "What is chocolate better than?" to provide a framework for designing an ideal positioning for chocolate. Results indicated that a credible product positioning would be chocolate that offers more satisfaction than most foods but not more than a romantic evening (see Table 13.2).

An expectation gap is calculated using importance and satisfaction ratings to identify ways to carve out a unique positioning in the marketplace by revealing meaningful consumer motivations (as shown in Table 13.3). This analysis matches product performance (satisfaction) to expectations (importance) to help target the product language that

Table 13.2 Chocolate Is Better Than…?

	Percent	
Answer	**Yes**	**No**
A cocktail	64	36
Exercise	63	37
Sports event	59	41
Most food	59	42
Shopping	56	47
Ice cream	56	43
Nothing	55	44
A massage	52	58
A good book	39	61
A good movie	39	61
Jewelry	35	65
Day at the beach	32	68
Sex	28	72

Table 13.3 Importance to Satisfaction Scores

	Importance > Satisfaction	**Target (%)**	**Satisfaction > Importance (%)**
To treat myself		110	
As an indulgent treat		113	
I don't feel guilty eating		112	
To satisfy a craving		111	
Has a taste I love			
Is my favorite food			120
To fill an emotional need			126
Has some health benefits			132

is relevant to consumers. In this study, only one attribute—"has a taste I love"—was on target, because the importance of this attribute was equal to the consumers' current product's satisfaction rating. Attributes that were less important, though satisfied by current products, were "a favorite food," "met an emotional need," and "has health benefits."

Postproduct analysis resulted in one product achieving the highest overall liking score. Milk chocolate ratings were higher than dark chocolate. The pre- and postproduct experience provided another satisfaction to importance gap analysis to explore consumer expectations (see Table 13.4). In the pretest, "Creamy smooth texture" was identified as the most critical attribute. In the post-test, the winning milk chocolate has a comparable importance to satisfaction rating, thereby being on target. Post-test satisfaction scores fell below pretest scores and did not meet consumer needs for "chocolate intensity," "the way it melts in the mouth," and "lingering flavor." None of the products were considered to have a sophisticated taste for the overall group, yet when user of milk and dark chocolate are segmented, all dark chocolate products were seen as sophisticated.

Insights from the Internet research were that

1. The illusion of chocolate for indulgence is greater than the satisfaction. The preproduct analysis demonstrated the importance of emotional and intellectual drivers in the chocolate category, and post-test scores indicate that the chocolate experience may not meet consumer's expectations for satisfying cravings.
2. Texture, including rate of melt, are the defining sensory properties in chocolate and are more important than flavor. Texture is tied to the ratings for sophistication and indulgence.
3. Brand name is a driving factor in chocolate selection; however, brand recognition was not apparent in this study.

Table 13.4 Comparison of Importance and Satisfaction Scores with Actual Product Satisfaction

	Pretest		Post-Test					
			Satisfaction					
Attribute	**Importance**	**Satisfaction**	**Dove Milk**	**Dove Dark**	**Godiva Milk**	**Godiva Dark**	**Valrhona Dark**	**Lindt Milk**
Has a creamy smooth texture	3.85	3.98	3.85	3.60	3.34	3.08	3.06	3.32
The intensity of the chocolate flavor	3.71	3.98	3.62	3.67	3.17	3.37	3.45	2.71
The way it melts in your mouth	3.61	4.01	3.78	3.58	3.40	3.10	2.99	3.38
Has a lingering flavor	3.03	3.62	3.53	3.36	3.10	3.11	3.20	2.93
The sophisticated taste	2.81	3.82	3.19	3.42	2.78	2.89	3.09	2.36
The flavor is dark chocolate	2.39	3.70	2.47	4.02	2.98	3.84	3.97	1.79

13.7 USING OTHER SENSORY METHODS TO UNCOVER INSIGHTS

13.7.1 Relating Affective and Descriptive Data

Product development professionals handling both the R&D and marketing aspects of a product cycle recognize that the consumer's response in terms of overall acceptance and purchase intent is the bottom line in the decision to go or not go with a product or concept (Beausire et al., 1988).

Despite the recognition of the need for affective data, the product development team is generally unsure about what the consumer means when asked about actual sensory perceptions. When a consumer rates a product as "too dry" or "not enough chocolate taste," is he really responding to perceived moistness/dryness or perceived chocolate flavor? Or is he responding to words that are associated in his mind with goodness or badness in the product? Too many researchers are taking the consumer's response at face value (as the researcher uses the sensory terms) and these researchers end up "fixing" attributes that may not be broken.

One key to decoding consumer diagnostics and consumer acceptance is to measure the perceived sensory properties of a product using a more objective sensory tool (Shepherd et al., 1988). The trained descriptive or expert panel provides a thumbprint or spectrum of product sensory properties. This sensory documentation constitutes a list of real attribute characteristics or differences among products that can be used both to design relevant questionnaires and to interpret the resulting consumer data after the test is completed. By relating consumer data to panel data—and, when possible, to ingredient and processing variables—or to instrumental or chemical analyses, the researcher can discover the relationships between product attributes and the ultimate bottom line: consumer acceptance.

When data are available for several samples (15–30) that span a range of intensities for several attributes (see the hand and body lotion example in Appendixes 13.1 and 13.2), it is possible to study relationships in the data using the statistical methods described in 15, Section 15.2. Figure 13.5 shows four examples. Graph A shows how consumer overall acceptance varies with the intensity of a descriptive panel attribute (e.g., color intensity); this allows the researcher to understand the effect of different intensities of a characteristic and to identify acceptable limits. In graph B, the abscissa depicts the intensity of an undesirable attribute, for example, an off flavor, and the ordinate is consumer acceptance of flavor; the steep slope indicates a strong effect on liking for one facet of the product. From the type of relationship in graph C, the researcher can learn how consumers use certain words relative to the more technically precise descriptive terms; note that the descriptive panel's rating for crispness correlates well with the consumers rating, but the latter rises less steeply. Finally, graph D relates two consumer ratings, showing the range of intensities of an attribute that the consumer finds acceptable. Such a relationship is tantamount to a "just about right" assessment.

The data relationships in Figure 13.5 are univariate. Consumer data often shows interaction between several variables (products, subjects, and one or more attributes). This type of data requires multivariate statistical methods such as principal component regression (PCR) or partial least squares (PLS) (see Muñoz et al., 1996 and Chapter 15).

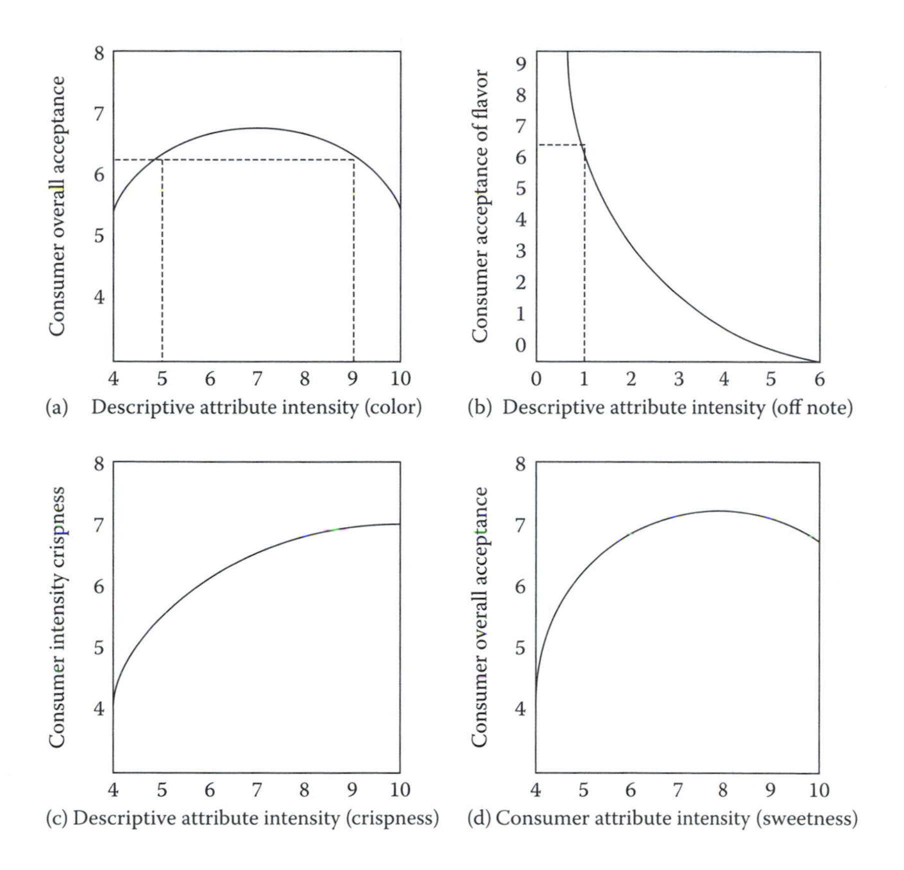

Figure 13.5 Examples of data relationships extracted from a consumer study. (a) consumer overall acceptance vs. descriptive attribute intensity (color intensity); (b) consumer acceptance for flavor vs. descriptive attribute intensity (flavor off note); (c) consumer intensity crispness vs. descriptive attribute intensity (crispness); (d) consumer overall acceptance vs. consumer attribute intensity (sweetness).

Case Study: Relating Consumer Qualitative Information with Descriptive Analysis Data

A manufacturer of sunscreens is developing a new body-lotion-sunscreen product. Results from the front end research indicate that there is a prime opportunity to sell a sunscreen that feels like a moisturizer but delivers the protection of a sunscreen. The manufacturer is looking for preliminary information and direction for the product developers.

The first step is to document a large array (15–30) of hand-lotion products in the marketplace from which a diverse subset of (4–8) are selected for discussion with consumers.

A trained Spectrum descriptive analysis panel documents the skinfeel properties of a large array (16–20) of lotions from the retail marketplace, thus defining precisely what the products feel like as they are dispensed in the hand and on the skin (see Chapter 12). Figure 13.6 shows the data range for the 16 samples for the initial rubout characteristics.

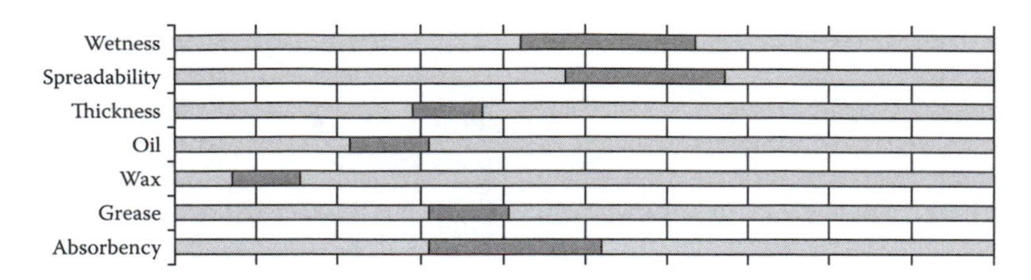

Figure 13.6 Data range graph for descriptive analysis rubout characteristics.

The descriptive panel results are analyzed using multivariate statistical techniques, such as principle component analysis. Maps of the data permit the researchers to look into/onto the whole space that encompasses the commercial hand lotions. From the map, a diverse subset of five lotions is selected for discussion with consumers.

Qualitative interviews are conducted with selected consumers. The consumers describe the optimum lotion-sunscreen as giving the skin a "soft, flexible, cushion, and hydrated" feeling. These terms are used by the consumers to describe the samples that the descriptive analysis panel describes as being higher in skin suppleness, lower in skin-texture visibility, and having more of a silicone feeling than oily or greasy. By considering the known range of sensory intensities from the original array, the sensory scientists develop a guideline for intensity, providing a development direction. This is illustrated in Figure 13.7. The product developers are now able to create prototypes for further testing, whether it is consumer acceptance, preference, or perception of efficacy.

13.7.2 Using Affective Data to Define Shelf-Life or Quality Control Limits

Chapter 17 describes quality control, and as noted in Section 12.6, a "modified" or short-version Spectrum descriptive analysis procedure can be used to define QA/QC or shelf-life limits. In a typical case, the first step is to send the fresh product out for an acceptability test in a typical user group. This initial questionnaire may contain additional questions asking the consumer to rate a few important attributes.

The product is also rated for acceptability and key attributes by the modified panel, and this evaluation is repeated at regular intervals during the shelf storage period, each time comparing the stored product with a control that may be the same product stored

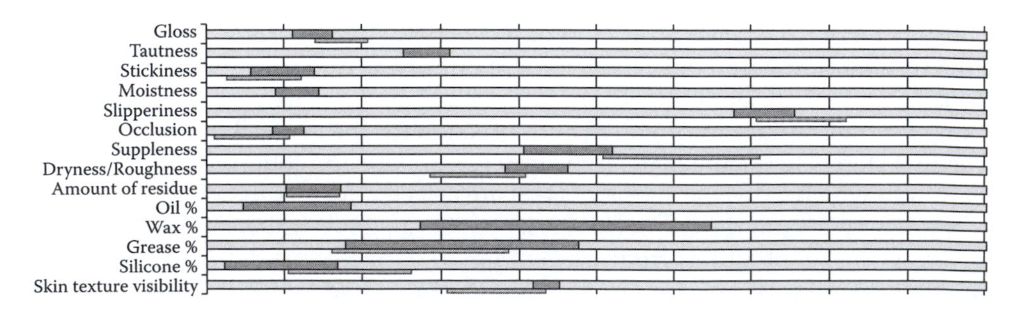

Figure 13.7 Suggested direction for afterfeel characteristics.

under conditions that inhibit perceptible deterioration (e.g., deep-freeze storage under nitrogen) or, if this is not possible, a fresh product of current production.

When a significant difference is found by the modified panel in overall difference from the control and/or in some major attribute(s), the samples are sent again to the user group to determine if the statistically significant difference is meaningful to the consumer. This is repeated as the difference grows with time of shelf storage. After the size of a panel difference can be related to what reduces consumer acceptance or preference, the internal panel can be used in the future to monitor regular production in shelf-life studies, with assurance that the results are predictive of consumer reaction.

Example 13.3: Shelf Life of Sesame Cracker

Problem/situation. A company wishes to define the shelf life of a new sesame cracker in terms of the "sell by" date that will be printed on packages on the day of production.

Project objective. To determine at what point during shelf storage the product will be considered "off," "stale," or "not fresh" by the consumer.

Test objective. (1) Use a research panel trained for the purpose of determining the key attributes of the product at various points during shelf storage and (2) submit the product to consumer acceptance tests (a) initially; (b) when the research panel first establishes a difference; and (c) at intervals thereafter until the consumers establish a difference.

Test design. Samples of a single batch of the sesame crackers were held for 2, 4, 6, 8, and 12 weeks under four different sets of conditions: *control*=near freezing in airtight containers; *ambient* = 70°F and 50% RH; *humid* = 85°F and 70% RH; and *hot* = 100°F and 30% RH.

Subjects. Twenty-five panelists from the R&D lab are selected for their ability to recognize the aromatics of stale sesame crackers, that is, the cardboard aromatic of the stale base cracker and the painty aromatic of oxidized oil from the seeds. Two hundred and fifty consumers must be users of snack crackers and are chosen demographically to represent the target population.

Sensory methods. The research panel used the questionnaire in Figure 13.8 and was trained to score the test samples on the seven line scales that represent key attributes of appearance, flavor, and texture related to the shelf life of crackers and sesame seeds. Research panelists also received a sample marked "control" with instructions to use the last line of the form as a difference-from-control test (see Section 7.8). The panelists were informed that these samples were part of a shelf-life study and that occasional test samples would consist of freshly prepared "control product" (such information reduces the tendency of panelists in shelf-life testing to anticipate more and more degradation in products).

On each occasion, the consumers received two successive coded samples (the test product and the control, in random order) each, along with the score sheet in Figure 13.9, which they completed immediately and returned to the interviewer.

Analyze results. The initial acceptance test, in which the 250 consumers received two fresh samples, provided a baseline rating of 7.2 for both, and the accompanying attribute ratings indicated that the crackers were perceived to be fresh and crisp.

The same two identical samples were rated 3.2 (out of 15) on the difference-from-control scale by the research panel. The 2- and 4-week samples showed no significant differences. At the 6-week point, the "humid" sample received a difference-from-control rating of 5.9,

Evaluation of Sesame Cracker

Instructions

1. Evaluate the cracker for appearance, flavor, and texture by placing a mark on each line below.

Appearance

Surface color	Light	Dark

Flavor

Toasted wheat	None	Strong
Sesame seed	None	Strong
Cardboard	None	Strong
Painty	None	Strong

Texture

Hardness	Soft	Hard
Crispness	Soggy	Crisp

2. Compare the cracker with the control and indicate the amount of difference between them by placing a mark on the line below:

No difference ———————————— Very different

Comments ____________________________________

____

____

Name ______________________________ Date ________

Figure 13.8 Research panel score sheet showing attribute rating and difference rating for Example 13.3: Shelf life of sesame cracker.

which was significantly different from 3.2. In addition, the "humid" sample was rated 4.2 in cardboard flavor (against 0 for the fresh control) and 5.1 in crispness (against 8.3 for the fresh control), both significant differences by ANOVA.

The 6-week "humid" samples were then tested by the consumers and were rated 6.7 on acceptance, against 7.1 for the control ($p < 0.05$). The rating for "fresh toasted flavor" also showed a significant drop.

The product researcher decided to conduct consumer tests with the other two test samples ("ambient" and "hot") as soon as the difference-from-control ratings by the research panel exceeded 5.0. Subsequent tests showed that consumers were only sensitive to differences that were rated 5.5 or above by the research panel. All further shelf-life testing on sesame crackers used the 5.5 difference-from-control rating as the critical point above which differences were not only statistically significant but also potentially meaningful to the consumer.

Sesame Cracker

Instructions

1. Overall evaluation. Place a mark in the box which you feel best describes how you like the product:

☐	☐	☐	☐	☐	☐	☐	☐	☐
Like extremely	Like very much	Like moderately	Like slightly	Neither like nor dislike	Dislike slightly	Dislike moderately	Dislike very much	Like extremely

2. Indicate by placing a mark how you feel the product rates in each category below:

Appearance

Color	☐ Light	☐	☐	☐	☐	☐ Dark

Flavor

Salty	☐ Not at all salty	☐	☐	☐	☐	☐ Very salty
Sesame flavor	☐ No sesame flavor	☐	☐	☐	☐	☐ Strong flavor
Fresh toasted flavor	☐ Stale, not fresh	☐	☐	☐	☐	☐ Very fresh

Texture

Crispness	☐ Soggy	☐	☐	☐	☐	☐ Crisp
Aftertaste	☐ Unpleasant	☐	☐	☐	☐	☐ Pleasant

Comments ____

____

____

Name ______________________________ Date ________

Figure 13.9 Consumer score sheet for Example 13.3: Shelf life of sesame cracker.

13.7.3 Rapid Prototype Development

There is an ongoing endeavor to identify and implement approaches to testing that would provide rapid feedback to product development and allow for a shortened development cycle. Various approaches exist that are easily implemented; however, it is suggested that the findings be validated prior to either large-scale market research or product launch. Loosely defined, rapid prototype development employs quantitative and qualitative techniques to collect consumer input, feedback, and insights during the product development process, following an accelerated timetable. Requirements for effective rapid prototyping include the need for actionable information and rapid feedback; it is also required to be iterative, able to handle multiple samples, of low to moderate cost, and of smaller scale.

The quantitative and qualitative techniques employed can be executed independently of each other or in combination. Three scenarios for rapid prototyping and the testing plan are as follows.

In Scenario 1, a group of target respondents are recruited to participate in focus groups on toothbrushes over a period of time. Two to four groups are conducted at one time point with respondents returning to participate in three to four successive rounds of testing. Each round focuses on either a new facet of the toothbrush such as the number, length, and stiffness of the bristles, or the size and shape of the handle. The groups focus on sensory properties with numerous stimuli presented to represent ranges of intensities for various attributes. Product developers, marketers, and sensory professionals viewing from the back room listen for sensory cues for product improvement. Ideas are taken back to the laboratory for creation of new prototypes.

Scenario 2 involves quantitative testing where a small number of consumers ($n = 50–75$) are either recruited from the mall or prerecruited to participate in small-scale taste tests on sweet and savory crackers. Respondents taste a series of five to six crackers that represent different levels of sweet impression and savory character. At the end of the tasting, short one-on-one interviews are conducted that allow respondents to verbalize their thoughts on the crackers. Utilizing electronic data-collection techniques, information is turned around rapidly and reviewed with the comments from the interviews. Product development is able to modify the prototypes that highlight consumer's response. This process is repeated three to four more times on a shortened cycle until measurable improvements are found in the products.

Scenario 3 uses community narratives or story telling as an approach. There is a desire to create a new or improved teen beverage that provides energy, nutritional value, and replaces lost nutrients and can be consumed during practice and games. A group of 10–15 teens who participate in sports are recruited for a 2-month testing program. One day per week, the teens go to a facility for 2 hours. The 2-hour block of time is divided into thirds, with one-third spent meeting or congregating to talk about their needs and wants, the second third spent exercising, and the final third spent tasting and discussing the products. This method develops a sense of community, allowing the teens to build off each other and provide feedback in a real-world setting.

Utilizing rapid prototype development means talking to consumers on a regular, ongoing basis; this allows them to provide constant feedback. Although it is a more hands-on approach, it provides direct feedback and the ability to clarify responses in a rapid manner.

APPENDIX 13.1 SCREENERS FOR CONSUMER STUDIES—FOCUS GROUP, CLT, AND HOME USE TEST (HUT)

Screener

Hand and Body Lotion

General: For Qualitative (Focus Group) or Quantitative (CLT or HUT)

Name ____________________ Date __________

Phone ____________________ Time __________

(Day) (Evening)

Street ____________________

City __________ State __________ Zip __________

Interviewer __________ Location __________

Appointment:

Date __________

Time __________

Introduction to respondent:

Hello, I'm __________ of __________, a national survey research firm. We are conducting a survey; do you have a few minutes to answer some questions?

If no: Ok, thank you for your time.

If yes: That would be great. If you qualify at the end of the survey, you will be asked to participate in a study. We will make an appointment for you to come in at that time.

So, let's begin.

Broad Questions

1. Record the gender of the respondent:

Male	( )	Terminate or continue based on quota
Female	( )	Terminate or continue based on quota

2. In the past 3 months, have you yourself participated in a survey, panel discussion, or consumer test?

Yes	( )	Terminate and tally
No	( )	Continue

3. Do you or does any member of your immediate family work for any of the following types of businesses? (Read List)

	Yes	No
Advertising agency or television	( )	( )
A marketing research firm	( )	( )
A public relations firm	( )	( )
Scientific research or related field	( )	( )
A company that retails, wholesales or manufactures personal care products	( )	( )
A cosmetic discount store	( )	( )

Terminate if yes, don't know, or refuse to answer to any of the questions

4. For classification purposes, please tell me which of the following best describes your age. (Recruit a mix)

Under 25 years	( )	Terminate or continue
25–34 years	( )	As quotas are filled
35–45 years	( )	
46–55 years	( )	
56+ years	( )	

5. Which of the following income brackets best describes your total household income? (Recruit a mix)

Under $30,000	( )	Terminate or continue
$30,000–55,000	( )	as quotas are filled
$55,000–80,000	( )	
$80,000–100,000	( )	
Over $100,000	( )	

Specific Questions

6. Which of the following items have you yourself purchased and used on a regular basis in the past 6 months? (Mark all that apply)

Hand & body lotion	( )	Terminate if not checked
Laundry detergent	( )	
Pretzels	( )	
Facial tissue	( )	
Soda	( )	

7. Do you have any skin allergies or sensitivities to

Bar soaps?	( )
Laundry detergents?	( )

Fragranced hand & body lotions? ( ) Terminate if checked
Shampoos? ( )
Nonfragranced hand & body lotions? ( ) Terminate if checked

8. How often do you apply hand & body lotion during a day?

None to 3 times ( )
4–6 times ( ) Terminate or continue based on quota
7–10 times ( )
More than 10 times ( )

9. Which brand of hand & body lotion do you use most often?

Brand A ( )
Brand B ( ) Must be checked to continue
Brand C ( )

10. I am going to read you a series of statements; tell me whether you strongly agree, agree, neither agree nor disagree, disagree, or strongly disagree with the following statements.

Focus Group Only	**Strongly Agree**	**Agree**	**Neither/ Nor**	**Disagree**	**Strongly Disagree**
a. Care of my skin is very important to me.	◊	◊	◊	◊	◊
b. I eat foods that offer vitamins and nutrients for my skin.	◊	◊	◊	◊	◊
c. I use a hand & body lotion that contains vitamins and minerals to nourish my skin.	◊	◊	◊	◊	◊
d. It is more important to apply hand & body lotion in the winter than in the summer.	◊	◊	◊	◊	◊
e. The fragrance of hand & body	◊	◊	◊	◊	◊

I would like your reaction to a few statements [read list]
I am comfortable expressing my opinions and beliefs. ___yes ___no
I enjoy group discussions in which everyone expresses their opinions. ___yes ___no
If asked to describe something, I can usually do so in detail. ___yes ___no

[To qualify respondents must answer "yes" on each statement]

My next question is somewhat different from the others I have asked so far, but please give me your best answer. If you could have dinner with anyone, who would it be, why would you choose them, and what would you talk about?

Who? ____

Why? ____

What? ____

> Note: This question is to screen the articulation of the respondent. Listen for the manner in which the respondent answers this question, not the content of his/her answer. We need respondents who can express themselves clearly and easily verbalize their thoughts on abstract concepts.

Invitations

Focus Group

Our company is inviting men and women such as you to participate in a market research study on __________ at _____________. The focus group discussion will last approximately 75 min and as compensation you will be paid $ _____ for your time and input. You will not receive payment if you are not present when the session begins or if you are unable to attend the entire session. Would you be willing to participate?

Yes ( ) No ( ) Time: __________

Central Location Test (CLT)

I would like to invite you to participate in an interesting study we are conducting at our office on ________. Would you be willing to come to our office to try several hand & body lotions over a 2-day period? Each day you would give your opinion of 4 different lotions. Each session will last approximately 45 min. For your time and participation you will receive $________. Would you be willing to participate?

Yes ( ) No ( ) Time: __________

Home Use Test (HUT)

I would like to invite you to participate in an interesting market research in-home use study of hand & body lotions. Over the next month, you will be asked to use two different hand & body lotions. You will be asked to pick up the first product from our facility at _________ and use it at home for 14 days. During that time, you will be asked to answer questions and give us your reactions and comments in a diary that will be provided to you. At the end of the 14 days, you will bring the product and diary to the facility and be given a second sample to use for 14 days. During that time, you will again be asked to answer questions and give us your reactions and comments in a diary that will be provided to you. For your time and participation, you will receive $________.

Would you be willing to participate?

Yes ( ) No ( )

APPENDIX 13.2 DISCUSSION GUIDE—GROUP OR ONE-ON-ONE INTERVIEWS

(Simple) Discussion Guide

Nurturing Hand & Body Lotion

Group 1 25–34 years; women; use hand & body lotion daily
Group 2 35–44 years; women; use hand & body lotion daily
Group 3 45–54 years; women; use hand & body lotion only when needed
Group 4 55–64 years; women; use hand & body lotion daily

- Purpose/introduction/warm-up/ground rules (15 min)
    - Thank everyone for participating; very interested in hearing what everyone has to say; there are no wrong answers; interested in everyone's opinions
    - Discuss rules: one person at a time; wait to be recognized; video tape for documentation and notes; no cell phones; location of facilities; length of discussion; consideration of others in room; confidentiality
    - Purpose of group discussion—to better understand use and wants of product
    - Around room intros: tell name, age, occupation, type of skin, skin concerns, and what kind of skin treatment used and how often
- Introduce and review concept (10 min): *A hand & body lotion that renews your skin by releasing nurturing vitamins and minerals with every use.*
- Reaction to concept (20 min)
    - Discussion to probe reaction to concept
        - On the paper in front of you, write three words that would describe the ideal product characteristics based on this concept
    - Probes:
        - What does concept say to you?
            - Expected performance
            - Meaning of "nurturing"
            - How it makes you feel
            - Perceived benefits; overall, from vitamins and minerals
            - When product would be used
            - What else would provide such benefits?
- Product sort and selection criteria (15 min)
    - Look at the collection of hand & body lotions (8–10 products) on the table
        - How would you group or categorize these products? Select a member of the group to take the lead. (*Observe the process and then probe on line up decision, placement, etc.*)
        - Define if any product is better, special, or different from the others. Is there a product on the table that best matches the concept?

Possible probes:

- Usual routine
- Types of products normally purchase

- Brands
- Quality
- Necessity versus indulgence
- Value
- Price points
- Additional expectations

APPENDIX 13.3 QUESTIONNAIRES FOR CONSUMER STUDIES

QUESTIONNAIRES FOR CONSUMER STUDIES

A. Candy Bar Questionnaire

Name ____________________

Product # ________________

Candy Bar

- Please rinse your mouth before starting.
- Evaluate the product in front of you by looking at it and tasting it.
- Considering *ALL* characteristics (*APPEARANCE,FLAVOR,* and *TEXTURE*) indicate your overall opinion by checking one box [√].

☐ ☐ ☐ ☐ ☐ ☐ ☐ ☐ ☐ ☐ ☐

Dislike extremely | Neither like nor dislike (nl/nd) | Like extremely

- Comments: Please indicate WHAT in particular you liked or disliked about this product. (USE WORDS NOT SENTENCES.)

LIKED	DISLIKED
____________________	____________________
____________________	____________________
____________________	____________________
____________________	____________________

1. Candy Bar Liking Questions

Please retaste the product as needed and indicate how much you LIKE or DISLIKE the following. *Check* the box that represents your response [√].

Overall appearance

☐ ☐ ☐ ☐ ☐ ☐ ☐ ☐ ☐ ☐ ☐

Dislike extremely | nl/nd | Like extremely

Overall flavor

☐ ☐ ☐ ☐ ☐ ☐ ☐ ☐ ☐ ☐ ☐

Dislike extremely | nl/nd | Like extremely

Overall texture

☐ ☐ ☐ ☐ ☐ ☐ ☐ ☐ ☐ ☐ ☐

Dislike extremely | nl/nd | Like extremely

2. Candy Bar Specific Evaluation

Retaste the product as needed and check the box for your response [✓] for both questions (LIKING and INTENSITY LEVEL) for each characteristic.

	Liking	Intensity/Level
Appearance		
Color	☐ ☐ ☐ ☐ ☐ ☐ ☐ ☐ ☐ Dislike extremely — nl/nd — Like extremely	☐ ☐ ☐ ☐ ☐ ☐ ☐ ☐ ☐ Light — Dark
Color uniformity	☐ ☐ ☐ ☐ ☐ ☐ ☐ ☐ ☐ Dislike extremely — nl/nd — Like extremely	☐ ☐ ☐ ☐ ☐ ☐ ☐ ☐ ☐ Non-uniform — Uniform
Amount of broken blisters	☐ ☐ ☐ ☐ ☐ ☐ ☐ ☐ ☐ Dislike extremely — nl/nd — Like extremely	☐ ☐ ☐ ☐ ☐ ☐ ☐ ☐ ☐ None — Many
Flavor		
Chocolate flavor	☐ ☐ ☐ ☐ ☐ ☐ ☐ ☐ ☐ Dislike extremely — nl/nd — Like extremely	☐ ☐ ☐ ☐ ☐ ☐ ☐ ☐ ☐ None — High
Peanut flavor	☐ ☐ ☐ ☐ ☐ ☐ ☐ ☐ ☐ Dislike extremely — nl/nd — Like extremely	☐ ☐ ☐ ☐ ☐ ☐ ☐ ☐ ☐ None — High
Roasted/toasted flavor	☐ ☐ ☐ ☐ ☐ ☐ ☐ ☐ ☐ Dislike extremely — nl/nd — Like extremely	☐ ☐ ☐ ☐ ☐ ☐ ☐ ☐ ☐ None — High
Sweetness	☐ ☐ ☐ ☐ ☐ ☐ ☐ ☐ ☐ Dislike extremely — nl/nd — Like extremely	☐ ☐ ☐ ☐ ☐ ☐ ☐ ☐ ☐ None — High

2. Candy Bar Specific Evaluation (Continued)

Texture

Attribute	Liking scale	Intensity scale
Firmness of whole bar	☐ ☐ ☐ ☐ ☐ ☐ ☐ ☐ ☐ Dislike extremely … nI/nd … Like extremely	☐ ☐ ☐ ☐ ☐ ☐ ☐ ☐ ☐ Soft … Firm
Crunchiness of nuts	☐ ☐ ☐ ☐ ☐ ☐ ☐ ☐ ☐ Dislike extremely … nI/nd … Like extremely	☐ ☐ ☐ ☐ ☐ ☐ ☐ ☐ ☐ Not crunchy … Crunchy
Rate of melt	☐ ☐ ☐ ☐ ☐ ☐ ☐ ☐ ☐ Dislike extremely … nI/nd … Like extremely	☐ ☐ ☐ ☐ ☐ ☐ ☐ ☐ ☐ Slow melt … Fast melt
Chalky mouth coating	☐ ☐ ☐ ☐ ☐ ☐ ☐ ☐ ☐ Dislike extremely … nI/nd … Like extremely	☐ ☐ ☐ ☐ ☐ ☐ ☐ ☐ ☐ None … Chalky

Raise your hand when finished. Thank you!

B. Paper Napkins Questionnaire

Name ____________________

Product # _______________

Paper Table Napkins

- Please be sure your hands are clean before starting.
- Evaluate the product in front of you.
- LOOK at this napkin, OPEN AND FEEL it, and answer the following questions.

Overall opinion

Please indicate how much you liked or disliked this product overall (considering ALL APPEARANCE, TACTILE/FEEL CHARACTERISTICS).*Circle* one of the numbers below ⊗ to express your overall opinion.

0	1	2	3	4	5	6	7	8	9	10
Dislike extremely					Neither like nor dislike (nl/nd)					Like extremely

- **Comments:** Please indicate what in particular you liked or disliked about this product. (use words not sentences, and be as specific as possible.)

LIKED	DISLIKED
____________________	____________________
____________________	____________________
____________________	____________________

1. Paper Table Napkins Liking Questions

Please retest the product as needed and indicate howmuch you LIKE or DISLIKE the following. *Circle* the number that represents your response ⊗.

Overall appearance

0	1	2	3	4	5	6	7	8	9	10
Dislike extremely					nl/nd					Like extremely

Overall texture

0	1	2	3	4	5	6	7	8	9	10
Dislike extremely					nl/nd					Like extremely

2. Paper Table Napkins Specific Evaluation

Retest the product as needed and *circle* your response for both questions (LIKING and INTENSITY LEVEL) for each characteristic.

	Liking	Intensity/level
Surface gloss	0 1 2 3 4 5 6 7 8 9 10 Dislike extremely — nl/nd — Like extremely	0 1 2 3 4 5 6 7 8 9 10 Dull finish — Glossy finish
Color/whiteness	0 1 2 3 4 5 6 7 8 9 10 Dislike extremely — nl/nd — Like extremely	0 1 2 3 4 5 6 7 8 9 10 Gray color — Bright color
Surface embossing	0 1 2 3 4 5 6 7 8 9 10 Dislike extremely — nl/nd — Like extremely	0 1 2 3 4 5 6 7 8 9 10 Not embossed — Very embossed
Specks in surface	0 1 2 3 4 5 6 7 8 9 10 Dislike extremely — nl/nd — Like extremely	0 1 2 3 4 5 6 7 8 9 10 No specks — Many specks
Stiffness	0 1 2 3 4 5 6 7 8 9 10 Dislike extremely — nl/nd — Like extremely	0 1 2 3 4 5 6 7 8 9 10 Not stiff — Very stiff
Smoothness of surface	0 1 2 3 4 5 6 7 8 9 10 Dislike extremely — nl/nd — Like extremely	0 1 2 3 4 5 6 7 8 9 10 Rough/not smooth — Very smooth
Body	0 1 2 3 4 5 6 7 8 9 10 Dislike extremely — nl/nd — Like extremely	0 1 2 3 4 5 6 7 8 9 10 Flimsy — Full bodied
Softness	0 1 2 3 4 5 6 7 8 9 10 Dislike extremely — nl/nd — Like extremely	0 1 2 3 4 5 6 7 8 9 10 Not soft — Very soft

Indicate to the test supervisor that you have completed this questionnaire. Thank you!

APPENDIX 13.4 PROTOCOL DESIGN FOR CONSUMER STUDIES

A. Protocol Design Format Worksheets

1. Product Screening

1. Test objective ______________________________
____
____

2. Sample selection
   a. Variables ______________________________
   ____
   ____
   b. Products/brands ______________________________
   ____
   ____
   ____

3. Reasons ______________________________
____
____

2. Sample Information

Sample conditions

1. Sample source ______________________________
____
____
   Age ______________________________
   Place ______________________________
   ____
   Code ______________________________
   ____
   Packaging condition ______________________________
   ____

2. Sample holding
____
____
____

3. Other
____
____

3. Sample Preparation

Total amount ____________________
Other ingredients ____________________
Temperature (storage or preparation) ____________________
Preparation/reconstitution time ____________________
Holding time ____________________
Containers ____________________
Other ____________________
____
Special instructions ____________________
____

4. Sample Presentation

Amount ____________________
Containers/utensils ____________________
Coding ____________________
____
Serving size ____________________
Temperature ____________________
Presentation procedure ____________________
____
Order ____________________
____
____

5. Subjects

Age range ____________________
Sex ____________________
Product usage ____________________
____
Frequency of product consumption ____________________
Availability ____________________

B. Protocol Design Example: Candy Bars

1. Product Screening

1. Test objective
   To determine the relative acceptance and attribute diagnostics for candy bars with different chocolate to peanut ratios and with some roast differences in peanuts
2. Sample selection
   a. Variables *Amount of standard 1050 coating on bar; amount of peanuts by weight; degree of roast color in peanuts*
   b. Products/brands *Screen 18 to 22 prototypes (experimental design) and 2 competitors;have descriptive data available to identify products with little or no differences from one another; choose 12 to 15 bars to test*
3. Reasons
   14 selected samples demonstrate differences in peanut/choclate balance and roast flavor intensity and crunchiness of nut pieces

2. Sample Information

Sample Conditions

1. Sample source *Trial run prototype samples (3 oz); competitors from same age carefully stored lots*
   Age *3 months old*
   Place *Lancaster production; competitors from midwest distribution*
   Code *Ours L432-439; competition A4192, 7425S*
   Packaging condition *All samples over wrapped in white foil wrappers(732 equipment Lancaster)*
2. Sample Holding
   Hold all foil wrapped samples for 3 weeks prior to test in boxes of 24 overwrapped in cellophane, at65°, in 50% RHstoreroom prior to shipping to test site
3. Other
   Ship all samples by truck in styrofoam chests to Indianapolis and Syracuse for test

3. Sample Preparation

Total amount *250 bars of each to each test site (150 needed)*
Other ingredient *None*
Temperature (storage or preparation) *Keep at 65 to 75°F*
Preparation/reconstitution time *None*
Holding time *None*
Containers *Use plastic plates*
Other *Leave bars wrapped until just before presentation to subject; discard any broken, split, or pitted samples*
Special instructions *Do not handle bars any more than a few seconds to prevent melting and damage*

4. Sample Presentation

Amount *Each subject to get one full bar of each product*
Containers/utensils *Plastic plates*
Coding *Three-digit codes; see attached sheets for each subject*
Serving size *One bar per subject*
Temperature *65 to 75°F*
Presentation procedure *Place sample in middle of coded 6 in. plastic plate*
Order *See attached sheet for codes and order for each subject [Such a sheet is not included here, but should be prepared based on the experimental design used.]*

5. Subjects

Age range *50% 12 to 25 years; 50% 25 to 55 years*
Sex *50% male; 50% female*
Product usage *Has eaten a chocolate coated candy bar within the last month*
Frequency of product consumption *5 or more bars/years*
Availability *Afternoons—3 to 5 or evening—7 to 9*

REFERENCES

Amerine, M. A., R. M. Pangborn, and E. G. Roessler (1965). *Principles of Sensory Evaluation of Food*. New York: Academic Press.

ASTM MNL63 (2009). *Just-About-Right (JAR) Scales: Design, Usage, Benefits, and Risks*. West Conshohocken, PA: ASTM International. www.astm.org.

ASTM E1958-12 (2012). *Standard Guide for Sensory Claim Substantiation*. West Conshohocken, PA: ASTM International. www.astm.org.

Barker, L. (1982). *The Psychobiology of Human Food Selection*. Westport, CT: AVI Publishing.

Beausire, R. L. W., J. P. Norback, and A. J. Maurer (1988). Development of an acceptability constraint for a linear programming model in food formulation. *J Sens Stud* 3(2): 137.

Carr, B. T. (1989). An integrated system for consumer-guided product optimization. In L.S. Wu (Ed.), *Product Testing with Consumer for Research Guidance, ASTM STP 1035* (pp. 41–53). Philadelphia: ASTM.

Casey, M. A., and R. A. Krueger (1994). Focus group interviewing. In H. J. H. MacFie and D. M. H. Thomson (Eds.), *Measurement of Food Preferences* (pp. 77–96). London: Blackie Academic and Professional.

Civille, G. V., A. Muñoz, and E. Chambers IV (1987). Consumer testing considerations. In *Consumer Testing. Course Notes*. Chatham, NJ: Sensory Spectrum.

Gacula, M. C. Jr. (1993). *Design and Analysis of Sensory Optimization*. Westport, CT: Food Nutrition Press.

Institute of Food Technologists (IFT) (1979). *Sensory Evaluation Short Course*. Chicago: IFT.

Kroll, B. J. 1990. Evaluation rating scales for sensory testing with children. *Food Technol* 44(11): 78–86.

Krueger, R. A. 1988. *Focus Groups. A Practical Guide for Applied Research*. Newbury Park, CA: Sage Publications.

Lawless, H. T., and H. Heymann (2010). *Sensory Evaluation of Food. Principles and Practices* (2nd edn.). New York: Chapman & Hall.

Meilgaard, M. C. (1992). Basics of consumer testing with beer in North America. In *Proceedings of the Annual Meeting of the Institute of Brewing, Australia & New Zealand Section* (pp. 37–47), Melbourne. See also *The New Brewer* 9(6): 20–25.

Meiselman, H. L. (1984). Consumer studies of food habits. In J. R. Piggott (Ed.), *Sensory Analysis of Foods*. London: Elsevier Applied Science.

Moskowitz, H. R. (1983). *Product Testing and Sensory Evaluation of Foods. Marketing and R&D Approaches.* Westport, CT: Food and Nutrition Press.

Moskowitz, H. R., A. Gofman, B. Itty, R. Katz, M. Manchaiah, and Z. Ma (2001). Rapid, inexpensive, actionable, concept generation and optimization: The use and promise of self-authoring conjoint analysis for the food service industry. *Food Ser Technol* 1: 149–68.

Muñoz, A. M., E. Chambers IV, and S. Hummer (1996). A multifaceted category research study: How to understand a product category and its consumer responses. *J Sens Stud* 11: 261–94.

Nestrud, M. A. (2011). A graph theoretic approach to food combination problems. Dissertation, Cornell University.

Nestrud, M. A., J. M. Ennis, C. M. Fayle, D. M. Ennis, and H. T. Lawless (2011). Validating a graph theoretical screening approach to food item combinations. *J Sens Stud* 26: 331–8.

Resurreccion, A. V. A. (1998). *Consumer Sensory Testing for Product Development*. Gaithersburg, MD: Aspen Publishers.

Schaefer, E. E (Ed.). (1979). *ASTM Manual on Consumer Sensory Evaluation*, ASTM Special Technical Publication 682. Philadelphia: ASTM International.

Shepherd, R., N. M. Griffiths, and K. Smith (1988). The relationship between consumer preference and trained panel responses. *J Sens Stud* 3: 19.

Stapleton, L. (2014). Digital real time data collection: Being a savvy user. Presentation at SSP Virtual Meeting, 2014.

Stone, H., and J. Sidel (2004). *Sensory Evaluation Practices*, (3rd edn.). New York: Academic Press/ Elsevier.

Szczesniak, A. S., E. Z. Skinner, and B. J. Loew (1975). Consumer textile profile method. *J Food Sci* 40: 1253–6.

Vickers, Z. (1988). Sensory specific satiety in lemonade using a just right scale for sweetness. *J Sens Stud* 3(1): 1–8.

Wu, L. S., and A. D. Gelinas (Eds.) (1989). *Product Testing with Consumers for Research Guidance*, Vol. 1, ASTM Standard Technical Publications STP 1035. Philadelphia: ASTM International.

Wu, L. S., and A. D. Gelinas (Eds.) (1992). *Product Testing with Consumers for Research Guidance*, Vol. 2, ASTM Standard Technical Publications STP 1155. Philadelphia: ASTM International.

14

Basic Statistical Methods

14.1 INTRODUCTION

The goal of applied statistics is to draw some conclusion about a population based on the information contained in a sample from that population. The types of conclusions fall into two general categories: estimates and inferences. Furthermore, the size and manner in which a sample is drawn from a population affects the precision and accuracy of the resulting estimates and inferences. These issues are addressed in the experimental design of a sensory study. This chapter presents the concepts and techniques of estimation, inference, and experimental design as they relate to some of the more fundamental statistical methods used in sensory evaluation. The topics are presented with a minimum of theoretical detail. Those interested in pursuing this area further are encouraged to read Gacula and Singh (1984), O'Mahony (1986), and Smith (1988) or, for more theoretically advanced presentations, Cochran and Cox (1957) and Snedecor and Cochran (1980).

Several definitions presented at this point will make the discussion that follows easier to understand. A population is the entire collection of elements of interest. The population of interest in sensory analysis varies from study to study. In some cases, the population may be people (e.g., consumers of a particular food), whereas in other cases, it may be products (e.g., batches of corn syrup). An element or unit from the population might be a particular consumer or a particular batch of syrup. Measurements taken on elements from a population may be discrete, that is, take on only specific values (such as a preference for brand A), or continuous, that is, take on any value on a continuum (such as the intensity of sweetness). The values that the measurements take on are governed by a probability distribution, usually expressed in the form of a mathematical equation that relates the occurrence of a specific value to the probability of that occurrence. Associated with the distribution are certain fixed quantities called *parameters*. The values of the parameters provide information about the population. For continuous distributions, for instance, the mean (μ) locates the center of the measurements. The standard deviation (σ) measures the dispersion or "spread" of the measurements about the mean. For discrete distributions, the proportion of the population that possesses a certain characteristic is of interest. For example, the population proportion (p) of a binomial distribution might summarize the distribution of preferences for two products.

Only in the rarest of circumstances is it possible to conduct a census of the population and directly compute the exact values of the population parameters. More typically, a subset of the elements of the population, called a *sample*, is collected, and the measurements of interest are made on each element in the sample. Mathematical functions of these measurements, called *statistics*, are used to approximate the unknown values of the population parameters. The value of a statistic is called an *estimate*.

Often, a researcher is interested in determining if a population possesses a specific characteristic (e.g., more people prefer product A than product B). There are risks associated with drawing conclusions about the population as a whole when the only information available is that contained in a sample. Formal procedures, called *tests of hypotheses*, set limits on the probabilities of drawing incorrect conclusions. Then, based on the actual outcome of an experiment, the researcher's risks are constrained within these known limits. Tests of hypotheses are a type of statistical inference that give sensory researchers greater assurance that correct decisions will be made.

The amount of information required to draw sound statistical conclusions depends on several factors (e.g., the level of risk the researcher is willing to assume, the required precision of the information, the inherent variability of the population being studied, etc.). These issues need to be addressed, and a plan of action, called *the experimental design*, should be developed before a study is undertaken. The experimental design, based on both technical and common-sense principles, will ensure that the experimental resources are focused on the critical issues in the study, that the correct information is collected, and that no excessive sampling of people or products occurs.

The remainder of the chapter is devoted to the further development of the ideas just presented. Section 14.2 presents some basic techniques for summarizing data in tabular and graphical forms. Section 14.3 combines estimation with some fundamental concepts of probability to present some methods for testing statistical hypotheses. Section 14.4 presents an introduction to the application of the Thurstonian model to sensory evaluation. The Thurstonian model provides an alternative approach for measuring differences among samples and provides unique insights on how the assessors are performing their evaluations. Section 14.5 covers the most commonly used experimental designs in sensory studies, including techniques for improving the sensitivity of panels for detecting differences among products. The basic techniques for calculating probabilities from some common distributions are presented in an appendix (see Section 14.6).

14.2 SUMMARIZING SENSORY DATA

The data from sensory panel evaluations should be summarized in both graphs and tables before formal statistical analyses (i.e., tests of hypotheses, etc.) are undertaken. Examination of the graphs and tables may reveal features of the data that would be lost in the computation of test statistics and probabilities. In fact, features revealed in the tables and graphs may indicate that standard statistical analysis procedures would be inappropriate for the data at hand.

Whenever a reasonably large number of observations are available, the first step of any data analysis should be to develop the frequency distribution of responses (see Figure 14.1).

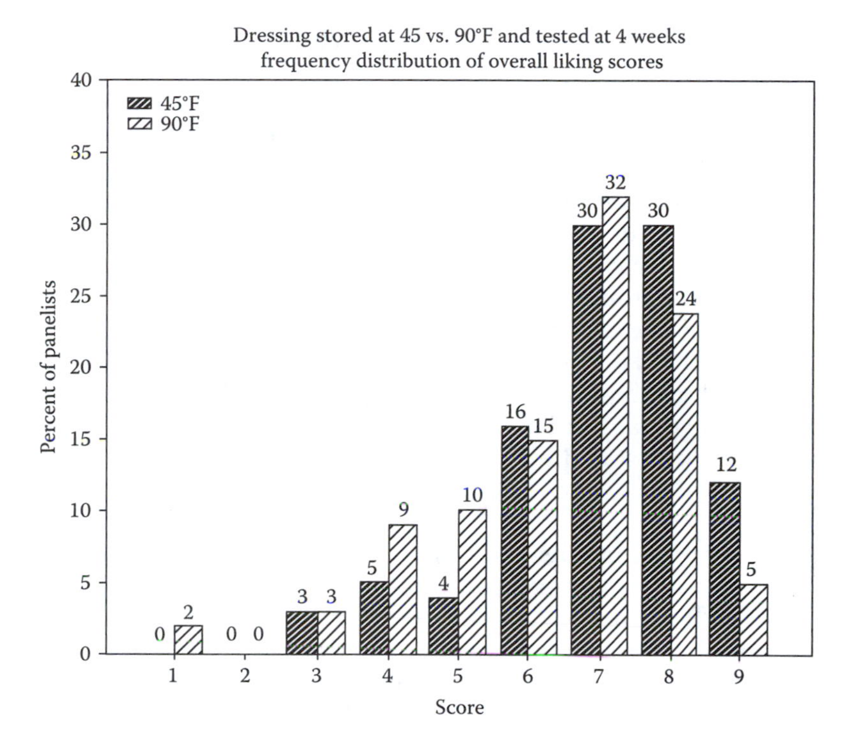

Figure 14.1 Histogram (with frequencies) of the overall liking scores for two samples of salad dressing.

Then, a basic set of summary statistics should be calculated. Included in the basic set would be the arithmetic or sample mean, $\bar{x}$, for estimating the center (or central tendency) of the distribution of responses; and the sample standard deviation, s, for estimating the spread (or dispersion) of the data around the mean. The sample mean is calculated as

$$\bar{x} = \left(\sum_{i=1}^{n} x_i \,/\, n \right) = (x_1 + x_2 + \cdots + x_n)\,/\,n, \tag{14.1}$$

where Σ represents the sum function. The subscript (i = 1) and superscript (n) indicate the range over which the summing is to be done. Equation 14.1 indicates that the sum is taken over all n elements in the sample. The sample standard deviation is calculated as

$$s = \sqrt{\left[\sum_{i=1}^{n} x_i^2 - \left(\sum_{i=1}^{n} x_i \right)^2 \Bigg/ n \right] \Bigg/ (n-1)}\,. \tag{14.2}$$

These basic statistics can sometimes be misleading. Instances where they should be used with caution include cases where the data are multimodal (i.e., several groups of data clustered at different locations on the response scale) or where there are extreme values (i.e., outliers) in the data.

Multimodal data may indicate the presence of several subpopulations with different mean values. In such situations, the sample mean of all the data may be meaningless, and, as a result, so might the sample standard deviation (since s measures the spread around the mean). Multimodal data should be examined further to determine if there is a way to break up the entire set into unimodal subgroups (e.g., by sex, age, geography, plant, batch, etc.). Separate sets of summary statistics could then be calculated within each subgroup. If it is not possible to break up the entire set, then the researcher must determine which summary statistics are still meaningful. For instance, the median divides the data in half with 50% of the observations falling below the median and 50% falling above it. This may be a meaningful way to identify the center of a set of multimodal data. Similarly, the spread of the data might be measured by the difference between the first and third quartiles of the responses (i.e., the points below which 25% and 75% of the values fall, respectively). This difference is called the *interquartile range.*

The sample mean, $\bar{x}$, is sensitive to the presence of extreme values in the data. The median is less sensitive to extreme values, so it could again be used in place of the sample mean as the summary measure of the center of the data. Another option is a robust estimator of central tendency called the *trimmed mean*. The trimmed mean is calculated in the same way as the sample mean but after a specific proportion (e.g., 5%) of the highest and lowest data values have been eliminated. Various computerized statistical analysis packages routinely compute a variety of measures of central tendency and dispersion.

Many statistical analysis procedures assume that the data are normally distributed. If the raw data used to calculate $\bar{x}$ are normally distributed, then so is $\bar{x}$. In fact, even if the raw data are not distributed as normal random variables, $\bar{x}$ is still approximately normal, provided that the sample size is greater than 25 or so. The mean of the distribution of $\bar{x}$ is the same as the mean of the distribution of the raw data, that is, μ, and if σ is the standard deviation of the raw data, then $\sigma/\sqrt{n}$ is the standard deviation of $\bar{x}$. $\sigma/\sqrt{n}$ is called the *standard error of the mean*. Notice that as the sample size n increases, the standard error of the mean decreases. Therefore, as the sample size becomes larger, $\bar{x}$ is increasingly likely to take on a value close to the true value of μ. The standard error (SE) of the mean is estimated by $SE = s/\sqrt{n}$, where s is the sample standard deviation calculated in Equation 14.2.

14.2.1 Summary Analysis of Data in the Form of Ratings

The overall liking responses of 30 individuals in each of four cities are presented in Table 14.1. The frequency distributions of the responses are presented tabularly in Table 14.2 and graphically, using simple bar plots, in Figure 14.2. There is no strong indication of multimodal behavior within a city. The summary statistics for these data are presented in Table 14.3. The box-and-whisker plots (see Danzart, 1986) in Figure 14.3 provide additional information about the distribution of ratings from city to city and, possibly, raise some minor concern about extreme observations.

Table 14.1 Data from a Multicity Monadic Consumer Test

Attribute: Overall Liking[a]				
Respondent	**Atlanta**	**Boston**	**Chicago**	**Denver**
1	12.6	10.4	7.9	10.3
2	9.8	10.4	7.8	11.7
3	8.6	8.9	6.3	11.5
4	9.8	8.0	11.1	9.9
5	15.0	10.4	5.5	11.7
6	12.7	11.0	6.5	10.3
7	12.8	7.4	8.8	11.6
8	9.5	10.5	5.2	12.1
9	12.4	9.2	7.8	11.6
10	9.6	9.2	7.6	12.3
11	9.2	9.8	6.3	12.4
12	7.1	9.1	7.1	10.5
13	9.9	9.7	8.0	12.4
14	12.4	10.3	5.7	14.4
15	8.7	9.1	5.5	11.1
16	11.9	10.3	5.2	9.9
17	9.9	11.7	7.2	11.9
18	11.3	9.8	8.0	8.8
19	10.4	10.2	9.1	12.3
20	11.8	9.5	8.4	8.6
21	11.5	12.4	4.0	11.9
22	8.9	9.5	6.9	9.3
23	11.4	12.9	6.6	10.0
24	6.9	11.1	7.4	10.2
25	8.8	13.3	7.3	10.8
26	11.6	12.9	7.5	12.7
27	11.3	11.4	9.1	11.1
28	9.7	9.0	6.9	11.9
29	10.0	10.1	8.4	10.2
30	11.2	11.2	6.1	10.1

[a] Measured on a 15 cm unstructured line scale.

14.2.2 Estimating the Proportion of a Population That Possesses a Particular Characteristic

The statistic used to estimate the population proportion p of a binomial distribution is $\hat{p}$, where

$$\hat{p} = \frac{\text{Number of "successes"}}{\text{Number of trials}}. \tag{14.3}$$

Table 14.2 Frequency Distributions from the Multicity Consumer Test Data in Table 14.1

Attribute: Overall Liking				
	Frequencies in			
Category Midpoint[a]	**Atlanta**	**Boston**	**Chicago**	**Denver**
1	0	0	0	0
2	0	0	0	0
3	0	0	0	0
4	0	0	1	0
5	0	0	2	0
6	0	0	6	0
7	2	1	8	0
8	0	1	9	0
9	5	6	3	3
10	9	12	0	8
11	4	5	1	4
12	6	2	0	13
13	3	3	0	1
14	0	0	0	1
15	1	0	0	0

[a] For example, in Atlanta, nine people responded with an overall liking rating between 9.5 and 10.4.

Suppose that 150 consumers participate in a preference test between two samples, A and B. Furthermore, suppose that 86 of the participants say that they prefer sample A. Preference for sample A was defined as a success before the test was conducted, so from Equation 14.3, $\hat{p} = 86/150 = 0.573$. That is, 0.573 or 57.3% of consumers preferred sample A. If a multicity test had been conducted, the estimated preferences for sample A could be represented graphically using a bar chart such as that in Figure 14.4.

14.2.3 Confidence Intervals on μ and p

The previously calculated single-valued statistics, called *point estimates*, provide no information as to their own precision. Confidence intervals supply this missing information. A confidence interval is a range of values within which the true value of a parameter lies with a known probability. Confidence intervals allow the researcher to determine if the point estimates are sufficiently precise to meet the needs of an investigation.

Three types of confidence intervals are presented: the one-tailed upper confidence interval, the one-tailed lower confidence interval, and the two-tailed confidence interval. The equations for calculating these intervals for both μ and p are presented in Table 14.4. In general, two-tailed confidence intervals are most useful, but if the analyst is only interested in an average value that is either "too big" or "too small," then the appropriate one-tailed confidence interval should be used.

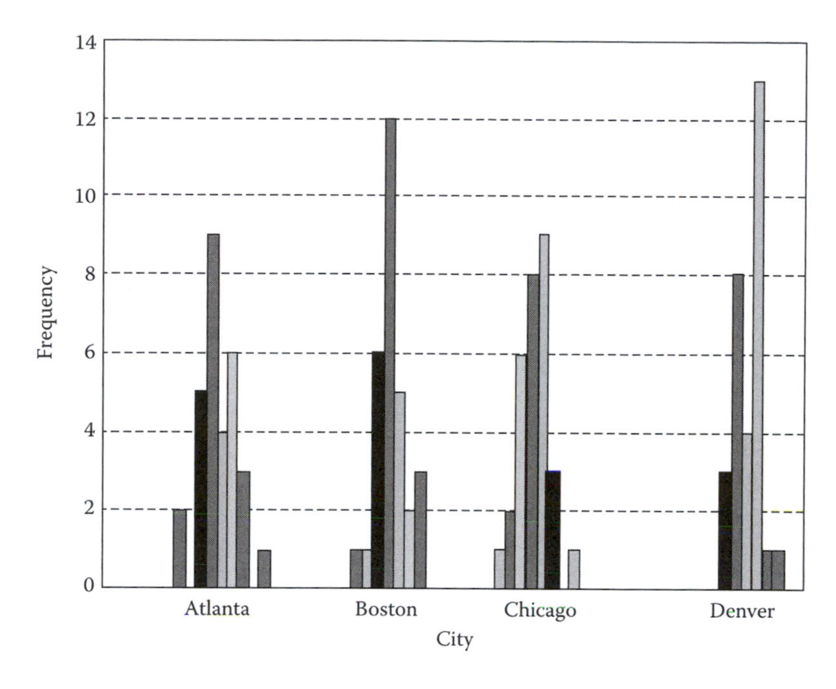

Figure 14.2 Histograms of the overall liking scores from the multicity consumer test data in Table 14.1.

Table 14.3 Summary Statistics from the Multicity Consumer Test Data in Table 14.1

Attribute: Overall Liking							
	City	**n**	**Mean**	**Median**	**Trimmed Mean**	**Standard Deviation**	**Standard Error**
Overall liking	Atlanta	30	10.557	10.200	10.573	1.793	0.327
	Boston	30	10.290	10.250	10.273	1.401	0.256
	Chicago	30	7.173	7.250	7.146	1.448	0.264
	Denver	30	11.117	11.300	11.115	1.276	0.233
	City		**Min**	**Max**		**Q1**	**Q3**
Overall liking	Atlanta		6.900	15.000		9.425	11.825
	Boston		7.400	13.300		9.200	11.125
	Chicago		4.000	11.100		6.250	8.000
	Denver		8.600	14.400		10.175	11.950

The quantities $t_{\alpha,n-1}$ and $t_{\alpha/2,n-1}$ in Table 14.4 are t-statistics. The quantity α measures the level of confidence. For instance, if $\alpha = 0.05$, then the confidence interval is a $100(1 - \alpha)\% = 95\%$ confidence interval. The quantity $(n - 1)$ in Table 14.4 is a parameter associated with the t-distribution called *degrees of freedom*. The value of t depends on the value of α and the number of degrees of freedom $(n - 1)$. Critical values of t are presented in Table 19.3.

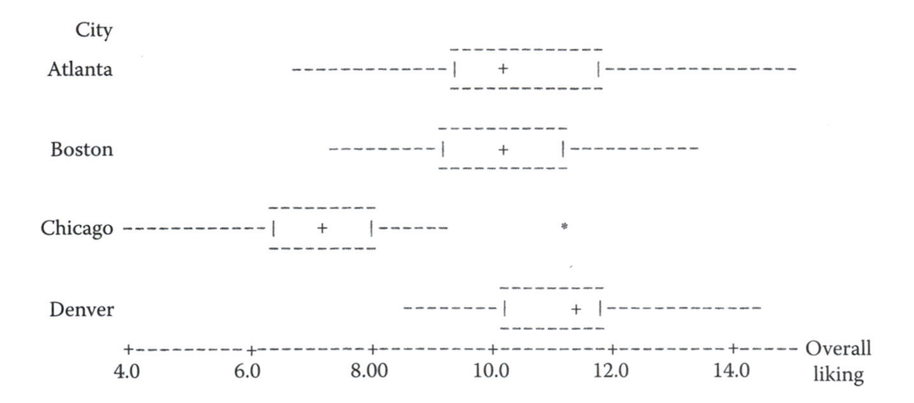

Figure 14.3 Box-and-whisker plots of the overall liking scores from the multicity consumer test data in Table 14.1.

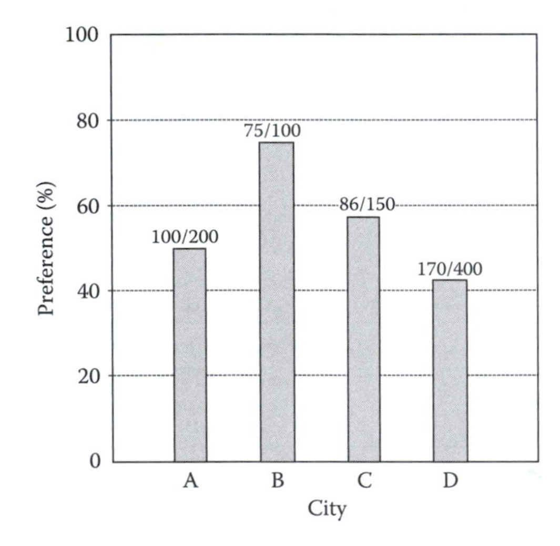

Figure 14.4 Bar chart of the preference results of a two-sample study conducted in four cities showing the relative difference from city to city. Actual preference results and total respondent base are included for each city.

The quantity z in Table 14.4 is the critical value of a standard normal variable. (The standard normal distribution has mean $\mu = 0$ and standard deviation $\sigma = 1$.) Critical values of z for some commonly used levels of α are presented in the last row of Table 19.3 (i.e., the row corresponding to ∞ degrees of freedom).

Consider the overall liking data presented in Table 14.1. The sample mean intensity for Atlanta was $\bar{x} = 10.56$ and the sample standard deviation of the data was $s = 1.79$. To construct a lower, one-tailed, 95% confidence interval on the value of the population mean, one uses Tables 14.4 and 19.3 to obtain

$$\bar{x} - t_{\alpha,n-1}s/\sqrt{n},$$

Table 14.4 Computational Forms for Confidence Intervals

	Parameter	
Type of Interval	μ	p
One-tailed upper	$\bar{x}+t_{\alpha,n-1}s/\sqrt{n}$	$\hat{p}+z_{\alpha}\sqrt{\hat{p}(1-\hat{p})/n}$
One-tailed lower	$\bar{x}-t_{\alpha,n-1}s/\sqrt{n}$	$\hat{p}-z_{\alpha}\sqrt{\hat{p}(1-\hat{p})/n}$
Two-tailed	$\bar{x}\pm t_{\alpha/2,n-1}s/\sqrt{n}$	$\hat{p}\pm z_{\alpha/2}\sqrt{\hat{p}(1-\hat{p})/n}$

where $\alpha = 0.05$ and $n = 30$, so $t_{\alpha,n-1}$ is $t_{0.05,29} = 1.699$, yielding

$$10.56-1.699(1.79)/\sqrt{30}=10.56-0.56=10.00.$$

The limit is interpreted to mean that the researcher is 95% sure that the true value of the mean overall liking rating in Atlanta is no less than 10.00.

A two-tailed 95% confidence interval on the mean is calculated as

$$\bar{x}\pm t_{\alpha/2,n-1s/\sqrt{n}},$$

where $\alpha = 0.05$ and $n = 30$, so $t_{\alpha/2,n-1}$ is $t_{0.025,29} = 2.045$, yielding

$$10.56\pm 2.045(1.79)/\sqrt{30}=10.56\pm 0.67 \text{ or } (9.89, 11.23).$$

That is, the researcher is 95% sure that the true value of the mean overall liking rating in Atlanta lies somewhere between 9.89 and 11.23. In Figure 14.5, the sample means and their associated 95% confidence intervals are presented for the overall liking data of each of the four cities presented in Table 14.1. The analyst can now begin to formulate some ideas about differences in average overall liking that may exist among the cities.

Consider the consumer preference test discussed before, where 86 of the 150 ($\hat{p} = 0.573$) consumers preferred sample A. To construct a 95% confidence interval (two-tailed) on the true value of the population proportion, p, one uses Table 14.4 and Table 19.3 to obtain

$$\hat{p}\pm z_{\alpha/2}\sqrt{\hat{p}(1-\hat{p})/n},$$

where $n = 150$, $\alpha = 0.05$, so $z_{\alpha/2} = t_{\alpha/2,\infty} = 1.96$, yielding

$$0.573\pm 1.96\sqrt{(0.573)(0.427)/150} \text{ or } (0.494, 0.652).$$

The researcher may conclude, with 95% confidence, that the true proportion of the population that prefers sample A lies between 49.4% and 65.2%. Confidence intervals on proportions can also be depicted graphically as in Figure 14.6, where 95% two-tailed confidence intervals have been added to the data summarized in Figure 14.4.

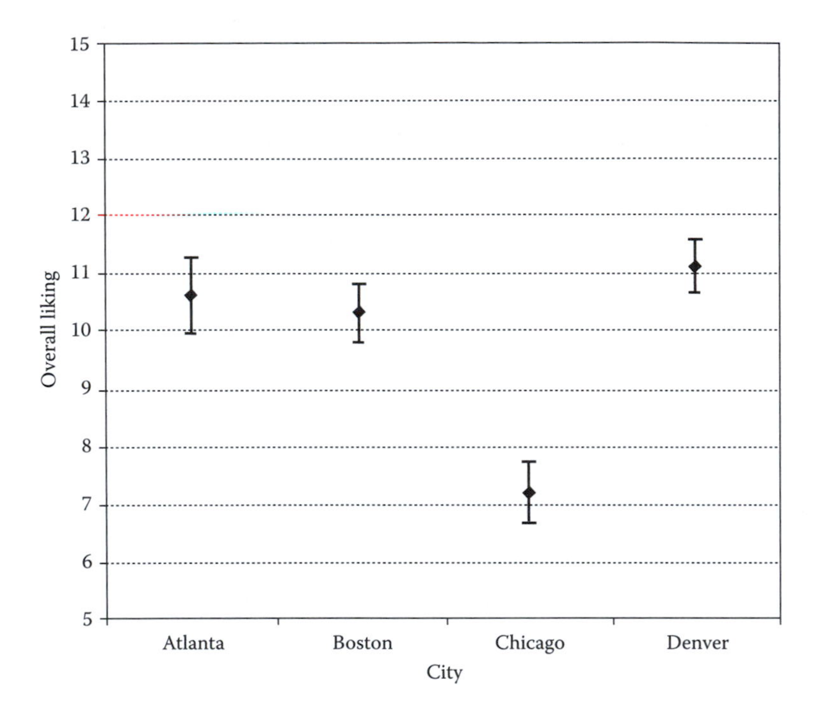

Figure 14.5 Average overall liking scores with 95% confidence intervals from the multicity consumer test data in Table 14.1. Note the large degree of overlap among Atlanta, Boston, and Denver compared to the much lower average value for Chicago.

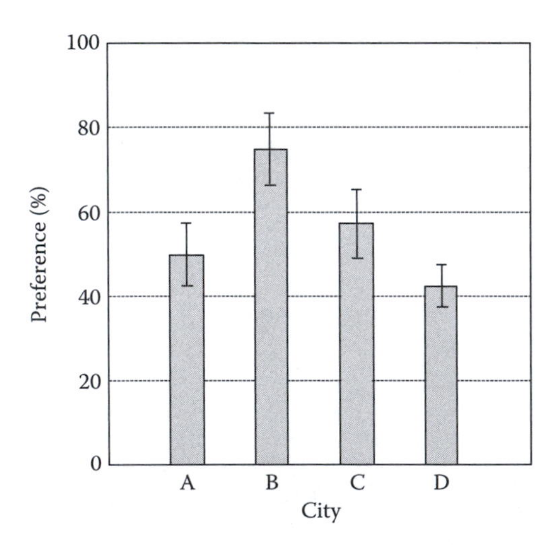

Figure 14.6 Bar chart of the preference results including 95% confidence intervals of a two-sample study conducted in four cities. Confidence intervals that overlap 50% indicate that no significant preference exists in that city ($\alpha = 0.05$). Confidence intervals from two cities that do not overlap indicate roughly that the two cities differ in their degree of preference for the product.

14.2.4 Other Interval Estimates

Confidence intervals state a range of values that have a known probability of containing the true value of a population parameter. The researcher may not always want to draw such a conclusion. There exist other types of statistical interval estimates.

For instance, a prediction interval is a range of values that has a known probability of containing the average value of *k* future observations. The researcher may choose $k = 1$ to calculate an interval that has a known probability of containing the next observed value of some response (e.g., being 95% confident that the perceived saltiness of the next batch of potato chips will lie between 7.2 and 10.4). Two-sided prediction intervals are calculated as

$$\bar{x} \pm t_{a/2,n-1} s \sqrt{(1/k)+(1/n)}\ .$$

Another statistical interval, called *a tolerance interval*, is a range of values that has a known probability of containing a specified proportion of the population. An example of a one-sided tolerance interval is that the researcher is 95% sure that 90% of all batches have firmness ratings less than 6.3. Two-sided tolerance intervals can also be computed (see Dixon and Massey, 1969).

14.2.5 Data Transformations

At times, a researcher may want to transform the scale of measurement from the one used to collect the data to a more meaningful scale for presentation. This is easy to carry out for a transformation called a *linear transformation*. If the original variable, x, is transformed to a new variable, y, using $y = a(x) + b$, then y is a linear transformation of x. Linear transformations are limited to multiplying the original variable by a constant, a, and/or adding a constant, b. Raising the original variable to a power and taking its logarithm, sine, inverse, and so on are all nonlinear transformations. If x has mean value μ and standard deviation σ, then the mean and standard deviation of y are $a\mu + b$ and $a\sigma$, respectively. These equations for computing the mean and standard deviation of the transformed variable y apply only to linear transformations. The sample mean, y, and sample standard deviation, s_y, are obtained by substituting $\bar{x}$ for μ and s_x for σ.

An example of this data transformation technique occurs in tests for overall differences such as triangle, duo–trio, and two-out-of-five tests where the original measurement is the proportion of correct responses, p_c. Using the triangle test as an example, p_c can be transformed to the proportion of the population that can distinguish the samples, p_d, by using $p_d = 1.5(p_c) - 0.5$. The expression for p_d is obtained by inverting the equation for the probability of obtaining a correct answer in a triangle test, $p_c = 1(p_d) + (1/3)(1 - p_d)$; that is, the probability of a correct answer is the probability of selecting a distinguisher, p_d (who will always give a correct answer), plus the probability of selecting a nondistinguisher $(1 - p_d)$, and having that person guess correctly (which has a probability of 1/3). Notice that when there are no perceptual differences between the samples in a triangle test, the expected proportion of correct answers is $p_c = 1/3$, which transforms to the expected proportion of distinguisher $p_d = 0$ (i.e., everyone is guessing).

In a triangle test involving n respondents, if x people correctly select the odd sample, then the estimated value of p_c is $\hat{p}_c = x/n$ and the estimated standard deviation of p_c is $s_c = \sqrt{\hat{p}_c(1-\hat{p}_c)/n}$. The estimated proportion of distinguishers is then $\hat{p}_d = 1.5(x/n) - 0.5$, with an estimated standard deviation of $s_d = 1.5s_c$. These transformations are applied in several places in Chapter 7.

These data transformations are particularly useful in the unified approach to discrimination testing discussed in Chapter 7. Confidence intervals can be constructed on the proportion of distinguishers in the population of panelists, p_d, using

$$\text{Lower confidence limit: } \hat{p}_d - z_\alpha s_d, \text{ and}$$
$$\text{Upper confidence limit: } \hat{p}_d + z_\beta s_d,$$

where $\hat{p}_d$ is the estimate of the proportion of distinguishers, s_d is the sample standard deviation of the proportion of distinguishers, and z_α and z_β are the α and β critical values from the standard normal distribution. The quantities $\hat{p}_d$ and s_d are obtained from $\hat{p}_c$ and s_c using the following transformations:

Method	$\hat{p}_d$	s_d
Triangle test	$1.5\hat{p}_c$–0.5	$1.5s_c$
Duo–trio and paired comparison	$2\hat{p}_c - 1$	$2s_c$
Two-out-of-five	$(10/9)\hat{p}_c$–$(1/9)$	$(10/9)s_c$

If the lower confidence limit is zero or less, then the null hypothesis of no perceptible difference cannot be rejected (at the $1 - \alpha$ confidence level). If the lower confidence limit is greater than zero, then the samples are perceptibly different. If the upper confidence limit is less than the proportion of distinguishers that the researcher wants to be able to detect, p_{max}, then the products are sufficiently similar (at the $1 - \beta$ confidence level). If the upper confidence limit is greater than p_{max}, then the samples are not sufficiently similar. (See Chapter 7 for examples using these confidence intervals.)

14.3 STATISTICAL HYPOTHESIS TESTING

Often, the objective of an investigation is to determine if it is reasonable to assume that the unknown value of a parameter is equal to some specified value or possibly that the unknown values of two parameters are equal to each other. In the face of the incomplete and variable information contained in a sample, statistical decisions of this type are made using hypothesis testing. The process of statistical hypothesis testing is summarized by the following five steps:

1. The objective of the investigation is stated in mathematical terms, called the *null hypothesis* (H_0), (e.g., H_0: $\mu = 8$).

2. Based on the prior interest of the researcher, another mathematical statement, called the *alternative hypothesis* (H_a) is formulated (e.g., H_a: $\mu > 8$, H_a: $\mu < 8$, or H_a: $\mu \neq 8$).
3. A random sample of elements from the population is collected and the measurement of interest is taken on each element of the sample.
4. The value of the statistic used to estimate the parameter of interest is calculated.
5. Based on the assumed probability distribution of the measurements and the null hypothesis assumption, H_0, the probability that the statistic takes on the value calculated in step 4 is computed. If this probability is smaller than some predetermined value (α), the null hypothesis is rejected in favor of the alternative hypothesis.

14.3.1 Statistical Hypotheses

In most sensory studies, statistical hypotheses specify the value of some parameter in a probability distribution, such as the mean μ or the population proportion p. The null hypothesis is determined by the objective of the investigation and serves as the baseline condition that is assumed to exist prior to running the experiment. The value specified in the null hypothesis is used to calculate the test statistic (and resulting p-value) in the hypothesis test. The alternative hypothesis is developed based on the prior interest of the investigator. For example, if a company is replacing one of the raw ingredients in its current product with a less expensive ingredient from an alternate supplier, the sensory analyst's only interest going into the study would be to determine with a high level of confidence that the product made with the less expensive ingredient is not less preferred than the company's current product. The null hypothesis and the alternative hypothesis for this investigation are

$$\mathrm{H}_0 : p_{\text{current}} = p_{\text{less expensive}}$$

$$\text{vs.}$$

$$\mathrm{H}_a : p_{\text{current}} > p_{\text{less expensive}}$$

where p_i is the proportion of the population that prefers product i. Both the null and the alternative hypotheses must be specified before the test is conducted. If the alternative hypothesis is formulated after reviewing the data, the results of the statistical tests are too often biased in favor of rejecting the null hypothesis.

14.3.2 One-Sided and Two-Sided Hypotheses

There are two types of alternative hypotheses: one-sided alternatives and two-sided alternatives. Some examples of situations leading to one-sided and two-sided alternatives are

One-Sided	Two-Sided
Confirm that a test brew is more bitter	Decide which test brew is more bitter
Confirm that a test product is preferred to the control	Decide which test product is preferred
In general, whenever H_a has the form A is more (less) than B, where both A and B are specified	In general, whenever H_a has the form A is different from B

Researchers often have trouble deciding whether the alternative hypothesis is one sided or two sided. General rules that work for one person may misguide others. There are no statistical criteria for deciding if an alternative hypothesis should be one sided or two sided. The form of the alternative hypothesis is determined by the prior interest of the researcher. If the researcher is only interested in determining if two samples are different, then the alternative hypothesis is two sided. If, on the other hand, the researcher wants to test for a specific difference between two samples, that is, one sample (specified) is more preferred or more sweet, and so on, than another sample, then the alternative hypothesis is one sided. Most alternatives are two sided, unless the researcher states that a specific type of difference is of interest before the study is conducted.

A point of confusion may arise regarding one-sided versus two-sided alternatives because in several common sensory testing situations, one-tailed tests statistics are used to test two-sided alternatives. For example, in a triangle test, the null hypothesis is only rejected for large numbers of correct selections (i.e., a one-tailed test criterion). However, the alternative hypothesis is two sided (i.e., H_a: the samples are perceivably different). Similar situations arise when χ^2 and F-tests are performed.

In practice, researchers should express their interests (i.e., the null and alternative hypotheses) in their own words. If the researcher's interests are clearly stated, it is easy to decide whether the alternative hypothesis is one sided or two sided. If not, then further probing is necessary. The sensory analyst should report the results of the study in terms of the researcher's stated interests (one sided or two sided), regardless of whether the appropriate statistical method is one tailed or two tailed.

14.3.3 Type I and Type II Errors

In testing statistical hypotheses, some conclusion is drawn. The conclusion may be correct or incorrect. There are two ways in which an incorrect conclusion may be drawn. First, a researcher may conclude that the null hypothesis is false when, in fact, it is true (e.g., that a difference exists when it does not). Such an error is called a *type I error.* Second, a researcher may conclude that the null hypothesis is true, or more correctly that the null hypothesis cannot be rejected, when, in fact, it is false (e.g., failing to detect a difference that exists). Such an error is called a *type II error* (see Figure 14.7). The practical implications of type I and type II errors are presented in Figure 14.8.

The probabilities of making type I and type II errors are specified before the investigation is conducted. These probabilities are used to determine the required sample size for the study (see, for example, Snedecor and Cochran 1980: 10). The probability of making a type I error is equal to α. The probability of making a type II error is equal to β. Although α and β are probabilities (i.e., numbers), it is currently a common practice to use type I error and α-error (as well as type II error and β-error) interchangeably. This somewhat casual use of terminology causes little confusion in practice.

The complementary value of type II error, that is, $1 - \beta$, is called the *power* of the statistical test. Power is simply the probability that the test will detect a given-sized departure from the null hypothesis (and, therefore, correctly reject the false null hypothesis). In discrimination testing, for example, the null hypothesis is H_0: $p_d = 0\%$. Departures from the null hypothesis are measured as values of $p_d > 0\%$. Suppose a researcher is

Truth \ Decision	Reject H_0	Do not reject H_0
H_0 true	Type I error Pr [type I error] = α	Correct decision
H_0 false	Correct decision	Type II error Pr [type II error] = β

Figure 14.7 Type I and type II errors of size α and β.

conducting a duo–trio test with 40 assessors and is testing at the $\alpha = 0.05$ level of significance. If the true proportion of distinguishers in the population of assessors is $p_d = 25\%$, then the power of the test is $1 - \beta = 0.44$—that is, the test, as designed, has a 44% chance of rejecting the null hypothesis at the $\alpha = 0.05$ level when 25% of the population can distinguish the samples. The power of a statistical test is affected by the size of the departure from the null hypothesis (i.e., p_d), the size of the type I error (α-risk) and the number of assessors, n.

14.3.4 Examples: Tests on Means, Standard Deviations, and Proportions

This section presents procedures for conducting routine tests of hypotheses on means and standard deviations of normal distributions and on the population proportion (or probability of success) from binomial distributions.

Example 14.1: Testing that the Mean of a Distribution Is Equal to a Specified Value

Suppose in the consumer test example in Section 14.2.1 that the sensory analyst wanted to test if the average overall liking of the sample for Chicago was six or greater than six. The mathematical forms of the null hypothesis and alternative hypothesis are

$$H_0 : \mu = 6$$

vs.

$$H_a : \mu > 6$$

The alternative hypothesis is one-sided.

Truth	Reject H_0	Do not reject H_0
H_0 is true	Type I error • Substitution takes place when it should not. • New product promotion done on same product as before. • Franchise in trouble due to loss of consumer confidence.	Correct decision
H_0 is false	Correct decision	Type II error • Substitution does not take place when it should. • Candidate sample is missed. • Money, effort and time are lost. • We "missed the boat."

(a) In testing for a difference

Truth	Reject H_0	Do not reject H_0
H_0 is true	Type I error • Substitution does not take place when it should. • Candidate sample is missed. • Money, effort and time are lost. • We "missed the boat."	Correct decision
H_0 is false	Correct decision	• Substitution takes place when it should not. • New product promotion done on same product as before. • Franchise in trouble due to loss of consumer confidence.

(b) In testing for similarity

Figure 14.8 The practical implications of type I and type II errors.

The statistical procedure used to test this hypothesis is a one-tailed, one-sample *t*-test. The form of the test statistic is:

$$t = (\bar{x} - \mu_{H_0}) / (s / \sqrt{n}). \tag{14.4}$$

The values of $\bar{x}$ and s are calculated in Table 14.3. Substituting into Equation 14.4 yields

$$t = (7.17 - 6) / (1.45 / \sqrt{30}) = 4.42. \tag{14.5}$$

This value of t is compared to the upper-α critical value of a *t*-distribution with $(n - 1)$ degrees of freedom (denoted as $t_{\alpha,n-1}$). The value of $t_{\alpha,n-1}$ marks the point in the *t*-distribution (with $(n - 1)$ degrees of freedom), for which the probability of observing any larger value of t is α. If the value obtained in Equation 14.5 is greater than $t_{\alpha,n-1}$, then the null hypothesis is rejected at the α-level of significance. Suppose the sensory analyst decides to control the type I error at 5% (i.e., $\alpha = 0.05$). Then, from the row of Table 19.3 corresponding to 29 degrees of freedom, the value of $t_{0.05,29} = 1.699$; therefore, the sensory analyst rejects the null hypothesis assumption that $\mu = 6$ in favor of the alternative hypothesis that $\mu > 6$ at the 5% significance level.

If this alternative hypothesis had been H_a: $\mu \neq 6$ (i.e., a two-sided alternative), then the null hypothesis would be rejected for absolute values of t (in Equation 14.5) greater than $t_{\alpha/2,n-1}$, that is, reject if $|t| > t_{0.025,29} = 2.045$ (from Table 19.3).

Example 14.2: Comparing Two Means—Paired-Sample Case

Sensory analysts often compare two samples by having a single panel evaluate both samples. When each member of the panel evaluates both samples, the paired *t*-test is the appropriate statistical method to use. In general, the null hypothesis can specify any difference of interest (i.e., H_0: $\delta = \mu_1 - \mu_2 = \delta_0$; setting $\delta_0 = 0$ is equivalent to testing H_0: $\mu_1 = \mu_2$). The alternative hypothesis can be two-sided (i.e., H_a: $\delta \neq \delta_0$) or one-sided (H_a: $\delta > \delta_0$ or H_a: $\delta < \delta_0$). In either case, the form of the paired *t*-statistic is

$$t = \frac{\bar{\delta} - \delta_0}{s_\delta / \sqrt{n}}, \tag{14.6}$$

where $\bar{\delta}$ is the average of the differences between the two samples and s_δ is the sample standard deviation of the differences. Consider the data in Table 14.5 that summarizes the scores of the panel on a single attribute. The analyst wants to test whether the average rating of sample 1 is more than two units greater than the average rating for sample 2. The null hypothesis in this case is H_0: $\delta \leq 2$ versus the alternative hypothesis H_a: $\delta > 2$. The test statistic is calculated as

$$t = \frac{2.54 - 2.00}{0.61 / \sqrt{10}} = 2.79,$$

where $n = 10$ is used as the sample size because there are 10 judges, each contributing one difference to the data set. The null hypothesis is rejected if this value of t exceeds the upper-α critical value of the *t*-distribution with $(n - 1)$ degrees of freedom (i.e., $t_{\alpha,n-1}$).

The analyst decides to set $\alpha = 0.05$ and finds in Table 19.3 that $t_{0.05,9} = 1.833$. The value of $t = 2.79$ is greater than 1.833, so the analyst rejects the null hypothesis and concludes,

Table 14.5 Data and Summary Statistics for the Paired t-test in Example 14.2

Judge	Sample 1	Sample 2	Difference
1	7.3	5.7	1.6
2	8.4	5.2	3.2
3	8.7	5.9	2.8
4	7.6	5.3	2.3
5	8.0	6.1	1.9
6	7.1	4.3	2.8
7	8.0	5.7	2.3
8	7.5	3.8	3.7
9	6.9	4.5	2.4
10	7.4	5.0	2.4
			$\bar{\delta} = 2.54$
			$s_\delta = 0.61$

at the 5% significance level, that the average rating for sample 1 is more than two units greater than the average rating for sample 2.

Example 14.3: Comparing Two Means—Independent (or Two-Sample) Case

Suppose that a sensory analyst has trained two descriptive panels at different times and that the analyst now wants to merge the two groups. The analyst wants a high level of confidence that the two groups will score samples with equivalent ratings before merging the groups and treating them as one panel.

The sensory analyst conducts several attribute panels to ensure that the two groups are similar. For each attribute considered, the analyst presents samples of the same product to all panelists and records their scores and the group to which they belong. The data from one of the studies is presented in Table 14.6. The null hypothesis for this test is H_0: $\mu_1 = \mu_2$ (or, equivalently, H_0: $\mu_1 - \mu_2 = 0$). The alternative hypothesis is H_a: $\mu_1 \neq \mu_2$ (i.e., a two-sided alternative).

The test statistic used to test the hypothesis is a two-sample t-test. The form of the test statistic is

$$t = \frac{(\bar{x}_1 - \bar{x}_2) - \delta}{\sqrt{\frac{(n_1-1)s_1^2 + (n_2-1)s_2^2}{n_1+n_2-2}}\sqrt{\frac{1}{n_1}+\frac{1}{n_2}}}, \tag{14.7}$$

where δ_0 is the difference specified in the null hypothesis ($\delta_0 = 0$ in the present example). Substituting the values from Table 14.6 into Equation 14.7 yields

$$t = \frac{(6.557 - 6.778) - 0}{\sqrt{\frac{(7-1)(0.580)^2 + (9-1)(0.460)^2}{7+9-2}}\sqrt{\frac{1}{7}+\frac{1}{9}}} = \frac{-0.221}{\sqrt{0.265}\sqrt{0.254}} = -0.85.$$

Table 14.6 Data and Summary Statistics for the Two-Sample *t*-test in Example 14.3

Group 1		Group 2	
Judge	**Score**	**Judge**	**Score**
1	6.2	1	6.7
2	7.5	2	7.6
3	5.9	3	6.3
4	6.8	4	7.2
5	6.5	5	6.7
6	6.0	6	6.5
7	7.0	7	7.0
		8	6.9
		9	6.1
$n_1 = 7$		$n_2 = 9$	
$x_1 = 6.557$		$x_2 = 6.778$	
$s_1 = 0.580$		$s_2 = 0.460$	

The value of $t = -0.85$ is compared to the critical value of a *t*-distribution at the $\alpha/2$ significance level (because the alternative hypothesis is two sided) with $(n_1 + n_2 - 2)$ degrees of freedom. For the present example (using $\alpha = 0.05$) $t_{0.025,14} = 2.145$ from Table 19.3. The null hypothesis is rejected if the absolute value (i.e., disregard the sign) of t is greater than 2.145. Because the absolute value of $t = -0.85$ (i.e., $|t| = 0.85$) is less than $t_{0.025,14} = 2.145$, the sensory analyst does not reject the null hypothesis and concludes that, on average, the two groups report similar ratings for this attribute.

Example 14.4: Comparing Standard Deviations from Two Normal Populations

The sensory analyst in Example 14.3 should also be concerned that the variability of the scores of the two groups is the same. To test that the variability of the two groups is equal, the analyst compares their standard deviations. The null hypothesis for this test is H_0: $\sigma_1 = \sigma_2$. The alternative hypothesis is H_a: $\sigma_1 \neq \sigma_2$ (i.e., a two-sided alternative). The test statistic used to test this hypothesis is

$$F = \frac{s^2_{\text{Larger}}}{s^2_{\text{Smaller}}}, \tag{14.8}$$

where s^2_{Larger} is the square of the larger of the two sample standard deviations and s^2_{Smaller} is the square of the smaller sample standard deviation. In Table 14.6, group 1 has the larger sample standard deviation, so $s^2_{\text{Larger}} = s_1^2$ and $s^2_{\text{Smaller}} = s_2^2$ for this example. The value of F in Equation 14.8 is then

$$F = (0.58)^2 / (0.46)^2 = 1.59\,.$$

The value of F is compared to the upper $\alpha/2$ critical value of an F-distribution with $(n_1 - 1)$ and $(n_2 - 1)$ degrees of freedom. (The numerator degrees of freedom are $(n_1 - 1)$

because $s^2_{\text{Larger}} = s^2_1$ for this example. If s^2_{Larger} had been s^2_2, then the degrees of freedom would be $(n_2 - 1)$ and $(n_1 - 1)$.) Using a significance level of $\alpha = 0.05$, the value of $F_{0.025,6,8}$ is found in Table 19.6 to be 4.65. The null hypothesis is rejected if $F > F_{\alpha/2,(n_1-1),(n_2-1)}$. Because $F = 1.59 < F_{0.025,6,8} = 4.65$, the null hypothesis is not rejected at the 5% significance level. The sensory analyst concludes that there is not sufficient reason to believe the two groups differ in the variability of their scoring on this attribute.

This is another example of a two-sided alternative that is tested using a one-tailed statistical test. The criterion for two-sided alternatives is to reject the null hypothesis if the value of F in Equation 14.8 exceeds $F_{\alpha/2,df_1,df_2}$ where df_1 and df_2 are the numerator and denominator degrees of freedom, respectively. Equation 14.8 is still used for one-sided alternatives (i.e., H_a: $s_1 > s_2$), but the criterion becomes "reject the null hypothesis if $F > F_{\alpha,df_1,df_2}$."

Example 14.5: Testing That the Population Proportion Is Equal to a Specified Value

Suppose that two samples (A and B) are compared in a preference test. The objective of the test is to determine if either sample is preferred by more than 50% of the population. The sensory analyst collects a random sample of $n = 200$ people; presents the two samples to each person in a balanced, random order; and asks each person which sample they prefer. For those respondents who refuse to state a preference, the "no preference" responses are divided equally among the two samples. It is found that 125 of the people said they preferred sample A. The estimated proportion of the population that prefer sample A is then $\hat{p}_A = 125/200 = 62.5\%$ by Equation 14.3.

The sensory analyst arbitrarily picks "preference for sample A" as a "success" and tests the hypothesis H_0: $P_A = 50\%$ versus the alternative H_a: $P_A \neq 50\%$. The analyst chooses to test this hypothesis at the $\alpha = 0.01$ significance level using the appropriate z-test:

$$z = \frac{\hat{p} - p_0}{\sqrt{(p_0)(1-p_0)/n}} \text{ for } \hat{p} \text{ and } p_0 \text{ proportions,}$$

or

$$z = \frac{\hat{p} - p_0}{\sqrt{(p_0)(100-p_0)/n}} \text{ for } \hat{p} \text{ and } p_0 \text{ percentages,} \tag{14.9}$$

where $\hat{p}$ and p_0 are the observed and hypothesized values of P, respectively. Substituting the observed and hypothesized values into Equation 14.9 yields

$$z = (62.5 - 50.0)/\sqrt{(50)(100-50)/200} = 3.54.$$

This value of z is compared to the critical value of a standard normal distribution. For two-sided alternatives, the absolute value of z is compared to $z_{\alpha/2} = t_{\alpha/2,\infty}$ (for one-sided alternatives, the value of z is compared to $z_\alpha = t_{\alpha,\infty}$) using Table 19.3. The value of $z_{0.005} = t_{0.005,\infty} = 2.576$. Because $z = 3.54$ is greater than 2.576, the null hypothesis is rejected and the analyst concludes at the 1% significance level that sample A is preferred by more than 50% of the population.

Table 14.7 Results of a Two Region Preference Test in Example 14.6

	Preference		
Region	**Product A**	**Product B**	**Total**
1	125	75	200
2	102	98	200
Total	227	173	400

Example 14.6: Comparing Two Population Proportions

Example 14.5 will be extended to take regional preferences into consideration. Suppose a company wishes to introduce a new product (A) into two regions and wants to know if the product is equally preferred over its prime competitor's product (B) in both regions. The sensory analyst conducts a 200-person preference test in each region and obtains the results shown in Table 14.7.

The null hypothesis in this example is H_0: $P_1 = P_2$ versus the alternative hypothesis H_a: $P_1 \neq P_2$, where P_i is the proportion of the population in region i that prefers product A. This hypothesis is tested using a z-test of the form

$$z = (\hat{p}_1 - \hat{p}_2) / SE, \tag{14.10}$$

where $\hat{p}_i$ is the estimated proportion of region i that prefers product A and SE is the estimated standard error of the difference between the two proportions, which is calculated as

$$SE = \sqrt{\hat{p}(1-\hat{p})(1/n_1 + 1/n_2)},$$

where $\hat{p}$, the pooled estimate of the population proportion, is: $\hat{p} = (n_1\hat{p}_1 + n_2\hat{p}_2)/(n_1 + n_2)$.

Substituting the values from Table 14.7 into the equations above:

$$\hat{p} = \frac{200 * 62.5 + 200 * 51.0}{200 + 200} = 56.75,$$

$$SE = \sqrt{56.75(100 - 56.75)\left(\frac{1}{200} + \frac{1}{200}\right)} = 4.954,$$

$$\text{and } z = (62.5 - 51.0) / 4.954 = 2.32.$$

The value of z in Equation 14.10 is compared to the upper-$\alpha/2$ critical value of the standard normal distribution. If the analyst chooses $\alpha = 0.10$ (i.e., 10% significance level), then the critical value $Z_{0.10/2} = t_{0.10/2,\infty} = 1.645$ (from Table 19.3). Because $z = 2.32 > Z_{0.10/2} = 1.645$, the analyst concludes at the 10% significance level that product A is not equally preferred over product B in both regions. Regional formulations may have to be considered.

Alternatively, the p-value associated with the test statistic, z, can be calculated using a spreadsheet function for the standard normal distribution, such as Excel's NORM.S.DIST. The p-value for the example is calculated as P = 2*(1-NORM.S.DIST(2.32,TRUE)) = 0.0203. Since the calculated p-value is less than the significance level chosen for the test, the analyst concludes at there is a significant difference between the regions. The multiplier of 2 in the calculation of the p-value is included because the example is a two-sided test. For a one-sided test, the multiplier of 2 would not be included in the calculation.

14.3.5 Calculating Sample Sizes in Discrimination Tests

The sample size required for a discrimination test is a function of the test sensitivity parameters, α, β, and p_d, or in the case of directional difference tests, p_{max}. Tables 19.7, 19.9, 19.11, and 19.13 can be used to find sample sizes for commonly chosen values of the parameters. Alternatively, researchers can use a spreadsheet to perform the necessary calculations. A test sensitivity analyzer has been developed in Microsoft Excel to allow researchers to study how various choices of α, β, and p_d (or p_{max}) affect the sample size and the number of correct responses necessary to claim that a difference exists or that the samples are similar (see Figure 14.9). The test sensitivity analyzer does this indirectly by letting the researcher choose values for the same size, n; the number of correct responses, x; and the maximum allowable proportion of distinguishers, p_d. (Although p_d is not meaningful in a directional difference test, the value of p_{max} is computed based on the value entered for p_d.) The test sensitivity analyzer then computes values for α and β. By adjusting the values of n, x, and p_d, the researcher can find the set of values that provides the best compromise between test sensitivity and available resources.

The binomial distribution, on which discrimination tests are based, is a discrete probability distribution. Only integer values for the sample size, n, and the number of correct responses, x, are valid. Small changes in n and x can have large impacts on the probabilities α and β, particularly for small values of n. Generally, it is not possible to select values of n, x, and p_d (or p_{max}) that yield values for α and β that are exactly equal to their target values. Instead, the researcher must select values for n, x, and p_d (or p_{max}) that yield values for α and β that are close to their targets.

As illustrated in Figure 14.9, the researcher wants to conduct a duo–trio test for similarity with the following target sensitivity values: $\alpha = 0.25$, $\beta = 0.10$, and $p_d = 25\%$. Strictly speaking, the values of n and x should be chosen so that both α and β are no greater than their target values. However, the researcher only has access to 60 assessors. Setting $n = 60$, the researcher finds that $x = 33$ correct responses yields values for α and β that are quite close to their targets, although the value for $\alpha = 0.26$ is slightly larger than desired. By adjusting n and x, the researcher finds that $n = 67$ assessors with $x = 37$ correct responses would be needed to yield values for α and β that are either at or below their targets. The researcher decides that the 60-assessor test is adequate, given that that is the maximum number of assessors available and that the α-risk is only 1% greater than the target value.

The test sensitivity analyzer is a useful tool for planning discrimination tests. Researchers can quickly run a variety of scenarios with different values of n, x, and p_d (or p_{max}) to observe the resulting impacts on α-risk and β-risk, selecting the values that offer the best compromise solution. The test sensitivity analyzer can be programmed in Excel by

INPUTS				OUTPUT			
Number of Respondents	Number of Correct Responses	Probability of a Correct Guess	Proportion Distinguishers	p_{max} or Probability of a Correct Response @ p_d	Type I Error	Type II Error	Power
n	x	p_0	p_d	p_{max}	α-risk	β-risk	1-β
60	33	0.50	0.25	0.625	0.2595	0.0923	0.9077

Interpretation:

33 or more correct responses is evidence of a difference at the α = 0.26 level of significance.

32 or fewer correct responses indicates that you can be 91% sure that no more than 25% of the panelists can detect a difference -- that is, evidence of similarity relative to p_d = 25% at the β = 0.09 level of significance.

Instructions:

1. Make entries in Row 4 of Columns A through D only!
   a. Enter the number of respondents (n) in Cell A4.
   b. Enter the number of correct responses (x) in Cell B4.
   c. Enter the probability of a correct guess (p_0) in Cell C4. (e.g., for the Triangle test C4 = 1/3, for the Duo-trio test C4 = 1/2, etc.)
   d. Enter the proportion of distinguishers (p_d) you want to be able to detect in D4.

2. Evaluate the results in Row 4, Columns F through H to decide if the test has adequate sensitivity.
   a. In testing for difference, choose small values for α-risk. (Adjust n and x to achieve the desired sensitivity (i.e., α-risk).)
   b. In testing for similarity choose small values for β-risk. (Adjust n, x and p_d to achieve the desired sensitivity (i.e., balance between p_d and β-risk).) (Do not choose values of x that are less than what would be expected by chance alone (e.g., $n/3$ for a Triangle test, $n/2$ for a Duo-trio, etc.). Increase n when necessary.)
   c. When testing for difference and similarity simultaneously, choose acceptably small values for both α-risk and β-risk. (Adjust n, x and p_d to achieve the desired sensitivity (i.e., balance between p_d, β-risk, and α-risk)).
   d. When using a two-sided directional difference test, double he computed value of α-risk to account for the two-sided nature of the test.

Figure 14.9 Test sensitivity analyzer illustrating the values of n, x, and p_d for a duo–trio test with target values of α = 0.25 and β = 0.10. Note that the α-risk is slightly greater than the target value specified.

making the entries in the cells indicated in Table 14.8. The explanatory text is entered in the appropriate cells, using the fonts and sizes necessary to achieve the desired visual effect.

In testing for similarity or in the unified approach, the number of correct responses, x, should not be chosen to be less than the number that would be expected by chance alone (e.g., $n/3$ in a triangle test, $n/2$ in a duo–trio test, etc.). Such values correspond to negative values for the proportion of distinguishers ($p_d < 0$). This is a logical impossibility that should not be used as the decision criterion in a test.

In using the test sensitivity analyzer for two-sided directional difference tests, researchers must remember to double the computed value for α-risk to account for the two-sided nature of the test.

Table 14.8 Excel Programming Information for Test Sensitivity Analyzer

Cell	Entry
E4	=D4+C4(1-D4) (can be hidden if desired)
F4	=1−BINOM.DIST(B4−1,A4,C4,TRUE)
G4	=BINOM.DIST(B4−1,A4,E4,TRUE)
H4	=1−G4
A7	=B4
F7	=F4
A9	=B4−1
F9	=H4 (using % format)
B10	=D4 (using % format)
D11	=B10 (using % format)
F11	=G4

14.4 THURSTONIAN SCALING

Although the percent-distinguisher model is appealing for its ease of interpretation, it is not a theoretical model of human behavior. It is useful for planning discrimination tests but, in general, it oversimplifies the behavior of sensory assessors. If individuals are truly either distinguishers or nondistinguishers, then in replicate triangle tests involving the same two samples, for example, one group of assessors should have 100% correct responses (the distinguishers) and another group should have 1/3 correct responses (the nondistinguishers). This is not what happens in practice. A more elaborate model of human behavior is necessary to better understand the results observed in discrimination tests. Thurstonian scaling (Thurstone, 1927) is one such model. The following overview of Thurstonian scaling is drawn largely from Ennis (2001), which was later represented in ASTM (2003).

14.4.1 A Fundamental Measure of Sensory Differences

The Thurstonian model is based on two assumptions. The first assumption is that perceptions have a probabilistic component that follows a normal probability rule. The second assumption is that assessors can faithfully execute the decision rule associated with the sensory task they are asked to perform.

The Thurstonian model recognizes that perceptions vary when an assessor performs repeated evaluations of the same product. The variations may result from heterogeneity in the product, or from momentary physiological or psychological changes in the assessor, or from some combination of these. Regardless of the sources, perceptions vary from one evaluation to another. The Thurstonian model assumes that the changes in perception follow a normal probability distribution on a nonspecific dimension of sensory magnitude. Without loss of generality, the simplest Thurstonian model assumes that the distribution

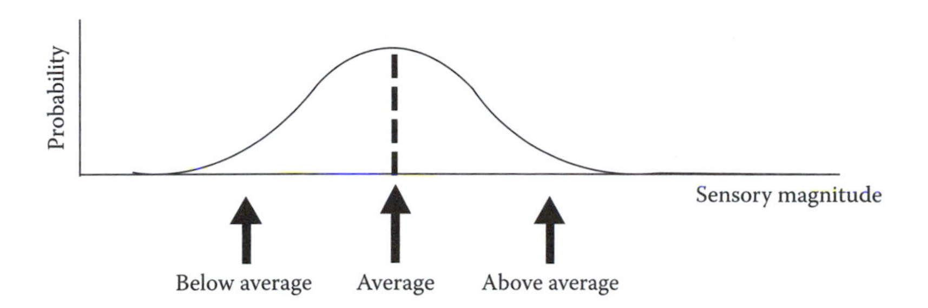

Figure 14.10 Normal distribution of sensory perceptions. An assessor's perceptions vary about the average of the product according to a normal probability distribution.

of perceptions fall about the sample's average magnitude with a standard deviation of one (see Figure 14.10).

It is the variation in perception that led assessors to confuse products that are "on the average" different. When two products have very different average sensory magnitudes, they will not be confused because there is no overlap in their perceptual distributions. For example, in Figure 14.11a, all of the perceived magnitudes of product B are higher than those of product A. However, when the difference between the products is small, the two distributions overlap, and it is possible for the products to be confused. In Figure 14.11b, although on the average product B is higher than product A, in a single evaluation, product B may be perceived to be closer to the average value of product A's distribution than it is to the average of its own distribution. The likelihood of this occurring is proportional to the distance between the averages of the two distributions. The distance between the two averages, called δ, is the fundamental measure of sensory difference in the Thurstonian model. δ is the number of standard deviations that separate the averages of the two distributions. If δ is small, the two products are similar. If δ is large, the two products are perceptibly different. The value of δ can be estimated using any forced-choice or category-scaling method. The statistic used to estimate δ is d'.

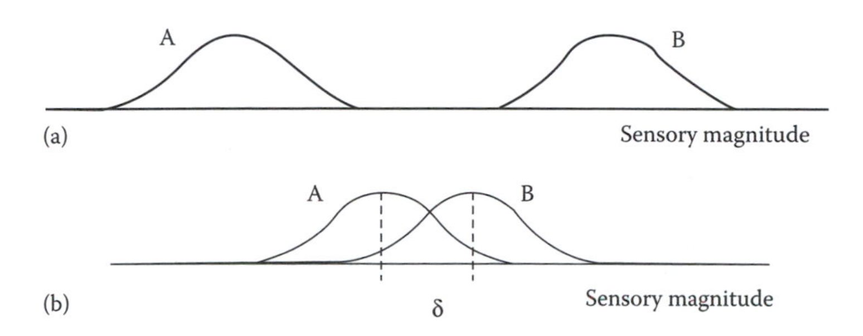

Figure 14.11 (a) Thurstonian representation of two products that are very different. (b) Thurstonian representation of two products that are confusable because their sensory distributions overlap. The degree of overlap is proportional to the distance between the average values of the distributions, δ.

14.4.2 Decision Rules in Sensory Discrimination Tests

The decision rules that assessors use to formulate their responses in forced-choice or category-scaling tests differ from one test method to another. For example, in a duo–trio test, the decision rule is to pick the coded sample that is perceived to be closer to the reference sample, whereas in a 2-AFC test, the decision rule is to pick the sample that is "stronger." Decision rules for several forced-choice tests are presented in Table 14.9. It is assumed that assessors apply the decision rules correctly when evaluating products. In other words, assessors will always give the correct answer based on what they perceive, even though the variability of perceptions may lead them to an incorrect answer with regard to the actual differences between the products.

An example using the triangle test will clarify this point. In the triangle test, an assessor is given three coded samples. Suppose two of the samples are from product A, and one of the samples is from product B. The assessor's task is to identify the "odd" sample—that is, to identify which one of the three samples is most different from the other two. The

Table 14.9 Decision Rules for 2-AFC, 3-AFC, Triangle, Duo–Trio, A/Not A and Same–Different Methods

Method	**Decision Rule**
2-AFC	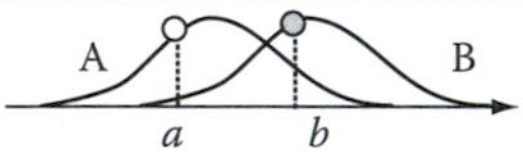 $P_c = P(b > a)$
3-AFC	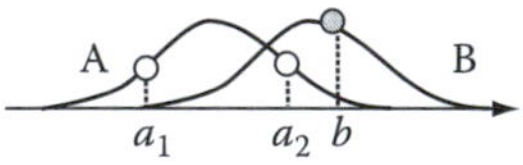 $P_c = P(b > a_1 \text{ and } b > a_2)$
Triangle	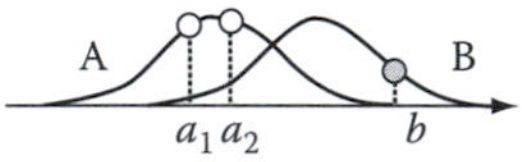 $P_c = P(\mid a_1 - a_2 \mid < \mid a_1 - b \mid \text{ and } \mid a_1 - a_2 \mid < \mid a_2 - b \mid)$
Duo–trio	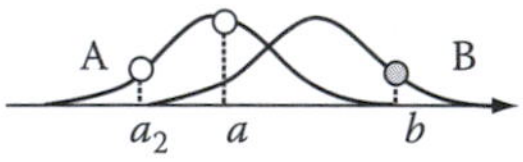 $P_c = P(\mid a_R - a \mid < \mid a_R - b \mid)$
A/Not A	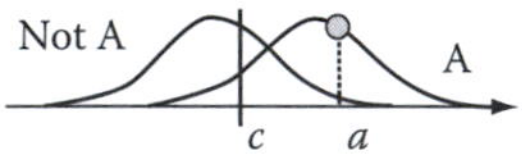 $P_a = P(a > c)$
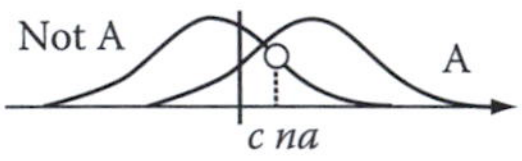	$P_{na} = P(na > c)$
Same–different	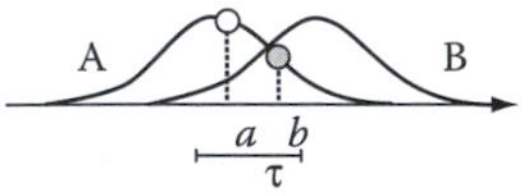 $P(S/U) = P(\mid b - a \mid < \tau)$
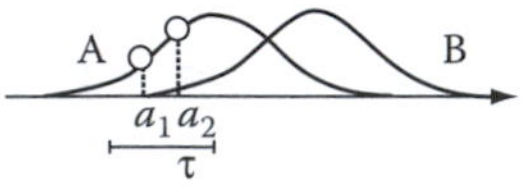	$P(S/M) = P(\mid x_1 - x_2 \mid < \tau), x = a \text{ or } b$

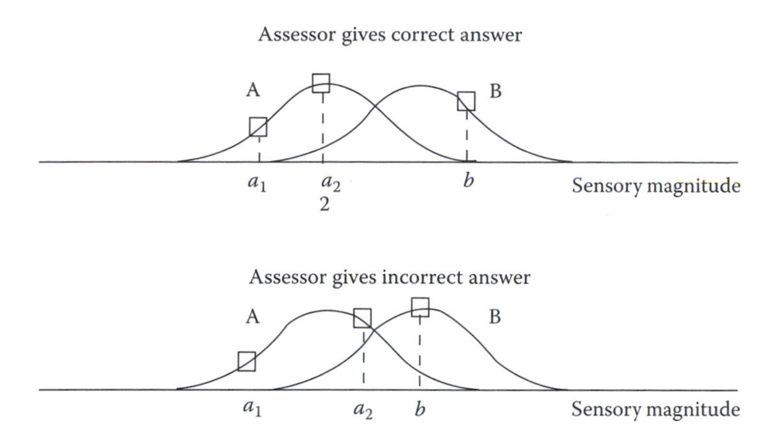

Figure 14.12 Correct and incorrect answers in a triangle test. In the top trial, the assessor correctly answers that *b* is the odd sample because, perceptually, both A_1 and A_2 are farther from the B than they are from each other. In the bottom trial, the assessor incorrectly answers that A_1 is the odd sample because perceptually the samples A_2 and B are farther from sample A_1 than they are from each other. In both cases, the assessor applied the decision rule correctly.

assessor will always select the sample that he or she perceives to be the "odd" sample on every trial of a triangle test.

Figure 14.12 presents the results of two possible trials of the triangle test. In the first trial, the assessor selects the B sample because perceptually it is farther from both of the A samples than the A samples are from each other. On this trial, the assessor gives a correct answer because the assessor's perceptions are consistent with the actual difference between the products. On the second trial, the assessor selects the A_1 sample because, perceptually, it is farther from the A_2 and B samples than the A_2 and B samples are from each other. On this trial, the assessor gives an incorrect answer because the assessor's perceptions are inconsistent with the actual difference between the products. On both trials, the assessor applied the decision rule correctly. However, due to the probabilistic nature of perception, in one trial the answer is correct, and in the other the answer is incorrect.

Ennis (1993) shows that because of differences in the decision rules, discrimination methods differ in their ability to detect a given-sized sensory difference, δ. In general, the more complex the decision rule, the more assessors are required to deliver the same statistical power from the test. For example, 2-AFC and 3-AFC tests are more powerful than the duo–trio and triangle test. Frijters (1979) used this approach to resolve the "paradox of discriminatory nondiscriminators."

14.4.3 Estimating the Value of δ

14.4.3.1 Forced-Choice Methods

The proportion of correct answers in a forced-choice test increases as the distance between the products increases. The proportion of correct answers can be combined with the decision rule of the test to estimate the value for δ. Tables of d', the statistic used to estimate δ, are widely available (see, for example, ASTM 2003, Elliott 1964, and Ennis 1993). In addition, ASTM (2003) and Bi et al. (1997) have published tables of the variance of d'. Knowing the variance of the estimate allows researchers to construct confidence intervals and statistical tests regarding the true value of δ. Software that computes d' and its associated statistics also is available, for example, from the Institute for Perception (Ennis, 2001).

The process for obtaining the value of d' is the same for all of the forced-choice tests that have a fixed, null-hypothesis probability of a correct response (2-AFC, 3-AFC, duo–trio, and triangle). For each of these tests, the value of d' is proportional to the observed proportion of correct responses. This is not the case for the "A"–"not A" and the same–different tests. Both of these test methods involve a placebo effect. For example, in the "A"–"not A" method, an assessor will tolerate some deviation around the average of the A distribution and will still consider the perception as coming from a sample of product A. When the perception falls too far from the average of the A distribution, the assessor will classify the sample as "not A." The boundary that the assessor uses to make this decision is called c. Similarly, for the same–different test, the maximum difference in the perceptions of the two samples that will still be classified as "same" is τ (see Table 14.9). These new parameters, called *cognitive criteria,* play a role in how the values of d' are obtained from each of the methods. In addition, the cognitive criteria can reveal the degree to which assessors vary in their tolerance of natural variation in perception.

14.4.3.2 Methods Using Scales

The Thurstonian model for category scales makes use of the c criterion from the "A"–"not A" test. Multiple c values are defined to mark the boundaries of the successive scale categories. For example, on a nine-point scale, there are eight c values, c_1 through c_8. To receive a rating of 5, the perceived intensity of a sample must be greater than c_4 boundary but less than c_5 boundary, as illustrated in Figure 14.13. The c values tend to not be equally spaced along the sensory dimension. Unequal spacing occurs due to biases in the assessors' behavior. For example, the neutral 5 category on a nine-point scale tends to be narrower than adjoining categories because assessors prefer to provide some positive or negative response, even if it is only weakly held.

In addition to providing d' estimates of the differences among samples, identifying the boundaries between categories reveals interesting information about how assessors are using the scales.

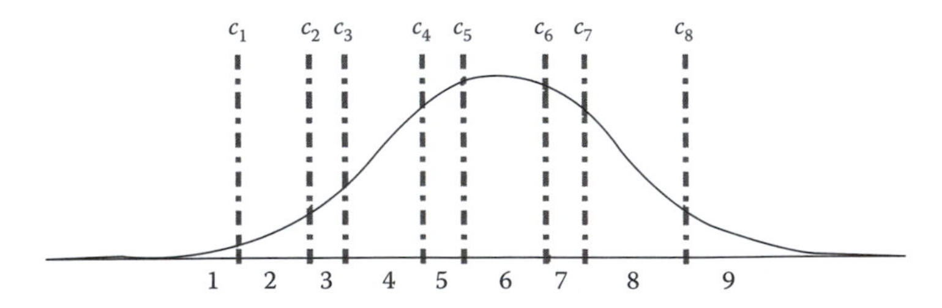

Figure 14.13 Multiple c criteria form the boundaries of the scale categories. In order to receive a rating of 5, the perceived intensity of a sample must be greater than c_4 but less than c_5.

The calculation of d' from full category scale data is too complicated to explain here. However, ASTM (2003) presents a rapid, table look-up approach that applies the technique used for the "A"–"not A" test to category scale data that has been collapsed into two categories (irrespective of how many categories there were on the physical scale used to collect the data). Detail is lost due to the collapsing, but the values of d' and its variance are still accurate.

14.4.4 Transitioning from Percent-Distinguisher Model to the Thurstonian Model for Planning Discrimination Tests

As stated earlier, the percent-distinguisher model is easy to use and understand, but it is oversimplistic. One of the outcomes that results from its simplicity is that it fails to distinguish the differences in performance of certain discrimination tests. Ennis (1993) used the Thurstonian model to demonstrate why the 3-AFC method is more power than the triangle test, while the percent-distinguisher model would indicate that the two methods are equally sensitive. The percent-distinguisher model considers only the guessing probability of a forced-choice discrimination test when assessing its statistical sensitivity. For example, based on the percent-distinguisher model, all tests with a guessing probability of 1/3 (e.g., triangle, 3-AFC, and the unspecified tetrad tests) would be equally sensitive at detecting perceptible differences among products. However, comparisons of the performance of these methods demonstrate that they are not equal. The Thurstonian model, which takes into consideration the different decision rules that are used from one test method to another, more accurately characterizes the relative sensitivities of different discrimination methods.

The percent-distinguisher model is used widely, especially in similarity testing. Product developers and other sensory clients have become familiar and comfortable with the model, so it cannot simply be abandoned. However, given its limitations and the availability of a better alternative, it is time to consider transitioning from the percent-distinguisher model to the Thurstonian model when planning sensory discrimination tests. The largest barrier to this transition is converting from the familiar test sensitivity parameter, p_d, to the less familiar Thurstonian measure of difference, δ. Jesionka et al. (2014) remove the barrier by providing tables that convert from p_d to δ for several of the most commonly used discrimination test methods (Table 14.10). Organizations that have chosen a particular value for p_d in their standard testing can now easily select the value of δ that corresponds to that of p_d for whatever test method they are using.

Another benefit of the transition from percent distinguishers to the Thurstonian model is that researchers now will be able to accurately calculate sample sizes for test methods with the same guessing probabilities but different decision rules. For example, if a researcher typically performs triangle tests using $n = 60$, $p_d = 25\%$, $\alpha = 5\%$, and $\beta = 20\%$, Table 14.10 can be used to find that a triangle test $p_d = 25\%$ corresponds to $\delta = 1.466$. The same level of sensitivity (i.e., $\delta = 1.466$, $\alpha = 5\%$, and $\beta = 20\%$) can be achieved in an unspecified tetrad test with a sample size of $n = 22$ and in a 3-AFC test with a sample size of $n = 11$. The percent-distinguisher model would indicate that $n = 60$ is required for all three methods.

Table 14.10 δ as a function of Proportion of Distinguishers for the 2-AFC, Duo–Trio, 3-AFC, Triangle and Tetrad Tests

Proportion Distinguishers	2-AFC	Duo–Trio	3-AFC	Triangle	Tetrad
0.00	0.000	0.000	0.000	0.000	0.000
0.05	0.089	0.530	0.116	0.612	0.431
0.10	0.178	0.761	0.229	0.879	0.618
0.15	0.267	0.949	0.339	1.095	0.768
0.20	0.358	1.115	0.448	1.287	0.900
0.25	0.451	1.271	0.557	1.466	1.022
0.30	0.545	1.421	0.665	1.638	1.138
0.35	0.642	1.568	0.774	1.807	1.251
0.40	0.742	1.715	0.885	1.976	1.363
0.45	0.845	1.865	0.999	2.146	1.474
0.50	0.954	2.020	1.116	2.321	1.588
0.55	1.068	2.182	1.238	2.504	1.705
0.60	1.190	2.355	1.366	2.696	1.826
0.65	1.322	2.542	1.504	2.903	1.955
0.70	1.466	2.749	1.652	3.129	2.095
0.75	1.627	2.985	1.817	3.381	2.248
0.80	1.812	3.263	2.005	3.674	2.424
0.85	2.036	3.608	2.230	4.028	2.633
0.90	2.326	4.072	2.520	4.492	2.902
0.95	2.772	4.814	2.961	5.213	3.315

14.5 STATISTICAL DESIGN OF SENSORY PANEL STUDIES

In this section, experimental designs that are commonly used in sensory evaluation are presented. The discussion is structured to avoid much of the confusion that often surrounds this topic. In Section 14.5.1, independent replications of an experiment are distinguished from multiple observations of a single sample. It is shown that confusing replications with multiple observations, which results directly from failing to recognize the population of interest, can lead to the incorrect use of measurement error in place of experimental error in the statistical analysis of sensory data. When this occurs, samples are often declared to be significantly different when they are not. In Section 14.5.2, the most commonly used designs for sensory panel studies are presented. These include randomized (complete) block designs, balanced incomplete block designs, Latin-square designs, and split-plot designs.

14.5.1 Sampling: Replication versus Multiple Observations

The fundamental intent of the statistical analysis of a designed experiment is to generate an accurate and precise estimate of the experimental error. All tests of hypotheses and confidence statements are based on this. Experimental error is the unexplainable, natural variability of the population being studied. Experimental error is expressed quantitatively as the variance or as the standard deviation of the population. One measurement taken on one unit from a population provides no means for estimating experimental error. In fact, multiple observations of the same unit provide no means for estimating experimental error, either. The differences among the multiple observations taken on a single unit result from measurement error. Several units from the same population need to be sampled to develop a valid estimate of experimental error. The measurements taken on different units are called *replications*. It is the unit-to-unit (or "rep-to-rep") differences that contain the information about the variability of the population (i.e., experimental error).

A common objective of sensory studies is to differentiate products based on differences in the perceived intensities of some attributes. If only a single sample (batch, jar, preparation, etc.) of each product is evaluated, there is no way to estimate the experimental error of the population of products. Often, measurement error, that is, judge-to-judge variability, is substituted for experimental error in the statistical analysis of sensory panel data. This is a very dangerous mistake because ignoring experimental error and replacing it with measurement error can lead an analyst to falsely conclude that significant differences exist among the products when, in fact, no such differences exist. Evaluating a single batch of product ignores the batch-to-batch differences that may contribute substantially to product variability. Just as repeated measurements of one individual's height tell us nothing about person-to-person differences, repeated evaluations of a single sample (regardless of the size of the panel) tell us nothing about product batch-to-batch variability.

Measurement error is real (as sensory professionals are well aware). However, measurement error cannot be casually substituted for experimental error without incurring the great risk of obtaining misleading results from statistical analyses. If in a taste test, the contents of one jar of mayonnaise are divided into 20 servings and presented to panelists, or a single preparation of a sweetener solution is poured into 20 cups and served, then the results of the test are equivalent to the repeated measurements of an individual's height. The variability estimate obtained from the study estimates measurement error. It is not a measure of the product variability (the valid experimental error) because the independent replicates (e.g., different batches) of the product were not presented. The only legitimate conclusion that could be drawn from such a study is whether the panelists were able to detect differences among the particular samples they evaluated. This is not the same as concluding that the products are different because there is no way to assess how constant any of the observed differences would be in future evaluations of different batches of the same products.

To avoid confusing independent replications of a treatment with multiple observations, the sensory analyst must have a clear understanding of what population is being studied. If the objective of a study is to compare several brands of a product, then several units from each brand must be evaluated. If an ingredient is known to be extremely uniform, then at the very least, separate preparations of samples with that ingredient should be served to each judge. (For extremely uniform products, the major source of variability may well be the preparation-to-preparation differences).

Suggesting that only one sample be taken from each jar of product or that each serving be prepared separately is undeniably more inconvenient than taking multiple observations on a single jar. However, the sensory analyst must compare this inconvenience to the price paid when, for instance, a new product fails in the market because a prototype formulation was falsely declared to be significantly superior to a current formulation based on the evaluation of a single batch of each product.

14.5.2 Blocking an Experimental Design

The blocking structure of an experimental design is a description of how the treatments are applied to the experimental material. To understand blocking structure, two concepts must be understood: the *block* and the *experimental unit.* A block is simply a group of homogeneous experimental material. Theoretically, any unit within a block will yield the same response to the application of a given treatment. The level of the response may vary from block to block, but the difference between any two treatments applied within a block is constant for all blocks. The experimental material within a block is divided into small groups called *experimental units.* An experimental unit is that portion of the total experimental material to which a treatment is independently applied.

The sensitivity of a study is increased by taking into account the block-to-block variability that is known to exist prior to running the experiment. If the treatments are applied appropriately, the block effects can be separated from the treatment effects and from the experimental error, thus providing "clean" reads of the treatment effects while simultaneously reducing the unexplained variability in the study.

In more familiar terms, in sensory tests, the experimental material is the large group of evaluations performed by the judges. The evaluations are typically arranged into blocks according to judge, in recognition of the fact that, due to differing thresholds for instance, judges may use different parts of the rating scale to express their perceptions. It is assumed that the size of the perceived difference between any two samples is the same from judge to judge. Within each judge (i.e., block), a single evaluation is the experimental unit. The treatments, which can be thought of as products at this point, must be independently applied at each evaluation. This is accomplished through such techniques as randomized orders of presentation, sequential monadic presentations, and wash-out periods of sufficient duration to allow the respondent to return to some baseline level of perception (constant for all evaluations).

14.5.2.1 Completely Randomized Designs

The simplest blocking structure is the completely randomized design (CRD). In a CRD, all of the experimental material is homogeneous; that is, a CRD consists of one large block of experimental units. CRDs are used, for example, when a single product is being evaluated at several locations by distinct groups of respondents (e.g., a monadic, multicity consumer test). In such cases, the significance of the differences due to location is determined in light of the variability that occurs within each location.

The overall liking data in Table 14.1 conform to a CRD. The box-and-whisker plots presented in Figure 14.3 and the confidence intervals in Figure 14.5 suggest that some city-to-city differences may exist. The average liking response can be used to summarize city-to-city differences. Analysis of variance (ANOVA) is used to determine if the observed differences in average liking are statistically significant. The ANOVA table for these data is presented in Table 14.11.

Table 14.11 ANOVA Table for a Completely Randomized Design for the Multicity Consumer Test Data in Table 14.1

Source of Variability	Degrees of Freedom	Sum of Squares	Mean Square	F
Total	119	541.56		
City	3	283.34	94.45	42.43
Error	116	258.22	2.23	

The F-ratio for cities in Table 14.11 is highly significant ($F_{0.01,3,116} = 3.96$), indicating that at least some of the observed differences among cities are real. To determine which of the averages are significantly different, another statistical method called a *multiple comparisons procedure* must be applied. For the present example, the multiple comparison technique called *Fisher's least significant difference* (LSD) is used. In general, the LSD value used to compare two averages, $\bar{x}_i$ and $\bar{x}_j$, is calculated as

$$\text{LSD}_\alpha = t_{\alpha/2,df_E}\sqrt{MS_E}\sqrt{(1/n_i)+(1/n_j)}\ , \tag{14.11}$$

where $t_{\alpha/2,df_E}$ is the upper-α/2 critical value of a t-distribution with df_E degrees of freedom (i.e., the degrees of freedom for error from the ANOVA), MS_E is the mean square for error from the ANOVA, and n_i and n_j are the number of observations that went into the calculation of $\bar{x}_i$ and $\bar{x}_j$, respectively. If the sample sizes are the same for all $\bar{x}$'s, then Equation 14.11 reduces to

$$\text{LSD}_\alpha = t_{\alpha/2,df_E}\sqrt{2MS_E/n}\ , \tag{14.12}$$

where n is the common sample size. In the example, $n = 30$, so the $\text{LSD}_{0.05} = 1.96\sqrt{2(2.23)/30} = 0.76$. Any two samples whose means differ by more than 0.76 are significantly different at the 5% level. As shown in Table 14.12, Chicago has a significantly lower average value than the other three cities, and Boston has a significantly lower average value than Denver.

Completely randomized designs are seldom used in multisample studies involving sensory panels because it is inefficient to have each panelist evaluate only a single sample, and yet it is recognized that different panelists might use different parts of the rating scales to express their perceptions. More elaborate panel designs are needed for such studies. Four of the most commonly used designs for sensory panels are discussed in the remainder of this section.

Table 14.12 Average Overall Liking Scores from the Multicity Monadic Consumer Test Data in Table 14.1

City	Atlanta	Boston	Chicago	Denver
Mean rating	10.56BC	10.29B	7.17A	11.12C

Note: Means not followed by the same letter are significantly different at the 5% level.

14.5.3 Randomized (Complete) Block Designs

If the number of samples is sufficiently small that sensory fatigue is not a concern, then a randomized (complete) block design is appropriate. Panelists are the "blocks"; samples are the "treatments." Each panelist evaluates (either by rating or ranking) all of the samples (hence the term *complete block*).

A randomized block design is effective when the sensory analyst is confident that the panelists are consistent in rating the samples but recognizes that panelists might use different parts of the scale to express their perceptions. The analysis applied to data from a randomized block design takes into account this type of judge-to-judge difference, yielding a more accurate estimate of experimental error and thus more sensitive tests of hypotheses than would otherwise be available.

Independently replicated samples of the test products are presented to the panelists in a randomized order (using a separate randomization for each panelist). The data obtained from the panelists' evaluations can be arranged in a two-way table as in Table 14.13.

14.5.3.1 Randomized Block Analysis of Ratings

Data in the form of ratings from a randomized block design are analyzed by ANOVA. The form of the ANOVA table appropriate for a randomized block design is presented in Table 14.14. The null hypothesis is that the mean ratings for all of the samples are equal (H_0: $\mu_i = \mu_j$ for all samples i and j) versus the alternative hypothesis that the mean ratings of at least two of the samples are different (H_a: $\mu_i \neq \mu_j$ for some pair of distinct samples i and j). If the value of the F-statistic calculated in Table 14.14 exceeds the critical value of an F with

TABLE 14.13 Data Table for a Randomized (Complete) Block Design

	Treatments (Samples)						
Blocks (Judges)	**1**	**2**	**·**	**·**	**·**	***t***	**Row Total**
1	x_{11}	x_{12}	·	·	·	x_{1t}	$x_{1\bullet} = \sum_{j=1}^{t} x_{1j}$
2	x_{21}	x_{22}	·	·	·	x_{2t}	$x_{2\bullet} = \sum_{j=1}^{t} x_{2j}$
·						·	·
·						·	·
·						·	·
b	x_{b1}	x_{b2}	·	·	·	x_{bt}	$x_{b\bullet} = \sum_{i=j}^{t} x_{bj}$
Column total	$x_{\bullet 1} = \sum_{i=1}^{b} x_{i1}$	$x_{\bullet 2} = \sum_{i=1}^{b} x_{i2}$	·	·	·	$x_{\bullet t} = \sum_{i=1}^{b} x_{it}$	

Table 14.14 ANOVA Table for Randomized Block Designs Using Ratings

Source of Variability	Degrees of Freedom	Sum of Squares	Mean Square	F
Total	$bt-1$	SS_T		
Blocks (judges)	$b-1$	SS_J		
Samples	$df_s=t-1$	SS_s	$MS_s=SS_s/df_s$	MS_s/MS_E
Error	$df_E=(b-1)(t-1)$	SS_E	$MS_E=SS_E/df_E$	

$(t-1)$ and $(b-1)(t-1)$ degrees of freedom (see Table 19.6), then the null hypothesis is rejected in favor of the alternative hypothesis.

If the F-statistic in Table 14.14 is significant, then multiple comparison procedures are applied to determine which samples have significantly different average ratings. Fisher's LSD for randomized (complete) block designs is

$$\text{LSD} = t_{\alpha/2,df_E}\sqrt{2MS_E/b}\,, \tag{14.13}$$

where b is the number of blocks (typically judges) in the study and $t_{\alpha/2,df_E}$ and MS_E are as defined previously.

14.5.3.2 Randomized Block Analysis of Rank Data

If the data from a randomized block design are in the form of ranks, then a nonparametric analysis is performed using a Friedman-type statistic. The data are arranged as in Table 14.13, but instead of ratings, each row of the table contains the ranks assigned to the samples by each judge. The column totals at the bottom of Table 14.13 are the rank sums of the samples.

The Friedman-type statistic for rank data, which takes the place of the F-statistic in the analysis of ratings, is

$$\text{T} = \frac{12}{bt(t+1)}\sum_{j=1}^{t} x_{\cdot j}^2 - 3b(t+1)\,, \tag{14.14}$$

where b is the number of panelists, t is the number of samples, and $x_{\cdot j}$ is the rank sum of sample j (i.e., the column total for sample j in Table 14.13). The dot in $x_{\cdot j}$ indicates that summing has been done over the index replaced by the dot, that is, $x_{\cdot j} = \sum_{i=1}^{b} x_{ij}$.

The test procedure is to reject the null hypothesis of no sample differences at the α-level of significance if the value of T in Equation 14.14 exceeds $\chi^2_{\alpha,t-1}$, and to accept H_0; otherwise, where $\chi^2_{\alpha,t-1}$ is the upper-α percentile of the χ^2 distribution with $t-1$ degrees of freedom (see Table 19.5). The procedure assumes that a relatively large number of panelists participate in the study. It is reasonably accurate for studies involving 12 or more panelists.

If the χ^2-statistic is significant, then a multiple comparison procedure is performed to determine which of the samples differ significantly. The nonparametric analog to Fisher's LSD for rank sums from a randomized (complete) block design is:

$$\text{LSD}_{\text{rank}} = z_{\alpha/2}\sqrt{bt(t+1)/6} = t_{\alpha/2,\infty}\sqrt{bt(t+1)/6}\,. \tag{14.15}$$

Two samples are declared to be significantly different at the α-level if their rank sums differ by more than the value of LSD_{rank} in Equation 14.15.

If the panelists are permitted to assign equal ranks or ties to the samples, then a slightly more complicated form of the test statistic T' must be used (see Hollander and Wolfe, 1973). Assign the average of the tied ranks to each of the samples that could not be differentiated. For instance, in a four-sample test, if the middle two samples (normally of ranks 2 and 3) could not be differentiated, then assign both the samples the average rank of 2.5. Replace T in Equation 14.14 with

$$T' = \frac{12\sum_{j=1}^{t}(x_{\cdot j} - G/t)^2}{bt(t+1) - [1/(t-1)]\sum_{i=1}^{b}[(\sum_{j=1}^{g_i} t_{i,j}^3) - t]}, \tag{14.16}$$

where $G = bt(t + 1)/2$, g_i is the number of tied groups in block i, and $t_{i,j}$ is the number of samples in the jth tied group in block i. (Nontied samples are each counted as a separate group of size $t_{i,j} = 1$).

14.5.4 Balanced Incomplete Block Designs

Balanced incomplete block (BIB) designs allow sensory analysts to obtain consistent, reliable data from their panelists even when the total number of samples in the study is greater than the number that can be evaluated before sensory fatigue sets in. In BIB designs, the panelists evaluate only a portion of the total number of samples (notationally, each panelist evaluates k of the total of t samples, $k < t$). The specific set of k samples that a panelist evaluates is selected such that, in a single repetition of a BIB design, every sample is evaluated an equal number of times (denoted by r), and all pairs of samples are evaluated together an equal number of times (denoted by λ). The fact that r and λ are constant for all the samples in a BIB design ensures that each sample mean is estimated with equal precision and that all pair-wise comparisons between two sample means are equally sensitive. The number of blocks required to complete a single repetition of a BIB design is denoted by b. Table 14.15 illustrates a typical BIB layout. A list of BIB designs, such as the one presented by Cochran and Cox (1957), is very helpful in selecting a specific design for a study.

To obtain a sufficiently large number of total replications, the entire BIB design (b blocks) may have to be repeated several times. The number of repeats or repetitions of the fundamental design is denoted by p. The total number of blocks is then pb, yielding a total of pr replications for every sample, and a total of $p\lambda$ for the number of times every pair of samples occurs in the total BIB design.

Experience with nine-point category scales and unstructured line scales has shown that the total number of replications (pr) should be at least 18 to yield sufficiently precise estimates of the sample means. This is a general rule, suggested only to provide a starting point for determining how many panelists are required for a study. The total number of replications needed to ensure that meaningfully large differences among the samples are declared statistically significant is influenced by many factors: the products, panelist

Table 14.15 Data Table for a Balanced Incomplete Block Design ($t = 7, k = 3, b = 7, r = 3, \lambda = 1, p = 1$)

Sample Block	1	2	3	4	5	6	7	Block Total
1	X	X		X				B_1
2		X	X		X			B_2
3			X	X		X		B_3
4				X	X		X	B_4
5	X				X	X		B_5
6		X				X	X	B_6
7	X		X				X	B_7
Treatment total	R_1	R_2	R_3	R_4	R_5	R_6	R_7	G

Note: X, an individual observation; B_i, the sum of the observations in row i; R_j, the sum of the observations in column j; G, the sum of all of the observations.

acuity, panelists' level of training, and so on. Only experience and trial and error can answer the question of how many replications are needed for any given study.

There are two general approaches for administering a BIB design in a sensory study. First, if the number of blocks is relatively small (e.g., four or five), it may be possible to have a small number of panelists (p in all) return several times until each panelist has completed an entire repetition of the design. (The order of presentation of the blocks should be randomized separately for each panelist, as should the order of presentation of the samples within each block.) Second, for large values of b, the normal practice is to call upon a large number of panelists (pb in all) and to have each evaluate the samples in a single block. The block of samples that a particular panelist receives should be assigned at random. The order of presentation of the samples within each block should again be randomized in all cases.

14.5.4.1 BIB Analysis of Ratings

ANOVA is used to analyze BIB data in the form of ratings (see Table 14.16). As in the case of a randomized (complete) block design, the total variability is partitioned into the separate effects of blocks, samples, and errors. However, the formulas used to calculate the sum of squares in a BIB analysis are more complicated than for a randomized (complete) block analysis. The sensory analyst should ensure that the statistical package used to perform the analysis is capable of handling a BIB design. Otherwise, a program specifically developed to perform the BIB analysis is required.

The form of the ANOVA used to analyze BIB data depends on how the design is administered. If each panelist evaluates every block in the fundamental design, then the "panelist effect" can be partitioned out of the total variability (see Table 14.16a). If each panelist evaluates only one block of samples, then the panelist effect is confounded (or mixed up) with the block effect (see Table 14.16b). The panelist effect is accounted for in both cases, thus providing an uninflated estimate of experimental error regardless of which approach is used.

Table 14.16 ANOVA Tables for Balanced Incomplete Block Designs

Source of Variability	Degrees of Freedom	Sum of Square	Mean Square	*F*
(a) Each of *p* Panelists Evaluates All *b* Blocks				
Total	$tpr - 1$	SS_T		
Panelists	$p - 1$	SS_P		
Blocks (within panelists)	$p(b - 1)$	$SS_{B(P)}$		
Samples (adj. for blocks)	$df_S = t - 1$	SS_S	$MS_S = SS_S/df_S$	MS_S/MS_E
Error	$df_E = tpr - t - pb + 1$	SS_E	$MS_E = SS_E/df_E$	
(b) Each of *pb* Panelists Evaluates One Block				
Total	tpr-1	SS_T		
Blocks	pb-1	SS_B		
Samples (adj. for blocks)	$df_S = t - 1$	SS_S	$MS_S = SS_S/df_S$	MS_S/MS_E
Error	$df_E = tpr - t - pb + 1$	SS_E	$MS_E = SS_E/df_E$	

If the *F*-statistic in Table 14.16 exceeds the critical value of an *F* with the corresponding degrees of freedom, then the null hypothesis assumption of equivalent mean ratings among the samples is rejected. Fisher's LSD for BIB designs has the form

$$LSD = t_{\alpha/2,df_E}\sqrt{2MS_E/(pr)}\sqrt{[k(t-1)]/[(k-1)t]}, \tag{14.17}$$

where t is the total number of samples, k is the number of samples evaluated by each panelist during a single session, r is the number of times each sample is evaluated in the fundamental design (i.e., in one repetition of b blocks), and p is the number of times the fundamental design is repeated. MS_E and $t_{\alpha/2,df_E}$ are as defined before.

14.5.4.2 BIB Analysis of Rank Data

A Friedman-type statistic is applied to rank data arising from a BIB design. The form of the test statistic is

$$T = \frac{12}{p\lambda t(k+1)}\sum_{j=1}^{t} R_j^2 - \frac{3(k+1)pr^2}{\lambda}, \tag{14.18}$$

where t, k, r, λ, and p were defined previously and R_j is the rank sum of the jth sample (i.e., the value for sample j in the last row of Table 14.15) (see Durbin, 1951). Tables of critical values of T in Equation 14.18 are available for selected combinations of $t = 3 - 6$, $k = 2 - 5$, and $p = 1 - 7$ (see Skillings and Mack, 1981). However, in most sensory studies, the total number of blocks exceeds the values in the tables. For these situations, the test procedure is to reject the assumption of equivalency among the samples if T in Equation 14.18 exceeds the upper-α critical value of a χ^2-statistic with $(t - 1)$ degrees of freedom (see Table 19.5).

If the χ^2-statistic is significant, then a multiple comparison procedure is performed to determine which of the samples differ significantly. The nonparametric analog to Fisher's LSD for rank sums from a BIB design is

$$\begin{aligned} LSD_{rank} &= z_{\alpha/2}\sqrt{p(k+1)(rk-r+\lambda)/6} \\ &= t_{\alpha/2,\infty}\sqrt{p(k+1)(rk-r+\lambda)/6}. \end{aligned} \tag{14.19}$$

14.5.5 Latin-Square Designs

In randomized block and BIB designs, a single source of variability (i.e., judges) is recognized and compensated for before the sensory panel study is conducted. When two sources of variability are known to exist before the panel is run, then a Latin-square design should be used. For example, it is commonly recognized that panelists can vary in how they perceive attributes from session to session (i.e., a session effect) and according to the order in which they evaluate the samples (i.e., a context effect). A Latin-square design can be used to compensate for these two sources of variability and, as a result, yield more sensitive comparisons of the differences among the samples.

The number of samples must be small enough so that all of them can be evaluated in each session ($t \leq 5$, typically). Furthermore, each panelist must be able to return repeatedly for a number of sessions equal to the number of samples in the study (i.e., t represents the number of samples and the number of sessions in a Latin-square design). For each panelist, each sample is presented once in each session. Across the t-sessions, each sample is presented once in each serving position. As can be seen in Figure 14.14, the equal allocation of the samples to each serving order/session combination can be displayed in a square array, thus giving rise to the term *Latin square*.

A separate randomization of serving orders is used for each judge. This can be carried out, for example, by first randomly assigning the codes S1, S2, S3, S4, and S5 in Figure 14.14 to the samples for each judge separately. Then, again for each judge, a particular order (i.e., row of Figure 14.14) is randomly selected for each session.

ANOVA is used to analyze data from a Latin-square design (see Table 14.17). The total variability is partitioned into the separate effects of judges, panel sessions, order of evaluations, samples, and error. If the F-statistic in Table 14.17 exceeds the critical value of an F with the corresponding degrees of freedom, then the null hypothesis assumption of equivalent mean ratings among the samples is rejected. Fisher's LSD for Latin-square designs has the form

$$LSD = t_{\alpha/2,df_E}\sqrt{2MS_E/(pt)},$$

where p is the number of panelists.

14.5.6 Split-Plot Designs

In randomized block and BIB designs, panelists are treated as a blocking factor; that is, it is assumed that the panelists are an identifiable source of variability that is known to exist before the study is run and, therefore, should be compensated for in the design of

Session \ Serving order	1	2	3	4	5
1	S_1	S_2	S_3	S_4	S_5
2	S_5	S_1	S_2	S_3	S_4
3	S_4	S_5	S_1	S_2	S_3
4	S_3	S_4	S_5	S_1	S_2
5	S_2	S_3	S_4	S_5	S_1

Figure 14.14 The Latin-square arrangement of serving orders by sessions for one judge and a five-sample study. The sample codes S1, S2,…,S5 are assigned randomly to the samples in the study. The serving orders and session orders are randomly permuted for each judge individually.

Table 14.17 ANOVA Table for Latin-Square Designs Using Ratings

Source of Variability	Degrees of Freedom	Sum of Square	Mean Squares	F
Total	$pt^2 - 1$	SS_T		
Judges	$p - 1$	SS_J		
Sessions (within judge)	$p(t - 1)$	SS_P		
Order (within judge)	$p(t - 1)$	SS_O		
Samples	$df_S = t - 1$	SS_S	$MS_S = SS_S/df_S$	MS_S/MS_E
Error	$df_E = (pt - p - 1)(t - 1)$	SS_E	$MS_E = SS_E/df_E$	

the panel. In ANOVA, the effects of environmental factors (e.g., judges) and treatment factors (e.g., products) are assumed to be additive. In practice, this assumption implies that although panelists may use different parts of the sensory rating scales to express their perceptions, the size and direction of the differences among the samples are perceived and reported in the same way by all of the panelists. Of course, the data actually collected in a study diverge slightly from the assumed pattern due to experimental error. Another way of stating this assumption is that there is no "interaction" between blocks and treatments (e.g., judges and samples) in a randomized block or BIB design. For a group of highly trained, motivated, and "calibrated" panelists, the assumption of no interaction between judges and samples is reasonable. However, during training, for instance, the sensory analyst may doubt the validity of this assumption. Split-plot designs are used to determine if a judge-by-sample interaction is present.

In split-plot designs, judges are treated as a second experimental treatment along with the samples. A group of b panelists are presented with t samples (in a separately randomized order for each panelist) in each of at least two panels ($p \geq 2$). The p panels are the blocks or "replicates" of the experimental design. Randomly selected batches or independent preparations of the samples are used for each panel. This is the first layer of randomization in a split-plot study. Then the panelists receive their specific sets of samples (arranged in a randomized order based on their arrival times at each panel). Due to the sequential nature of the randomization scheme, where first one treatment factor (samples) is randomized within replicates and then a second treatment factor (judges) is randomized within the first treatment factor (i.e., samples), a split-plot design is appropriate.

14.5.6.1 Split-Plot Analysis of Ratings

A special form of ANOVA is used to analyze data from a split-plot design (see Table 14.18). The sample effect is called the *whole-plot effect*. Judges and the judge-by-sample interaction are called the *subplot effects*. Separate error terms are used to test for the significance of whole-plot and subplot effects (because of the sequential nature of the randomization scheme described previously).

The whole-plot error term (Error (A) in Table 14.18) is calculated in the same way as a panel-by-sample interaction term would be, if one existed. The F_1-statistic in Table 14.18 is used to test for a significant sample effect. If the value of F_1 is larger than the upper-α critical value of the F-distribution with $(t - 1)$ and $(p - 1)(t - 1)$ degrees of freedom, then it is concluded that there are significant differences among the average values of the samples.

The F_2 and F_3 statistics in Table 14.18 are used to test for the significance of the subplot effects, judges, and the judge-by-sample interaction, respectively. The denominator of both F_2 and F_3 is the subplot error term $MS_{E(B)}$. If F_3 exceeds the upper-α critical value of the F-distribution with $(b - 1)(t - 1)$ and $t(p - 1)(b - 1)$ degrees of freedom, then a significant judge-by-sample interaction exists. The significance of the interaction indicates that the judges are expressing their perceptions of the differences among the samples in different ways. Judge-by-sample interactions result from insufficient training, confusion over the definition of an attribute, or lack of

Table 14.18 ANOVA Table for Split-Plot Designs Using Ratings

Source of Variability	Degrees of Freedom	Sum of Squares	Mean Square	F
Total	$pbt - 1$	SS_T		
Panel	$p - 1$	SS_P		
Samples	$t - 1$	SS_S	$MS_S = SS_S/(t - 1)$	$F_1 = MS_S/MS_{E(A)}$
Error(A)	$df_{E(A)} = (p - 1)(t - 1)$	$SS_{E(A)}$	$MS_{E(A)} = SS_{E(A)}/df_{E(A)}$	
Judges	$b - 1$	SS_J	$MS_J = SS_J/(b-1)$	$F_2 = MS_S/MS_{E(B)}$
Judge-by-sample	$df_{JS}=(b - 1)(t - 1)$	SS_{JS}	$MS_{JS} = SS_{JS}/df_{JS}$	$F_3 = MS_{JS}/MS_{E(B)}$
Error(B)	$df_{E(B)} = t(p - 1)(b - 1)$	$SS_{E(B)}$	$MS_{E(B)} = SS_{E(B)}/df_{E(B)}$	

familiarity with the rating technique. When a significant judge-by-sample interaction exists, it is meaningless to examine the overall sample effect (tested by F_1 in Table 14.18) because the presence of an interaction indicates that the pattern of differences among the samples depends on which judge or judges are being considered. Tables of individual judges' mean ratings and plots of the judge-by-sample means should be examined to determine which judges are causing the interaction to be significant (see Chapter 10, Section 10.5.1).

If F_3 is not significant but F_2 is, then an overall judge effect is present. A significant judge effect confirms that the judges are using different parts of the rating scale to express their perceptions. This is not of as great a concern as a significant judge-by-sample interaction. However, depending on the magnitude of the differences among the judges, it may indicate that the panel needs to be recalibrated through the use of references.

If F_3 is not significant but F_1 is, then an overall sample effect is present. To determine which of the samples differ significantly, use Fisher's LSD for split-plot designs:

$$\text{LSD} = t_{\alpha/2,df_{E(A)}}\sqrt{2MS_{E(A)}/(pb)}, \tag{14.21}$$

where p is the number of independently replicated panels and $df_{E(A)}$ and $MS_{E(A)}$ are the degrees of freedom and mean square for Error(A), respectively.

14.5.7 A Simultaneous Multiple Comparison Procedure

Thus far, we have used only Fisher's LSD multiple comparison procedure to determine which samples differ significantly in a designed sensory panel study. There are, in fact, two classes of multiple comparison procedures. The first class, including Fisher's LSD, controls the comparison-wise error rate; that is, the type I error (of size α) applies each time a comparison of means or rank sums is made. Procedures that control the comparison-wise error rate are called one-at-a-time multiple comparison procedures. The second class controls the experiment-wise error rate; that is, the type I error applies to all of the comparisons among means or rank sums simultaneously. Procedures that control the experiment-wise error rate are called simultaneous multiple comparison procedures.

Tukey's honestly significant difference (HSD) is a simultaneous multiple comparison procedure. Tukey's HSD can be applied regardless of the outcome of the overall test for differences among the samples. The general form of Tukey's HSD for the equal sample-size case for ratings data is

$$\text{HSD} = q_{\alpha,t,df_E}\sqrt{MS_E/n}, \tag{14.22}$$

where q_{α,t,df_E} is the upper-α critical value of the studentized range distribution with df_E degrees of freedom (see Table 19.4) for comparing t sample means. As with the LSD, df_E and MS_E are the degrees of freedom and the mean square for error from the ANOVA, respectively, (Error (A) in the split-plot ANOVA); n is the sample size common to all the means being compared. For randomized (complete) block designs, $n = b$; for split-plot designs, $n = p$. Tukey's HSD for BIB designs has the form

$$\text{HSD} = q_{\alpha,t,df_E}\sqrt{MS_E/(pr)}\sqrt{[k(t-1)]/[(k-1)t]}\ . \tag{14.23}$$

The nonparametric analog to Tukey's HSD for rank sums is

$$\text{HSD}_{\text{rank}} = q_{\alpha,t,\infty}\sqrt{bt(t+1)/12} \tag{14.24}$$

for randomized (complete) block designs, and

$$\text{HSD}_{\text{rank}} = q_{\alpha,t,\infty}\sqrt{p(k+1)(rk-r+\lambda)/12} \tag{14.25}$$

for BIB designs.

APPENDIX 14.1 PROBABILITY

The purpose of this section is to present the techniques for calculating probabilities based on some commonly used probability distributions. The techniques are the foundation for the statistical estimation and inference that were discussed in Sections 14.2 and 14.3 as well as for the more advanced topics discussed in Chapter 15.

Normal Distribution

The normal distribution is among the most commonly used distributions in probability and statistics. The form of the normal distribution function is

$$f(x) = \frac{1}{\left(\sqrt{2\pi\sigma}\right)}\exp\left[-\frac{(x-\mu)^2}{2\sigma^2}\right],$$

where exp is the exponential function with base e. The parameters of the normal distribution are the mean $\mu(-\infty < \mu < \infty)$ and the standard deviation $\sigma(\sigma > 0)$. The normal distribution is symmetric about μ, that is, $f(x - \mu) = f(\mu - x)$. The mean μ measures the central location of the distribution. The standard deviation, σ measures the dispersion or "spread" of the normal distribution about the mean. For small values of σ, the graph of the distribution is narrow and peaked; for large values of σ, the graph is wide and flat (see Figure 14.15). As with all continuous probability distributions, the total area under the curve is equal to one, regardless of the values of the parameters.

Let x be a random variable having a normal distribution with mean μ and standard deviation σ (often abbreviated as $x \sim \text{n}(\mu, \sigma)$). Define the variable z as

$$z = (x-\mu)/\sigma. \tag{14.26}$$

The random variable z also has a normal distribution. The mean of z is zero and its standard deviation is one (i.e., $z \sim n(0, 1)$). z is said to have a standard normal distribution, or often z is called a *standard normal deviate*. Given the values of μ and σ for a normal random variable x and a table of standard normal probabilities (see Table 19.2), it is possible to calculate various probabilities of interest.

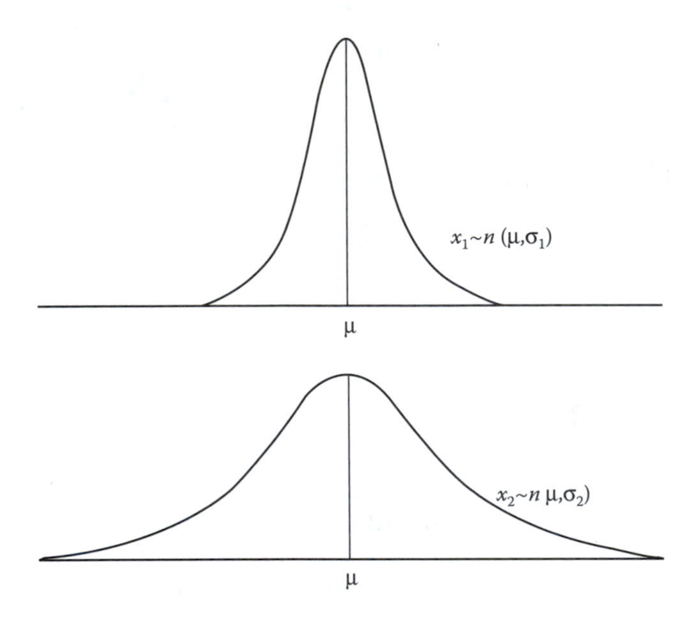

Figure 14.15 A comparison of two normal distributions with the same mean but with $\sigma_1 < \sigma_2$.

Example 14.7: Calculating Normal Probabilities on an Interval

Consider the problem of calculating the probability that a normal random variable x with mean $\mu = 50$ and standard deviation $\sigma = 5$ takes on a value between 50 and 60 (notationally, $\Pr[50 < x < 60]$). The first step in solving the problem is to "standardize" x using (Equation 14.26):

$$\Pr[50 < x < 60] = \Pr[(50 - 50)/5 < (x - 50)/5 < (60 - 50)/5]$$
$$= \Pr[0 < z < 2].$$

Table 19.2 gives the probabilities of a standard normal deviate taking on a value from zero (i.e., its mean) to some specified number. Therefore, consulting the row corresponding to 2.0 and the column corresponding to 0.00 in Table 19.2, the analyst finds that the probability sought is equal to 0.4772 (see Figure 14.16).

Alternatively, the probability can be calculated using a spreadsheet function such as Excel's NORM.DIST. To find the probability that the normal random variable falls between 50 and 60 is computed in Excel by entering the following in the spreadsheet:

1. In cell A1 enter = NORM.DIST(60, 50, 5, TRUE)
2. In cell A2 enter = NORM.DIST(50, 50, 5, TRUE)
3. In cell A3 enter = A1–A2

The desired probability, 0.4772, is displayed in cell A3.

Next, consider the problem of finding $\Pr[45 < x < 50]$, where, as before, $x \sim n(50, 5)$. Standardizing:

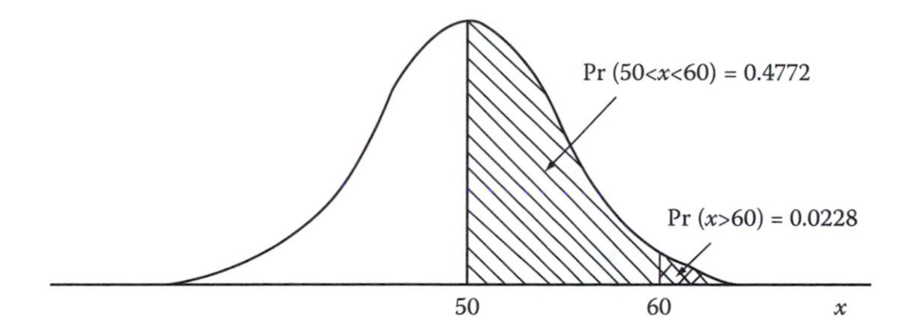

Figure 14.16 A graphical depiction of calculating normal probabilities on an interval and in the tail of the distribution.

$$\Pr[45 < x < 50] = \Pr[(45-50)/5 < (x-50)/5 < (50-50)/5]$$
$$= \Pr[-1 < z < 0].$$

Because the standard normal distribution is symmetric about its mean, zero, it follows that $\Pr[-c < z < 0] = \Pr[0 < z < c]$ for any constant c. Therefore, by Table 19.2,

$$\Pr[-1 < z < 0] = \Pr[0 < z < 1] = 0.3413\,. \tag{14.27}$$

Therefore, $\Pr[45 < x < 50] = 0.3413$.

The same probability can be computed in Excel by entering the following in the spreadsheet:

1. In cell A1 enter = NORM.DIST(50, 50, 5, TRUE)
2. In cell A2 enter = NORM.DIST(45, 50, 5, TRUE)
3. In cell A3 enter = A1–A2

The desired probability, 0.3413, is displayed in cell A3.

Finally, consider $\Pr[45 < x < 60]$ for the same random variable $x \sim n(50, 5)$. This problem is solved as follows:

$$\Pr[45 < x < 60] = \Pr[-1 < z < 2] \text{ (Standardizing by [14.26])}$$
$$= \Pr[-1 < z < 0] + \Pr[0 < z < 2]$$
$$= \Pr[0 < z < 1] + \Pr[0 < z < 2] \text{ (by [14.27])}$$
$$= 0.3413 + 0.4772$$
$$= 0.8185$$

The same probability can be computed in Excel by entering the following in the spreadsheet:

1. In cell A1 enter = NORM.DIST(60, 50, 5, TRUE)
2. In cell A2 enter = NORM.DIST(45, 50, 5, TRUE)
3. In cell A3 enter = A1–A2

The desired probability, 0.8185, is displayed in cell A3.

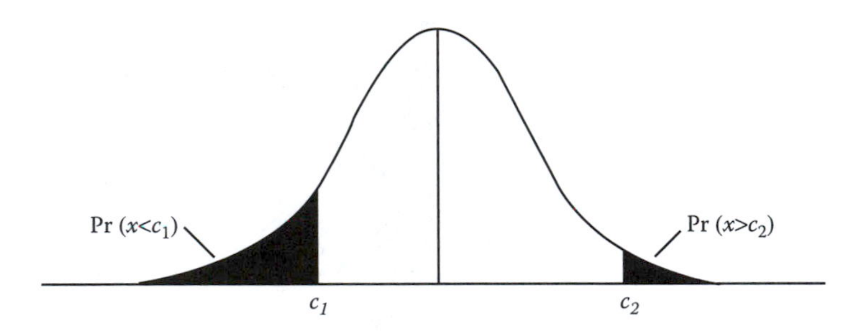

Figure 14.17 Tail probabilities of a normal distribution.

Example 14.8: Calculating Normal Tail Probabilities

Tail probabilities are associated with the areas under the probability curve at the extremes of the distribution (see Figure 14.17). Notationally, tail probabilities are stated as $\Pr[x > c]$ or $\Pr[x < c]$ for some constant c. Tail probabilities are widely used in testing statistical hypotheses.

Consider the problem of finding $\Pr[x > 60]$, where $x \sim n(50, 5)$. Noting that the total area (i.e., probability) under any probability curve is one, it follows from the symmetry of the normal distribution that $\Pr[x < \mu] = \Pr[x > \mu] = 0.50$. Therefore,

$$\Pr[x > 60] = \Pr\left[\frac{x-50}{5} > \frac{60-50}{5}\right]$$

$$= \Pr[z > 2] \text{ (by [14.26])}$$

$$= 0.50 - \Pr[0 < z < 2]$$

$$= 0.50 - 0.4772(\text{from Example14.7})$$

$$= 0.0228$$

(See the crosshatched area in Figure 14.16 for an understanding of the third step.)

The same probability can be computed in Excel by entering the following in the spreadsheet:

1. In cell A1 enter = 1 – NORM.DIST(60, 50, 5, TRUE)

The desired probability, 0.0228, is displayed in cell A1.

14.6.2 Binomial Distribution

The binomial distribution function is

$$\Pr[x = k] = b[k] = \binom{n}{k} p^k (1-p)^{n-k} \tag{14.28}$$

for $k = 0,1,2,\ldots, n$; $n > 0$ and an integer; and $0 \le p \le 1$.

The parameters of the binomial distribution are n = the number of trials; and p = the probability of "success" on any trial. The choice of what constitutes a success on each trial is arbitrary. For instance, in a two-sample preference test (A vs. B), preference for A could constitute a success or preference for B could constitute a success. Regardless, $k_i = 1$ for $I = 1, 2, \ldots, n$ if the result of the ith trial is a success; $k_i = 0$, otherwise. In Equation 14.28, $k = \Sigma_{i=1}^{n} k_i$ is the total number of successes in n trials. Exact binomial probabilities can be calculated using spreadsheet functions such as Excel's BINOM.DIST. Approximate binomial probabilities can be calculated using the normal approximation to the binomial.

Example 14.9: Calculating Exact Binomial Probabilities

Suppose $n = 16$ assessors participate in a two-out-of-five difference test. The probability of correctly selecting the two odd samples from among the five follows a binomial distribution with probability of success, $p = 0.10$ (when there is no perceptible difference among the samples). To find the probability that exactly two ($k = 2$) of the assessors make the correct selections (i.e., exactly 2 successes in 16 trials), enter the following in a cell in an Excel spreadsheet: = BINOM.DIST(2, 16, 0.10, FALSE). The response displayed will be 0.2745, which is the desired probability.

To find the probability that between two and six assessors (inclusive) make the correct selections, one notes that

$$\Pr[2 \le x \le 6] = \Pr[x \le 6] - \Pr[x \le 1], \tag{14.29}$$

which is computed in Excel by entering the following in the spreadsheet:

4. In cell A1 enter = BINOM.DIST(6, 16, 0.10, TRUE)
5. In cell A2 enter = BINOM.DIST(1, 16, 0.10, TRUE)
6. In cell A3 enter = A1–A2

The desired probability, 0.4848, is displayed in cell A3.

There are two approaches for calculating tail probabilities using spreadsheet functions, depending on whether you want to compute the probability in the lower or upper tail of the distribution. To compute probabilities in the lower tail of the distribution (e.g., that less than three assessors make the correct selections), use the following technique:

$$\Pr[x < 3] = \Pr[x \le 2]. \tag{14.30}$$

Therefore, enter the following in a cell in an Excel spreadsheet: =BINOM.DIST(2, 16, 0.10, TRUE), and the resulting probability, 0.7892, will be displayed. On the other hand, consider the probability that at least three (i.e., three or more) assessors make the correct selections—that is,

$$\Pr[x \ge 3] = 1 - \Pr[x < 3] = 1 - \Pr[x \le 2]. \tag{14.31}$$

Therefore, enter the following in a cell in an Excel spreadsheet: =1–BINOM.DIST(2, 16, 0.10, TRUE), and the resulting probability, 0.2108, will be displayed.

Example 14.10: The Normal Approximation to the Binomial

When a computerized spreadsheet with a binomial probability function is not available, approximate binomial probabilities can be calculated using the normal distribution. To use the methods of Section 14.5.1, one needs to know the values of μ and σ. For the number of successes, these are

$$\mu = np$$

$$\sigma = \sqrt{np(1-p)}. \tag{14.32}$$

Let $n = 36$ and $p = 1/3$ and consider the problem of calculating the probability of at least 16 successes. For Equation 14.32, one computes:

$$\mu = (36)(1/3) = 12$$

$$\sigma = \sqrt{(36)(1/3)(1-1/3)}$$

$$= \sqrt{8} = 2.828.$$

Therefore, using the methods of Example 14.8:

$$\Pr[x \geq 16] = \Pr[(x-12)/2.828 \geq (16-12)/2.828]$$

$$= \Pr[z \geq 1.41]$$

$$= 0.5 - \Pr[0 < z < 1.41]$$

$$= 0.50 - 0.4207$$

$$= 0.0793.$$

One can also use the normal approximation to the binomial to calculate probabilities associated with the proportion of successes. For this case,

$$\mu = p$$

$$\sigma = \sqrt{p(1-p)/n}. \tag{14.33}$$

In most sensory evaluation tests, the number of trials is large enough so that the normal approximation gives adequately accurate results. A common general rule is that the normal approximation to the binomial is sufficiently accurate if both $np > 5$ and $n(1 - p) > 5$; that is, for the normal approximation to be reasonably accurate, the sample size n should be sufficiently large so that one would expect to see at least five successes and at least five failures in the sample results.

REFERENCES

ASTM (2003). Standard practice for estimating Thurstonian discriminal distances. In *Standard Practice E2262-03*. West Conshohocken, PA: ASTM International.

Bi, J., D. M. Ennis, and M. O'Mahony (1997). How to estimate and use the variance of d′ from difference tests. *J Sens Stud* 12: 87–104.

Cochran, W. G., and G. M. Cox (1957). *Experimental Designs* (2nd edn.). New York: Wiley.

Danzart, M. (1986). Univariate procedures. In J. R. Piggott, (ed.), *Statistical Procedures in Food Research* (pp. 19–60). Essex: Elsevier Applied Science.

Dixon, W. J., and F. J. Massey (1969). *Introduction to Statistical Analysis*. New York: McGraw-Hill.

Durbin, J. (1951). Incomplete blocks in ranking experiments. *Br J Math Stat Psychol* 4: 85–92.

Elliot, P. B. (1964). Tables of d′. In J. A. Swet (ed.), *Signal Detection and Recognition by Human Observers* (pp. 651–84). New York: Wiley.

Ennis, D. M. (1993). The power of sensory discrimination methods. *J Sens Stud* 8: 353–70.

Ennis, D. M. (2001). *IFPrograms User Guide, Version 7*. Richmond, VA: The Institute for Perception.

Frijters, J. E. R. (1979). The paradox of discriminatory nondiscriminators resolved. *Chem Sens Flavor* 4: 355–8.

Gagula, M. C., and J. Singh (1984). *Statistical Methods in Food and Consumer Research*. Orlando, FL: Academic Press.

Hollander, M., and D. A. Wolfe (1973). *Nonparametric Statistical Methods*. New York: Wiley.

Jesionka, V., B. Rousseau, and J. M. Ennis (2014). Transitioning from proportion of discriminators to a more meaningful measure of sensory difference. *Food Qual Prefer* 32: 77–82.

O'Mahony, M. (1986). *Sensory Evaluation of Food: Statistical Methods and Procedure*. New York: Marcel Dekker.

Skillings, H. H., and G. A. Mack (1981). On the use of a Friedman type statistic in balanced and unbalanced block designs. *Technometrics* 23: 171–7.

Smith, G. L. (1988). Statistical analysis of sensory data. In J. R. Piggott, (ed.), *Sensory Analysis of Foods* (pp. 335–79). Essex, UK: Elsevier Applied Science.

Snedecor, G. W., and W. G. Cochran (1980). *Statistical Methods*. Ames, IA: Iowa State University Press.

Thurstone, L. L. (1927). A law of comparative judgment. *Psycholog Rev* 34: 273–86.

15

Advanced Statistical Methods

15.1 INTRODUCTION

The basic statistical techniques presented in Chapter 14 are all that would be required to analyze the results of most sensory tests. However, when the objectives of the study go beyond simple estimation or discrimination, then more sophisticated statistical methods may need to be applied. This chapter presents some of the more common of these advanced techniques. The computational complexity of the methods makes hand calculation impractical. It is assumed that the reader has access to computer resources capable of performing the necessary calculations.

Sensory studies seldom include only a single response variable. More often, many variables are measured on each sample and often one of the goals of the study is to determine how the different *multivariate* measurements relate to each other. Approaches for studying multivariate data relationships are presented in Section 15.2. First, correlation analysis, principal components analysis (PCA), and cluster analysis are discussed. These techniques are used to study sets of multivariate data in which all of the variables are of equal status. Second, regression analysis, principal component regression, partial least-squares, and discriminant analysis are presented. These methods apply when the variables in the data set can be classified as being either independent or dependent, with the goal of the analysis being to predict the value of the dependent variables using the independent variables. Section 15.3 presents the various approaches to preference mapping, in which all of the methods presented in Section 15.2 are applied to link consumer acceptance to the sensory properties of a group of test products. In Section 15.4, experimental plans for systematically studying the individual and combined effects of more than one experimental variable are presented. The discussion includes factorial experiments, fractional factorials (or *screening studies*), and response surface methodology (or *product optimization studies*).

15.2 DATA RELATIONSHIPS

The need to determine if relationships exist among different variables often arises in sensory evaluation. The manner and degree in which different descriptive attributes increase or decrease together, the similarity among consumers' liking patterns for various products, and the ability to predict the value of a perceived attribute based on the age of a product are three examples of this type of problem.

The various statistical methods that exist for drawing relationships among variables can be divided into two groups. The first group of methods handles data sets in which all of the variables are independent, in the sense that they are all equally important with no one or few variables being viewed as being driven by the others (e.g., a set of descriptive flavor attributes). The second group of methods is applied to data sets that contain both dependent and independent variables. These are data sets in which one or more of the variables are of special or greater interest relative to some others (e.g., overall liking vs. descriptive attribute ratings). Because the methods in both of these groups deal with more than one variable at a time, they are members of the class of multivariate statistical methods.

15.2.1 All Independent Variables

When all of the variables are viewed as being equally important, the goal of the statistical analysis is to determine the nature and degree of relationships among the variables, to determine if groups of related variables exist, or to determine if distinct groups of observations exist.

15.2.1.1 Correlation Analysis

The simplest of multivariate techniques, correlation analysis, is used for measuring the strength of the linear relationship between two variables. The strength of the relationship between attributes X and Y, for instance, is summarized in the correlation coefficient r, where

$$r = \frac{\Sigma(y_i - \bar{y})(x_i - \bar{x})}{\sqrt{\Sigma(y_i - \bar{y})^2 \Sigma(x_i - \bar{x})^2}}.$$

The value of r lies between –1 and +1. A value of –1 indicates a perfect inverse linear relationship (i.e., one variable decreases as the other increases) while a value of +1 indicates a perfect direct linear relationship (i.e., both variables either increase or decrease together). A value near zero implies that little linear relationship exists between the two variables. A strong correlation does not imply causality; that is, neither variable can automatically be assumed to be "driving" the other, but rather that the two co-vary to some degree.

Correlation coefficients are summary measures. An analyst should always examine the scatterplots of the paired variables before deciding if the value of r is an adequate summary of the relationship. A strong linear trend among relatively evenly spaced observations, as in Figure 15.1a, is safely summarized by the correlation coefficient, as is an unpatterned spread of observations spanning the ranges of both variables, as in Figure 15.1b. Some relationships may have high values of r but are clearly better summarized in nonlinear terms, as in Figure 15.1c. In other cases, patterns that are clearly apparent visually may have very low values of r, as in Figure 15.1d, e. Conversely, the correlation coefficient may be misleadingly large when distinct groups of observations (with no internal correlation) are present,

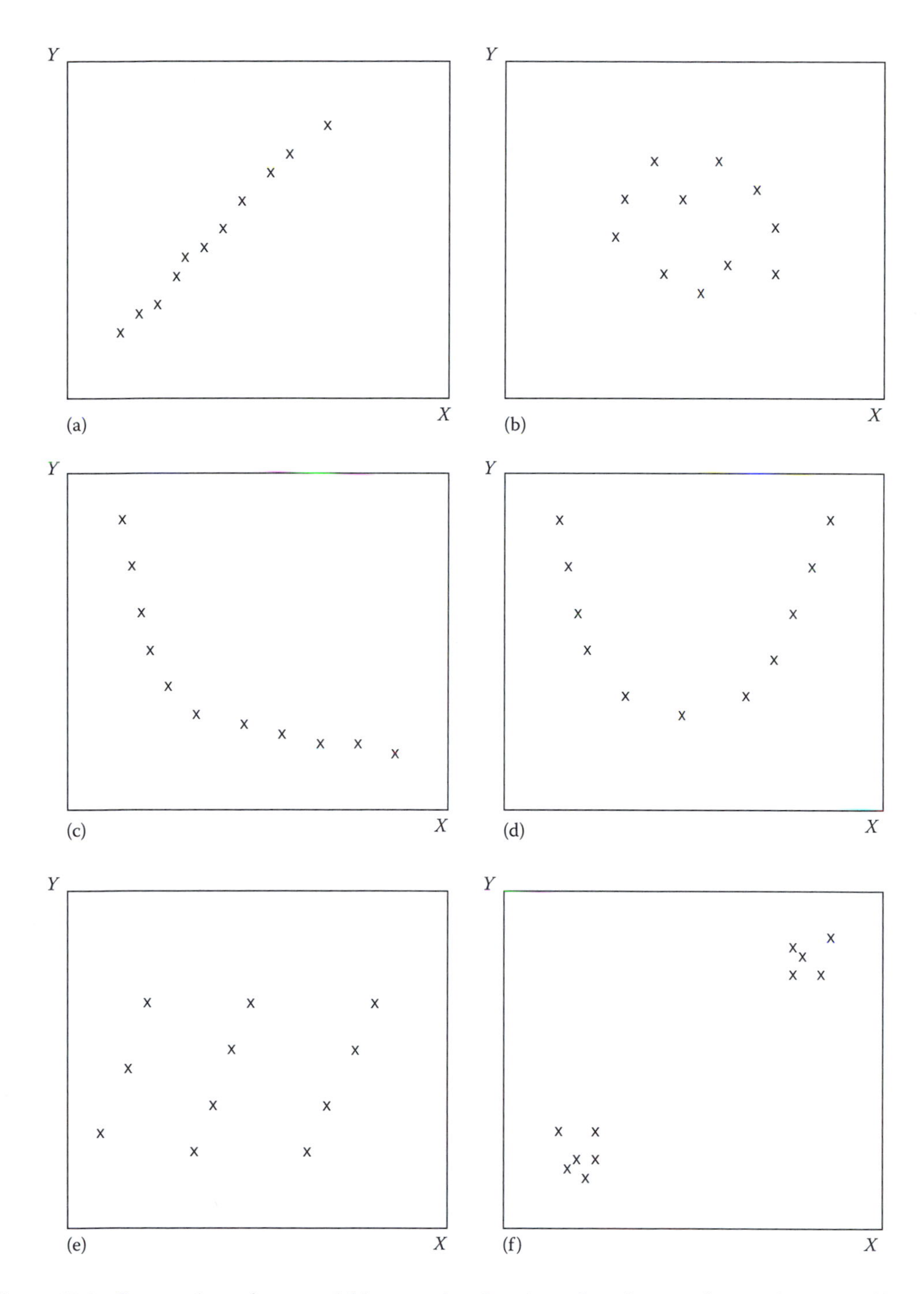

Figure 15.1 Scatterplots of two variables, x and y, showing when the sample correlation coefficient, r, is a good summary measure (i.e., plots a and b) and when it is not (i.e., plots c through f).

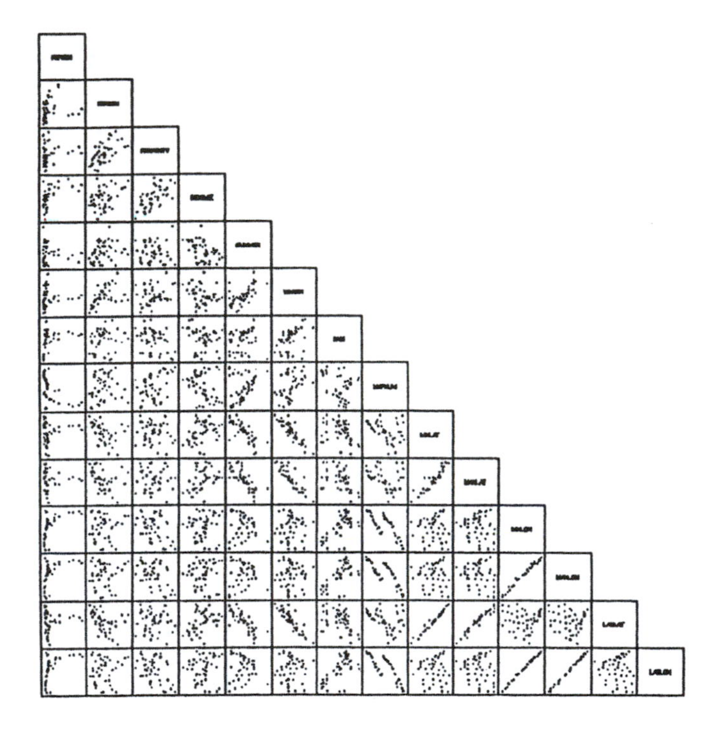

Figure 15.2 Scatterplot matrix of multiple responses useful for identifying correlated or otherwise related pairs of variables.

as in Figure 15.1f. A scatterplot matrix (see Figure 15.2) allows all of the pairwise plots of the data to be displayed in a compact format.

Correlation analysis can be used to identify groups of responses that vary in similar ways, possibly distinct from other such groups. Also, correlation analysis can be used to determine the strength of the relationship between data arising from different sources (e.g., consumer ratings and descriptive data from a trained panel, descriptive attribute ratings, and instrumental measurements, etc.). Chapter 13, Section 13.7, contains examples of these types of relationships.

Patterns of correlation may vary across product categories, regionally, or from one market segment to another. Care should be taken to ensure that the data to which correlation analysis is applied arise from a single population and not a blend of heterogeneous ones.

15.2.1.2 Principal Components Analysis

An initial correlation analysis might identify one or more groups of variables that are highly correlated with each other (and not highly correlated with variables from other groups). This suggests that variables in each group contain related information and that possibly a smaller number of unobserved (or *latent*) variables would provide an adequate summary of the total variability. PCA is the statistical technique used to identify the smallest number of latent variables, called *principal components*, that explain the greatest amount of observed variability. It is often possible to explain as much as 75%–90% of the total variability in a data set consisting of 25–30 variables with as few as two to three principal components.

Computer programs that extract the principal components from a set of multivariate data are widely available, so theoretical and computational details are not included in the following discussion. Those interested in a more analytical discussion of PCA are referred to Piggott and Sharman (1986).

PCA analyzes the correlation structure of a group of multivariate observations and identifies the axis along which the maximum variability in the data occurs. This axis is called the *first principal component*. The *second principal component* is the axis along which the greatest amount of remaining variability lays subject to the constraint that the axes must be perpendicular (at right angles) to each other (i.e., orthogonal or uncorrelated). Each additional principal component is selected to be orthogonal to all others and such that each successive principal component explains as much of the remaining unexplained variability as possible. The number of principal components can never be larger than the number of observed variables and, in practice, is often much less. The process of extracting principal components ends either when a prespecified amount of the total variability has been explained (this quantity is always included in the computer output of a PCA) or when extraction of another principal component would add only trivially to the explained variability. This situation is depicted graphically by the flattening out of the scree plot (Cattell, 1966) in Figure 15.3.

The direction of the axis defined by each principal component, y_i, is expressed as a linear combination of the observed variables, x_j, as in

$$y_i = a_{i1}x_1 + a_{i2}x_2 + \cdots + a_{ip}x_p. \tag{15.1}$$

The coefficients, a_{ij}, are called *weights* or *loading factors*. They measure the importance of the original variables on each principal component. Like correlation coefficients, a_{ij} takes

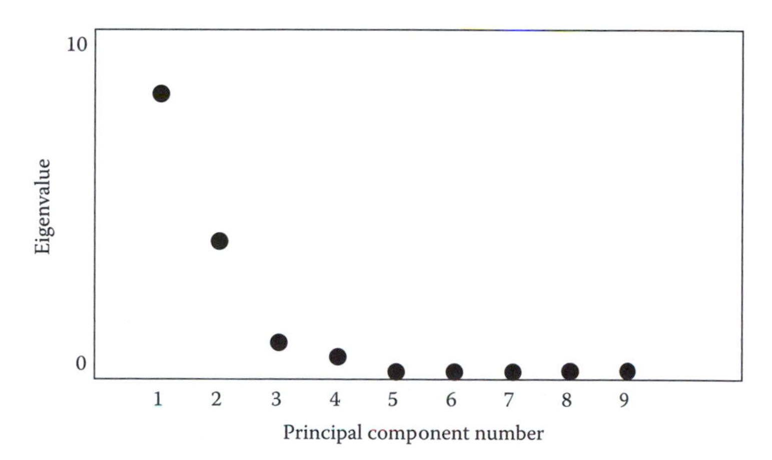

Figure 15.3 A scree plot of the variability explained by each principal component used to determine how many principal components should be retained in a study. Only those principal components that are extracted before the plot flattens out are retained.

on values between –1 and +1. A value close to –1 or +1 indicates that the corresponding variable has a large influence on the value of the principal component; values close to zero indicate that the corresponding variable has little influence on the principal component. Typically, groups of highly correlated observed variables segregate themselves into non-overlapping groups predominantly associated with a specific principal component.

Examination of the loading factors reveals how the observed variables group together and may lead to a meaningful interpretation of the type of variability being summarized by each principal component. To further aid in interpretation, the principal component axes are sometimes rotated to increase their alignment with the axes of the original variables. After rotation, the first principal component will no longer lie in the direction of maximum variability, followed by the second, and so on, but the advantage gained by having a small number of interpretable latent variables offsets this effect.

In addition to depicting the associations among the original variables, PCA can be used to display the relative "locations" of the samples. A plot of the principal component scores for a set of products can reveal groupings and polarizations of the samples that would not be as readily apparent in an examination of the larger number of original variables. Cooper et al. (1989) used PCA to depict in two dimensions the relationship of 16 orange juice products originally evaluated on ten attributes (see Figure 15.4). In their analysis, the first two principal components explained 79% of the original variability. Piggott and Sharman (1986) present additional examples. Powers (1988) presents numerous references of the application of PCA in descriptive analysis.

PCA provides a way to summarize data collected on a large number of variables in fewer dimensions. It is tempting to ask if it is necessary to continue to evaluate all of the original variables as opposed to only a few "representative" ones. The number of original variables studied should not be reduced based on PCA results. As seen in Equation 15.1, each of the original variables is included in the computation of each principal component. Retaining only a small group of representative variables on a sensory ballot ignores the multivariate nature of the effects of the original variables, would not allow for future verification of the stability of the principal components, and could lead to misleading results in future evaluations.

15.2.1.3 Multidimensional Scaling

One of the primary benefits of PCA is the visual characterization of the similarities and differences among the test products that are evaluated by a sensory panel. The "perceptual map" obtained from a PCA illustrates both the similarities and differences among the products and the attributes that account for the differences along each sensory dimension (i.e., each principal component). The raw data for a principal component analysis are often the average intensity ratings of descriptive sensory attributes collected on a set of products. The degree to which the PCA provides an accurate summary depends on how well the attributes represent all of the sensory properties of the products. PCA maps based on incomplete sets of attributes may not accurately portray the overall differences among the products.

Multidimensional scaling (MDS) is a data analysis technique that generates perceptual maps of a set of stimuli based directly on their degree of similarity or dissimilarity, without regard for the attributes that the products possess. In a classic example of MDS,

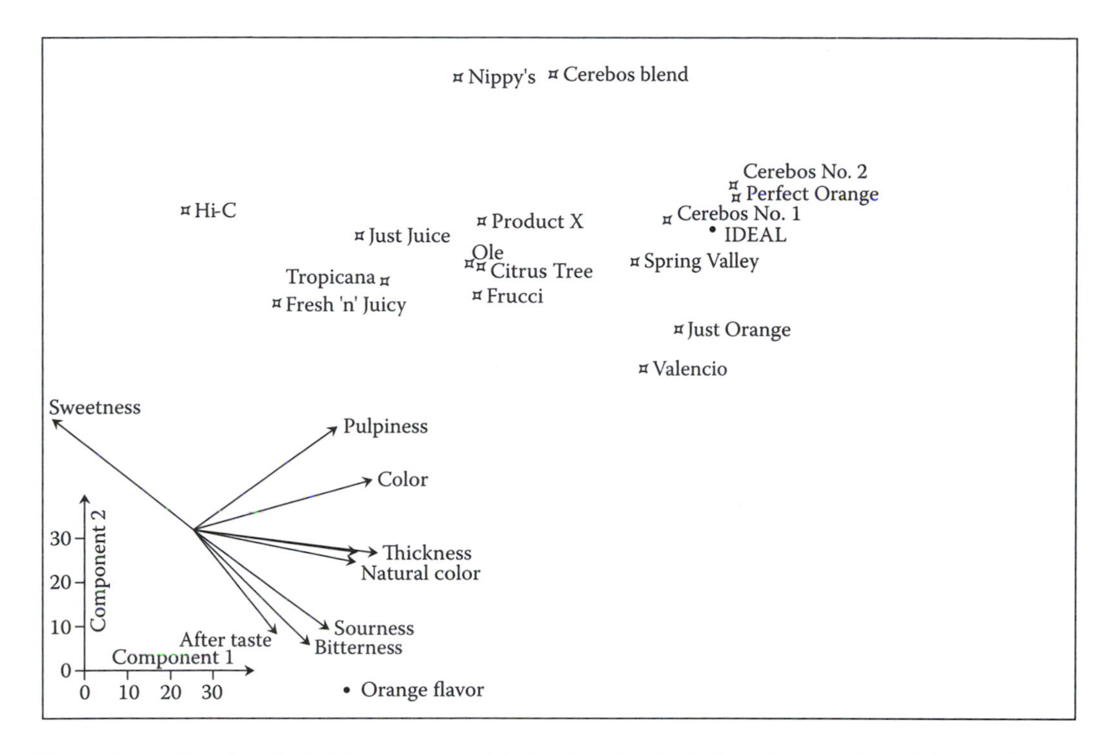

Figure 15.4 Results of a PCA on orange drinks showing both the relationships of the products to each other and the associations among the original descriptive attributes. The plot of the products is offset from the plot of the attributes to aid the visual presentation while maintaining the relative directions and magnitudes. For example, Hi-C is high in sweet and low in sour and bitter, whereas Cerebos No. 2 is less sweet and relatively high in pulp, color. (From H.R. Cooper, et al. (1989). *Product Testing with Consumers for Research Guidance*, L.S. Wu, ed., Philadelphia, PA: ASTM International. With permission.)

Kruskal and Wish (1978) accurately illustrate the relative positions of ten US cities based only on the distances between each pair of cities (see Table 15.1). The map (Figure 15.5) is obtained from the distances between the cities as opposed to their longitudinal and latitudinal coordinates. In the same way, MDS can be used in sensory evaluation to generate accurate visual summaries of the relative positions of a set of stimuli based on a panel's assessment of overall similarity or difference. Products that are similar to each other will be positioned close together on the map, while products that are different from each other will be positioned farther apart on the map. Intensity ratings on a prespecified set of sensory attributes are not required.

There are a variety of MDS procedures available in commercial software packages. They differ in the details of the algorithms used to fit the model. Theoretical and computational details are not included in the following discussion; interested readers are encouraged to study the details of the procedures available in the software package they will be using.

Table 15.1 Inter-city Distances for Multidimensional Scaling (MDS) Analysis

City	Atlanta	Chicago	Denver	Houston	LA	Miami	NY	SF	Seattle	DC
Atlanta	0	587	1212	701	1936	604	748	2139	2182	543
Chicago	587	0	920	940	1745	1188	713	1858	1737	597
Denver	1212	920	0	879	831	1726	1631	949	1021	1494
Houston	701	940	879	0	1374	968	1420	1645	1891	1220
LA	1936	1745	831	1374	0	2339	2451	347	959	2300
Miami	604	1188	1726	968	2339	0	1092	2594	2734	923
NY	748	713	1631	1420	2451	1092	0	2571	2408	205
SF	2139	1858	949	1645	347	2594	2571	0	678	2442
Seattle	2182	1737	1021	1891	959	2734	2408	678	0	2329
DC	543	597	1494	1220	2300	923	205	2442	2329	0

All MDS procedures have several characteristics in common. First, all use similar approaches to measure the quality of the fit of the model. The most common measure is called stress. The lower the stress, the better the fit of the model. Although stress can be calculated in several ways, all approaches measure the difference between the observed differences among the stimuli in the raw data and the fitted differences between the coordinates of the stimuli in the model. Stress always decreases as the number of dimensions in the model increase, so stress can be used to decide how many dimensions should be fit. Also, stress can be calculated in two general contexts—metric and nonmetric. In a metric MDS model, stress is assessed based on the proportional differences between the observed and fitted differences among the stimuli. In a nonmetric MDS model, stress is assessed based on the rank order of the differences between the observed and fitted differences among the stimuli.

In sensory evaluation, MDS models are often fit at the assessor level; then, the individual models are aggregated using any of a number of approaches to arrive at a consensus model. Fitting MDS models to the individual assessors allows researchers to assess the degree to which the assessors agree on the similarities and differences among the test products, whether or not some of the assessors are "outliers" in the sense that their fitted model is extremely different from those of the other assessors, or if there are clusters of assessors that share similar fitted models within a cluster but different models from assessors in other clusters. In developing the consensus model, assessors are often weighted based on how similar they are to the other assessors in the study. Assessors who are substantially different from the general trends receive a lower weight and thus have less impact on the consensus model. Separate consensus models, weighted or unweighted, can be obtained for each cluster of assessors identified in the panel.

As can be seen in Figure 15.5, the relative positions of the stimuli are accurately illustrated, but the orientation of the axes on the map may not match the actual orientation of the objects. Correlation and regression techniques can be used to interpret the dimensions of the MDS model. For example, if a collection of beverages were located on a MDS map, and sets of instrumental and sensory descriptive data were available on the same samples, the instrumental and sensory data can easily be added to the map to aid in interpretation. A simple approach to accomplish this is to scale the MDS axes so that all of the stimuli fall between –1 and +1. Then perform a correlation analysis between the instrumental and sensory data and the coordinates of the stimuli on the MDS axes. Plot the instrumental and sensory data on the map using their correlations with each axis—that is, for example, plot a sensory attribute on the MDS map using the correlation of the attribute with the first MDS dimension as that attribute's location on the x-axis and the correlation of the attribute with the second MDS dimension as that attribute's location on the y-axis. Since correlations fall between –1 and +1, the locations of the instrumental and sensory attributes will reveal the properties that best characterize each of the stimuli and will illustrate the relationships of the instrumental and sensory attributes to each other.

15.2.1.4 Cluster Analysis

In the same spirit that PCA identifies groups of attributes based on their degree of correlated behavior, the multivariate statistical method cluster analysis identifies groups of observations based on the degree of similarity among their ratings. The ratings may be

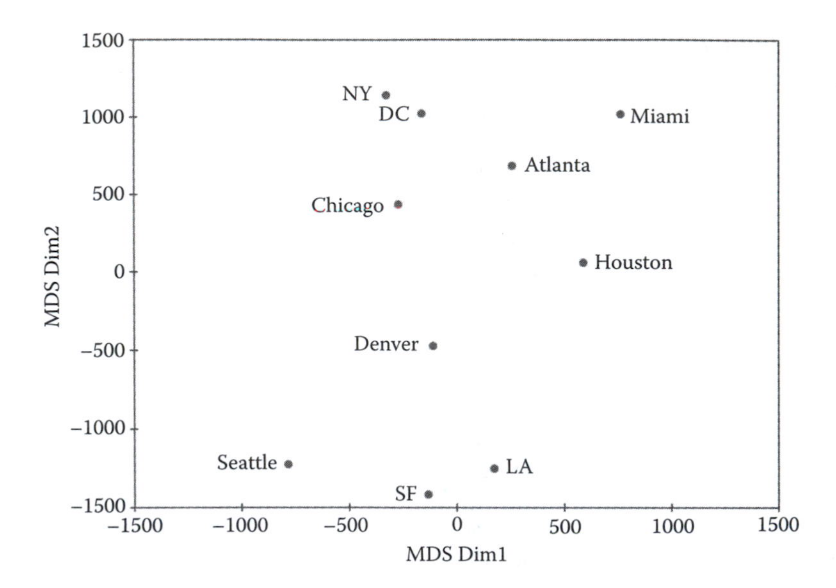

Figure 15.5 Map of the relative positions of US cities derived from their pairwise distances from each other using multidimensional scaling (MDS).

different attributes collected on a single sample or a single attribute collected on a variety of samples. There are a large number of cluster analysis algorithms in common use at present; therefore, no fair treatment of the computational details of cluster analysis could be presented in a general discussion such as this. Interested readers are referred to Jacobsen and Gunderson (1986) for their discussion of applied cluster analysis that includes a step-by-step example, a list of food science applications, and a list of texts and computer programs on the topic. Godwin et al. (1978) present an interesting application of cluster analysis in which sensory attributes and instrumental measurements are grouped based on their relation to concomitantly collected hedonic responses. Although not entirely statistically proper (attributes are not randomly sampled observations from some extant population), their approach is an interesting numerical technique for studying data relationships and should not be overlooked.

There are two classes of cluster analysis algorithms: the hierarchical and the nonhierarchical methods. The practical distinction between them is that after an observation is assigned to a cluster by a hierarchical method, it can never be moved to another cluster, whereas moving an observation from one cluster to another is possible in nonhierarchical methods.

Hierarchical methods proceed in one of two directions. In the more common approach, each observation is initially considered to be a cluster of size one, and the analysis successively merges the observations (or intermediate clusters of observations) until only one cluster exists. Alternatively, the analysis may begin by treating all the observations as belonging to one cluster and then proceed to break groups of observations apart until only single observations remain. The successive mergers or divisions are graphically depicted in a dendrogram (or tree diagram) (see Figure 15.6). The dendrogram charts the

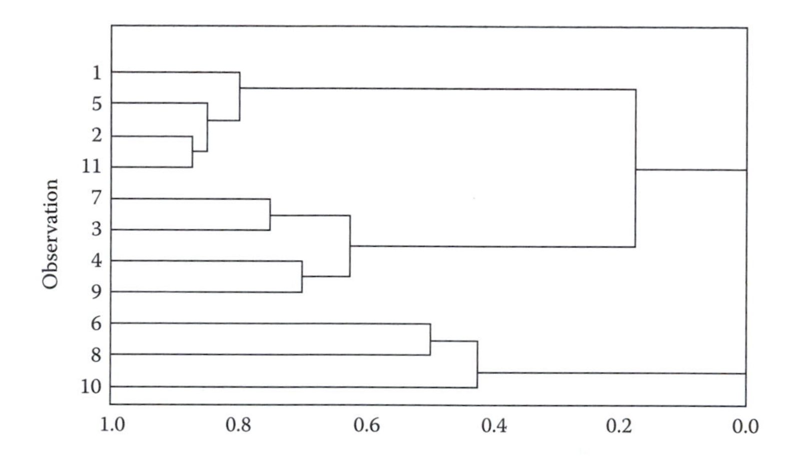

Figure 15.6 A dendrogram from a cluster analysis showing which observations are grouped together and the degree of separation among the clusters.

hierarchical structure of the observations, measures the degree of change in the clustering criterion, and is used to decide how many clusters truly exist.

The general difference between hierarchical cluster analysis algorithms is the way in which the distance (or linkage) among two clusters is measured. Commonly used algorithms include average linkage, centroid linkage, median linkage, furthest neighbor (or complete) linkage, nearest-neighbor (or single) linkage, and Ward's minimum variance linkage (see SAS, 1989). Ward's method uses an analysis of variance (ANOVA)-type sum of squares as a "distance" measure. Each approach has its advantages and disadvantages. None has emerged as a clear favorite for general use.

Nonhierarchical methods include the *k*-means method (MacQueen, 1967) and the fuzzy objective function (or FCV) method (Bezdek, 1981). Iterative mathematical techniques are used in both. For both, the user must indicate the number of clusters that are believed to exist. In the *k*-means method, each observation is assigned to a cluster based on its (Euclidean) distance from the center of the cluster. As more observations are added to a cluster, the center moves; thus, the assignments of the observations must be repeated until no further changes occur. FCV replaces the concept of *cluster membership* with *degree of membership*. The method assigns a membership weight, between 0 and 1, to each observation for every one of the prespecified number of clusters. Instead of reassigning observations to different clusters, adjustment of the membership weights continues until the convergence criteria are met (e.g., minimal shift in the locations of the centers of the clusters).

FCV offers the advantage of distinguishing observations that are strongly linked to a particular cluster (i.e., with membership weights close to +1.0) from those observations that have some association with more than one cluster (i.e., with membership weights nearly equal for two or more clusters). In addition, Jacobsen and Gunderson (1986) present a discussion of some approaches for determining the discriminatory importance of the original variables using an FCV clustering example of Norwegian beers based on gas-chromatographic data.

An application of cluster analysis that is particularly important in sensory acceptance testing is that of identifying groups of respondents that have different patterns of liking across products. While some respondents may favor an increasing intensity of some flavor note, others may find it objectionable. Merging such distinct groups may lead to a misunderstanding of the acceptability of a product because, in statistical terms, failing to recognize the clusters leads to computing the mean of a multimodal set of data. The center of such a set of observations may not represent any real group of respondents and thus is an inappropriate summary measure of overall liking (see Figure 15.7). Performing cluster analyses to discriminate patterns of liking may uncover groups of respondents that cross over demographic boundaries known to exist prior to running the study. As such, cluster analysis has an advantage over this classical approach to "segmentation."

Another important application of cluster analysis often arises in external preference mapping studies (see Section 15.3.2). The number of products that could be included in the study may exceed the resources available for consumer testing. If descriptive profiles are available for all of the potential products, the perceptual map of the products can be constructed using PCA. The resulting factor scores of the products can then be submitted to a cluster analysis to identify groups of similar products. If, for example, only 10 of the 18 possible products can be tested with consumers, a 10-cluster solution would be selected.

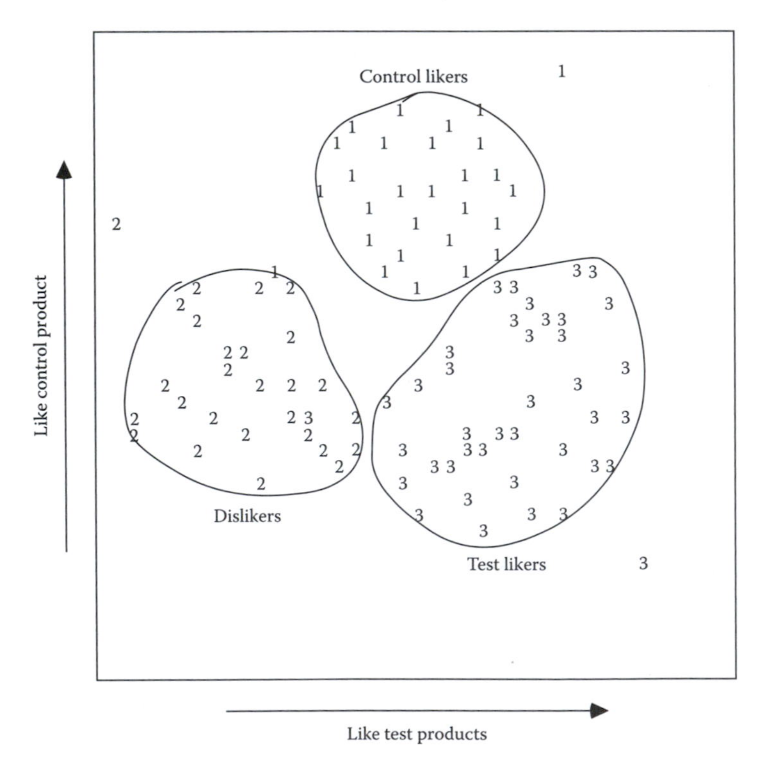

Figure 15.7 A plot of three clusters of respondents grouped by their patterns of overall liking for a group of yogurt products. The plot shows how the overall mean of the three groups would be a poor summary of the "average" liking for a product.

One product is selected from each cluster to represent the cluster in the consumer test. Using cluster analysis in this way ensures to as great a degree as possible that the range of sensory differences present in all 18 products is preserved among the 10 products that are tested with consumers.

After clusters of respondents are identified, correlation analysis, PCA, and/or regression analyses (to be discussed in the next section) can be performed to determine the similarity and differences among the clusters in how the perceived attributes relate to liking. Multiple products/varieties, "niche" marketing, or line extensions may be indicated. Lastly, demographic summaries of each cluster could be performed to determine if the members form a targetable population for marketing purposes.

15.2.2 Dependent and Independent Variables

For this set of methods, the values of some variable(s) are viewed as being dependent on the values of the other (independent) variables in the set. The statistical methods for such data either use the independent variables to predict the value of a continuous dependent variable, or they use the independent variables to group observations into particular categories of a discrete dependent variable. Even when both dependent and independent variables are present, a researcher should first apply the methods described in the previous section to uncover fundamental relationships among all the variables (using both correlation and PCA) and to determine if all the observations can be analyzed as a single group or if clusters exist that display distinctly different patterns of relationships (via cluster analysis). These preliminary analyses help to ensure that a meaningful and complete summary of the information contained in the data is obtained.

15.2.2.1 Regression Analysis

Predicting the value of one variable based on the values of one or more other variables has become commonplace. Consumer acceptability has been predicted by descriptive data or by formula and process values. Descriptive data values have been predicted by instrumental results. The perceived intensity of various responses has been predicted based on the intensity (or concentration) of a stimulus using either psychophysical models (e.g., Stevens' law) or by kinetic models (e.g., the Michaelis–Menten/Beidler equation). All of these examples use regression analysis to relate the value of a continuous dependent variable to the values of one or more independent variables.

Regression can be used simply to predict the value of a response or, not so simply, to determine what and how changes in one variable cause changes in another. By itself, regression analysis does not yield causal relationships. If a researcher comes prepared with hypotheses about the dynamics of a system, then regression analysis can be used to test the validity of the hypothesized relationships. In general, however, a highly accurate predictive model obtained by regression analysis is only just that. A highly accurate model does not imply that the independent variables drive the dependent variable. The researcher must provide the meaning behind data. It cannot be obtained from the numerical analysis procedures used to analyze the data.

Plotting data is essential to a successful regression analysis. For the same reasons noted in the discussion of correlation analysis, researchers could easily be misled by blindly applying computer programs to perform regression computations without first examining plots of the dependent variable(s) versus the independent variable(s) (see Figure 15.1). Other plots that are useful in determining the quality of the regression model are presented in the following subsections.

15.2.2.1.1 *Simple Linear Regression*

In simple linear regression, the value of a single dependent variable, y, is predicted using the value of a single independent variable, x, using a linear model of the form:

$$y = \beta_0 + \beta_1 x + \varepsilon , \qquad (15.2)$$

where β_0 and β_1 are parameters of the regression equation that will be estimated in the analysis and ε is the unexplained deviation between the observed value of y and its predicted value, called a *residual.*

The original units of measure do not have to be retained in simple linear regression. If examination of a plot of y versus x reveals a nonlinear relationship, it is often possible to transform either x or y, or both, to obtain a straight-line relationship. These transformed values of y and x can then be substituted into Equation 15.2 to obtain estimates of β_0 and β_1 (on the transformed scales). For example, the data in Figure 15.1c might be linearized by taking the logarithm of y.

The coefficients β_1 and β_0 are estimated by

$$\hat{\beta}_1 = \frac{\sum (x_i - \overline{x})(y_i - \overline{y})}{\sum (x_i - \overline{x})^2} \qquad (15.3)$$

and

$$\hat{\beta}_0 = \overline{y} - \hat{\beta}_1 \overline{x} . \qquad (15.4)$$

Based on the estimated regression coefficients, the predicted (or expected) value of the dependent variable, y, is

$$\hat{y}_i = \hat{\beta}_0 + \hat{\beta}_1 x_i . \qquad (15.5)$$

The estimates in Equations 15.3 and 15.4 are "best" in the sense that they minimize the sum of the squared differences between the observed and predicted values of y; that is, they minimize the sum of the squared residuals (i.e., SS_E):

$$SS_E = \sum (y_i - y)^2 = \sum \varepsilon_i^2 . \qquad (15.6)$$

This is what is meant when it is said that the regression equation was fit to the data using the *method of least squares.*

A fundamental criterion used to assess the quality of the regression equation is to determine if the fitted line results in a substantial reduction in the variability of the dependent variable. The variability of y around the line (i.e., vs. $\hat{y}$) is compared with the variability of y around its sample mean $\overline{y}$ (which is the original "expected" value of y) (see

Figure 15.8). This notion is formalized statistically by adding the assumption that the residuals of the regression analysis are normally distributed, independent of each other, all with the mean value of zero and the same variance, σ^2, that is, $\varepsilon \sim n(0, \sigma^2)$. ANOVA can then be used to determine if a significant reduction in unexplained variability is obtained by using least-squares estimates to predict y based on x. The F-ratio in the ANOVA table for simple linear regression, such as in Table 15.2, actually tests H_0: $\beta_1 = 0$ versus H_A: $\beta_1 \neq 0$, which is equivalent to the reduction in variability argument stated previously (if $\beta_1 = 0$ then the line is horizontal and $\hat{y} = \bar{y}$, so no reduction in variability could occur).

Other criteria are used to assess the quality of the regression equation. The coefficient of determination,

$$R^2 = 1 - \frac{SS_E}{SS_T}, \tag{15.7}$$

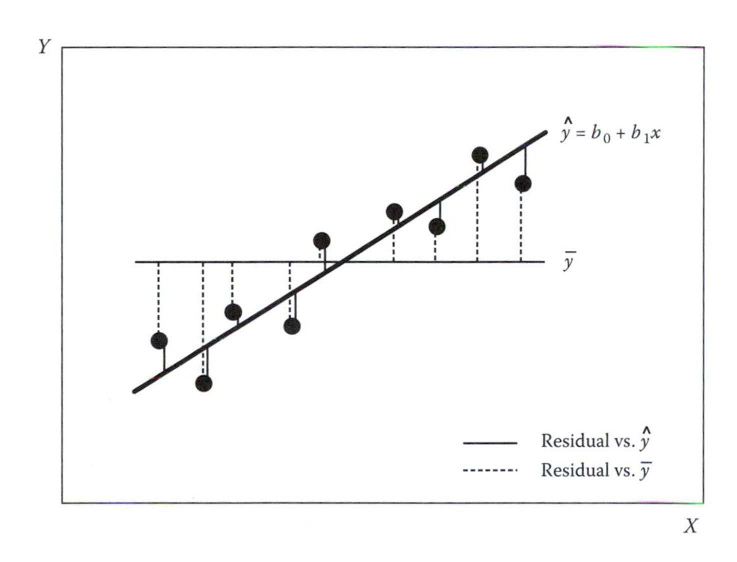

Figure 15.8 A comparison of the residuals from a fitted regression line with the residuals from $\bar{y}$ used to determine if the fitted line significantly reduces the amount of unexplained variability in the response. The reduction in the size of the residuals (i.e., distance from the "expected" value) between $\hat{y}$ vs. $\bar{y}$ shows that the regression line is a better summary of the data.

Table 15.2 ANOVA Table for Simple Linear Regression

Source of Variation	Degrees of Freedom	Sum of Squares	Mean Square	F
Total	$n - 1$	SS_T		
Regression	1	SS_{Reg}	$MS_{Reg} = SS_{Reg}$	MS_{Reg}/MS_E
Error	$df_E = n - 2$	SS_E	$MS_E = SS_E/df_E$	

summarizes the proportion of the total variability that is explained by using x to predict y. SS_E and SS_T in Equation 15.7 are the error and total ANOVA sums of squares from Table 15.2, respectively. In sensory evaluation, values of $R^2 > 0.75$ are generally considered to be acceptable. However, whether this is true depends on the intended use of the regression equation. Other criteria may be more informative. A confidence interval on β_1 can be constructed using

$$\hat{\beta}_1 \pm t_{\alpha/2,n-2}\sqrt{MS_E / SS_x}\,, \tag{15.8}$$

where $SS_x = \Sigma(x_i - x)^2$ and $t_{\alpha/2,n-2}$ is the upper-$\alpha/2$ critical value of Student's t-distribution with $n - 2$ degrees of freedom. The F-ratio from ANOVA only tells the analyst if β_1 is different from zero. The confidence interval in Equation 15.8 tells the analyst whether β_1 is estimated with sufficient precision to be useful in applications. The idea of a confidence interval on β_1 can be extended to confidence bands on the predicted value of y using

$$\hat{y}_0 \pm t_{\alpha/2,n-2}\sqrt{MS_E[(1/n)+(x_0-\bar{x})^2/SS_x]} \tag{15.9}$$

The confidence bands can be plotted along with the predicted values to provide a visual assessment of the quality of the fit (see Figure 15.9). If the confidence bands are too

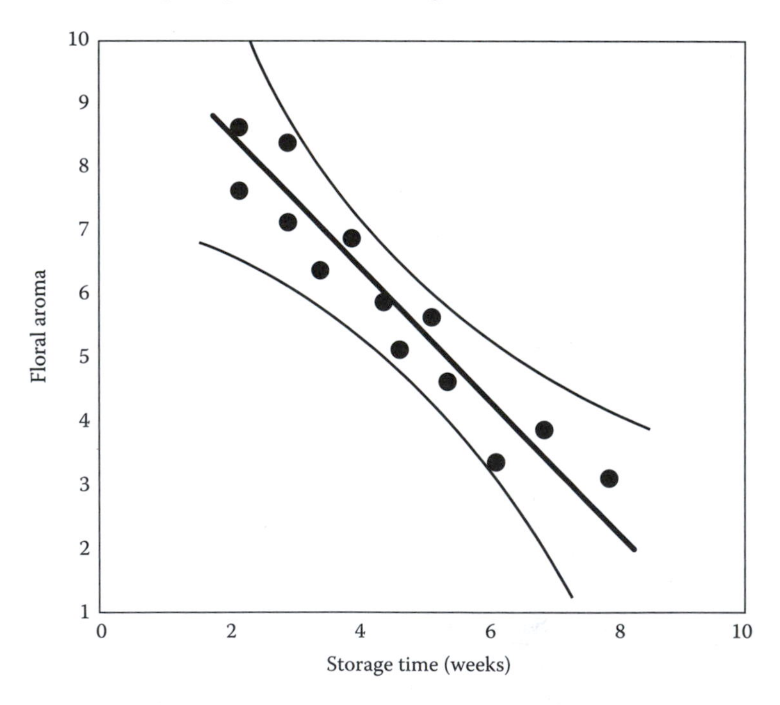

Figure 15.9 A fitted regression line with 95% confidence bands. The width of the bands provides a visual assessment of the quality of the fitted line. Narrow bands such as these indicate that the data are well fitted by the line, while wide bands indicate that a large amount of unexplained variability remains.

wide, then regardless of the F-ratio test or the value of R^2, the fitted simple linear regression equation is not sufficiently good.

Several possibilities exist to explain a poor-fitting regression equation. Most of these possibilities can be studied with plots of the residuals from the regression. The residuals, $\hat{\varepsilon}$, should be plotted versus the predicted values, $\hat{y}$, and versus the independent variable, x. The residuals should be randomly dispersed across the range of both the predicted values and the independent variable (see Figure 15.10a). Any apparent trends indicate that a simple linear regression is not sufficient and that a more complex relationship exists between x and y. Higher-order terms (e.g., x^2) may be needed (see Figure 15.10b) or data transformations may need to be performed (see Figure 15.10c). An individual point falling far from the rest in either the vertical or horizontal direction may be an outlier that is having an unreasonably large influence on the fit of the model (see Figure 15.10d). Such observations should be examined and, when appropriate, eliminated from the data.

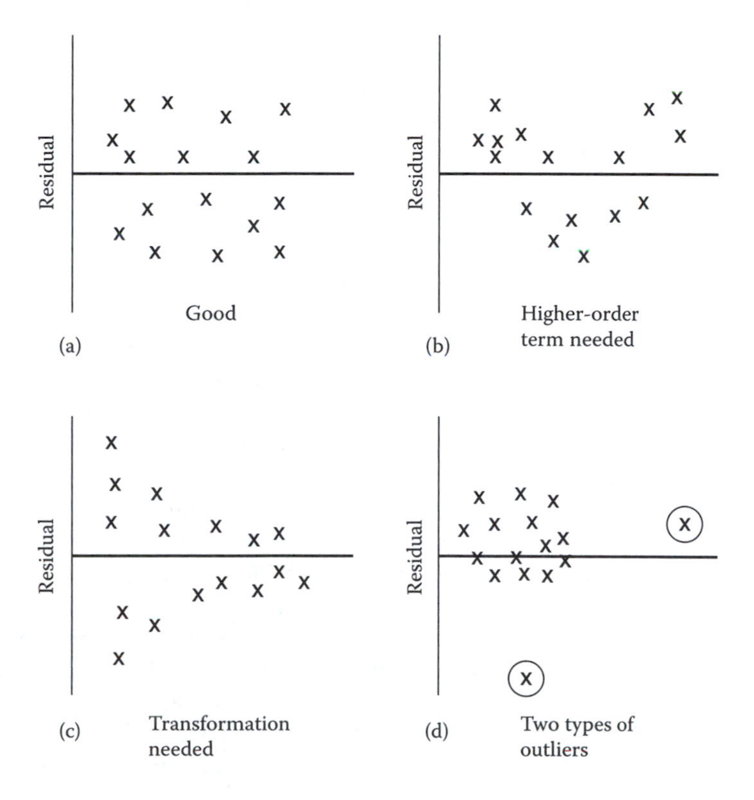

Figure 15.10 Plots of the residuals from a simple linear regression showing the desired, random arrangement in (a) and several undesirable patterns, i.e., plots (b), (c), and (d), along with their interpretation.

15.2.2.1.2 *Multiple Linear Regression*

Sometimes more than one independent variable is needed to obtain an acceptable prediction of a response, y. It may be that a polynomial in a single variable, x, is needed because the relationship between x and y is not a straight line, such as in

$$y = \beta_0 + \beta_1 x + \beta_2 x^2 + \beta_3 x^3 \,.$$

In other cases, the response may be influenced by more than one independent variable such as in

$$y = \beta_0 + \beta_1 x_1 + \beta_2 x_2 + \beta_3 x_3 \,,$$

or a combination of both cases may exist. Regardless, multiple regression analysis is a straightforward extension of simple linear regression that allows multiple independent variables to be included in the regression equation for y. An integral part of the analysis involves the assessment of the value of each term considered for inclusion in the model.

A pitfall associated with multiple regression is that now any relationships that exist among the independent variables will influence the resulting regression equation. *Multicolinearity* is the term used to describe situations in which two or more independent variables are highly correlated with each other. This mutual correlation will influence the values of the estimated coefficients, b_is, and could lead to incorrect conclusions about the importance of each term in the model. It is important that the correlation structure of the independent variables be studied before undertaking a multiple linear regression analysis. The correlation analysis will be more meaningful if it is accompanied by a scatter-plot matrix of the independent (and dependent) variables, like the one in Figure 15.2. A designed approach to multiple regression, called *response surface methodology* (RSM), avoids the problems of multicolinearity. RSM will be discussed in Section 15.3.

Not all of the independent variables that could be included in a multiple linear regression model may be needed. Some of the independent variables may be poor predictors of the response, or, due to multicolinearity, two or more independent variables may explain the same variability in the dependent variable. Several approaches for selecting variables to include in the model are available (see Draper and Smith, 1981). Most computer packages include more than one.

One "brute force" approach is the *all possible regressions* method. As the name implies, all possible subsets of the independent variables are considered, starting with all of the one-variable models, then all of the possible two-variable models, and so on. Computer output typically only presents a small number of the best models from each size group. Several criteria are used to determine which models are best in all possible regressions. The multiple R^2 (from Equation 15.7) is a common measure, with larger values being preferred. Another criterion is the size of the residual mean square, MS_E, from the ANOVA. MS_E is the estimated residual variance, so smaller values are desirable.

Associated with the residual mean square criteria is the adjusted R^2, where

$$R^2_{adj} = 1 - \frac{(n-1)MS_E}{SS_T} \,. \tag{15.10}$$

R^2_{adj} is interpreted in the same way as R^2. In multiple regression, the residual mean square will initially decrease with the addition of new terms into the model; R^2 will increase. Some studies reach a point where the further additions of new terms will result in an increase in the residual mean square (a bad sign), but R^2 will continue to increase (a good sign). When the residual mean square begins to increase, the adjusted R^2 will begin to decrease so that the two statistics agree qualitatively.

A final criterion commonly used in all possible regressions is Mallow's C_p statistic (1973):

$$C_p = \frac{SS_{E(p)}}{MS_{Full}} + 2p - n \tag{15.11}$$

where $SS_{E(p)}$ is the residual sum of squares from a model containing p terms and MS_{Full} is the residual mean square from the model containing all of the independent variables. Unlike R^2 and MS_E, C_p considers how good the full model is and uses this as a base of comparison to gauge the quality of a model containing only a subset of the independent variables. One drawback to C_p is that it can only be calculated when the number of observations in the data set is greater than the number of independent variables. For instance, if the number of descriptive attributes is greater than the number of samples, then Mallow's C_p criterion could not be used.

Another group of variable-selection procedures used in multiple regression is the forward inclusion, backward elimination, and stepwise selection procedures. These use similar criteria for deciding if an independent variable should be included in the model or not. Forward inclusion starts by adding to the model the independent variable that maximizes the reduction in the unexplained variability, measured by the residual sum of squares (SS_E). Additional terms are added based on the additional reduction in SS_E that occurs as a result of their inclusion. Computer packages use an "F-to-enter" statistic to determine if adding a particular term will result in a statistically significant reduction in the unexplained variability. The independent variable with the largest F-to-enter value is the next term added to the model. The F-to-enter value can be set to correspond to the analyst's desired significance level. For example, a value of 4.0 corresponds roughly to $\alpha = 0.05$ for data sets containing 30–50 observations. The forward inclusion procedure continues to add terms until none of the F-to-enter values are large enough (compared to the value set in the program).

Backward elimination starts with all of the independent variables in the model and proceeds to eliminate terms based on how little of the variability in the dependent variable they explain. Computer packages use an "F-to-remove" statistic to measure how unimportant a particular term is. The term with the smallest F-to-remove is the next one to be excluded from the model. Once again, the analyst can select the value for F-to-remove, and the procedure will continue to remove terms until none of the F-to-removes are too small (4.0 is also a good initial value for F-to-remove).

Stepwise selection is the recommended variable-selection procedure for building multiple linear regression models. Stepwise selection combines forward inclusion with backward elimination by allowing for either the addition or removal of a term at each step of the procedure (starting after two terms have been added). Terms are initially added to the model using the *F*-to-enter criteria from forward selection. Because of multicolinearity, the importance of each term in the model changes depending on which other terms are also in the model. A term in the model may become redundant as others are added. Stepwise selection allows for such a term to be removed using the *F*-to-remove criteria from backward elimination. The values of *F*-to-enter and *F*-to-remove are recomputed for each independent variable at each step of the procedure. The analysis ends when none of the statistics satisfy the values set by the analyst.

All of the diagnostics used to assess the quality of the model in simple linear regression should be used to determine the goodness of fit of the multiple linear regression model. These include R^2 (now also including R^2_{adj}), MS_E, confidence intervals on the individual coefficients, and, most importantly, plots of the residuals. The potential for missing nonlinear relationships or outliers is higher in multiple linear regression because of the difficulty of visualizing in more than three dimensions. Plots of the residuals versus the independent variables are the easiest way to explain problems and/or to determine if further improvements in the model are possible.

15.2.2.2 Principal Component Regression

A weakness of multiple linear regression is the manner in which it deals with correlated predictor variables (i.e., the *x*-variables), a problem called *multicolinearity*. As noted in the previous section, if two highly correlated predictor variables are included in the regression model, the size and even the sign of the slope coefficients can be misleading. The problem is overcome to some degree by using one of the variable-selection procedures, such as stepwise regression. However, the regression model that results from a stepwise procedure does not tell the whole story when it comes to identifying all of the predictor variables, *x* values, that are related to the predicted variable, *y*. For example, when using attribute intensity data from a descriptive panel to predict consumer acceptance, a particular attribute, such as sweet taste, may be highly related to acceptance, but it may not appear in the regression model because another attribute that is highly correlated with sweet taste, such as sweet aftertaste, may already be in the model. The stepwise procedure will not include both sweet taste and sweet aftertaste in the regression model precisely because they are highly correlated with each other (and, thus, are explaining the same variability in acceptance). This gives the researcher the incorrect impression that only sweet aftertaste, and not sweet taste, is important to acceptance. A more correct interpretation is that the term in the regression model, that is, sweet aftertaste, is, in effect, representing all of the descriptive attributes with which it is highly correlated. However, the regression model does not reveal which attributes are correlated. Thus, a stepwise regression procedure applied to correlated predictor variables does not, by itself, uncover all of the attributes that are "driving" acceptance.

Principal components regression (PCR) is a method that overcomes this weakness. PCR is a straightforward combination of principal components analysis (PCA) and regression. Continuing the example of using descriptive data to predict acceptance, a PCA is performed on the average attribute profiles of the samples in the study. A set of factor scores is obtained for each sample. The factor scores are used as the predictor variables, that is, the *x* values, in a regression analysis to predict consumer acceptance, *y*. The factors obtained from the PCA represent the underlying dimensions of sensory variability in the samples and are typically easy to interpret based on the factor loadings. Attributes with large positive or negative loadings (i.e., close to +1 or −1) on a single factor are the attributes that define the factor, so if the factor is found to be a significant driver of acceptance, the researcher knows that all of the attributes associated with that factor are, as a group, influencing acceptance. In addition, the factors obtained from PCA are not correlated with each other. Thus, the problem of multicolinearity is avoided. Popper et al. (1997) present an excellent example of PCR applied to the prediction of consumer acceptance of 12 honey-mustard salad dressings based on their descriptive profiles.

The factors identified with PCA are ordered according to how much of the variability in the original data each explains. The first factor accounts for the greatest amount of variability, the second factor accounts for the next greatest amount, and so on. It may not be the case that the factor that explains the most variability in the original x-variables is the factor that is most highly related to y. In fact, some factors may not be related to y at all and, therefore, do not need to be included in the PCR model. Stepwise regression can be used to generate a PCR model that includes only those factors which are statistically significant predictors of y. In the descriptive analysis example, the stepwise approach to PCR allows the researcher to identify which underlying dimensions of sensory variability (and all of their associated attributes) "drive" acceptance and which do not.

The PCR approach can be further extended by recognizing that the factor scores may be related to the response variable, y, in nonlinear and interactive ways. Nonlinear and interactive relationships can be accounted for in PCR by including the squares and the cross products of the factor scores, respectively, as variables in the regression, as is done in response surface methodology (see Section 15.3.3). This allows the researcher to identify the levels of the factor scores that are associated with, for example, the best liked product—that is, an optimum or "ideal" point.

15.2.2.3 Partial-Least Squares Regression

As noted above, PCA generates factors that may not be related to the independent variable of the regression, y. Partial least-squares (PLS) regression (see Martens and Martens, 1986) is a multivariate technique related to PCR that overcomes this weakness. Where PCA concentrates only on explaining the variability exhibited by the correlated predictor variables (x-variables), PLS derives factors (i.e., linear combinations of the x-variables) that (1) explain large portions of the variability in the x-variables and (2) simultaneously correlate to as great an extent as possible with the dependent variable y. While the PLS factors may not explain as much of the variability in the x values as would the PCA factors, PLS ensures that each factor identified has maximal predictive power on y.

PLS has two other important advantages over PCR. First, it generates graphical output that clearly illustrates the relationships both among the predictor variables, x values, and between the predictor variables and y. Second, PLS readily extends to simultaneously predicting more than one dependent variable. When a single dependent variable is predicted, the analysis is called *PLS1*. When several dependent variables are predicted, the analysis is called *PLS2*. The graphical relationships in PLS2 can be used to illustrate, for example, how consumer vocabulary relates to descriptive attribute ratings (see Figure 15.11; Muñoz and Chambers, 1993; Popper et al., 1997).

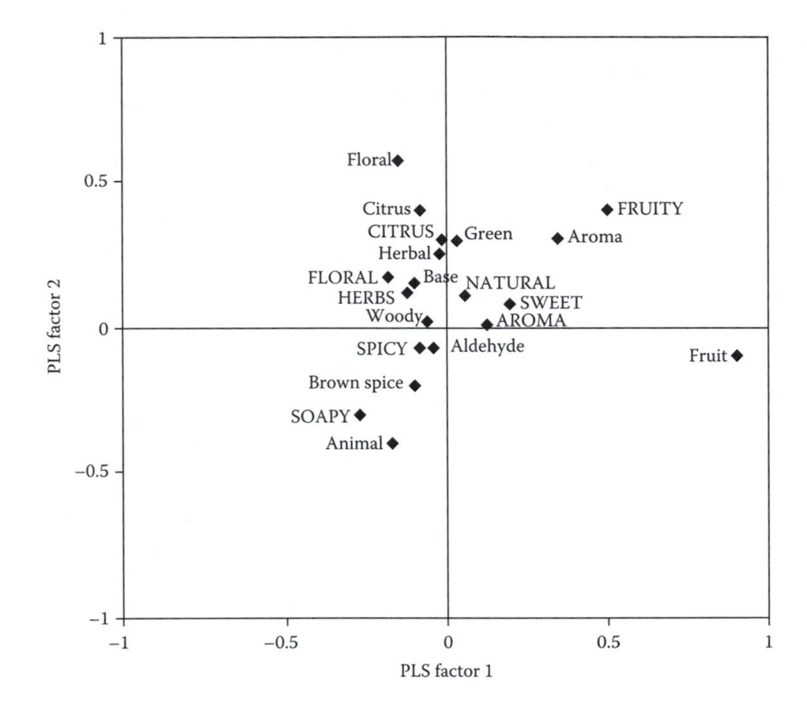

Figure 15.11 Use of graphical relationships in PLS2 to illustrate the relationship between consumer vocabulary and descriptive attribute ratings. Terms in upper case are consumer responses. Terms in lower case are descriptive attributes. Note that consumer and descriptive vocabularies do not agree in all cases.

The computations performed in a PLS regression are beyond the scope of this discussion. Several computer programs that perform PLS analyses are available (Pirouette, 2006; Unscrambler, 2006; SAS, 2004). The programs and the growing number of publications with examples of PLS applied to sensory data make this technique increasingly accessible to interested researchers.

15.2.2.4 Discriminant Analysis

Discriminant analysis is a multivariate technique that is used to classify items into preexisting categories (defined by a discrete dependent variable). A mathematical function is developed using the set of continuous independent variables that best discriminate among the categories from which the items arise. For instance, descriptive attribute data might be used to classify a finished product as being "acceptable" or "unacceptable" from a quality control perspective or descriptive and/or instrumental measures might be used to determine the source (e.g., country or manufacturer) of a raw ingredient.

Discriminant analysis is similar to several of the multivariate techniques that have already been discussed. In one sense, discriminant analysis is similar to regression in that a group of continuous independent variables is used to predict the "value" of a dependent variable. In regression analysis, the value is the magnitude of a continuous dependent variable that is predicted using a *regression equation*. In discriminant analysis, the value is the category of a discrete dependent variable that is predicted using a *discriminant function*. In another sense, discriminant analysis is similar to PCA in that the correlated nature of the independent variables is considered in developing new axes (i.e., weighted linear combinations of the original variables). In PCA, the axes are chosen to successively explain the maximum amount of variability. In discriminant analysis, the axes are chosen to maximize the differences between the centers of the discrete categories of the dependent variable. A simple graphical depiction of discriminant analysis is presented in Figure 15.12, in which acceptable and unacceptable samples of product are displayed in a plot of two descriptive attributes: staleness and crispiness. The discriminant function defines the new axis, *D*, on which the difference between the means of the two groups is maximized.

If the dependent variable contains only two categories, then only a single discriminant function is needed. If the dependent variable contains more than two categories, then more than one discriminant function may be needed to accurately classify the observations. The number of possible discriminant functions is one less than the number of categories. Regardless, the linear combination(s) of the original variables that best separate the categories is the one that maximizes the ratio

$$\frac{\text{Variance between category means}}{\text{Variance within categories}}.$$

In addition to this ratio, the quality of the discriminant function is measured by the proportion of the items that it correctly classifies. This evaluation can be carried out by using the same observations that were used to build the discriminant function. However, when sufficient data are available, it is preferable to withhold some of the observations from the model building analysis and use them only after the fact to verify that the

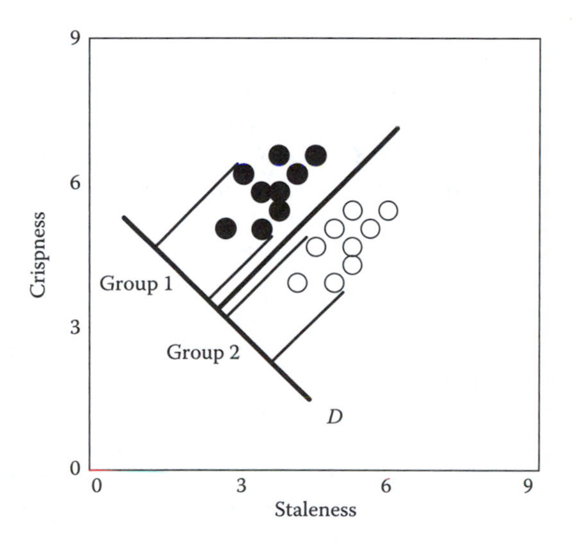

Figure 15.12 A graphical depiction of discriminant analysis with the samples plotted in two of the original attributes (staleness and crispness) and with the "axis of maximum discrimination," *D*.

discriminant function performs the classification task satisfactorily. As such, the verification process is more objective.

Not all of the original independent variables may be needed to accurately classify each item. As in regression analysis, there are four commonly used variable-selection criteria: forward inclusion, backward elimination, stepwise, and all-possible-functions. In each, the criterion for determining the value of a variable is the degree to which it contributes to the discrimination among the categories. Powers and Ware (1986) present a summary of a stepwise discriminant analysis in which six blue-cheese products were best categorized by using only 14 of the original 28 profile attributes.

The Powers and Ware reference just cited is a comprehensive discussion of applied discriminant analysis in sensory evaluation. Two alternatives to linear discriminant analysis, canonical discriminant analysis and nearest-neighbor discriminant analysis, are discussed. A variety of industrial applications, relations to other multivariate techniques, and relevant computer software concerns are also presented. Analysts interested in performing discriminant analysis are encouraged to familiarize themselves with the material presented there.

15.3 PREFERENCE MAPPING

Preference mapping is a collection of multivariate techniques for illustrating the relationships between sensory (and sometimes instrumental) data and consumer acceptance. Preference mapping studies go by many names, including category appraisals, competitive assessments, product space mappings, and key drivers analyses. The variety of names can lead to confusion. For example, whereas one organization may refer to preference mapping studies as category appraisals, another might reserve the term category appraisals for studies that track marketing information, such as share of sales, advertising initiatives, new line extensions, and so on. Whereas some perceive the term preference mapping as being too technical, when discussing this approach, it is important to use the term that colleagues understand appropriately and not confuse with some other test method.

Another source of confusion is that there are several varieties of preference maps. The two major varieties are internal preference mapping and external preference mapping. In this chapter, PLS is presented as a third variety of preference mapping. A feature that discriminates the three approaches is the information that is used to locate the test products on the maps. Internal preference mapping uses consumer acceptance ratings to locate the products on the maps. External preference mapping uses sensory descriptive attribute ratings to locate the products on the maps. PLS mapping uses both the consumer and the sensory data to locate the products on the maps.

To keep the difference between internal and external straight, it is helpful to recall that marketing researchers, conducting acceptance tests, coined the term preference mapping. The consumer acceptance ratings are "internal" to the studies they conduct; therefore, the method that uses the consumer ratings to locate the samples on the map is *internal preference mapping*. To marketing researchers, sensory attribute ratings are an "external" source of information, so the method that uses the descriptive attribute ratings to locate the samples on the map is *external preference mapping*. Internal, external, and PLS are three broad categories of preference mapping. There is a good deal of variety in the details of the methods that fall within each of the three categories. Therefore, once again, it is important to confirm prior to running a preference mapping study that the approach you intend to use is the same one that your colleagues expect.

A single data set is used to illustrate the three approaches to preferences mapping. The data consist of 30 sensory attributes evaluated on 10 prepared meal products. One hundred consumers evaluated the same 10 products for overall acceptance. Each consumer evaluated all 10 products.

In Section 15.3.1, the consumer data is used to develop an internal preference map. The map is then extended through a creative application of correlation analysis to illustrate the relationship between the sensory data and acceptance. In Section 15.3.2, the external preference map is developed using PCA and regression analysis, as discussed in the previous section. Also included in Section 15.3.2 will be an application of *preference segmentation*, in which cluster analysis is used to identify segments of consumers with similar liking patterns for the ten test products. It is important to identify homogeneous groups of consumers before running external preference mapping on average acceptance ratings. Averaging across groups with different preferences can mask the sensory attributes that drive consumer acceptance. Lastly, in Section 15.3.3, PLS mapping is applied to the combined sensory and consumer data, including the preference segments identified in Section 15.3.2.

15.3.1 Internal Preference Mapping

Internal preference mapping is an application of PCA, discussed in Section 15.2. The test products form the rows of the data set. The overall liking ratings of the consumers form the columns (see Table 15.3). Because PCA requires complete data, consumer tests in which each respondent evaluates all of the products are preferred in order to avoid the excessive imputation of missing values that would be required if incomplete serving designs were used. The factor scores of the products and the factor loadings of the respondents from a two-dimensional PCA solution are plotted on the same graph.

Table 15.3 Format of Input Data for Internal Preference Mapping

Product	Resp1	Resp2	Resp3	...	Resp100
A	6	8	9	...	8
B	3	7	7	...	1
C	8	8	9	...	7
D	2	7	9	...	3
E	3	7	7	...	3
F	2	6	6	...	2
G	8	9	8	...	7
H	2	8	9	...	5
I	7	8	9	...	4
J	5	9	9	...	7

Products form the rows of the data set. The overall liking ratings of each respondent form the columns.

Constructing the map is a simple, four-step process:

1. Plot the factor scores of the products and the factor loadings of the respondents on the same graph (see Figure 15.13a). The factor scores of the products need to be rescaled to fall between –1 and +1 so that they cover the same range as the factor loadings of the respondents. For each dimension, this is accomplished by dividing the original factor scores by the largest factor score (ignoring the sign) on the dimension.
2. Delete the respondents whose liking data are not well fitted to the map (see Figure 15.13b). These respondents can be identified in one of two ways. If a factor analysis procedure, such as PROC FACTOR in SAS (2005), was used to fit the PCA model, the output includes communality statistics for each respondent. Like the R^2 value from a regression analysis, communalities are the percent of the variability in the respondent's liking data that is being explained by the PCA model. Delete the respondents with communalities less than a preselected cutoff (typically, somewhere between 0.50 and 0.75). Alternatively, delete respondents who fall too close to the origin (i.e., the (0, 0) point) on the map. A respondent's distance from the origin is the square root of the communality. Compute the square of each respondent's distance from the origin by summing the squares of their factor loadings. Delete respondents whose squared distance from the origin is less than a preselected cutoff (again, typically, somewhere between 0.50 and 0.75).
3. Rescale the remaining respondents to fall equally far from the origin of the plot (see Figure 15.13c). The point that represents each respondent is rescaled to fall a unit distance from the origin of the plot. To do this, divide both factor loadings of each respondent by the distance that respondent falls from the origin (i.e., Distance = $\mathrm{SQRT}(F_1^2 + F_2^2)$, where F_1 and F_2 are the respondent's factor loadings). This step is not required but is commonly used in practice.

   Internal preference maps are self-segmenting. Respondents tend to form multiple clusters on the map. A group of respondents who fall close to each other on the map form a segment. The test products that are closest to them are, in general, the ones they like the most. The products that fall on the opposite side of the map are the ones they like the least. For example, in Figure 15.13c, there is one segment in the lower right quadrant representing those who most like products A, C, G, and H; and most dislike products B, E, and F. There is a more dispersed segment in the upper right quadrant representing those who also like products A, C, and G the most; are more accepting of products B, E, and F; and dislike product H the most.
4. Incorporate the sensory descriptive information by correlating the attribute intensities of the products with their factor scores and plotting the resulting correlation coefficients on the map (see Figure 15.13d). This creative use of correlation analysis was first proposed by McEwan (1998). Each product has a full set of attribute intensities and two factor scores, one for factor 1 (the horizontal axis of the map) and one for factor 2 (the vertical axis of the map). For each attribute, use the correlation of the attribute ratings with factor 1 as the x-coordinate and the correlation of the attribute ratings with factor 2 as the y-coordinate.

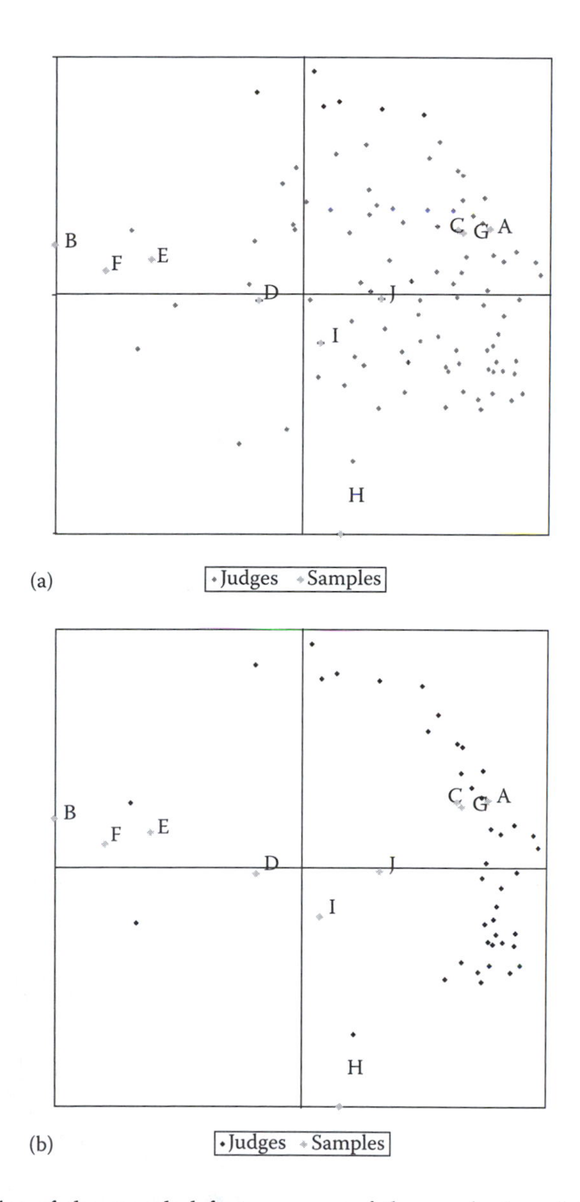

Figure 15.13 (a) A plot of the rescaled factor scores of the products and the factor loadings of all respondents. (b) A plot of the rescaled factor scores of the products and the factor loadings of respondents with communalities greater than 0.50—that is, the respondents for which the model explains at least 50% of the variability in their liking ratings.

The orientation of the attributes with each other reveals their internal correlation structure. Attributes that fall close to each other on the map are positively correlated with each other. Attributes that fall on opposite sides of the map are negatively correlated with each other. Attributes that fall at nearly right angles to each other are not correlated.

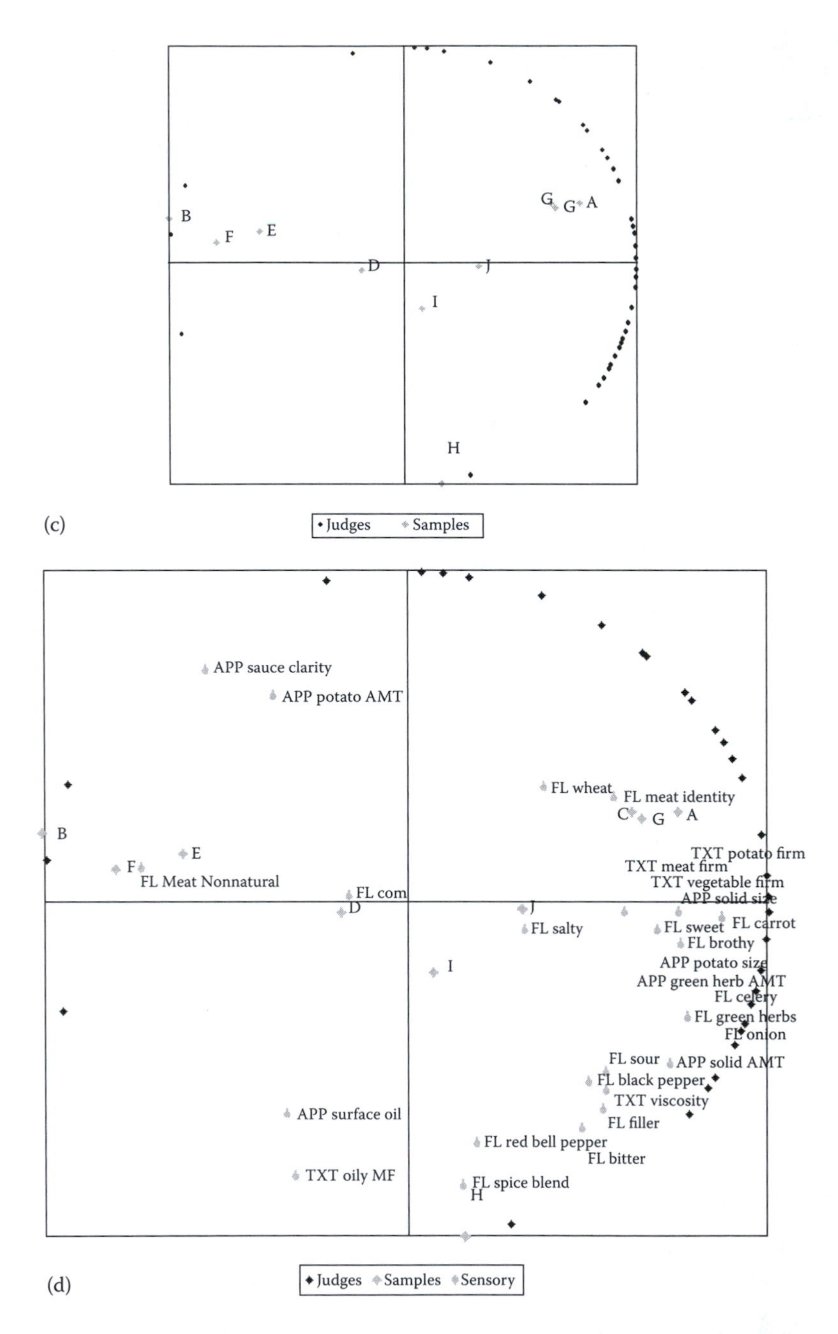

Figure 15.13 (Continued) (c) A plot of the rescaled factor scores of the products and the factor loadings of respondents rescaled to fall equidistant from the origin. This optional step sometimes makes it easier to identify clusters of respondents. (d) The final internal preference map including sensory attributes. The coordinates of the sensory attributes are their correlations with the factor scores of the samples from each of the two dimensions of the map.

More importantly, the location of the attributes to the samples and to the clusters of respondents reveals the positive and negative drivers of acceptance. Attributes that fall close to a cluster of respondents are positive drivers for that cluster. Higher intensities are preferred over the lower intensities of such attributes. Attributes that fall on the opposite side of the map are negative drivers (lower intensities are preferred). Attributes that fall at right angles to the cluster are not key drivers for that segment. For example, in Figure 15.13d, the respondents represented in the segment in the lower right quadrant prefer high intensities of green herbs, onion, celery, and carrot flavors. They dislike a sauce with high clarity, an appearance dominated by potatoes, and a nonnatural meat flavor.

It is important to understand that in all forms of preference mapping, when it is said that consumers prefer *low* or *high* intensities of attributes, it should be interpreted as low or high in the range of intensities exhibited by the products in the study. Low or high in preference mapping does not mean the extremes of the rating scale used to evaluate the products. Attribute intensities for products from the same category may range over only 2–5 units on a 15-point scale. There can be a large difference in acceptance for a product with, for example, a 2.5 intensity as opposed to one with a 4.5 intensity on a key attribute.

15.3.2 External Preference Mapping

An external preference map is more complicated to construct than an internal preference map. Three statistical methods are used in the analysis. PCA (discussed in Section 15.2.1.2) is used to develop a perceptual map of the product space based only on the sensory characteristics of the products. Cluster analysis (discussed in Section 15.2.1.3) is used to identify preference segments among the consumers. A preference segment represents a group of respondents who exhibit similar patterns of liking across the products in the test but whose pattern differs in some meaningful way from respondents in another preference segment. Lastly, regression analysis (discussed in Section 15.2.2.1) is used to link the sensory information to consumer acceptance through models that use the factor scores of the samples from the perceptual map to predict the acceptance ratings of the products from the total respondent base and any preference segments that were identified in the cluster analysis (see Figure 15.14).

15.3.2.1 Constructing the Perceptual Map of the Product Space

The perceptual map of the product space is obtained from the PCA of the product-by-attribute data. The products form the rows (observations) of the dataset and the sensory attributes form the columns (variables). In keeping with the idea that a map is being created, the principal components obtained from the analysis are called the *key sensory dimensions of the perceptual space*.

Before conducting the PCA, it must be decided if all of the products and all of the attributes should be included in the analysis. If only one product in the study possesses suprathreshold intensities of one or more attributes, the PCA is likely to create a factor dedicated entirely to distinguishing that product from all of the others. The creation of one or more of these *single-sample dimensions* has the negative side effect of masking otherwise meaningful differences among the other products in the study. For example, including one pepperoni pizza in a study with ten plain cheese pizzas might obscure some subtle but

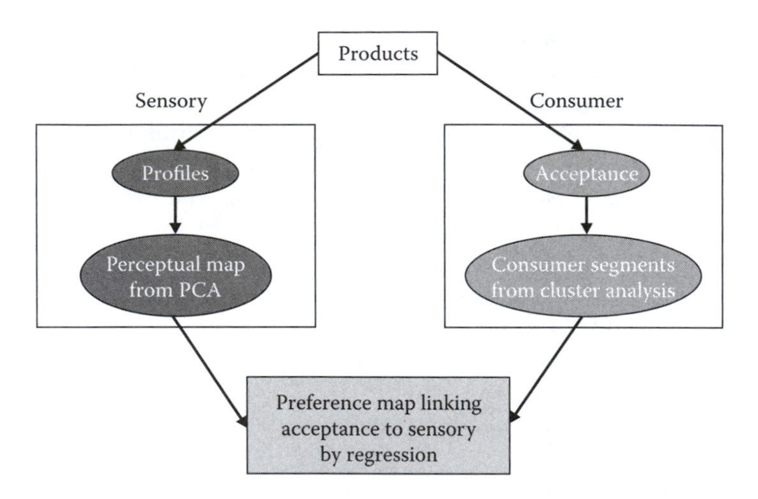

Figure 15.14 Schematic diagram of an external preference mapping analysis. Products are submitted for both sensory descriptive and consumer evaluations. Sensory results are summarized on a perceptual map. Consumer acceptance ratings are submitted to a cluster analysis to identify preference segments. The sensory and consumer information are linked using regression analysis.

meaningful differences among the cheese pizzas. More in-depth information about the attributes that drive liking may be obtained if the pepperoni pizza is eliminated and the category of interest is redefined to be plain cheese pizza. If only one attribute is involved, another option is to eliminate the attribute, especially if prior research has shown that it does not play a significant role in acceptance. Alternatively, the preference mapping study could be conducted twice. The first run would include all of the products with special interest paid to the effect of the pepperoni-related attributes. The second run would consist of only the plain cheese pizza product to focus more specifically on the attributes that drive liking in that group.

The same consideration should be paid to the possible elimination of attributes. If the intensity of an attribute is the same for all of the products in the test, or if the intensity varies over a trivially small range of values (e.g., range $\leq$ 0.5 units on a 15-point scale), the attribute should be dropped from the analysis. Similarly, if the intensity of an attribute is subthreshold for all of products, the attribute should be dropped. This is especially important if the PCA is conducted using the correlation matrix as opposed to the covariance matrix of the responses. Large but spurious correlations can occur with attributes that exhibit only a small range of values. A correlation-based PCA cannot distinguish between attributes with trivial ranges of intensities and those with meaningfully large ranges when it computes loading factors and factor scores. Attributes with trivially small ranges may appear to play key roles in defining the perceptual space of the products. By itself, this point seems to support the use of the covariance matrix in the PCA because the use of the covariance matrix down-weights attributes that exhibit little variability. However, it is widely recognized that small differences in certain attributes (e.g., off-notes) can have large impacts on acceptance. A covariance-based PCA could down-weight these attributes to the point that their true importance to acceptance is lost. Using the correlation matrix after eliminating attributes with trivially small ranges of intensities preserves the importance of these attributes. Because of this, correlation-based PCAs are recommended for the perceptual mapping step of an external preference mapping analysis.

Any elimination of products or attributes needs to be carried out with caution. Whenever possible, conduct the analyses with all products and attributes, then repeat the analysis with larger and larger groups of products and attributes eliminated. The multiple analyses often reveal interesting insights concerning the unique products and attributes, as well as clearer understanding about what is driving liking in the category as a whole.

A preliminary screening of the candidate products can be used to determine how well the products fill the sensory space of the category. The researcher may decide to eliminate extreme products, or products that possess unique attributes, to obtain a more uniform coverage of the new, more narrowly defined category. Alternatively, additional products or specifically formulated prototypes could be added to the study to fill any "gaps" in the sensory space. This screening also provides some context for the researcher when the data are analyzed and mined for information and insights.

For the prepared meals data, the attributes potato flavor, red bell pepper flavor, and sweetness were eliminated because of trivial variability. Corn flavor was eliminated because only sample D had a suprathreshold intensity, and preliminary correlation analyses revealed that corn flavor had no significant impact on acceptance. The remaining 26 attributes were submitted to a PCA.

The first decision that must be made in the development of the perceptual map is how many dimensions (i.e., factors) to include on the map. Three criteria are used to make this decision. The first is to determine when there are no more large big drops in the magnitudes of the eigenvalues. Eigenvalues measure the amount of variability each dimension explains. If adding another dimension to the map does not substantially increase the amount of variability being explained, it may be time to stop. This criterion is best assessed graphically through the use of a scree plot. When the eigenvalues in the plot begin to flatten out, it is time to quit adding dimensions (see Figure 15.15). The second criterion is to stop once at least 75% of the total variability in the data has been explained. The third

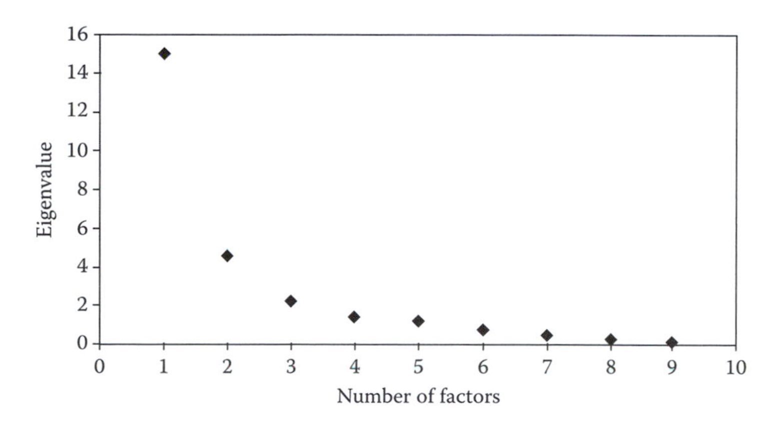

Figure 15.15 A scree plot of the eigenvalues from the PCA can help decide how many factors to include on the perceptual map.

criterion is to stop adding dimensions when the individual eigenvalues fall below 1.0. All standard PCA output includes these two pieces of information (see Table 15.4).

Examination of Figure 15.15 reveals that the eigenvalues begin to flatten out at the third dimension. Examination of Table 15.4 reveals that 75% of the variability is explained by just two dimensions and that the individual eigenvalues fall below 1.0 at six dimensions. It was decided to fit two-, three- and four-dimensional solutions and to examine the resulting factor loadings to see which made the most sense from the sensory and product points of view. After reviewing the results with the sensory analyst and the product developer, a three-dimensional solution was chosen.

The three dimensions can be interpreted by examining the factor loading of the attributes on each dimension. Factor loadings are similar to correlation coefficients. They range in value from –1 to +1. Values close to either extreme indicate a strong association between the attribute and the dimension. Small values (e.g., <±0.6) indicate a weak association. Focusing on the larger factor loadings helps in the interpretation of each dimension.

Examination of the factor loadings in Table 15.5 reveals that the first sensory dimension deals, in general, with the overall "wholesomeness" of the products. On the first dimension, the products range from those that are high in nonnatural meat flavor to those that are high in meat identity, with lots of large and firm meat, potato, and vegetable pieces and high intensities of green herb, carrot, and celery flavors.

The second sensory dimension is a combination of sauce appearance and texture and spiciness. Products range from those with thin and clear sauces to products with viscous sauces. The products with viscous sauces also tend to be high in spicy/black pepper character. For brevity, the second dimension will be called "Sauce: Clear to Viscous." The third sensory dimension clearly captures the differences in the perceived oiliness of the products.

This brings up an important point of caution regarding perceptual mapping. The dimensions that emerge on the map are based on the correlations that exist among the sensory attributes that are included in the analysis. Some of these correlations may be inherent in the product category. For example, it may be natural in this category for the thinner

Table 15.4 Summary of Eigenvalues and Explained Variability from the PCA

Factor	Eigenvalue	Variability Explained (%)	Cumulative (%)
1	14.98	58	58
2	4.59	17	75
3	2.22	9	84
4	1.41	5	89
5	1.23	5	94
6	0.76	3	97
7	0.44	2	99
8	0.24	1	100
9	0.12	0	100

Seventy-five percent of the variability is explained by the first two factors. Eigenvalues are greater than one up to factor 5.

Table 15.5 Factor Loadings of the Sensory Attributes Are Arranged in Decreasing Order by Factor

Attribute	Wholesomeness	Sauce: Clear to Viscous	Oiliness
FL meat identity	0.90	—	—
FL brothy	0.87	—	—
TXT meat firm	0.84	—	—
APP potato size	0.82	—	—
APP green herb AMT	0.80	—	—
FL carrot	0.79	—	—
APP solid size	0.79	—	—
TXT potato firm	0.78	—	—
FL celery	0.77	0.62	—
FL wheat	0.75	—	—
TXT vegetable firm	0.74	—	—
FL green herbs	0.72	0.64	—
FL meat nonnatural	−0.86	—	—
TXT viscosity	—	0.89	—
FL black pepper	—	0.89	—
FL sour	—	0.87	—
FL filler	—	0.86	—
FL spice blend	—	0.85	—
FL bitter	—	0.85	—
APP solid AMT	—	0.75	—
FL onion	0.65	0.70	—
APP sauce clarity	—	−0.92	—
APP surface oil	—	—	0.92
TXT oily MF	—	—	0.88
FL salty	—	—	—
APP potato AMT	—	—	—

Only loadings greater than ±0.6 are displayed. Factor loadings are similar to correlation coefficients. Values close to ±1.0 are important. The attributes with large loadings on a factor help to interpret the sensory variability that the factor is explaining.

sauces to have higher clarity and for the thicker sauces to be more translucent or opaque. However, some of the correlations may be strictly coincidental, resulting only from the specific set of products that were included in the analysis. For example, thick, translucent sauces do not have to be high in spicy and black pepper notes. When interpreting the sensory dimensions, it is important to distinguish between the relationships that are inherent to the product category and those that are coincidental to the products that were included in the study.

Naming the sensory dimensions has both good and bad effects. Names aid in interpretation and add a comforting level of familiarity to the results. Referring to the perceptual map in terms of DIM1, DIM2, and so on often turns off the less technical members of the project team, and consequently the information is not used as fully as it could be. However, names put boundaries on the dimensions. If an important attribute is not mentioned in the name of the dimension, users of the information may forget that it plays any role at all. To have the broadest appeal to both the technical and the nontechnical users of the perceptual mapping results, naming the dimensions is preferred. However, it is important to stress that the names are not comprehensive summaries of all of the attributes involved. A table like Table 15.5, in which the factor loadings (>±0.6) of the attributes have been replaced with their signs, is a good way to illustrate all of the attributes that play significant roles on the sensory dimensions.

The relationships among the attributes and the products can now be illustrated on the perceptual map. Figure 15.16 shows the relationships of the attributes and the test products on the first two sensory dimensions. It can be seen, for example, that product B is high in sauce clarity and low in sauce viscosity and spiciness, whereas products H and I are higher in sauce viscosity and spicy flavor and lower in sauce clarity. Products D, E, and F are high in nonnatural meat flavor and low in size and firmness of meat, potato, and

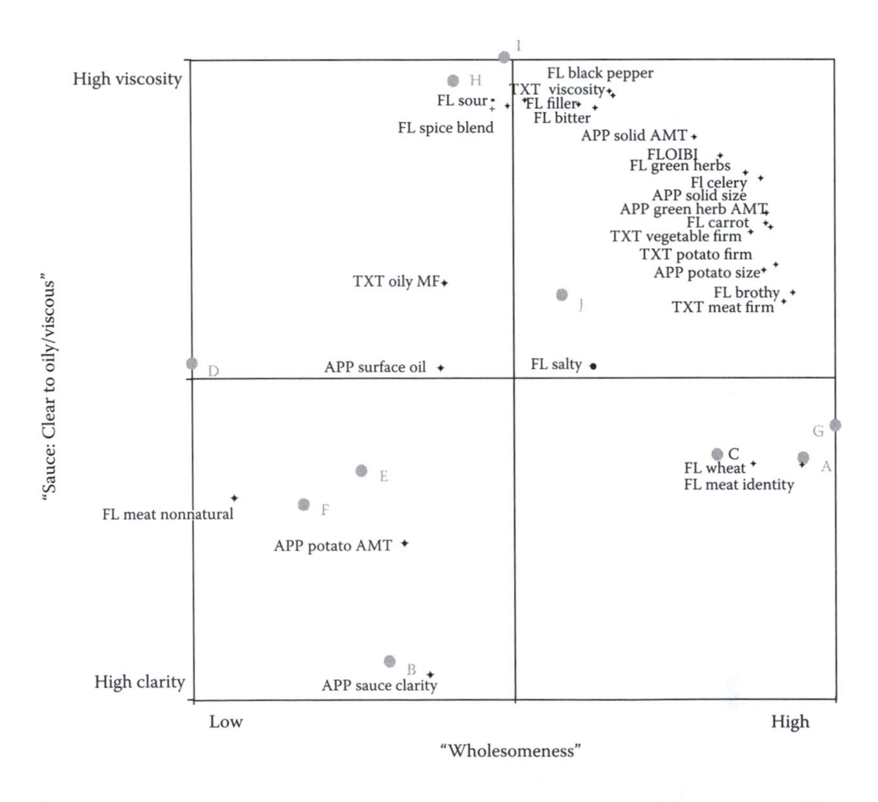

Figure 15.16 The perceptual map of the first two dimensions of the prepared meals study. Products A, C, and G are high in meat identity and wheat flavors. Product B is high in sauce clarity and low in viscous texture and in sour, bitter, spicy, and black pepper flavors. Products H and I have the opposite set of characteristics from product B.

vegetable pieces; and in vegetable flavors. Products A, C, and G are high in "meat identity" and wheat flavors, as well as size and firmness of meat, potato, and vegetable pieces; and vegetable flavors.

15.3.2.2 Identifying Preference Segments

Now that the perceptual map of the product space has been developed, the sensory information can be linked to consumer acceptance. In a typical preference mapping study, the acceptance ratings of the consumers are averaged across the total base of respondents as well as within various subgroups based on demographics and on attitudinal and usage patterns (e.g., by age or gender, among exercise enthusiasts or sedentary individuals, among heavy or light category users, etc.). Understanding the liking patterns in various demographic, attitudinal, and usage segments is important for positioning products and for identifying niche opportunities in defined markets. However, when looking into segments of these types, the unspoken assumption is that everyone in the segment has the same preferences for the products in the category. This may not be true and, as a result, the true preferences of individuals may be masked by averaging their liking ratings with those who have different product preferences. Therefore, before performing the analysis that links the sensory and consumer information, it is important to ensure that the average liking ratings of the products come from groups of consumers with similar preferences. Cluster analysis is the tool that is used to accomplish this task.

As discussed in the previous section, several methods are available to perform cluster analyses, and there are many variations within each of the major methods. In the analysis of the prepared meals data, hierarchical clustering using Ward's method was applied. The data that were analyzed were the overall liking ratings of the consumers. The respondents formed the rows (observations) of the data set; the products formed the columns (variables). The liking ratings were centered by subtracting each respondent's average liking rating from each of his or her individual ratings. Centering removes scale-usage effects from the raw data. For example, two respondents may have the same preferences for the test products, but one respondent tends to use the middle part of the scale (4, 5, and 6) to rate the products, whereas the other respondent uses the high end of the scale (7, 8, and 9) to rate the products. Centering the liking ratings allows the cluster analysis to group the respondents based on their patterns of liking ratings across the products rather than on their absolute levels.

The dendrogram in Figure 15.17 suggests that either two or four preference segments are present. The final decision on how many preference segments to include on the preference map needs to balance the internal homogeneity of the segments against their size. Averaging liking ratings from fewer than 25 to 30 respondents should be avoided because averages from such small groups lack precision. In the present example, two of the four segments have fewer than 25 respondents in them. For that reason, two preference segments were chosen.

The difference between the segments is illustrated in the graph of their average overall liking ratings (see Figure 15.18). Segment 1 exhibits a wide range of average ratings. Respondents in segment 1 like products A, C, and G the most and like products B, E, and F the least. Respondents in segment 2 exhibit a narrower range of liking ratings than those

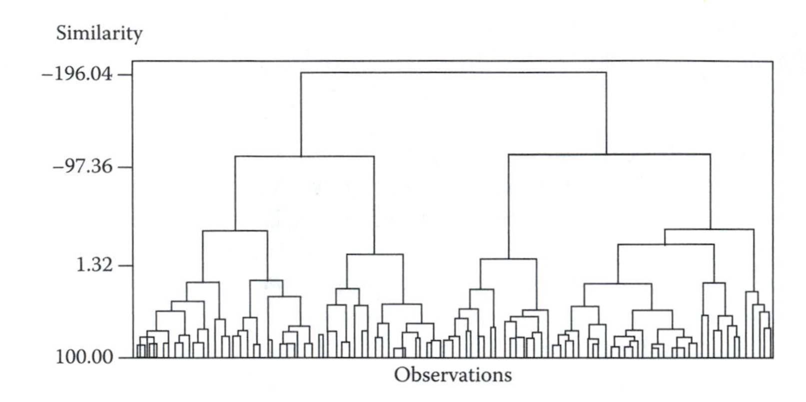

Figure 15.17 The dendrogram of the consumers' centered overall liking ratings suggests either two or four segments. The two segment solution was chosen because two of the segments in the four-segment solution have fewer than 25 respondents in them.

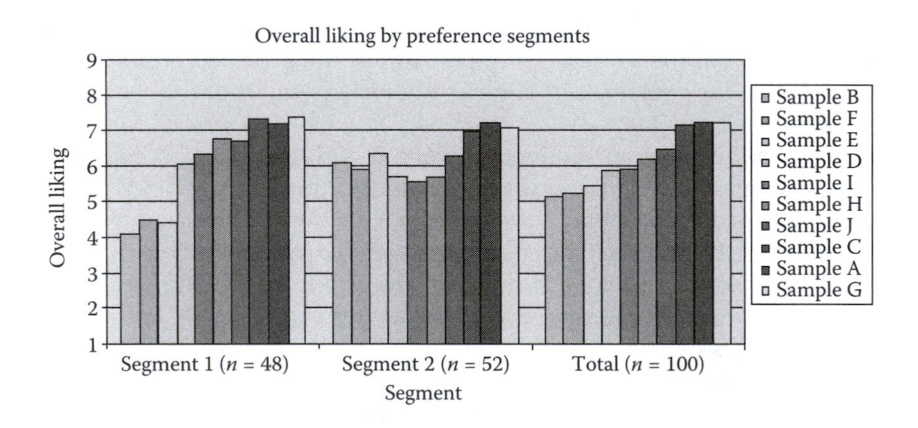

Figure 15.18 The average liking ratings of the products by preference segment reveals that both segments like products A, C, and G most. Segment 1 dislikes products B, E, and F. Segment 2 does not dislike any of the products but likes products D, H, and I the least.

in segment 1. Respondents in segment 2 also like products A, C, and G the most but, unlike segment 1, they like products D, H, and I the least.

15.3.2.3 From Perceptual Map to Preference Map

The final step in the development of an external preference map is to fit regression equations to the average overall liking ratings of the total respondent base and all of the consumer segments of interest in the study (demographic, attitudinal, usage, and preference segments). The independent variables (i.e., the predictors) are the factor scores of the test products obtained from the PCA. Both the linear and quadratic forms of the factor scores are included in the regression analysis. Including the quadratic terms in the regression model creates the opportunity to identify an intermediate point on the sensory dimension that is predicted to be more well-liked than either extreme. Because of this, regression models that include quadratic terms are called *ideal point* models. Regression models that include only the linear terms are called *vector models* because they can only point in the direction of increasing liking.

A variable-selection procedure such as stepwise regression or backward elimination is used to identify the sensory dimensions that have a significant relationship to overall liking. When the number of products in the test is sufficiently large, backward elimination is preferred to stepwise regression because it gives all of the predictors an equal chance of ending up in the final model (Anderson and Whitcomb, 2005).

The results of the regression analysis for the total respondent base are presented in Figure 15.19. Each line on the graph represents a sensory dimension that has a significant impact on overall liking. Steep lines have large impacts on liking; flatter lines have smaller impacts. Lines that slope up indicate that higher levels on the sensory dimension are more well-liked than lower levels. Conversely, lines that slope down indicate that lower levels are more well-liked. Curved lines indicate that an intermediate point is most well-liked. To generate a line on the graph, hold all but one of the sensory dimensions constant and plot the changes in overall liking that result from varying the remaining dimension from its low to its high level. Repeat the process for all of the significant sensory dimensions. Figure 15.19 reveals that the "wholesomeness" dimension has a strong positive impact on overall liking—higher levels are preferred. The "sauce: clear to viscous" dimension also has a significant impact on overall liking. A medium-high level of this dimension is most well-liked. The line for the "oiliness" dimension is flat, indicating that this dimension has no significant impact on liking.

The predicted liking ratings for any point on the preference map can be illustrated in a contour plot (discussed in Section 15.4) (see Figure 15.20). Also included in the figure is the convex hull formed by the test products. The convex hull represents the limits

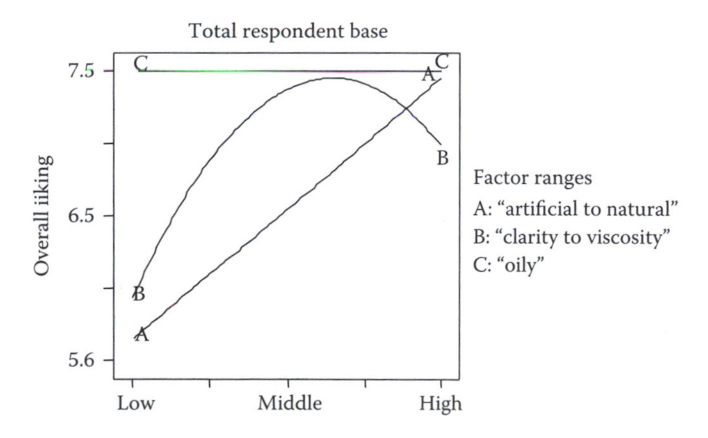

Figure 15.19 Perturbation chart of the key-drivers model for the total respondent base. "Wholesomeness" (A) and "Sauce: Clear to Viscous" (B) are both equally important to overall liking. "Oily" (C) does not have a significant impact on liking. The high level of "Wholesomeness" and the medium-high level of "Sauce: Clear to Viscous" are preferred.

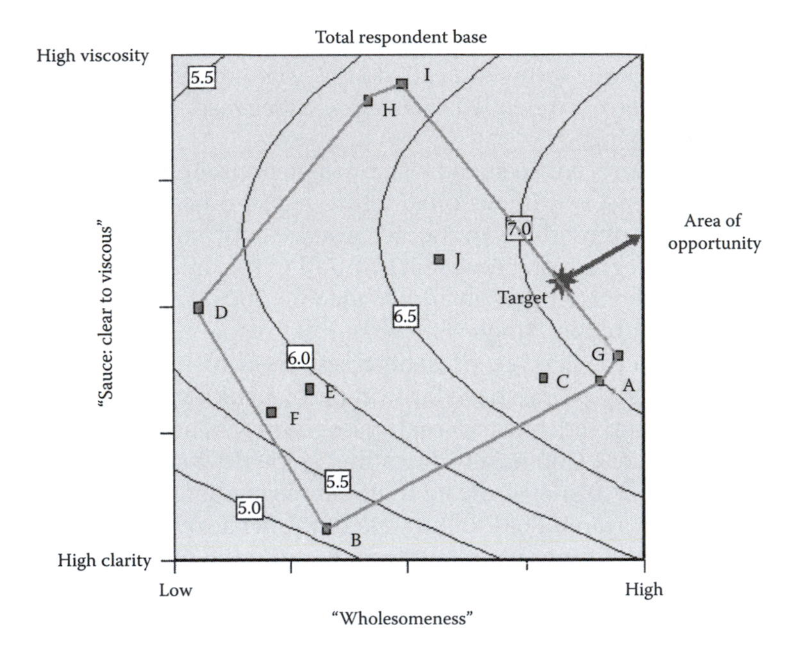

Figure 15.20 A contour plot of the first two dimensions of the preference map from the prepared meals study. Any point on a line is predicted to have an overall liking rating of the value indicated. Overall liking is maximized at high levels of "Wholesomeness" and medium-high levels of "Sauce: Clear to Viscous." To stay within the confines of the product space, the target product lies on the convex hull. The trends indicate the higher levels of both dimensions may be more well-liked.

of the product space. Predictions of points that fall outside of the product space are extrapolations and should be viewed with caution. Although the regression model may indicate that moving farther in a certain direction should have a positive impact on liking because there are no data for points outside of the product space, there is no way to tell how far the trend continues. The predicted values for points outside of the product space could be quite unreliable.

The point that is predicted to be most well-liked is identified in Figure 15.20 as the "target" product. When the target lies on the edge of the product space, as it does in this example, the direction of increasing liking that is indicated by the regression model can be denoted as an area of opportunity. No predicted liking values are given, but the analysis does indicate that this area may deserve some additional exploration.

15.3.2.4 Reverse Engineering the Profile of the Target Product

At the same time as the regression models for the overall liking data are developed, regression models that use the factor scores of the samples to predict the original sensory attributes are built. Only the linear terms are included in the models, and no variable-selection procedures are applied, so each sensory attribute is predicted by all of the sensory dimensions. The factor scores that correspond to the target product are plugged into the models

Table 15.6 Profile of Target Product Determined by Reverse Engineering

Sensory Attribute	Target Profile
FL meat identity	4.9
FL brothy	4.9
TXT meat firm	6.2
APP potato size	10.8
APP green herb AMT	6.1
APP solid size	8.5
TXT potato firm	4.3
FL carrot	3.0
FL celery	2.9
FL wheat	3.7
TXT vegetable firm	2.7
FL green herbs	3.3
FL meat nonnatural	2.3
TXT viscosity (sauce)	3.1
FL black pepper	1.7
FL spice blend	0.2
FL bitter	2.8
FL sour	1.7
FL filler	4.4
APP solid AMT (no potato)	7.1
FL onion	2.4
APP sauce clarity	4.1
APP surface oil	8.8
TXT oily MF	5.6
FL salty	7.4
APP potato AMT	7.4
FL corn	0.4
FL potato	0.0
FL red bell pepper	0.1
FL sweet	0.7

for the individual sensory attributes to obtain the sensory profile that is predicted to correspond to the target product (see Table 15.6).

15.3.2.5 External Preference Mapping of Individual Respondents

Another approach to external preference mapping is to fit individual regression models to each respondent's overall liking data. As in internal preference mapping, respondents with poor fitting models can be dropped from further analyses. For the remaining respondents, an action standard is defined to represent "satisfaction." For example, a respondent could be said to be satisfied with any point on the preference map with a predicted liking

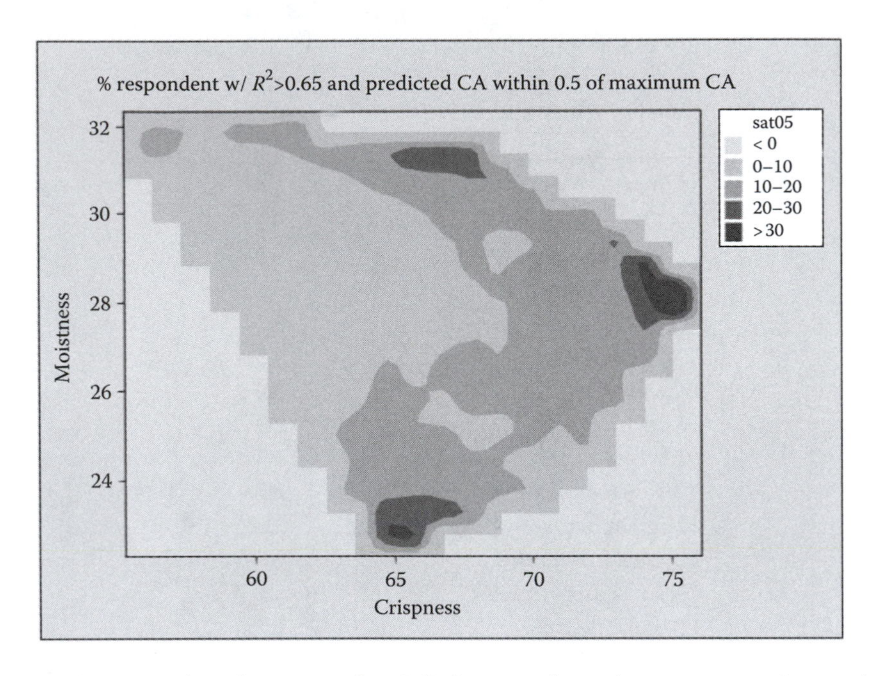

Figure 15.21 Contour plot of percent of satisfied respondents (among respondents whose individual models had an $R^2 > 0.65$). Note multiple regions of high satisfaction at mid crispness/low moistness, high crispness/mid moistness, and mid crispness/high moistness.

rating of 5.0 or more. Alternatively, a respondent could be said to be satisfied with any point on the map that has a predicted liking rating within 0.5 units of his or her maximum predicted liking rating. The percentage of respondents who are satisfied with every point on the preference map is then plotted on a contour plot such as in Figure 15.21. An advantage of this approach is that, like internal preference mapping, it is self-segmenting. Multiple target products can be identified as separate points on the map that correspond to areas of high satisfaction. An advantage that this approach does not share with internal preference mapping is that each respondent's data can be fitted using an ideal point model, so interior points on the map can be identified as the point of maximum overall liking. (Internal preference mapping is, in a sense, a vector model, in that it can only point in the direction of increasing liking.) The disadvantage of the method is that the models of the individual respondent's data are often poor. Either many respondents are dropped from the analysis or a very liberal definition of an acceptable model needs to be used to keep a large proportion of the respondents in the analysis.

15.3.3 Partial Least-Squares Mapping

PLS mapping is a direct application of partial least-squares regression, described in Section 15.2.2.3. When PLS is used as a preference-mapping tool, the dependent (y) variables in the PLS model are the overall liking ratings of the consumers and the independent (x) variables are the sensory attribute ratings. Because PLS can handle multiple dependent variables in the same model, the overall liking ratings of the total respondent base, as well as those of any consumer segments of interest, can be fitted in a single analysis. This is helpful for determining the similarities and differences in the attributes that drive liking among the segments.

The PLS map for the prepared meals data is presented in Figure 15.22. The overall liking ratings of the total respondent base, as well as those of the two preference segments presented in Figure 15.18, were the dependent variables in the PLS analysis. The same sensory attributes that were used in the external preference map were the independent variables in the PLS model. Products A, C, and G appear in the same quadrant as the points for consumer preference segments, segment 1 and segment 2, indicating that these products were well liked by both segments. Products B, E, and F fall on the opposite side of the map from segment 1, indicating that these products were not well liked by that segment of consumers. Conversely, products H and I fall on the opposite side of the map from segment 2, indicating that these products were the least liked among consumers in that segment.

The distance between the points for segment 1 and segment 2 in Figure 15.22 indicate different sensory attributes drive liking in the two segments. The sensory attributes that fall close to the point plotted for each segment are positive drivers for that segment. The attributes

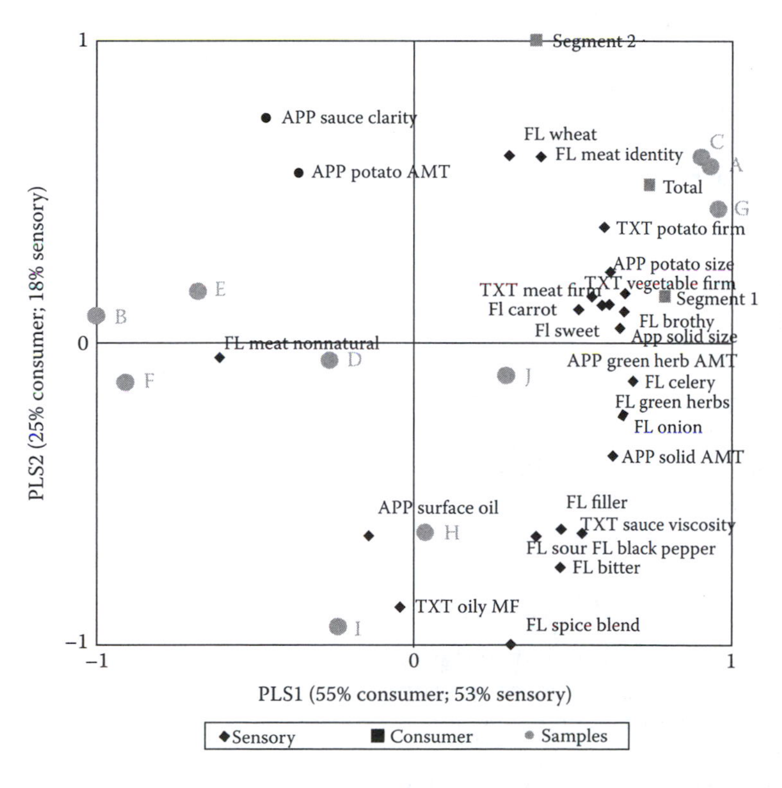

Figure 15.22 PLS map of prepared meals showing distribution of test products and difference between preference segments. Both segments like products A, C, and G. Segment 1 dislikes products B, E, and F. Segment 2 dislikes products H and I.

that fall on the opposite side of the map from the plotted point are negative drivers. Inspection of Figure 15.23a and b reveals the differences in the key drivers between the two segments. Segment 1 prefers products with lots of large and firm meat, potato, and vegetable pieces; high intensities of meat identity, green herb, carrot, onions, and celery flavors; and low intensity of nonnatural meat flavor. Segment 2, on the other hand, prefers lots of large firm potatoes in a clear, low viscosity sauce; high meat-identity flavor; low oiliness; and low spiciness (especially black pepper). These findings agree strongly with the results from both the internal and external preference maps.

In this way, the researcher can try different statistical methods to mine the data in an effort to confirm the results and look for any additional information about products and consumers.

Using different statistical methods to mine the data allows researchers to cross validate the primary results of a study and to uncover additional information about both products and consumers.

15.4 TREATMENT STRUCTURE OF AN EXPERIMENTAL DESIGN

In the experimental designs discussed in Chapter 14, the treatments (or products) were viewed as a set of qualitatively distinct objects having no particular association among themselves. Such designs are said to have a one-way treatment structure. One-way experiments commonly occur toward the end of a research program when the objective is to decide which product should be selected for further development.

In many experimental situations, however, the focus of the research is not on the specific samples but rather on the effects of some factor or factors that have been applied to the samples. For instance, a researcher may be interested in the effects that different flours and sugars have on the flavor and texture of a specific cake recipe, or he may be interested in the effects that cooking time and temperature have on the flavor and appearance of a prepared meat. In situations such as these, there are specific plans available that provide highly precise and comprehensive comparisons of the effects of the factors, while at the same time minimize the total amount of experimental material required to perform the study.

Two "multiway" treatment structures are discussed in this section. They are the factorial treatment structure (often called *factorial experiments*) and the response surface treatment structure (often called *response surface methodology* or *RSM*).

15.4.1 Factorial Treatment Structures

Researchers are often interested in studying the effects that two or more factors have on a set of responses. Factorial treatment structures are the most efficient way to perform such studies. In a factorial experiment, specific levels for each of several factors are defined. A single replication of a factorial experiment consists of all possible combinations of the levels of the factors. For example, a brewer may be interested in comparing the effects of two kettle boiling times on the hop aroma of his beer. Furthermore, if the brewer is currently using two varieties of hops, he may not be sure if the two varieties respond similarly to

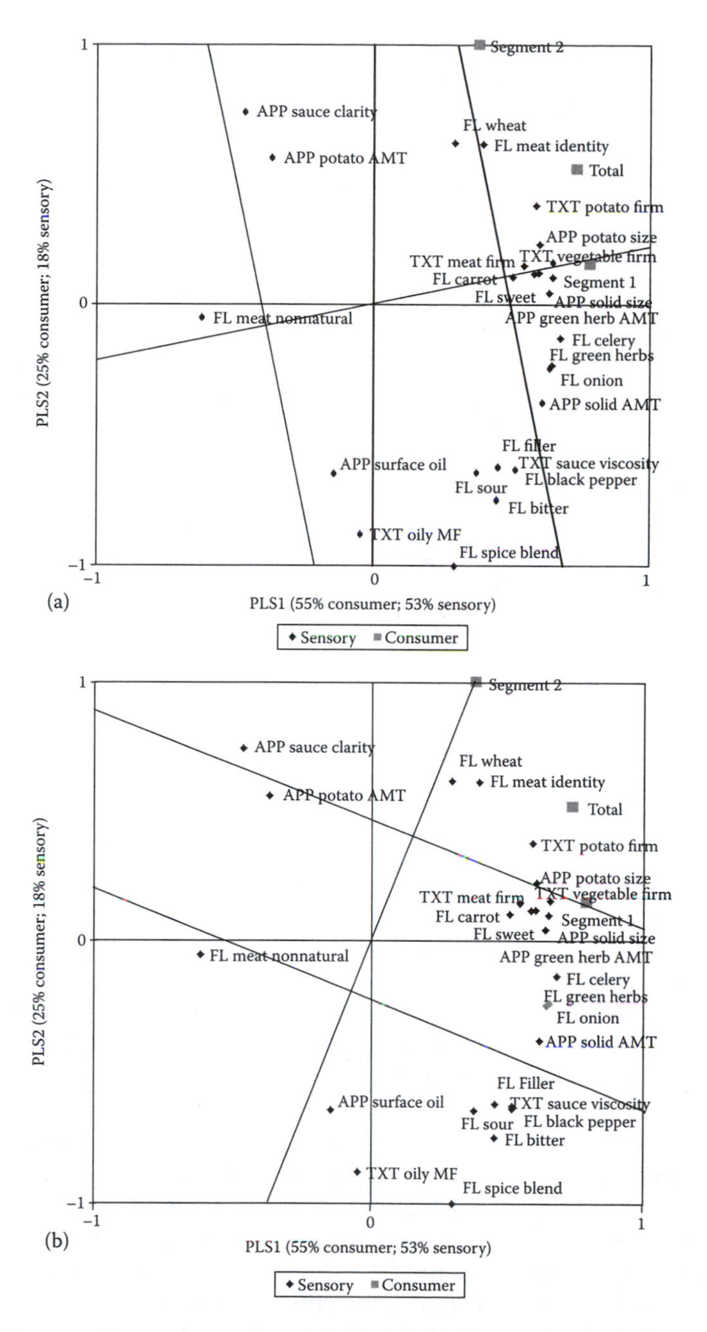

Figure 15.23 PLS maps illustrating positive and negative drivers by segment. Chart (a) illustrates that segment 1 prefers lots of large and firm meat, potato, and vegetable pieces; high intensities of meat identity and of green herb, carrots, onion, and celery flavors; and low intensity of nonnatural meat flavor. Chart (b) illustrates that segment 2 prefers lots of large firm potatoes in a clear, low viscosity sauce; high meat identity flavor; low oiliness; and low spiciness (especially black pepper).

Table 15.7 Factorial Treatment Structure for Two Factors Each Having Two Levels

		Hop Variety	
		A	**B**
Kettle boiling	Low (1)	$T_{1A} = 6$	$T_{1B} = 13$
Time	High (2)	$T_{2A} = 12$	$T_{2B} = 7$

changes in kettle boiling time. Combining the two levels of the first factor (kettle boiling time) with the two levels of the second factor (variety of hops) yields four distinct treatment combinations that form a single replication of a factorial experiment (see Table 15.7). The experimental variables in a factorial experiment may be quantitative (e.g., boiling time) or qualitative (e.g., variety of hops). Any combination of quantitative and qualitative factors may be run in the same factorial experiment.

An *effect* of a factor is the change (or difference) in the response that results from a change in the level of the factor. The effects of individual factors are called *main effects*. For example, if the entries in Table 15.7 represent the average hop aroma rating of the four beer samples, the main effect due to boiling time is

$$\frac{(T_{1A} - T_{2A}) + (T_{1B} - T_{2B})}{2}. \tag{15.12}$$

Similarly, the main effect due to variety of hops is

$$\frac{(T_{1A} - T_{1B}) + (T_{2A} - T_{2B})}{2}. \tag{15.13}$$

In some studies, the effect of one factor depends on the level of a second factor. When this occurs, there is said to be an *interaction* between the two factors. Suppose for the beer brewed with hop variety A that the hop aroma rating increased when the kettle boiling time was increased, but that hop aroma decreased for the same change in boiling time when the beer was brewed with hop variety B (see Table 15.7). There is an interaction between kettle boiling time and variety of hops because the effect of boiling time depends on which variety of hops is being used.

Graphs can be used to illustrate interactions. Figure 15.24a illustrates the interaction between boiling time and variety. The points on the graph are the average hop aroma ratings of the four experimental conditions presented in Table 15.8. The interaction between the two factors is indicated by the lack of parallelism between the two lines. If there were no interaction between the two factors, the lines would be nearly parallel (deviating only due to experimental error) as in Figure 15.24b. Researchers must be very cautious in interpreting main effects in the presence of interactions. Consider the data in Table 15.8 that illustrates the "boiling time by hop variety" interaction. Applying Equation 15.12 yields an estimated main effect due to boiling time of $[(6 - 12) + (13 - 7)]/2 = 0$, which indicates that boiling time has no effect. However, Figure 15.24a clearly shows that for each variety of hops there is a substantial effect due to boiling time. Because the separate variety effects are opposite, they cancel each other in calculating the main effect due to boiling time.

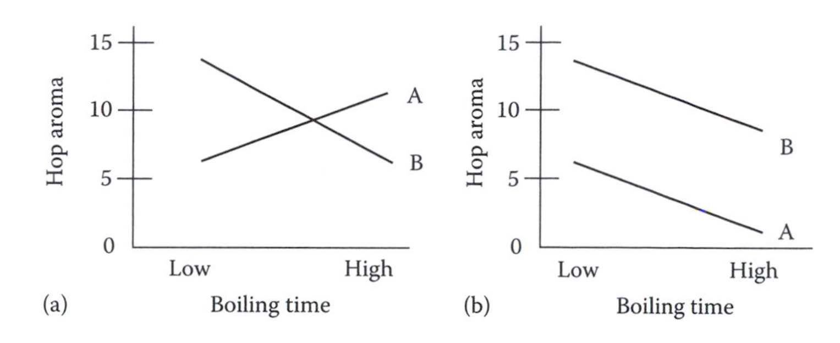

Figure 15.24 Plots of the mean hop aroma response illustrating (a) interaction and (b) no interaction between the factors in the study.

Table 15.8 ANOVA Table for a Factorial Experiment

Source of Variation	Degrees of Freedom	Sum of Squares	Mean Square	F
Total	$rab - 1$	SS_T		
A	$a - 1$	SS_A	$MS_A = SS_A/(a - 1)$	$F_A = MS_A/M_{SE}$
B	$b - 1$	SS_B	$MS_B = SS_B/(b - 1)$	$F_B = MS_B/M_{SE}$
AB	$df_{AB} = (a - 1)(- 1)$	SS_{AB}	$MS_{AB} = SS_{AB}/df_{AB}$	$F_{AB} = MS_{AB}/M_{SE}$
Error	$df_E = {}_{ab}(r - 1)$	SS_E	$MS_E = SS_E/df_E$	

Note: Factor A has a levels, factor B has b levels, and the entire experiment is replicated r times. The samples are prepared according to a completely randomized blocking structure.

In the presence of an interaction, the effect of one factor can only be meaningfully studied by holding the level of the second factor fixed.

Researchers sometimes use an alternative to factorial treatment structures, called one-at-a-time treatment structures, in the false belief that they are economizing the study. Suppose in the beer brewing example that the brewer had only prepared three samples: the low boiling time/variety A point T_{1A}, the low boiling time/variety B point T_{1B}, and the high boiling time/variety B point T_{2B}. (The high boiling time/variety A treatment combination T_{2A} is omitted.) Because only three samples are prepared, it would appear that the one-at-a-time approach is more economical than the full factorial approach. This is not true, however, if one considers the precision of the estimates of the main effects. Only one difference due to boiling time is available to estimate the main effect of boiling time in the one-at-a-time study (i.e., $T_{1B} - T_{2B}$). The same is true for the variety effect (i.e., $T_{1A} - T_{2B}$). Equations 15.12 and 15.13 show that, for the factorial treatment structure, two differences are available for estimating each effect. The entire one-at-a-time experiment would have to be replicated twice, yielding six experimental points, to obtain estimates of the main effects that are as precise as those obtained from the four points in the factorial experiment.

Another advantage that factorial treatment structures have over one-at-a-time experiments is the ability to detect interactions. If the high-temperature/variety A observation

$T_{2A} = 12$ were omitted from the data in Table 15.7 (as in the one-at-a-time study), one would observe that beer brewed at the high boiling time has less hop aroma than beer brewed at the low boiling time and that beer brewed with hop variety A has less hop aroma than beer brewed with hop variety B. The most obvious conclusion would be that beer brewed at the high boiling time using hop variety A would have the least hop aroma of all. The complete data in Table 15.7 and the plot of the interaction in Figure 15.24a show this would be an incorrect conclusion.

The recommended procedure for applying factorial treatment structures in sensory evaluation is as follows. Prepare at least two independent replications of the full factorial experiment. Submit the resulting samples for panel evaluation using the appropriate blocking structure as described in Chapter 14, Section 14.5. Take the mean responses from the analysis of the panel data and use them as raw data in an ANOVA. The output of the ANOVA includes tests for main effects and interactions among the experimental factors (see Table 15.8). This procedure avoids confusing the measurement error, obtained from the analysis of the panel data, with the true experimental error, which can only be obtained from the differences among the independently replicated treatment combinations.

15.4.2 Fractional Factorials and Screening Studies

Early in a research program, many variables are proposed as possibly having meaningful effects on the important responses. To execute an efficient research plan, experimenters need an approach that will allow them to screen out the influential variables from those that have little or no impact on the responses. This determination must be done with a minimum amount of work so that sufficient resources exist at the end of the program to do the necessary fine-tuning and "finishing" work on the final prototype. There is a class of experimental plans called *fractional factorials*, which allow researchers to screen for the effects of many variables simultaneously with a minimum number of experimental samples.

As the number of experimental variables grows in a factorial experiment, each main effect is estimated by an increasing number of "hidden replications." For example, as noted in the previous section, in a 2 × 2 (or 2^2) factorial, each main effect is estimated by two differences (i.e., two hidden replications). In a 2^6 factorial (i.e., six factors, each with two levels), the number of hidden replications for estimating each main effect has grown to 32. This may be excessive. A single replication of a 2^6 factorial consists of 64 experimental samples. If interest is primarily focused on identifying individual experimental variables with significant main effects, then the number of hidden replications could safely be reduced to 16 or even 8 without excessively sacrificing sensitivity. The number of experimental samples would be concurrently reduced to 32 or 16, thus yielding a manageable experiment. Figure 15.25 shows that the number of samples in a 2^3 factorial can be cut in half from eight to four while still providing two differences for estimating each main effect.

15.4.2.1 Constructing Fractional Factorials

Most screening studies are performed by selecting two levels, low and high, for each experimental variable. The various treatment combinations of low and high levels make up the experimental design; that is, the treatment combinations define the levels of the

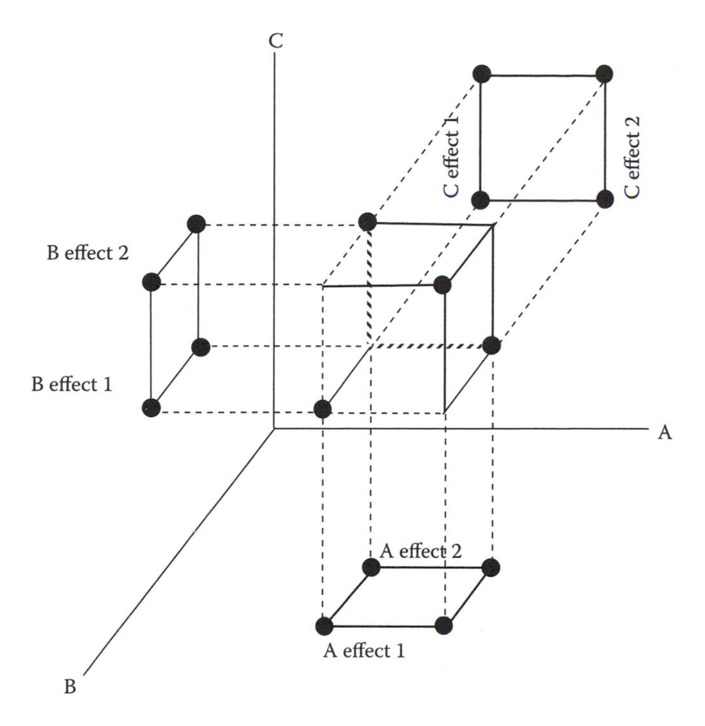

Figure 15.25 A graphical display of a ½-replicate fractional factorial of a 2^3 experiment showing by projection that two differences remain for estimating each main effect even though the total experiment has been reduced from eight to four samples.

experimental variables that should be used to produce each of the experimental samples. A convenient notation has been developed to identify the levels of the factors in each treatment combination. The high level of a variable A is denoted by the lower case *a*, the high level of B by *b*, and so on. The low level of a variable is denoted by the absence of the lower-case letter. For example, in a 2^3 factorial, the treatment combination high-A, high-B, high-C would be denoted as *abc*; the treatment combination low-A, high-B, high-C would be denoted as *bc*; and the treatment combination low-A, low-B, high-C would be denoted by *c*. The symbol used to represent the combination of all factors at their low levels is (1).

The eight treatment combinations that make up a single replication of a 2^3 factorial experiment are presented in the first column of Table 15.9. The remaining columns contain the signs of the coefficients that would be used to estimate each of the factorial effects. (The coefficients are either –1 or +1; therefore, only the sign is needed.) The treatment combinations are grouped by the sign of the coefficient for estimating the three-way interaction ABC. The two groups formed in this way are each ½-replications of a 2^3 factorial. Either of the two groups of four treatment combinations could be selected for use in a screening study. Cochran and Cox (1957) present plans for fractional factorial experiments for both 2^n and 3^n experiments where the number of factors, n, is as large as eight.

By choosing the treatment combinations that have the same sign for the coefficients of the three-way interaction, any ability to estimate the magnitude of this effect has been

Table 15.9 Factorial Effects in a 23 Factorial Experiment Arranged as Two ½-Replicate Fractional Factorial Experiments (ABC+ and ABC−)

Treatment Combination	Factorial Effects						
	A	B	C	AB	AC	BC	ABC
a	+	−	−	−	−	+	+
b	−	+	−	−	+	−	+
c	−	−	+	+	−	−	+
abc	+	+	+	+	+	+	+
(*I*)	−	−	−	+	+	+	−
ab	+	+	−	+	−	−	−
ac	+	−	+	−	+	−	−
bc	−	+	+	−	−	+	−

sacrificed. ABC is called the *defining contrast* because it is the criterion that was used to split the factorial into two ½-replications.

Notice in Table 15.9 that within each group there are two +s and two −s for estimating each effect. These are the hidden replications that remain even when only half of the full factorial is run. Suppose that the first group of four treatment combinations was selected to be run (i.e., the ABC+ group). Then the main effect of variable A would (apart from a divisor of 2) be estimated by

$$A = abc + a - b - c\,.$$

However, the estimate of the two-way interaction BC is also

$$BC = abc + a - b - c.$$

The main effect of A is said to be *aliased* with the two-way interaction BC, notationally denoted as A = BC. Similarly, B = AC and C = AB. In practical terms, if two factorial effects are aliased, then it is impossible to separate their individual impacts on the responses of interest. The apparent effect of A may be really due to A or due to BC, or possibly even due to a combination of the two.

The aliasing of main effects with interactions is the price paid for fractionalizing a factorial experiment. Typically, it is reasonable to assume that the magnitudes of the main effects are larger than the magnitudes of the interactions and, in such cases, fractional factorials can be used safely to screen for important experimental variables. If, however, large interactive effects are present, then a researcher may be misled into concluding that a variable has an important influence on the response when, in fact, it does not. This caution is not intended to frighten researchers away from using fractional factorials but rather only to make them aware of the issue, because it may serve to explain otherwise incongruous results that arise as a research program progresses.

15.4.2.2 Plackett–Burman Experiments

Fractional factorials are not the only plans that can be used to screen for influential variables. Plackett–Burman (1946) experiments are even more economical in the number of samples they require. The number of samples in a Plackett–Burman experiment is always a multiple of four. The number of experimental variables that can be screened with a Plackett–Burman experiment is, at most, one less than the number of samples (i.e., 4, 5, 6, or 7 variables can be screened with 8 samples; 8, 9, 10, or 11 variables can be screened with 12 samples; etc.) Box and Draper (1987) present the construction of Plackett–Burman experiments covering the range from 4 to 27 experimental factors (i.e., for studies involving 8–28 experimental samples).

15.4.2.3 Computer-Aided Optimal Fractional Designs

Another approach to generating fractional factorial designs is to use a computer-aided design of experiments (DOE) program (Carr, 2010). Many popular statistical packages, including Design Expert, JMP, Minitab, and SAS, have modules that perform this function. With the computer-aided approach, researchers first specify a candidate set of experimental samples (e.g., the full factorial design for the factors and levels they want to study). Next, researchers specify the statistical model that they want to fit to the data (e.g., all of the main effects of the experimental variables and, possibly, some interactions of interest). Lastly, the researchers specify the number, n, of experimental samples to include in the design. The number must be greater than the number of terms in the statistical model specified in the previous step. The computer-aided DOE program will then select the subset of size n samples from the candidate set that will do the "best" job of fitting the specified model to the data. A variety of optimality criteria can be applied to determine the "best" subset of samples (see Carr, 2010), D-Optimal being among the most widely used.

Computer-aided optimal DOE has several advantages over the traditional approaches to building fractional designs. Optimal designs save resources because they often require fewer samples than the traditional approaches. Optimal designs are more flexible than traditional approaches. Researchers can choose different numbers of levels for each experimental variable, as opposed to being restricted to, for example, two levels in the standard fractional factorial approach. Researchers can exclude infeasible samples from the candidate set so that only samples that can actually be produced will be included in the final design. Also, researchers have greater flexibility in specifying the effects they want to include in the statistical model that will be used in the analysis. The model can include linear and quadratic main effects for some or all of the experimental variables and specific interactive effects among the experimental variables.

15.4.2.4 Analysis of Screening Studies

Both fractional factorials and Plackett–Burman experiments can be analyzed by ANOVA. However, because of the small number of samples involved, it is sometimes impossible to compute F-ratios to test for the significance of the effects. This happens because in some screening experiments there are no degrees of freedom available for estimating experimental error. Regardless, even when the ANOVA computations can be performed, the tests are not very sensitive, so that the possibility of missing a real effect (i.e., a type II error) is relatively high.

A graphical technique for analyzing screening experiments allows the researcher more input into the decisions on which variables are affecting the response. The technique is motivated by the logic that if none of the variables have an impact on the response, then the values of their estimated effects are actually just random observations from a distribution (assumed to be normal) with a mean of zero. If these estimated effects are plotted against their corresponding normal random deviates, they should form a straight line (in the absence of any real effects). If, however, some of the variables affect the response, then the estimated effects are more than random observations. Real effects will fall off the line in the plot, either high and to the right (for positive effects) or low and to the left (for negative effects). The researcher can examine the *normal probability plot* of the estimated effects, such as that presented in Figure 15.26, to decide which variables actually affect the response.

Constructing a normal probability plot is a four-step process:

1. Estimate the effects of the experimental variables using ANOVA.
2. Rank the estimated effects in increasing order from $i = 1$ to n.

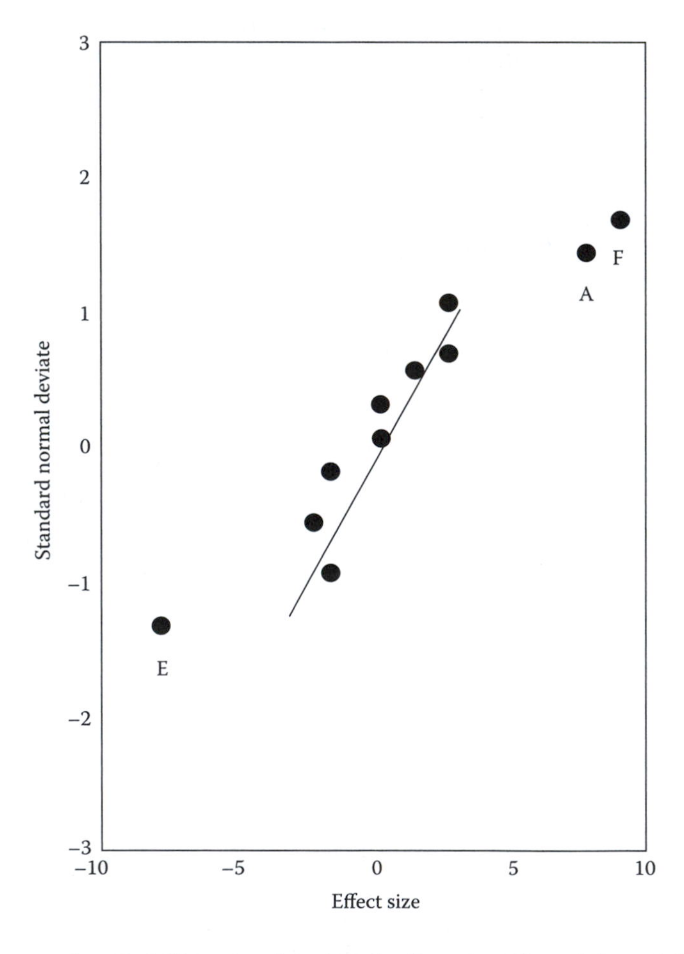

Figure 15.26 A normal probability plot showing the "nonsignificant" factorial effects falling on the line and the "significant" effects falling high and to the right, and low and to the left.

3. Pair the ordered estimates with the new variable $z = \Phi^{-1}[p]$, where $p = (3i - 1)/(3n + 1)$ and Φ^{-1} is the inverse of the standard normal distribution function. Many statistical computer packages contain a function for computing the value of z from p (sometimes called PROBIT).
4. Plot z versus the estimated effects using a standard plotting routine, fit (by eyeball) a straight line to the data, and look for points that fall high and to the right or low and to the left. These identify the "significant" variables.

15.4.3 Conjoint Analysis

Conjoint analysis is an especially useful application of fractional designs. Conjoint analysis routinely uses highly fractionated experimental designs, called orthogonal arrays, which contain the minimum number of experimental samples that will still allow the main effects of the experimental variables to be estimated. Conjoint analysis is especially useful for optimizing concepts and for generating models that permit market simulations. Because of the large number of experimental samples involved in a typical conjoint design, conjoint analysis is not widely used in product testing, but it can provide product developers with valuable preliminary information on features that consumers want in a product before prototype development begins.

In traditional conjoint analysis, respondents are presented with a description of a product. The description is comprised of a particular combination of levels of all of the experimental variables being studied in the design. The combination is called a product profile, and the method is called full profile conjoint analysis. Each respondent evaluates all of the product profiles included in the orthogonal array, typically assigning one rating related to purchase intent and another rating related to the believability of the profile. Each respondent's data are analyzed separately to estimate the impact that each level of each of the experimental variables has on the respondent's purchase intent and believability ratings. The impacts of the variable levels are called part-worths or partial utilities, and the sum of the part-worths of all of the components of the product profile is called the total utility of the profile. In conjoint analysis, it is assumed that a person will select the product with the highest total utility from a set of competing options. Since a separate model is developed for each respondent, market simulations can be run to estimate the market share each of a competing set of virtual products would be expected to garner. The components of the virtual products can be manipulated to determine the combination of features that will produce the highest market share against products already in the market (to the extent that the existing products can be accurately described by the levels of the experimental variables included in the study).

The number of profiles in a traditional conjoint design grows rapidly as the number of experimental variables and their levels increase. Also, traditional conjoint analysis has been criticized for being unrealistic because consumers do not actual assign ratings to individual profiles when making purchase decisions. Several variations on traditional conjoint analysis have been developed to address this shortcoming. For example, adaptive conjoint analysis was developed to reduce the number of profiles and the number of variables in each profile that a respondent needs to evaluate in the study. In adaptive conjoint analysis, each respondent provides information on the importance they place on each of the experimental variables in the study (called self-explicated information). This information, along with a computer-aided survey tool, is used to develop a custom set of profiles for each respondent to evaluate. The computerized survey tool uses the self-explicated information and the respondent's responses to early profiles to adapt the profiles each respondent evaluates in order to develop precise estimates of the partial utilities in as few evaluations as possible. Another variation on traditional conjoint analysis, called choice-based or discrete-choice conjoint, replaces the rating task with a choice task in which each respondent is presented with two or more profiles and is asked to select the one they would most likely buy. Proponents of the approach claim that the data collected more accurately represent a consumer's decision process when choosing a product to buy. The task can be extended to have the respondent also select the profile they would be least likely to buy. This approach, called best-worst or MaxDiff scaling, provides more information to estimate partial utilities than is available from the "pick the winner" approach. Both adaptive conjoint and choice-based methods create the opportunity to test larger numbers of experimental variables with more levels within each variable, providing more complete and realistic product profiles to be tested.

With the advent of hierarchical Bayesian estimation techniques, partial utilities can be obtained at the respondent level of adaptive and choice-based studies, so once again, market simulations and consumer segmentation analyses can be performed based on data obtained from the more economical methods.

15.4.4 Response Surface Methodology

The treatment structure known as response surface methodology (RSM) is essentially a designed regression analysis (see Montgomery, 1976; Giovanni, 1983). Unlike factorial treatment structures, where the objective is to determine if (and how) the factors influence the response, the objective of an RSM experiment is to predict the value of a response variable (called the *dependent variable*) based on the controlled values of the experimental factors (called *independent variables*). All of the factors in an RSM experiment must be quantitative.

RSM treatment structures provide an economical way to predict the value of one or more responses over a range of values of the independent variables. A set of samples (i.e., experimental points) is prepared under the conditions specified by the selected RSM treatment structure. The samples are analyzed by a sensory panel, and the resulting average responses are submitted to a stepwise regression analysis. The analysis yields a predictive equation that relates the value of the response(s) to the values of the independent variables. The predictive equation can be depicted graphically in a response surface plot as shown in Figure 15.27. Alternatively, the predicted relationship can be displayed in a *contour plot* as in Figure 15.28. Contour plots are easy to interpret. They allow the researcher to determine the predicted value of the response at any point inside the experimental region without requiring that a sample be prepared at that point.

Several classes of treatment structures can be used as RSM experiments. The most widely used class, discussed here, is very similar to a factorial experiment. One part of the plan consists of all possible combinations of the low and high levels of independent variables. (In a two-factor RSM experiment, this portion consists of the four points: [low, low], [low, high], [high, low], and [high, high].) This factorial portion of the RSM experiment is

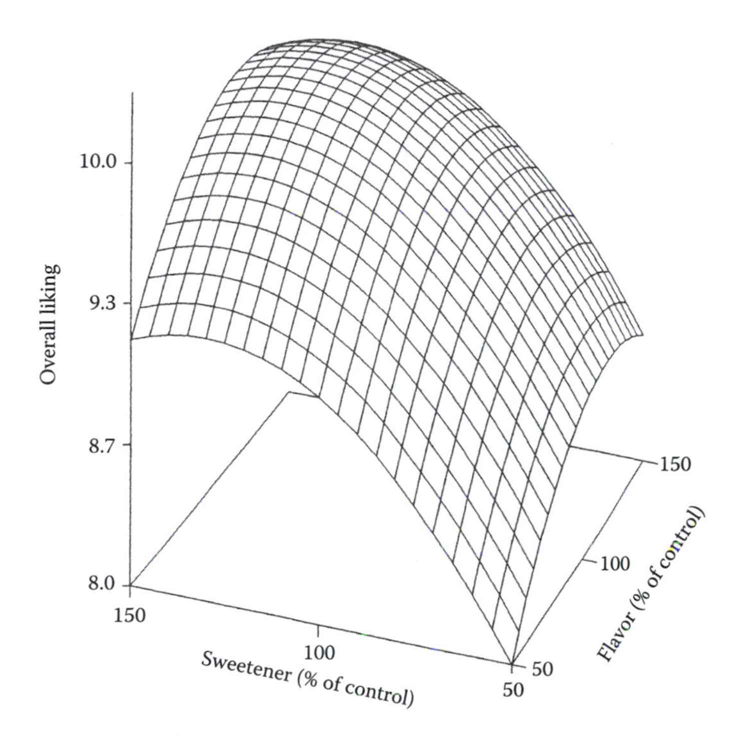

Figure 15.27 A response surface plot showing the predicted relationship between overall liking and the levels of sweetener and flavor in the product.

augmented by a center point (i.e., the point where all of the factors take on their average values, [low+high]/2). Typically, the center point is replicated several times (not less than three) to provide an independent estimate of experimental error (see Figure 15.29). The regular practice in an RSM experiment is to assign the low levels of all the factors the coded value of –1, the high levels the coded value of +1, and the center point the coded value of zero.

The treatment structure of an RSM experiment depicted in Figure 15.29 is called a *first-order RSM experiment*. The full regression equation that can be fit by the treatment structure has the form

$$y = \beta_0 + \beta_1 x_1 + \beta_2 x_2 + \cdots + \beta_k x_k, \tag{15.14}$$

where β_i is the coefficient of the regression equation to be estimated and x_i is the coded level of the *k*-factors in the experiment. First-order RSM experiments are used to identify general trends and to determine if the correct ranges have been selected for the independent variables. The first-order models are used early in a research program to identify the direction in which to shift the levels of the independent variables to affect a desirable change in the dependent variable (e.g., increase desirable response or decrease undesirable response).

First-order models may not be able to adequately predict the response if there is a complex relationship between the dependent variable and the independent variables.

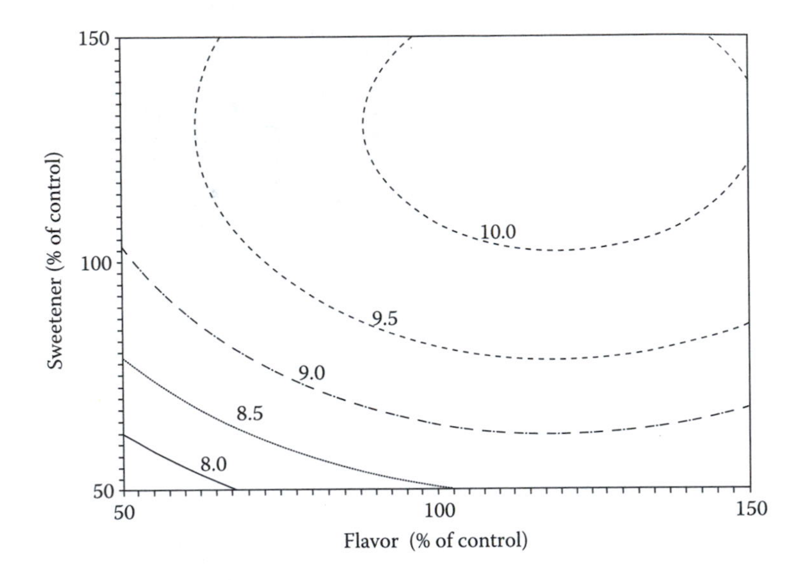

Figure 15.28 A contour plot of the predicted relationship between overall liking and the levels of sweetener and flavor in the product. Contour plots provide a quantitative assessment of the sensitivity of the product to changes in the levels of the ingredients.

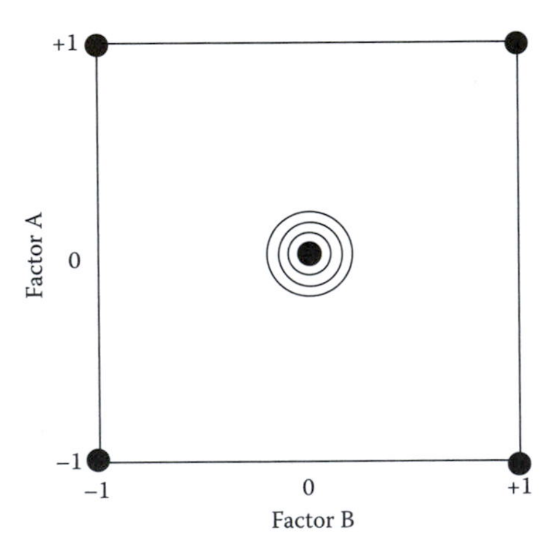

Figure 15.29 A two-factor, first-order RSM experiment. The figure illustrates the arrangement of the factorial and center points in an RSM experiment with two independent variables that permit estimation of a first-order regression model in Equation 15.14.

A second-order RSM treatment structure is required for these situations. The full regression model that can be fit to a second-order RSM treatment structure has the form

$$y = \beta_0 + \beta_1 x_1 + \beta_2 x_2 + \cdots + \beta_k x_k + \beta_{11} x_1^2 + \beta_{22} x_2^2 + \cdots + \beta_{kk} x_k^2 + \beta_{12} x_1 x_2 + \beta_{13} x_1 x_3 + \cdots + \beta_{k-1,k} x_{k-1} x_k. \quad (15.15)$$

The addition of the squared and cross-product terms in the model allows the predicted response surface to "bend" and "flex," resulting in an improved prediction of complex relationships.

A popular class of second-order RSM experiments is the central-composite, rotatable treatment structures. Central-composite experiments are developed by adding a set of axial or "star" points to a first-order RSM treatment structure (see Figure 15.30). There are $2k$ axial points in a k-factor RSM experiment. Using the normal −1, 0, +1 coding for the factor levels, the axial points are $(\pm\alpha, 0, \ldots, 0)$, $(0, \pm\alpha, 0, \ldots, 0)$, …, $(0, 0, \ldots, 0, \pm\alpha)$, where α is the distance from the axial point to the center of the experimental region (i.e., the center point). The value of α is $(F)^{1/4}$, where F is the number of noncenter factorial points in the first-order experiment. For example, in a two-factor experiment, $F = 4$ and $\alpha = (4)^{1/4} = 1.414$.

Second-order RSM models have several advantages over first-order models. As mentioned before, the second-order models are better able to fit complex relationships between the dependent variable and the independent variables. In addition, second-order models can be used to locate the predicted maximum or minimum value of a response in terms of the levels of the independent variables.

As mentioned in Section 15.4.2.3, computer-aided "optimal" DOE programs are available that will let researchers selected the test samples to include in their RSM design that will do the "best" job of fitting the RSM model to the data that result from the experiment. Researchers have the flexibility of specifying which linear and quadratic effects of the individual experimental variables to include in the model as well as the interactions among several experimental variables. The optimal designs often contain fewer experimental samples than the traditional RSM designs.

The recommended procedure for performing an RSM experiment is as follows (also see Carr, 1989): First, the experimental samples should be prepared according to the RSM plan. Second, perform a regular BIB analysis of the samples from the RSM treatment structure, ignoring the association among the samples. (The balanced incomplete block [BIB] blocking

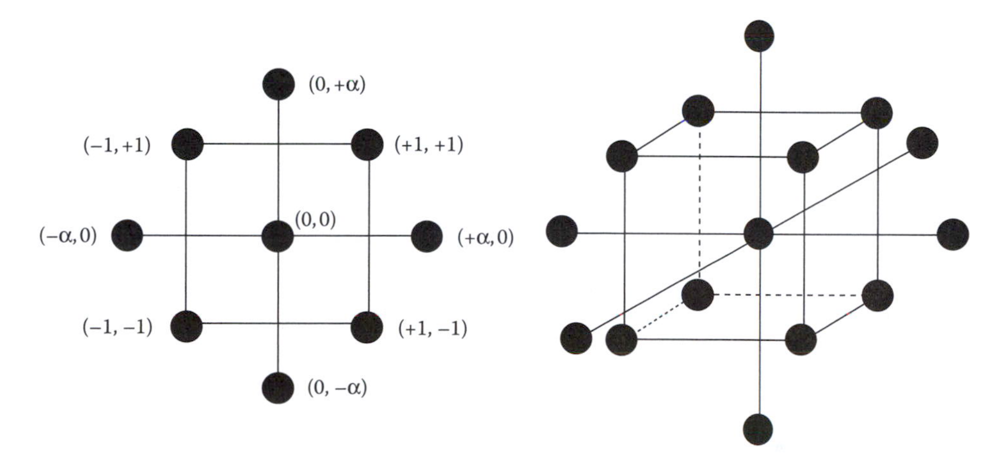

Figure 15.30 Central composite RSM experiments. The figures illustrate the arrangement of the factorial, axial, and center points in an RSM experiment with two and three independent variables that permit estimation of a second-order regression model as in Equation 15.15.

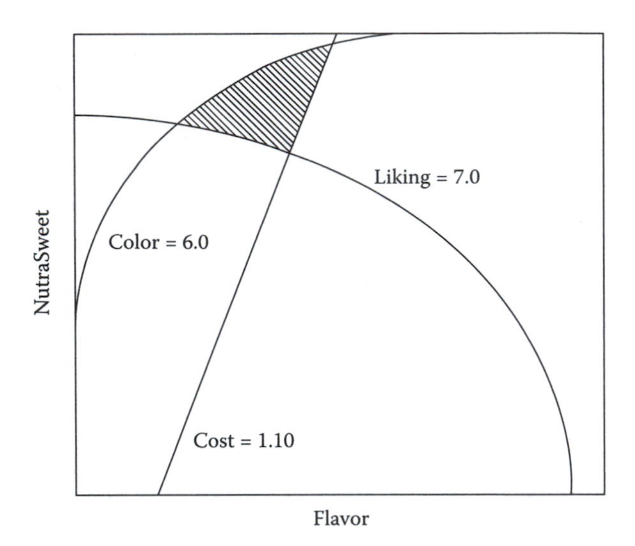

Figure 15.31 Overlaid contour plots of the critical limits for several response variables showing the region of formula levels predicted to satisfy all of the constraints simultaneously.

structure is suggested because there are normally too many samples in an RSM experiment to evaluate at one sitting. If, however, it is possible to evaluate all of the samples together, then a randomized [complete] block design can be used.) The only output of interest from the BIB analysis is the set of adjusted sample means. The significance (or lack of significance) of the overall test statistic is of no interest. Next, submit the sample means to a stepwise regression analysis to develop the predictive equation that relates the value of the response to the levels of the experimental factors. The predictive equation is then used to generate a contour plot that provides a graphical depiction of the effects of the factors on the response. If there is only one response, the region where the response takes on acceptable values (or attains a minimum or maximum value) is apparent in the contour plot. When several responses are being considered, the individual contour plots can be overlaid. Hopefully, a region where all of the responses take on acceptable values can be identified as in Figure 15.31.

REFERENCES

Anderson, M. J. and P. J. Whitcomb (2005). *RSM Simplified: Optimizing Processes Using Response Surface Methods for Design of Experiments*. Connecticut: Productivity.

Bezdek. J. (1981). *Pattern Recognition with Fuzzy Objective Function Algorithms*. New York: Plenum Press.

Box, G. E. P. and N. R. Draper. (1987). *Empirical Model Building and Response Surfaces*. New York: Wiley.

Carr, B. T. (1989). An integrated system for consumer-guided product optimization. In L. S. Wu (ed.), *Product Testing with Consumers for Research Guidance* (pp. 41–53). Philadelphia, PA: ASTM International.

Carr, B. T. (2010). Statistical design of experiments in the 21st century and implications for consumer product testing. In S. R. Jaeger and H. MacFie (eds.), *Consumer-Driven Innovation in Food and Personal Care Products* (pp. 427–69). Cambridge: Woodhead Publishing Limited.

Cattell, R. B. (1966). The screen test for the number of factors. *Multivariate Behavioral Research* 1: 245–76.
Cochran, W. G. and G. M. Cox (1957). *Experimental Designs* (2nd edn.). New York: Wiley.
Cooper, H. R., M. D. Earle, and C. M. Triggs (1989). Ratio of ideals—A new twist on an old idea. In L. S. Wu (ed.), *Product Testing with Consumers for Research Guidance* (pp. 54–63). Philadelphia, PA: ASTM International.
Draper, N. and H. Smith (1981). *Applied Regression Analysis* (2nd edn.). New York: Wiley.
Giovanni, M. (1983). Response surface methodology and product optimization. *Food Technol* 37(11): 41–3.
Godwin, D. R., R. E. Bargmann, and J. J. Powers (1978). Use of cluster analysis to evaluate sensory-objective relations of processed green beans. *J Food Sci* 43: 1229–30.
Jacobsen, T., and R. W. Gunderson (1986). Applied cluster analysis. In J.R. Piggott (ed.), *Statistical Procedures in Food Research* (pp. 361–408). Essex, UK: Elsevier Science.
Kruskal, J. B. and M. Wish (1978). *Multidimensional Scaling*. Newbury Park, CA: Sage.
MacQueen, J. B. (1967). Some methods for classification and analysis of multivariate observations. In *Proceedings of the Fifth Berkeley Symposium on Mathematical Statistics Probability* (Vol. 1, pp. 281–97). Berkeley: University of California Press.
Mallows, C. L. (1973). Some comments on cp. *Technometrics* 15: 661–5.
Martens, M. and H. Martens (1986). Partial least squares regression. In J. R. Piggott (ed.), *Statistical Procedures in Food Research* (pp. 293–359). Essex, UK: Elsevier Science.
McEwan, J. A., P. J. Earthy, and C. Ducher (1998). *Preference Mapping: A Review*. Gloucestershire, UK: Campden & Chorleywood Food Research Association; (Review, No. 6).
Montgomery, D. C. (1976). *Design and Analysis of Experiments*. New York: Wiley.
Muñoz, A. M. and E. Chambers IV (1993). Relating sensory measurements to consumer acceptance of meat products. *Food Technol* 47(11): 128–31, 134.
Piggott, J. R. and K. Sharman (1986). Methods to aid interpretation of multivariate data. In J. R. Piggott (ed.), *Statistical Procedures in Food Research* (pp. 181–232). Essex, UK: Elsevier Science.
Pirouette (2006). *Comprehensive Chemometrics Modeling Software*. Bothell, WA: Infometrix.
Plackett, R. L. and J. P. Burman (1946). The design of optimum multifactor experiments. *Biometika* 33: 305–25.
Popper, R., H. Heymann, and F. Rossi (1997). Three multivariate approaches to relating consumer to descriptive data. In A.nM. Muñoz (ed.), *Relating Consumer, Descriptive and Laboratory Data to Better Understand Consumer Responses, ASTM Manual 30* (pp. 39–61). Philadelphia, PA: ASTM International (American Society for Testing and Materials).
Powers, J. J. (1998). Descriptive methods of analysis. In J. R. Piggott (ed.), *Sensory Analysis of Foods* (2nd edn., pp. 187–256). Essex, UK: Elsevier Science.
Powers, J. J. and G. O. Ware (1986). Discriminant analysis. In J. R. Piggott (ed.), *Statistical Procedures in Food Research* (pp. 125–80). Essex, UK: Elsevier Science.
SAS (1989). *SAS/STAT User's Guide, Version 6* (Vol. 1, 4th edn). Cary, NC: The SAS Institute.
SAS (2004). SAS OnlineDoc® 9.1.3. Cary, NC: SAS Institute.
SAS (2005). SAS/STAT User's Guide 9.1.3. Cary, NC: SAS Institute.
Unscrambler. 2006. *Multivariate Statistical and Analytical Software*. Trondheim, Norway: CAMO.

16

Guidelines for Reporting Results

16.1 INTRODUCTION

16.1.1 Rationale

- Reports are a written record or documentation of the work performed. In that respect, they should have sufficient information to enable one to replicate the work done, capture the findings so they can be referred to, and add to the database of knowledge.
- Reports are a tool to disseminate information. The author must consider the intended audience; reports that cannot be interpreted waste the time of the reader. A major challenge for today's sensory analyst is to reach multiple audiences simultaneously.
- Reports are a tool to enable decision making, whether they are in the form of an article demonstrating a new sensory method, a report designed to offer strategic guidance for product development, or an informational tool to understand consumers and highlight opportunities for innovative new products.

16.1.2 Qualities of a Good Report

For the user of a sensory report, the most important considerations are how much confidence he or she can place in the report and whether or not the important information can be extracted. Many factors determine this (Larmond, 1981):

1. Reliability: Would similar results be obtained if the test were repeated with the same panelists? With different panelists?
2. Validity: How valid are the conclusions? Did the test measure what it was intended to measure?
3. Clarity: Are the results communicated? Is the format appropriate for the audience?

Because of the many opportunities for variability and bias resulting from the use of human subjects, reports of sensory tests must contain more detail than reports of physical or chemical measurements. The amount of information provided and the format it takes depends on the report's audience. There are three main types of reports:

- Technical reports
  This style of report is often used between sensory professionals and product developers within the same company. The conciseness of the report is ideal for busy professionals. The level of detail provided often assumes prior familiarity with the project objective or the product category.
- Business reports
  Business reports are written as an exchange between the sensory professional and a wider business audience. They contain the methodological and analytical details of a business technical report, but they are often structured differently. The summary information and conclusions may be given first and are then are followed by the technical details.
- Journal articles (intended for peers in the field of sensory science)
  Journal articles are the most formal reporting style. The minutia of the organization and style of the report are determined by the journal that publishes the article. Since these reports are intended for professionals across the field, they include comparisons between the current experiment and other published work. The high likelihood of readers having no familiarity with the work supports the convention of including extended background information and justifications for the experiment in the introduction.

16.2 ANATOMY OF THE REPORT

The recommendations below are based on Prell (1976), the Sensory Evaluation Division of the Institute of Food Technologists (1981), and industry practice. Applications of the suggested guidelines are illustrated in the examples at the end of this chapter.

16.2.1 Part 1: Summary or Abstract

What did the test teach us? It is an important courtesy to the user not to oblige him or her to hunt through pages of text in order to discover the essence of the results. The conclusion is obvious to the sensory analyst, and he or she should state it briefly and concisely in the opening summary or abstract.

The summary for business technical reports and journal articles should answer the four "whats":

- What was the objective?
- What was done?
- What were the results?
- What can be concluded?

Additionally, the following questions are gaining popularity for adding recommendations for future action to the executive summary or abstract:

- So what?
- Now what?

16.2.2 Part 2: Objectives and Introduction

As reiterated many times in this book, clearly written descriptions of the project objective and the test objective are fundamental to the success of any sensory experiment (see Sections 6.2 and 6.3).

16.2.3 Part 3: Materials and Methods

The "Materials and Methods" section for all report types should provide sufficient detail to allow the work to be repeated. Accepted methods should be cited by adequate references.

Journal articles, technical reports, and business reports should have the following components, although the level of detail and order in which they are presented may change depending on the needs of the audience.

1. Experimental design: Assuming that the objective was clearly stated previously, the text should now explain the "layout" of the experiment in terms of the objective. If there are major and minor objectives, the report should show how this is reflected in the design. If an advanced design is used (randomized complete block, balanced incomplete block, Latin square, etc.), it should be stated. Next, state the measurements made (e.g., sensory, physical, chemical), sample variables and level of the variables (where appropriate), number of replications, and limitations of the design (e.g., lots available for sampling, nature and number of samples evaluated in a test session). Include the efforts made to reduce the experimental error.
2. Sensory methods: When describing the sensory methods employed, use the terminology in this book (see Tables 6.2 through 6.5), which is the same as that of the Institute of Food Technologists (IFT, 1981). See also the guidelines recommended by the International Standards Organization (ISO, 2005).
3. The panel: The number of panelists for each experimental condition should be stated, as it influences the statistical significance of the results obtained. If too few panelists are used, large differences are required for statistical significance, whereas if too many are used (e.g., 1000 for a triangle test), statistical significance may result when the actual difference is too small to have practical meaning. Changes in the panel during the course of the experiments should be avoided, but if they do occur, they must be fully described. The extent of the training and the other techniques used to prepare the panelists for the current tests, including a full description of any reference standards used, are important information needed to judge the validity of the results. The composition of the panel (age, sex, etc.) should be described if any affective tests were part of the experiment.
4. Conditions of the test: The physical conditions of the test facility as well as the way samples are prepared and presented are important variables that influence both the reliability and the validity of the results. The report should contain the following information:
   a. Test area: The location of the test area (lab booth, mall, home, field service) should be stated, and any distractions present (odors, noise, heat, cold, lighting) should be described together with efforts made to minimize their influence.

b. Sample preparation: The equipment and methods of sample preparation should be described (time, temperature, any carrier used). Identify and describe raw materials and formulations if applicable.
c. Sample presentation: The description should enable the reader to judge the degree of bias likely to be contained in the results and may include any of the following conditions capable of influencing them:
   - Whether panelists work individually or as a group
   - Lighting used if different from normal/ambient
   - Sample quantity, container, utensils, temperature
   - Order of presentation (randomized, balanced)
   - Coding of sample containers, for example, three-digit random numbers
   - Any special instructions such as mouth rinsing; information about the identity of samples or variable under test; time intervals between samples; samples being swallowed or expectorated
   - Any other variable that could influence the results, for example, time of day, high or low humidity, age of samples, and so on

5. Statistical techniques: The manner in which the data reported were derived from actual test responses should be defined, for example, conversion of scores to ranks. The type of statistical analysis used and the degree to which underlying assumptions (e.g., normality) are met should be discussed, as should the null hypothesis and alternate hypothesis, if not trivial.

16.2.4 Part 4: Results and Discussion

For all report styles, results should be presented concisely in the form of tables and figures, and enough data should be given to justify conclusions. However, the same information should not be presented in both forms. Tabular data generally are more concise, except for trends and interactions, which may be easier to see from figures.

The results section should summarize the relevant collected data and the statistical analyses. All results should be shown, including those that run counter to the hypotheses. Reports of tests of significance (F, c2, t, r, etc.) should list the probability level, the degrees of freedom if applicable, the obtained value of the test statistic, and the direction of the effect.

Table 16.1 summarizes how each piece of a report might be customized based on the type of report needed.

16.3 GRAPHICAL PRESENTATION OF DATA

16.3.1 Introduction

For as long as people have collected and presented data, there have also been people who are confused by those same data. The business professional and student should consider including graphs and charts throughout their reports to clearly communicate the findings of their research. The successful graphic allows data to be presented in a clear and concise

Table 16.1 Summary of Report Components by Type of Report

	Technical Report	Journal Article	Business Report
Abstract/Summary	Since technical reports are often an exchange between two busy professionals, summaries should not exceed 100 words (Prell, 1976), but they should straightforwardly answer the "whats."	The abstract is often the only part of the paper readers will see. As such, the time investment that facilitates a good abstract is worthwhile. The following is a general guideline: Introduction and Justification, Objectives, Materials and Methods (briefly), Results (provide details), and Conclusion (typically the last sentence).	The business report executive summary is one of the most challenging parts of the report to write. At times, the sensory professional must condense over 100 slides of information to 1 bulleted list. As a result, this summary concentrates on overarching themes, conclusions, or insights without detailed minutia.
Objectives and Introduction	The technical report (if directed to the project leader) should state and explain the test objective; if the report covers a complete project, it should state and explain the project objective as well as the objective of each specific test in this part of the report.	The introduction should include a review, with references, of pertinent previous work. This should be followed by a brief definition of the problem. If the study is based on a hypothesis, this hypothesis should be made evident to the reader in the introduction. Subsequent sections of the report should provide the test of the hypothesis.	The objective should be stated clearly at the beginning. Technical language should be avoided to facilitate understanding of the project's purpose by a wide audience.

(*Continued*)

Table 16.1 (Continued) Summary of Report Components by Type of Report

	Technical Report and Journal Article	Business Report
Materials and Methods	The materials and methods section should provide sufficient detail to allow the work to be repeated. Accepted methods should be cited by adequate references. This section may justify the design of the current project with previous work.	The material and methods sections are more likely to be abbreviated and presented in bulleted lists. References to methodology of published literature are generally not included. Additionally, instead of the complete methodology being in one place, portions of the descriptions of the methodology may be placed in relevant sections of the report and may appear after the results for that section. For example, a report that has both descriptive and consumer information may take this flow: **Example Flow of Business Report** • Executive Summary for Descriptive and Consumer Results • Descriptive Results • Descriptive Methodology • Consumer Results • Consumer Methodology
Results and Discussion	The theoretical and practical significance of the results should be identified and related to previous knowledge. The discussion should begin by briefly stating whether the results support or fail to support any original hypothesis. The interpretation of data should be logically organized and should follow the design of the experiment. The results should be interpreted, compared, and contrasted (with limitations indicated), and the report should end with clear-cut conclusions.	The discussion should provide insights and conclusions regarding the research and recommendations for next steps. Emphasis is on the implications of the results in business decisions, rather than comparing the current study to previous published work.

way that minimizes the opportunity for misunderstanding and increases the amount of information that can be quickly and clearly understood. Graphical displays should show the data in a nondistracting form that expresses the story behind the data without manipulating the facts of the analysis. A good graphic can help minimize the amount of language needed to explain results.

Many different types of graphics may be utilized for this purpose; however, good design principals and graphing standards should always be employed to ensure that graphics complement and illuminate the data accurately and clearly. The failure to consider both the form and function of graphs may result in poorly communicated or, worse, entirely misunderstood data, which leads to tenuous decision making.

The topics presented in this chapter attempt to provide some very basic information that every scientist should use in reporting and presenting data. For further information on the role of graphs in data presentation, consider reading Tufte (2001).

To begin, a few definitions may be helpful. *Graphics* may be defined as or relating to the pictorial arts and are generally the part of the presentation devoid of text. *Tables* are a systematic arrangement of data usually in rows and columns for ready reference.

16.3.2 General Guidelines for Graphing Data

- Labels and titles: Ensure that all graphs have all axes clearly labeled. The title should succinctly describe the topic of the graph.
- Encodings: Be sure that all colors and symbols are clearly explained through the use of a legend or proper labeling.
- Usage: Reference your graphics in the text or your presentation. While a good graphic should be able to stand alone, use the graphic to help explain results and conclusions or tell the story.
- Scale use: Only meaningful differences should be highlighted by scale usage. Line and bar charts should not be cut off but instead should start at zero in order to show the true magnitude. Contrast Figure 16.1, showing the appropriate use of

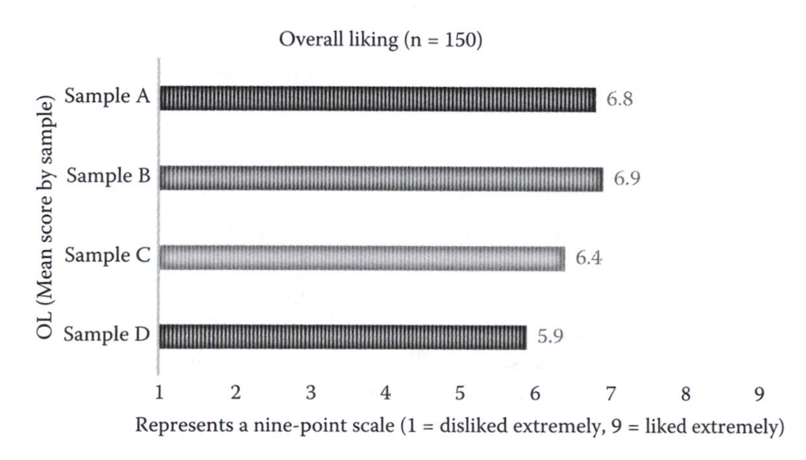

Figure 16.1 Appropriate use of axes.

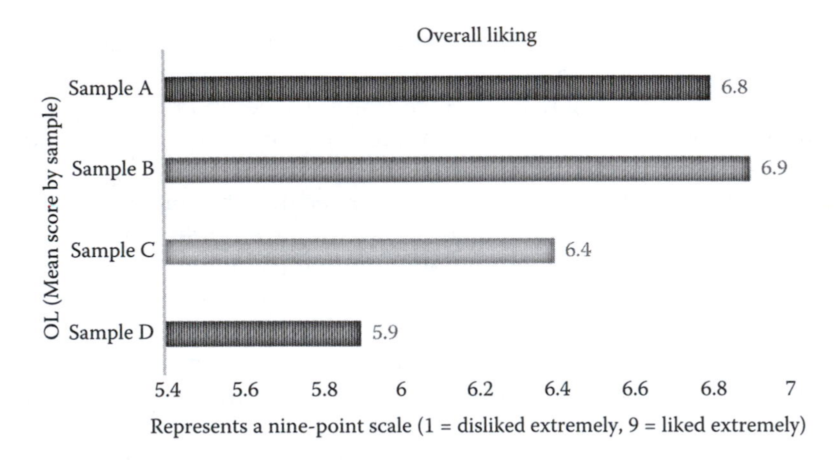

Figure 16.2 Inappropriate use of axes.

axes with Figure 16.2, showing the inappropriate use of axes. The inappropriate use of a truncated scale in Figure 16.2 causes the magnitude of the differences between samples to be amplified. In this case, a quick review of the bar graph might be misleading to some viewers.

- Color: Color helps distinguish between plotted variables. However, be mindful that color printing is often expensive. In the event that black-and-white printing is the only option, prepare by using different shades of black and different textures or lines to differentiate variables.

16.3.3 Appropriateness of Graphs

Initially consider the types of data intended for presentation. Will the data be used to compare products? Will they be used to explain a relationship between two or more variables? Will they be used to show trends over time? Answering these simple questions can guide the scientist to the most appropriate graph.

16.3.4 Common Graphs and Examples

Comparing two or more items can be done using a variety of charting types.
Bar Graphs: A bar graph is a good way to present data that compares items or ideas such as the performance across samples for a single attribute. This is frequently employed for consumer research testing to show the comparison of samples for a single attribute or question. Limit the number of bars used. Too many bars reduce the readability of the graphic. Also, consider how to order the bars. Should they be in order from the most-liked to the least-liked sample?

Histogram: For sensory data, histograms are commonly used to display frequency information such as the frequency and distribution of degree of differences scores across time. In consumer testing reports, the distribution of liking data across a categorical scale is often included to identify possible segmentation in liking.

Scatterplot: Scatterplots may be used to visualize patterns of the relationship between two sets of data. Scatterplots can visually show the strength and type of relationship (e.g., linear, quadratic, etc.) between the variables. Perceptual maps such as those in Figure 16.3 are scatterplots that display descriptive data processed through principal component analysis (PCA, discussed in more detail in Chapter 15). The maps represent the sensory space of a product category based on the factor scores and factor loadings generated from PCA. Two dimensions are presented at a time even if the PCA solution contains more than two dimensions. Products that are near each other are similar, and generally they are characterized by attributes within their immediate vicinity. Products that are far from each other are dissimilar. For example, in Figure 16.3, Brand B Regular and Brand E Kids are similar.

Pie chart: Pie charts show proportions. These are especially desirable for showing proportions of the population that prefer particular samples. Additionally, they can be used to show demographic information.

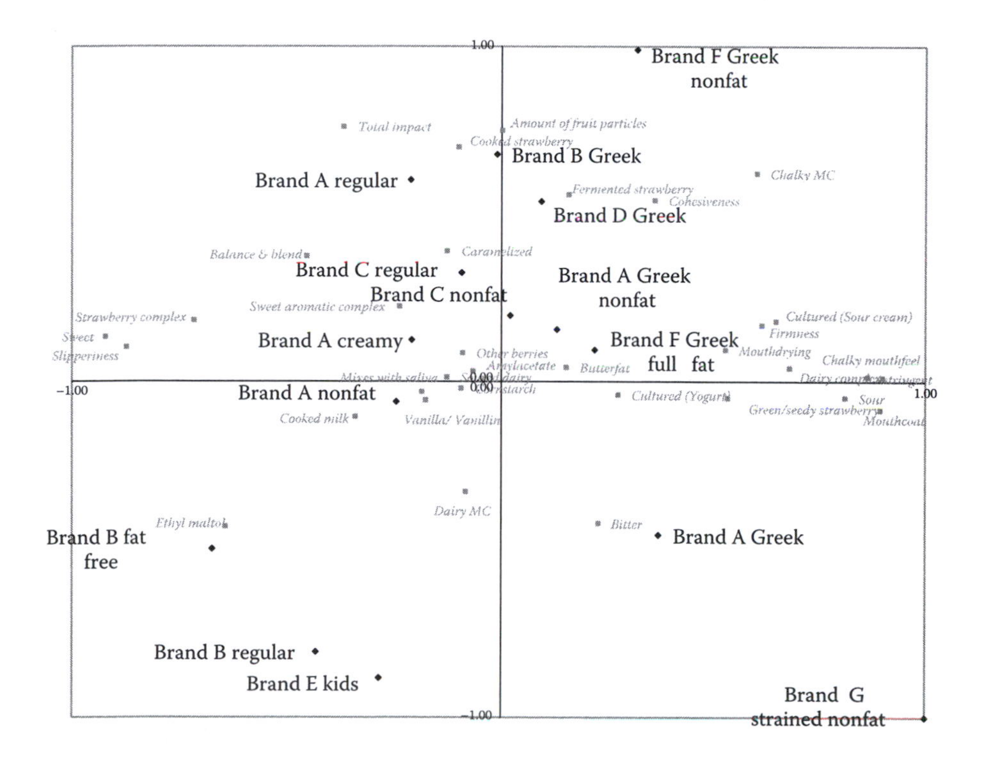

Figure 16.3 Perceptual map.

16.4 EXAMPLE REPORTS (TABLES 16.2 THROUGH 16.4)

Table 16.2 Example of Technical Report: Hop Character in Five Beers

	Summary
What was the objective? What was done? What were the results? What can be concluded?	To choose among five lots of hops on the basis of the amount of hop character they are likely to provide, pilot brews were made with hop samples 1, 2, 3, 4, and 5, costing $1.00, $1.20, $1.40, $1.60 and $1.80/lb, respectively; 20 trained members of the brewery panel judged each beer three times on a scale from 0 to 9. Sample 4 received a rating of 3.9, significantly higher than samples 2 and 5, at 3.0 and 2.9. Samples 1 and 3 were significantly lowest at 2.1 and 1.4. It can be concluded that hop samples 4 and 2 deliver more hop character per dollar than the remainder
	Objectives
Project objective, test objectives, agreed before the experiment	The brewery obtained representative lot samples from several suppliers. The project objective was to choose among the lots based on their ability to provide hop character. The test objectives were to (1) compare the five beers for degree of hop character on a meaningful scale and (2) obtain a measure of the reliability of the results
	Materials and Methods
Design which accomplishes objectives 1 and 2	Describe conditions of test: screening of samples, information to panel, panel area, sample presentation
Describe sensory tests used	Sensory evaluation—The tasters evaluated the amount of hop character on a scale of 0–9; reference standards were available as follows; synthetic hop character at 1.0 mg/L = 3.0 scale units, and at 3 mg/L = 6.0 scale units
Describe panel: number, training, etc.	The panel—20 panel members were selected on the basis of past performance evaluating hop character; all 20 panelists tested all three sets
Describe conditions of test: screening of samples, information to panel, panel area, sample presentation	Sample preparation and presentation—The test beers were stored at 12°C and evaluated 7–10 days after bottling. Samples were screened by two experienced tasters who found them representative of the type of beer with no differences in color, foam, or flavor other than hop character. Panel members were informed that samples were test brews with different hops, but the identity of individual samples was not disclosed. Members worked individually in booths and no discussion took place after the sessions. Sample portions of 70 mL were served at 12°C in clear 8-oz. glasses. The five samples were presented simultaneously in balanced, random order. Samples were swallowed
Statistical techniques	Statistical evaluation—Results were evaluated by split-plot analysis of variance

Present results concisely

Results and Discussion

The average results for the five beers are shown in Table 1 and the corresponding statistical analysis in Table 2. Sample 4 received a significantly higher rating for hop character (3.9) than the remaining samples

Give enough data to justify conclusions

TABLE 1
Average Hop Character Ratings for the Five Beer Samples

Sample	4	2	5	1	3
Mean	3.9[a]	3.0	2.9	2.1	1.4
Hops used, lb/bbl	0.36	0.38	0.34	0.32	0.35

[a] Sample is significantly different from other samples at the 5% significance level.

Give probability levels, degrees of freedom, obtained value of test statistic

TABLE 2
Split-Plot ANOVA of the Results

Source of Variation	Degrees of Freedom	Sum of Squares	Mean Squares	*F*
Total	299	975.64		
Replications	2	8.89		
Samples	4	221.52	55.38	41.88[a]
Error(A)	8	10.58	1.32	
Subjects	19	412.30	21.70	17.79[a]
Sample×Subject	76	89.81	1.18	0.97
Error(B)	190	232.53	1.22	

Note: Error(A) is calculated as would be the Rep × Sample interaction. Error(B) is calculated by subtraction.

[a] Significant at the 1% level.

(*Continued*)

Table 16.2 (Continued) Example of Technical Report: Hop Character in Five Beers

Interpret the data, following the design of the experiment	Samples 2 and 5, with nearly identical ratings of 3.0 and 2.9, had significantly less hop character than sample 4, but significantly more than samples 1 and 3. The statistical evaluation shows no significance for the subject-by-sample interaction ($F = 0.97$); it may therefore be assumed that the panelists were consistent in their ratings; the significance of the subject effect ($F = 17.79$) suggests that the panelists used different parts of the scale to express their perceptions; this is not uncommon; furthermore, when there is no interaction, the subject-to-subject differences are of secondary interest. The primary concern, the difference among samples, was evaluated using an HSD multiple comparison procedure; $HSD_{5\%} = 0.7$, which results in the differences shown by underscoring in Table 1. Variations in the amounts of hops used to obtain the BU level of 14 were small compared with the variations in perceived hop character intensity
	Conclusions
End with clear-cut conclusions	Of the five samples tested, sample 4 ($1.60/lb) produced a significantly higher level of hop character. Sample 2 ($1.20) merits consideration for less expensive beers

Note: This report covers the test described in Example 8.8.

Table 16.3 Example of Business Report: Descriptive Analysis of Chai Tea (slide presentation)

Intended Audience: Business personnel (technical and/or non-technical)
Purpose: To communicate technical sensory results to product developers and other business partners

	Cover Letter
In today's electronic-driven society, the cover letter is often written in the body of an e-mail, and the rest of the report is attached.	Date Client Location Dear Client, Attached please find the report for 2 chai teas that were submitted for evaluation. Please call if you or your colleagues have any questions or would like to schedule additional work. Signature
Slide 1:	**Title and Authors**
	Descriptive Analysis of Chai Tea Prepared for: ABC Tea Company Prepared by: Sensory Spectrum, Inc. May 2012
Slide 2:	**Background and Objectives**
Background information that justifies the project and the objectives.	Background: ABC Tea Company would like to understand sensory similarities and differences between two chai tea formulations – original/control & new. Due to a shortage of a specific ingredient, ABC Tea Company was forced to make a substitution in their current formulation. Consumer complaints rose when this ingredient substitution was made. They would like to understand the impact of that substitution on the flavor of their product. Objectives: 1. Provide descriptive profiles (appearance & flavor) on two samples of chai tea: 1. Original/Control 2. New (sample with substituted ingredient) 2. Provide Degree of Difference (DOD) testing to provide guidance to product development to clarify the possible causes of the consumer complaints and to recommend changes to the new product to change its flavor closer to the original (control)

(Continued)

Table 16.3 (Continued) Example of Business Report: Descriptive Analysis of Chai Tea (slide presentation)

Guidance	Slide content	
Slide 3:		**Executive Summary**
Simply state findings for key samples and objectives.		• Overall the new product is less complex and less blended in flavor than the control. • The new product lacks the overarching "brown" notes that blend the caramelized & brown spice notes with the tea notes (also brown in character). • The new product's brown spice notes are more singular chemical notes rather than the more complex, woody brown spices (typical of whole spices) seen in the control. • The new product has no common aromatic (i.e. "brown") to blend the tea's flavors leaving each of its flavor components to stand alone. In doing so, the spices do not blend with each other or with the tea notes and the overall character of the product becomes "spiky." • In contrast, the control tea is very "brown" in character (caramelized, browned/ tobacco tea notes, brown/woody spices). The brown notes overlap amongst the different flavor "buckets" and blend the tea, making it difficult to know where the "brown" of one flavor ends and another begins. The overall character of the control is very balanced & blended with no notes "sticking out."
Slide 4:		**Executive Summary (continued)**
In addition to the executive summary, presenting product differences in table format can make the report appeal to a wider audience. Some readers may prefer tabular format over bulleted lists. The inclusion of recommendations for driving business decisions are a primary distinction between business reports and journal articles.	**DOD = 7.0***	**To Move Closer to Control**
	Appearance*	• Make sample darker and brighter • Increase opacity
	Flavor	Brown spices should be more woody in character as is typical of ground whole spices
Additional slides:		**Results (Detailed)**
Detailed results should follow the executive summary. They offer deeper information to support recommendations.	See Figure 16.4.	

Include raw data with the written summaries. Attributes with significant or meaningful differences can be highlighted for emphasis. Alternatively, non-significant or non-meaningful data could be moved to an appendix.

Attribute definitions and detailed sample information could also be included in the appendices.

Appendix 1: Descriptive Analysis Methodology

- Prior to the evaluation, panelists reviewed references for brown spices and ginger
- The samples were analyzed for appearance & flavor by nine members of the Sensory Spectrum Food Panel, trained and experienced in each type of evaluation.
- The strength of each attribute was rated on the 15-point Spectrum Scale, where 0 = none and 15 = very strong.
  - This scale incorporates the ability to use tenths of a point and therefore has the potential of 150 scale differentiations.
  - The scale may be expanded beyond 15 points to include extreme ratings if necessary.
- The panelists evaluated each sample using the following procedure:
  - Each panelist received one sample for aromatics, basic tastes, and chemical feeling factors
  - Samples were evaluated between 140F and 150F
  - All samples were expectorated
  - Panelists recorded individual ratings, then each attribute was discussed until a consensus value was decided upon and recorded
  - After flavor was evaluated, the panel rated the appearance of a second set of samples.

Appendix 2: Degree of Difference Methodology

- The degree of difference scale is a 0.0 to 10.0 rating indicating how different a product is from a reference product or control, with 0.0 meaning no difference and 10.0 being extremely different
- The degree of difference rating quantifies the magnitude of the difference but is not directional
  - A degree of difference between 2.0 and 3.0 is considered low enough that a number of consumers would not be able to tell the two samples apart.
  - A degree of difference of 5.0 or greater is high enough that a number of consumers would be able to tell the samples apart.
  - A degree of difference of 4.0 is considered a questionable area where consumers may or may not be able to tell the difference.

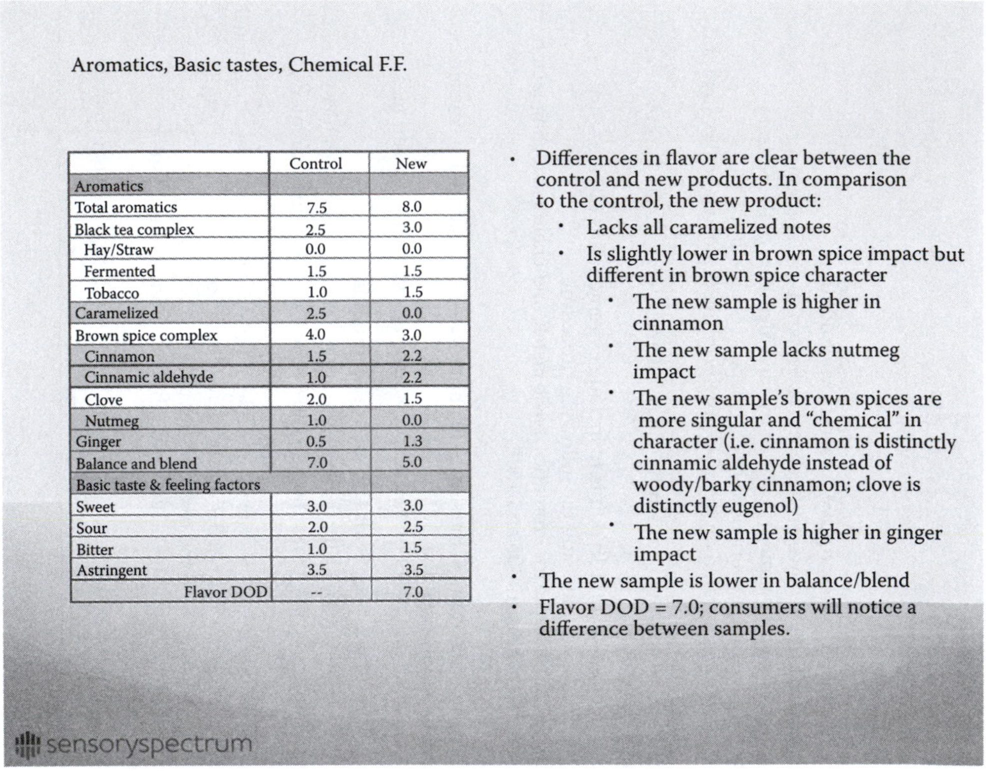

Aromatics, Basic tastes, Chemical F.F.

	Control	New
Aromatics		
Total aromatics	7.5	8.0
Black tea complex	2.5	3.0
Hay/Straw	0.0	0.0
Fermented	1.5	1.5
Tobacco	1.0	1.5
Caramelized	2.5	0.0
Brown spice complex	4.0	3.0
Cinnamon	1.5	2.2
Cinnamic aldehyde	1.0	2.2
Clove	2.0	1.5
Nutmeg	1.0	0.0
Ginger	0.5	1.3
Balance and blend	7.0	5.0
Basic taste & feeling factors		
Sweet	3.0	3.0
Sour	2.0	2.5
Bitter	1.0	1.5
Astringent	3.5	3.5
Flavor DOD	--	7.0

- Differences in flavor are clear between the control and new products. In comparison to the control, the new product:
  - Lacks all caramelized notes
  - Is slightly lower in brown spice impact but different in brown spice character
    - The new sample is higher in cinnamon
    - The new sample lacks nutmeg impact
    - The new sample's brown spices are more singular and "chemical" in character (i.e. cinnamon is distinctly cinnamic aldehyde instead of woody/barky cinnamon; clove is distinctly eugenol)
    - The new sample is higher in ginger impact
- The new sample is lower in balance/blend
- Flavor DOD = 7.0; consumers will notice a difference between samples.

sensoryspectrum

Figure 16.4 Attributes summary for chai tea descriptive analysis.

The outline below is condensed and paraphrased from a full published article (Lawless et al. 2012). Please refer to the original article to determine the appropriate level of detail for a journal. The paraphrased version below is for demonstration purposes.

Intended Audience: Peers in the sensory field

Purpose: To add to the body of scientific knowledge through reporting sensory results and comparing them to previous research

Table 16.4 Example of Journal Article Format: Optimizing Fruit Juices

	Title and Authors
What is a phrase that describes the experiment? Include relevant key words in the title to aid Internet search engines in finding the paper. In sensory evaluation, authors are typically ordered by contribution to the manuscript. Other fields of study or specific journals may practice another convention.	Consumer-based Optimization of Blackberry, Blueberry, and Concord Juice Blends Lydia J.R. Lawless, Renee T. Threlfall, Jean-François Meullenet, Luke R. Howard
	Abstract
What is the justification? What is the objective? What was done (briefly)? What are the results? What are the major conclusions?	Juices from fruits such as blackberry, blueberry, and Concord grapes show potential in filling consumer demand for healthier products. Four optimization methods (response surface, ideal point, desirability, and intuition) were tested with consumers for their ability to maximize sensory quality of juice blends while maintaining potential health properties. Ten juice blends were formulated based on a mixture design and tested with consumers; the data obtained were used to develop four optimized juice blends. A validation study was performed in which consumers evaluated the optimized juice blends and scored them for overall liking. The hedonic scores for three of the optimized juices achieved parity with 100% Concord juice. Information about anthocyanins increased purchase intent for the three juices that were higher in anthocyanins. Results indicate that informing consumers of potential health properties of juice could increase purchase intent of juice blends and therefore product success.

(*Continued*)

Table 16.4 (Continued) Example of Journal Article Format: Optimizing Fruit Juices

	Introduction
What other published work justifies the current experiment? What is the history of the experimental subject? In this example, potential health properties of the fruits, optimization methods, and scales (verbal nine-point, ideal point) should be discussed.	The population is increasingly interested in healthy eating. Juices from polyphenolic-rich fruits such as blackberry, blueberry, and Concord grapes are desirable candidates to pique this interest because polyphenolics are associated with potential health benefits (Timberlake and Henry 1988; Folts 1998; Goyarzu et al. 2004; Seeram et al. 2006). Using optimization methods to develop juice blends is advantageous because they identify solutions under specific constraints. Additionally, they are cost-effective and efficient. Many optimization methods have been developed including preference mapping and extensions thereof. Some beverage optimizations have incorporated a mixture design in which a treatment is represented by the relative proportions of its ingredients, and the proportions sum to one (Kumar et al. 2010; Dooley et al. 2012). Varying optimization methods may use differing scaling techniques to elicit data from consumers. The verbal 9-point hedonic scale has been widely used in consumer-based sensory research. Other scales such as the ideal point scale have been introduced more recently. The ideal point scale can elicit the distance between actual attribute intensities and the *ideal* intensities of attributes (Worch et al. 2010). The ideal point scale may be more appropriate for health-oriented products because consumers may have a different mental sensory target for products marketed as healthier.
What are the objectives (detailed)?	The objectives of this research were 1) execute a sensory test in which consumers evaluated 10 juice blends and then use the data to develop 4 optimized juice blends; 2) test the optimized juice blends in a validation study; and 3) determine the effect of polyphenolic information on purchase intent.
	Materials and Methods
Describe the process such that anyone could reproduce the results. Use subheadings to increase readability and ease of reference for the reader.	**Sample Preparation** Juice was produced at the University of Arkansas Enology Research Laboratory in Fayetteville. Fresh blackberries and blueberries and frozen Concord grape juice were sourced and processed according to previous studies, which simulated commercial juice production. The fresh fruit was frozen and then cooked. The must was pressed, and the resulting juice was filtered. Concord grape juice concentrate was reconstituted to 15% soluble solids.

Sample preparation: Who is the manufacturer? What are the storage conditions? How was the sample selected?	Juice formulations were determined by a ABCD mixture design and included 100% blueberry, 100% blackberry, 100% Concord, 50% blueberry + 50% blackberry, 50% blueberry + 50% Concord, 50% blackberry + 50% Concord, 67% blueberry + 17% blackberry + 17% Concord, 17% blueberry + 67% blackberry + 17% Concord, 17% blueberry + 17% blackberry + 67% Concord, and 33% blueberry + 33% blackberry + 33% Concord. Once juice was processed, it was bottled in jars and pasteurized. The jars were capped, cooled, and stored at 36°F.
If your experiment contained instrumental analysis in addition to the sensory evaluation, add details of the analysis in a separate section.	**Instrumental Analysis** Total anthocyanins and total phenolics analyses were performed on the juice samples to gauge their nutraceutical value. Soluble solids were also performed.
	Sensory Evaluation
Describe conditions of test: Place of evaluation Testing area Balloting (computer program or paper) Sample presentation Describe the type of subjects used (trained panel, semi-trained panel, or naïve consumers). If the panel is trained, describe their training. If the panel is composed of naïve consumers, indicate their recruitment criteria.	Consumer testing was performed in the University of Arkansas Sensory Service Center. Consumers evaluated samples in a partitioned white booth using a computer system. A balanced complete block design was used to determine presentation order. Panelists for the initial consumer study were recruited based on blackberry, blueberry, and Concord grape liking and juice consumption (minimum of 3 times per week). Panelists (n = 108) evaluated 5 juice products per day over a 2-day period for overall liking based on the 9-point hedonic scale and ideal intensities of the attributes using the ideal point scale. Panelists from the initial study were invited to return for the validation study to test the four optimized solutions as well as three formulations from the original study, which framed the optimized solutions. As in the initial study, consumers evaluated the samples for overall liking on the 9-point hedonic scale and the ideal intensities of the attributes on the ideal point scale. Consumers were asked to indicate their purchase intent for each product, and then they were given additional information about the anthocyanin content of each of the samples. Next, consumers were asked to rate their purchase intent again.

(Continued)

Table 16.4 (Continued) Example of Journal Article Format: Optimizing Fruit Juices

	Statistical Analysis
If the analysis is straightforward, a brief description is sufficient. If the analysis is innovative or complex, a more thorough description is warranted. Indicate the statistical sensitivity (e.g., α level).	Analysis of variance and Tukey's mean separation were performed on instrumental data with $\alpha = 0.05$. Mixture design data were statically analyzed with response surface, ideal point, and desirability techniques. An intuitive solution was also developed. In *response surface* analysis, an equation was developed for each consumer and mapped in a ternary plot. For each equation, a lower level was set at each consumer's average liking. Any solution that was calculated to be below average for each consumer was plotted as an unfeasible region for that consumer. Therefore, the overall feasible region represented juice formulations that were above average liking for the majority of consumers. The solution was chosen from this region of the sensory space. In *ideal point* analysis, the mean distance from the actual attribute intensity and the ideal intensity was calculated. Solutions were based on minimizing the distance from ideal. The *desirability* function represented the panelists' balanced and maximized overall liking scores. Each point within the ternary plot was designated a desirability score between 0 and 1 with 1 being the highest level of desirability possible. In the validation study, ANOVA and mean separation were performed on overall liking means, and matched pair analysis was performed on purchase intent data.
	Results and Discussion
Present results concisely. Start this section with a general summary and then move to more specific outcomes. As in Materials and Methods, use subheadings to organize the section. Use graphs to support reader understanding as shown in Lawless (2012). Compare results to results in similar studies. How do the results support your objectives?	Initial Consumer Study Juice blends high in Concord juice received the highest overall liking scores, followed by predominantly blueberry blends and then blackberry blends. Liking scores may have been driven by sensory attributes. Blackberry juice was bitter, which was potentially due to high total phenolics levels (Hoye and Ross, 2011). Previous research confirmed that sweetness and grape note were drivers of liking, and bitterness and blackberry note were detractors of overall liking (Lawless et al., 2012a). Four optimized solutions were generated from the four optimization techniques. Response surface optimized solution. In the response surface optimization, the majority of consumers were satisfied at the solution (18% blueberry + 19% blackberry + 63% Concord). Ideal point optimized solution. The contours for sweetness, sourness, and total anthocyanins were graphed in the ternary plot. The solution that minimized the distance from ideal for sweetness and sourness while simultaneously maximized total anthocyanins (within the desirable sensory region) was 9% blackberry + 20% blueberry + 71% Concord.

	Desirability Function Solution. The desirability function solution was 13% blackberry + 87% Concord, although when desirability was rounded to three significant digits, ranges of 84%–89% Concord, 0%–1% blueberry juice, and 11%–16% blackberry juice were equally desirable. Desirability decreased linearly across the plot surface.
	Validation Study All of the optimized solutions were well-liked in the validation study (6.6–7.5 on a 9-point scale). The desirability function solution had the highest overall liking mean, but it was not significantly different than the intuitive optimum, the ideal point solution, or 100% Concord. Although blends that were liked more than Concord were not created, several blends that were liked equally well and had more anthocyanins were generated. Purchase intent was measured before and after consumers were given information about the anthocyanin content in each of the juices. Consumers indicated that their purchase intent increased for the three samples highest in anthocyanins and decreased for the two samples lowest in anthocyanins. Future research could explore the role of health information delivery in label graphics and price in purchase-intent evaluations (Chrea et al., 2011). The current study supports previous research, which showed that consumers were interested in nutraceutical products, especially those related to physiological benefits (particularly cardiovascular health) (Siegrist et al., 2008; Ares et al., 2008). Consumers consistently express trade-offs between taste and health. The negative effect of off-flavors on overall liking can be mitigated with health claims (Tuorila and Cardello, 2002). In the same sense, in the current experiment, purchase intent increased for samples that were highest in anthocyanins even if they were not the most well-liked. Optimization methods that can accommodate consumer trade-offs between taste and health are needed to maximize consumer utility.
	Conclusions
The journal may or may not require a separate section for conclusions. If the journal does not require a separate section, write the conclusions in the last paragraph of the Results and Discussion section.	Information about anthocyanin content in juice blends increased consumer purchase intent for blends that were higher in anthocyanins, which indicates validating the potential health information and communicating it to consumers could increase market success for nutraceutical-rich juice blends.

(*Continued*)

Table 16.4 (Continued) Example of Journal Article Format: Optimizing Fruit Juices

	References
Follow the particular journal's style for references.	Ares, G., A. Gimenez, and A. Gambaro. 2008. Uruguayan consumers' perception of functional foods. *J Sens Stud* 23: 614–630. Chrea, C., L. Melo, G. Evans, C. Forde, C. Delahunty, and D.N. Cox. 2011. An investigation using three approaches to understand the influence of extrinsic product cues on consumer behavior: An example of Australian wines. *J Sens Stud* 26: 13–24. Dooley, L., R.T. Threlfall, and J.F. Meullenet. 2012. Optimization of blended wine quality through maximization of consumer liking. *Food Qual Prefer* 24: 40–47. Folts, J.D. 1998. Antithrombotic potential of grape juice and red wine for preventing heart attacks. *Pharm Biol* 36: 21–27. Kumar, S.B., R. Ravi, and G. Saraswathi. 2010. Optimization of fruit punch using mixture design. *J Food Sci* 75: S1–S7. Goyarzu, F., D.H. Malin, F.C. Lau, G. Taglialatela, W.D. Moon, R. Jennings, E. Moy, D. Moy, S. Lippold, and B. Shukitt-Hale. 2004. Blueberry supplemented diet: Effects on object recognition memory and nuclear factor-kappa B levels in aged rats. *Nutr Neurosci* 7: 75–83. Hoye, C., Jr and C.F. Ross. 2011. Total phenolic content, consumer acceptance, and instrumental analysis of bread made with grape seed flour. *J Food Sci* 76: S428–S436. Lawless, L.J.R., R.T. Threlfall, L.R. Howard, and J.F. Meullenet. 2012a. Sensory, compositional, and color properties of nutraceutical-rich juice blends. *Am J Enol Vitic* 63: 529–537 Seeram, N.P., L.S. Adams, Y. Zhang, R. Lee, D. Sand, H.S. Scheuller, and D. Heber. 2006. Blackberry, black raspberry, blueberry, cranberry, red raspberry, and strawberry extracts inhibit growth and stimulate apoptosis of human cancer cells in vitro. *J Agric Food Chem* 54: 9329–9339. Siegrist, M., N. Stampfli and H. Kastenholz, 2008. Consumers' willingness to buy functional foods. The influence of carrier, benefit and trust. *Appetite* 51: 526–529. Timberlake, C.F. and B.S. Henry, 1988. Anthocyanins as natural food colorants. *Prog Clin Biol Res* 280: 107–121. Tuorila, H., and A.V. Cardello. 2002. Consumer responses to an off-flavor in juice in the presence of specific health claims. *Food Qual Prefer* 13: 561–569. Worch, T., L. Sébastien, P. Punter, and P. Jérôme. 2010. Can we trust consumers' ideal? Study of the relationship between the consumers' preference and their ideals. 10th Sensometrics conference, Rotterdam, the Netherlands.

REFERENCES

IFT (1981). Guidelines for the preparation and review of papers reporting sensory evaluation data. Sensory Evaluation Division, Institute of Food Technologists. *Food Technol.* 35(11): 50.

ISO (2005). International Standard ISO 6658:2005. Sensory analysis—Methodology—General guidance. International Organization for Standardization. New York: American National Standards Institute.

Larmond, E. (1981). Better reports of sensory evaluation. *Tech Q Master Brew Assoc Am* 18: 7.

Lawless, L. J., R. T. Threlfall, J. Meullenet, and L. R. Howard (2012). Consumer-based optimization of blackberry, blueberry, and concord juice blends. *J. Sensory Stud.* 63: 529–537.

Prell, P. A. (1976). Preparation of reports and manuscripts which include sensory evaluation data. *Food Technol.* 30(11): 40.

Tufte, E. R. (2001). *The Visual Display of Quantitative Information* (2nd ed.) Graphics Press.

17

Sensory Evaluation in Quality Control (QC/Sensory)

17.1 INTRODUCTION

Sensory evaluation has played an integral role in quality control (QC) for decades. In its earliest stages, QC/sensory functions were performed by experts—perfumers, brewmasters, or winemakers. A product's quality was judged on a subjective basis by these individuals (Dove, 1947; Hinreiner, 1956; also see Section 1.2). Gradually, responsibility for quality evaluations transitioned to a larger pool of experts or to small panels of judges. Early evaluation methods focused directly on quality (poor to excellent) and to some extent on the product's characteristics (Platt, 1934; Dawson and Harris, 1951; Peryam and Shapiro, 1955; Amerine et al., 1959; Ough and Baker, 1961).

Formal sensory methods began to be used in the latter half of the twentieth century (Pangborn and Durkley, 1964), when attention began to be given to the selection and training of quality assessors, to the utilization of formal sensory methods including the use of reference standards, and to the use of formal data analysis methods to summarize and interpret the quality data. Over time, QC/sensory evolved along two equally important paths. One focused on the challenges of initiating and maintaining a QC/sensory program in an organization (Rutenbeck, 1985; Mastrian,1985; Stouffer, 1985; Carlton, 1985). The second focused on sensory methods that were practical and effective for an in-plant QC function (Dziezak, 1987; Chambers, 1990; Hubbard, 1990). Both paths were captured in the ASTM publication *Manual 14: The Role of Sensory Analysis in Quality Control* (1992).

The challenges involved in implementing and maintaining a QC/sensory program are substantial. Readers whose job functions include QC/sensory are encouraged to review the literature on the topic to become familiar with both the challenges and with the techniques that have proven to be effective in dealing with them (e.g., Munoz et al., 1991). This chapter will not deal with the establishment and management of QC/sensory programs. Instead, as an update of *Sensory Evaluation in Quality Control,* it will focus on the sensory methods that have proven to be effective tools for monitoring and maintaining high-quality products.

Many sensory methods have been proposed to support a QC function. Over time, three methods have emerged as the most widely used: attribute descriptive methods (Chambers, 1990), difference-from-control methods (Aust et al., 1985), and the in/out method (Sidel et al., 1983). The remainder of the chapter will be devoted to describing each method, its strengths and weaknesses, and the logistical issues that need to be addressed to implement each in an ongoing QC/sensory program.

17.2 ATTRIBUTE DESCRIPTIVE METHODS

Attribute descriptive methods are possibly the most powerful tools for assessing the quality of a product (Munoz et al., 1991). These methods focus on a small number of sensory attributes that have been demonstrated to impact consumer acceptance. The methods provide specific information on the nature and range of sensory variability that the product exhibits and set limits on an attribute-by-attribute basis for what constitutes the acceptable limits of variability. Because the sensory specifications are based on consumer acceptance, the attribute descriptive methods are typically applied only to finished products.

Attribute descriptive methods provide detailed information on what is varying perceptibly in the product, the magnitude of the variability, and the direction of variability relative to the product's sensory specifications. As such, attribute descriptive methods closely resemble many of the instrumental methods used in QC.

The major drawback of attribute descriptive methods is the high level of resources (human resources, cost, and time) required to implement and maintain them. Significant resources are required to identify the sensory attributes that are important to product quality and to set limits on the acceptable ranges of intensities for these attributes. This implementation step involves parallel descriptive and consumer evaluations of a large number of production samples and elaborate data analyses to identify the key attributes and their acceptable ranges. Program maintenance requires extensive and ongoing training of assessors, maintenance of a battery of relevant and up-to-date reference standards, and the availability of facilities where the evaluations can be performed in a reliable manner. As a result, the attribute descriptive methods tend to be used only for a company's most important brands and products. Applications of these methods for products that generate relatively small amounts of revenue would be difficult to justify.

The four steps in implementing an attribute descriptive QC program are

1. Establish sensory specifications
2. Train and operate the plant panels
3. Establish a data collection, analysis, and reporting system
4. Operate the ongoing program

17.2.1 Establishing Sensory Specifications

The first major step in the development of an attribute descriptive QC/sensory program is to develop a set of relevant sensory specifications. This is a two-step process involving the identification of a small number of sensory attributes (5–15) that influence the consumer's acceptability of the product and the establishment of ranges of intensities for each of these attributes within which the product maintains an acceptably high level of liking.

17.2.1.1 Initial Sample Screening

The process begins with the collection and screening of a large number of production samples. Sampling should be done in a manner that maximizes the probability that the products collected span the entire range of production variability. Reviewing the logistics of sample production is helpful. For example, when applicable, samples should be collected from multiple plants, across multiple days (spread out sufficiently to ensure the turn-over of raw materials), and from all product shifts and production lines, including multiple pick-ups per shift. It is easy to see that such an extensive sampling program will result in a very large number of samples. It also needs to be recognized that the number of units of production that need to be collected at this stage must be adequate for subsequent formal sensory descriptive evaluations and consumer testing. It is not unusual to collect more than 100 product samples, with the number of sampled units regularly totaling in the 1000s. Logistical issues involving such things as the amount of space available to store samples, how quickly the samples can be evaluated, and the number of samples that can be reliably evaluated before shelf-life issues arise all need to be considered in advance. There is no value in collecting more samples than can be appropriately stored and evaluated by the descriptive panel and consumers in an acceptably short period of time.

An initial screening of the production samples is useful for reducing the number of samples that move forward into descriptive evaluations and consumer testing. Samples that exhibit extreme sensory properties should be included in the next step of evaluations, while only a single representative of a group of highly similar samples needs to move forward into additional testing. This initial screening process often reduces the number of samples requiring additional evaluations by 50%–75%. The remaining samples are then submitted to formal sensory descriptive evaluation.

17.2.1.2 Sensory Descriptive Evaluations and Sample Selection for Consumer Testing

The descriptive evaluation should be conducted according to the methods described in Chapters 11 and 12. The goal of this step is to obtain a comprehensive description of all of the sensory properties of each sample and to fully characterize the ranges of variability exhibited by the complete set of samples (see Table 17.1 for an example involving 25 production samples of potato chips).

Once the production samples have been fully characterized and the ranges of variability for all of the attributes have been established, the "whittling down" process can begin. The goal is to arrive at the smallest set of products and attributes that adequately characterize the total variability of the product. Attributes that exhibit trivial ranges of variability do not need to be included in future ballots. However, what constitutes a trivial range needs to be decided in a very thoughtful manner. Off notes may exhibit small ranges of variability, but they may have substantial impacts on acceptability. Each attribute needs to be considered individually, but in many cases, the same rule of thumb for what constitutes a meaningful range of variability may apply. For example, setting the limit at a 0.5 unit range for products measured on a 15-unit scale may apply to many of the attributes on the ballot. The absence of statistically significant differences should not be used to define

Table 17.1 Summary Statistics of the Sensory Attributes for 25 Samples of Potato Chips (Appearance, Flavor, and Texture)

Attribute	Mean	Std. Dev.	Minimum	Maximum	Range
Color Int.	3.2	0.5	2.0	8.0	6.0
Even color	6.0	0.9	4.0	13.3	9.3
Blotches	0.0	0.0	0.0	0.0	0.0
Translucency	9.7	0.6	8.6	12.2	3.6
Even size	3.4	1.3	4.0	8.4	4.4
Even shape	3.6	0.8	1.0	6.1	5.1
Thickness	4.5	0.1	4.4	4.6	0.2
Bubbles	5.2	0.2	5.1	5.4	0.3
Folds	7.3	0.8	5.0	10.1	5.1
Potato complex	5.3	0.7	2.6	6.7	4.1
Raw	2.1	0.2	0.7	4.2	3.5
Cooked	2.0	0.0	2.0	2.0	0.0
Fried	3.3	0.5	2.0	5.0	3.0
Skins	0.7	0.1	0.6	0.8	0.2
Heated oil	1.5	0.2	1.2	1.6	0.4
Earthy	0.8	0.1	0.7	0.8	0.1
Cardboard	1.2	0.3	0.0	5.1	5.1
Painty	0.0	0.2	0.0	3.8	3.8
Salty	10.5	1.1	8.1	14.7	6.6
Sweet	4.3	0.1	4.2	4.3	0.1
Bitter	0.5	0.0	0.5	0.5	0.0
Astringent	5.0	0.0	5.0	5.0	0.0
Burn	1.3	0.5	0.0	3.2	3.2
Surface bumps	5.2	0.1	5.0	5.3	0.3
Oily surface	4.1	0.5	2.8	5.1	2.3
Hardness	7.3	0.7	4.7	9.4	4.7
Crisp/crunch	12.4	0.7	9.5	14.6	5.1
Denseness	8.9	0.6	6.1	10.0	3.9
Number of particles	11.5	0.2	11.2	11.6	0.4
Abras. of part.	4.5	0.1	4.2	4.6	0.4
Persist. of crisp	6.1	0.1	6.0	6.2	0.2
Mixes w/saliva	10.1	0.1	9.9	10.2	0.3
Cohesiveness	4.1	0.3	3.8	4.3	0.5
Grainy mass	6.5	0.0	6.5	6.5	0.0
Toothpack	3.5	0.5	1.5	4.7	3.2
Oily film	4.3	0.1	4.3	4.4	0.1

a trivially small range. Absence of a significant difference can occur for several reasons other than the absence of true differences among the samples. Small panel size and confusion over the attribute also may lead to an outcome of no significant difference. Familiarity with the product and sound sensory judgment should be used in combination with the statistical results to decide which attributes should remain on the ballot.

Once the attributes that will remain on the ballot have been determined, attention can shift to deciding which products will be tested with consumers. The goal of this step is to identify the smallest number of samples that fully represent the total sensory variability of the product. Samples that exhibit extreme attribute intensities are automatically included (see Figure 17.1a). Samples that possess unique combinations of attribute intensities should also be included (see Figure 17.1b). Once the unique samples have been identified, the remaining samples should be chosen to span the remaining sensory variability in a comprehensive and uniform manner. If only a small number of attributes remain on the ballot, the process of selecting samples representing the low, middle, and high ranges of intensities may be done easily with tables and graphs. However, when the number of attributes is too large for such a simple process, principal component analysis (PCA) (see Section 15.2.1.2) can be applied. PCA represents the variability of the products on all of their sensory attributes in a small number of latent sensory dimensions called principal components or factors. Each sensory dimension represents a bundle of correlated sensory attributes. As such, if the samples are chosen to represent the low, middle, and high ranges of the sensory dimensions, it is very likely that the resulting sample set will span the entire range of sensory variability at the individual attribute level (see Figure 17.1c).

17.2.1.3 Consumer Testing Production Samples

The selected samples now move on to be tested with consumers. The screening requirements and test design need to be given serious consideration on a case-by-case basis, but the most commonly used designs involve medium to heavy users participating in a central location test (CLT) using a balanced incomplete block serving protocol (see Section 14.5.4). Variations of this general design include testing in multiple geographic regions (for companies that seek regional instead of national/global consistency) and testing over multiple days to achieve a complete block serving protocol so that preference segments can be identified through the use of cluster analysis (for companies that are willing to consider stockkeeping unit (SKU) multiple SKU's of the same product to address the individual preferences of the segments) (see Section 15.2.1.4).

The questionnaire for the consumer test can be relatively short, consisting of acceptability questions—overall and by modality (aroma, appearance, texture, flavor, etc.), and of a small number of specific attribute intensity questions. The acceptability data will be used to set the sensory specifications. The consumer intensity data are used to confirm that consumers are perceiving differences among the products in a manner consistent with the intensity ratings obtained from the descriptive panel. Consumer and trained panel ratings do not always agree because of differences between consumer and descriptive panel terminology, among other things. However, it is reasonable to expect a high level of agreement on common, broadly understood characteristics such as light to dark appearance, soft to firm texture, and common tastes and flavors like salty, sweet, and minty. Lack of agreement on these "sanity check" attributes may indicate a problem with the execution of the consumer test or changes in the sensory properties of the test samples between the initial sensory descriptive evaluations and the consumer test. If it arises, the lack of agreement needs to be resolved before moving forward to setting the sensory specifications.

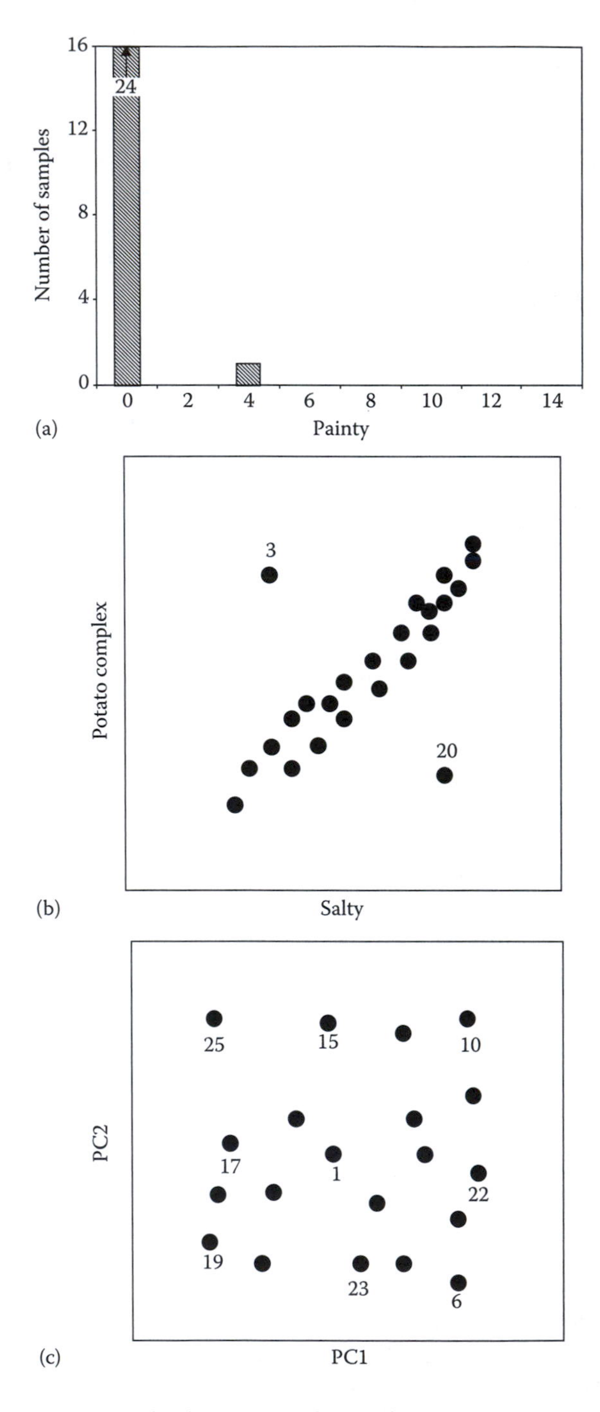

Figure 17.1 (a) One extreme sample, (b) two samples not following the general trend, and (c) samples selected to "fill" the sensory space of the remaining samples.

The results of the consumer test are summarized using the standard methods employed by the company, so that members of the QC/sensory team will be familiar with the results of the consumer test (e.g., average overall liking, top-two-box purchase intent, etc.). The key consumer acceptance measures will be combined with the sensory descriptive ratings of the test samples to identify the attributes that affect consumer acceptance and the ranges of attribute intensities that yield product with acceptably high liking ratings.

17.2.1.4 Establishing the Sensory Specifications

The relationships between consumer acceptance and the intensities of the sensory attributes lead to the sensory specifications for the product. Relatively simple statistical procedures and charting are sufficient to complete the task. Correlation analysis (Chapter 15) can be used to identify attributes that have a strong straight-line relationship with acceptance. The absence of a large correlation statistic does not always mean that there is no meaningful relationship between the sensory attribute and acceptance. Correlation measures only the strength of the straight-line relationship between two variables. Strong curvilinear relationships may exist between acceptance and an attribute that would be missed by the correlation analysis. That is why it is important to plot the acceptance versus intensity ratings of the study samples for each attribute on the final sensory descriptive ballot. Most graphing applications provide for the addition of curvilinear trend lines to the chart, so the general relationship between acceptance and the attributes intensities is readily apparent when viewing the charts (see Figure 17.2).

Three key patterns are likely to emerge on the charts:

1. There may be no systematic relationship between acceptance and the intensity of the sensory attribute (Figure 17.2a).
2. There may be a strong linear trend, either increasing or decreasing (Figure 17.2b).
3. There may be a strong curvilinear trend where intermediate intensities of the attribute have higher acceptability than either extreme (Figure 17.2c).

Each of the trends leads to a different type of sensory specification.

If there is no systematic relationship between acceptability and the intensity of an attribute, no sensory specification is needed for that attribute. For attributes that exhibit linear relationships with acceptability, one-sided specifications apply. Lower specification limits are set for attributes with increasing trends; upper specification limits are set for attributes with decreasing trends. For attributes that exhibit curvilinear relationships with acceptance, both lower and upper specification limits may be needed.

The process of setting the specification limits first requires that a minimal acceptance level be specified. The minimal acceptance level should fall between the lowest and highest acceptance ratings observed in the consumer test. If it does not, then either all of the production samples "pass" and no sensory specifications are required or all of the production samples "fail" and the organization needs to readjust its expectations of the consumers' opinion of their product. Assuming that the minimal acceptance level falls within the range of the consumers' acceptance ratings, then it can be used to establish the specification limits for sensory attributes that are systematically related to product acceptance.

A horizontal reference line at the minimal acceptance level can be added to the acceptance versus attribute intensity chart. The intensity or intensities that correspond to the

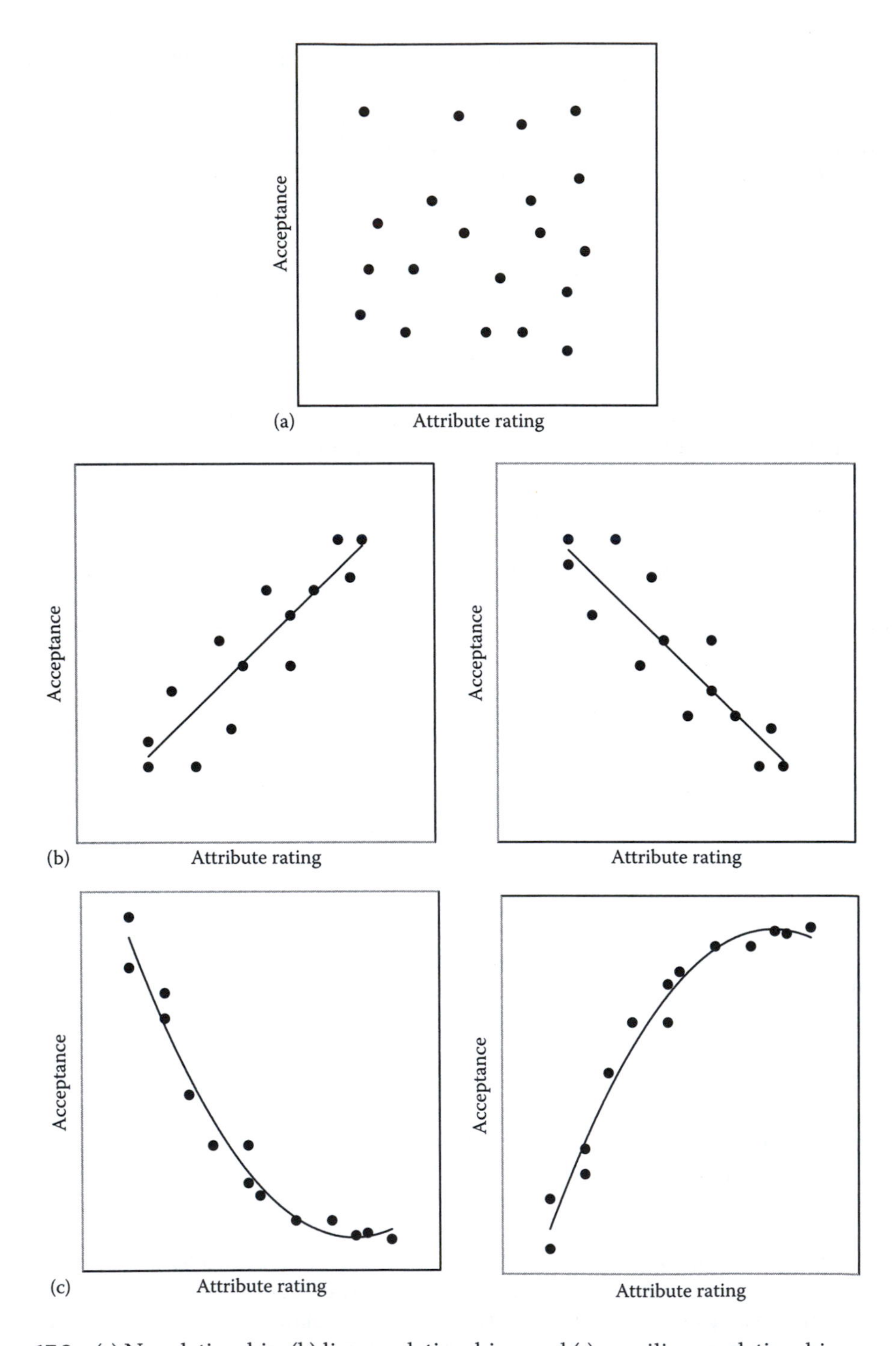

Figure 17.2 (a) No relationship, (b) linear relationships, and (c) curvilinear relationships.

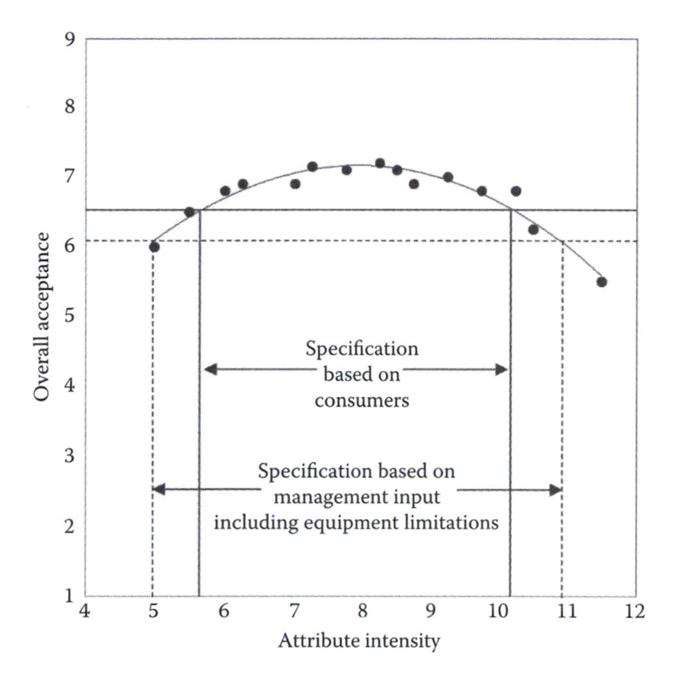

Figure 17.3 Establishment of specifications based on consumer responses alone and based on consumer responses and management input.

point(s) where the reference line crosses the trend line on the chart can be set as the tentative sensory specifications (see Figure 17.3). These tentative specifications should be reviewed by the QC/sensory team and adjusted as needed to account for any practical issues that influence the final decision, such as financial or production rate issues. The final set of consumer derived sensory specifications can then be published (See Table 17.2).

17.2.2 Implementing the In-Plant QC/Sensory Function

Once the QC/sensory specifications are developed, it is time to develop the in-plant resources that will be used to monitor the sensory quality of the products. The first step in this process is to identify an in-plant sensory coordinator who will oversee the operation of the in-plant QC/sensory panel and coordinate with the other QC functions to communicate the results of the panel's evaluations. Secondly, candidates for QC/sensory assessors need to be identified, screened, trained, and put to work.

QC/sensory assessors can be drawn from company employees or from nonemployees who are willing to work part time on the panel. There are pros and cons to both approaches, and the best solution will be different from one company to another. The availability of employees to participate on the panel, over and above their "normal jobs," is one issue. The cost of bringing in a pool of part-time employees is another issue. For more information on descriptive panels, see Chapter 10.

Once a pool of assessors is identified, the QC/sensory panel is developed using the practices described in Chapter 10. The candidates need to be screened for basic acuity and

Table 17.2 Final Sensory Specifications for Potato Chips

Attribute	Score
Appearance	
Color intensity	3.5–6.0
Even color	6.0–12.0
Even size	4.0–8.5
Flavor	
Fried potato	3.0–5.0
Cardboard	0.0–1.5
Painty	0.0–1.0
Salty	8.0–12.5
Texture	
Hardness	6.0–9.5
Crisp/crunch	10.0–15.0
Denseness	7.0–10.0

for their ability to be contributing members of the panel. The assessors who pass the initial screening then need to be trained on the attributes that will be evaluated as part of the QC/sensory evaluations and to be calibrated to the intensity scales that will be used. This process typically involves exposing the assessors to three samples per attribute, where the samples span the range of low to middle to high attribute intensities that the product samples are likely to exhibit. During this time, the external reference standards that will be needed during the ongoing operation of the program can be identified.

17.2.3 Product Sampling, Data Analysis and Reporting

When possible, the samples that the QC/sensory panel evaluates should be drawn from production at the same time as other QC samples are acquired. A common practice is to draw three samples per shift or per lot—early, middle, and late in the production lot.

The data analyses associated with the sensory descriptive program are straightforward. For each attribute, the panel's average intensity rating is calculated for each sample collected. If three samples are collected per lot, the averages are computed for each sample separately (see Table 17.3). This allows for the calculation of both an overall lot average and a measure of within-lot variability. The most common QC summary measures are the sample mean, $\bar{x}$, and the sample range, R, which is simply the difference between the highest and lowest intensities of the multiple samples that are collected within each lot (see Table 17.3).

The panel coordinator should tabulate the average intensities for all of the attributes for the lot and compare the averages to the sensory specifications. Any lots with attribute intensities that fall outside of the QC/sensory specification are reported to management (see Table 17.3).

Table 17.3 Summary Analysis of QC/Sensory Attributes

	Sample from Lot 011592A2				
Attribute	**Early**	**Middle**	**Late**	**Mean**	**Range**
Color intensity	4.6	4.2	5.0	4.6	0.8
Even color	4.8	4.6	4.2	4.5*	0.6
Even size	4.1	5.2	6.2	5.2	2.1
Fried potato	3.6	3.3	3.9	3.6	0.6
Cardboard	5.0	3.3	4.6	4.3**	1.7
Painty	0.0	0.0	0.0	0.0	0.0
Salty	10.5	11.8	12.3	11.5	1.8
Hardness	7.0	7.5	7.9	7.5	0.9
Crisp/Crunch	12.5	13.1	14.0	13.2	1.5
Denseness	7.4	7.1	7.5	7.3	0.4

* = Below low specification limit
** = Above upper specification limit

If the organization has implemented a statistical process control (SPC) program, the lot average and ranges can be plotted on control charts to monitor the changes in sensory attribute intensities from lot to lot over time. Standard SPC rules for identifying when assignable causes of variability are present can be applied to the QC/sensory data in the same way they would be applied to any other quantitative QC measure (see Figure 17.4).

The reporting of QC/sensory results should be integrated into the standard QC reporting system, so that the QC/sensory results can be considered along with all of the other QC measurements to determine the final disposition of a production lot.

17.3 DIFFERENCE-FROM-CONTROL METHODS

The difference-from-control (DFC) method (also known as the degree of difference method) (see Section 7.8) is widely used in QC/sensory. Implementations of the DFC method vary widely, from simply rating the overall difference from control to rating overall difference, rating differences on specific attributes, and collecting diagnostic information on the nature of the differences. The DFC method can be applied to finished products, raw materials, and intermediate products such as unflavored bases and so on. The scale used to rate difference from control also offers flexibility. It can be designed as a simple unidirectional scale ranging, for example, from 0 = "Not at All Different" to 4 = "Extremely Different". The scale also can be designed to provide directional information related to the attribute being analyzed—for example, −4 = "Extremely Weaker," 0 = "Not at All Different," to 4 = "Extremely Stronger" (see Figure 17.5). Training the panel to use the DFC method is less involved than the training for the attribute descriptive method. Panelists need to be familiarized with the control product and calibrated to the DFC scale through exposure to a variety of samples that have been manipulated to exhibit specific differences from the control product. The potential simplicity, high flexibility, and ease of panel training related

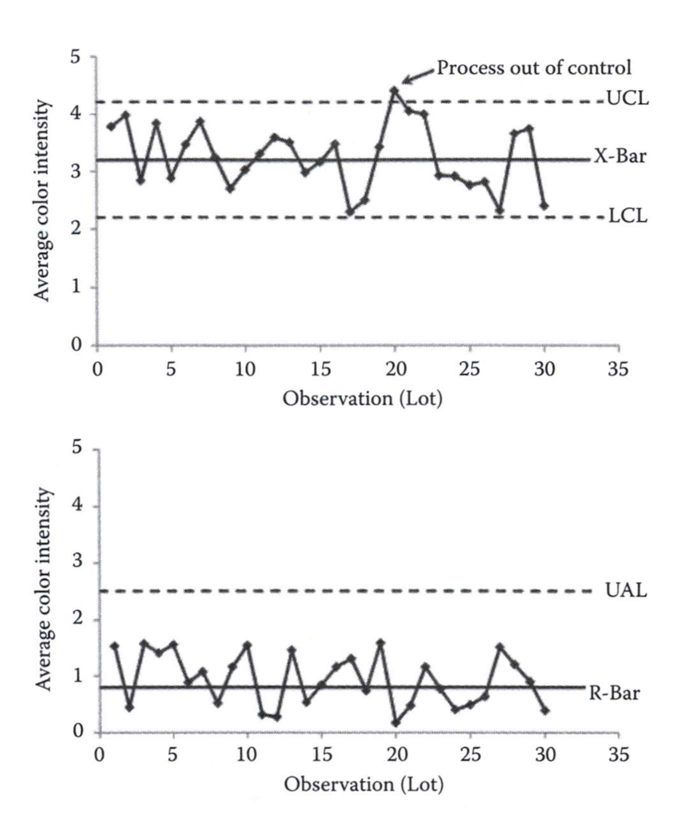

Figure 17.4 X-bar chart and R-chart of color intensity showing that Lot 20 is "out of control" because the lot average falls above the upper control limit (UCL) of the chart.

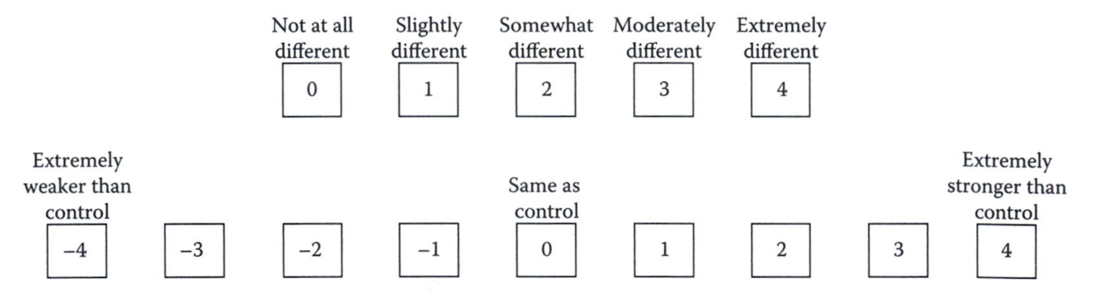

Figure 17.5 Examples of unidirectional and bidirectional difference-from-control scales.

to the DFC method are three of its primary advantages. There are two key disadvantages of the DFC method. The first is that the overall difference rating does not provide specific guidance on the nature of the difference that could be used to identify its source. Even the inclusion of a number of attribute scales does not guarantee that all sources of difference will be identified. The second potential disadvantage is that a sample that represents the control product must always be available. This means that either the control product must be easy to produce or that a control can be stored for extended periods of time without noticeable changes in its sensory properties.

The four steps in implementing a difference-from-control QC program are the same as those associated with the attribute descriptive approach:

1. Establish sensory specifications
2. Train and operate the plant panels
3. Establish a data collection, analysis, and reporting system
4. Operate the ongoing program

17.3.1 Establishing Sensory Specifications

The process of establishing QC/sensory specification in the DFC method is similar to the process used in the attribute descriptive method. A large number of samples are obtained from production. The samples are screened to eliminate redundancy and to arrive at a small number of samples that represent the full range of sensory differences that occur during production. This smaller set of samples is rated by a trained panel using the DFC scale that has been chosen by the company. The samples may also be evaluated for their descriptive properties to provide valuable diagnostic information regarding the magnitude of the DFC ratings and the characteristics that are associated with them. This small set of samples may be as few as five or six for a product that exhibits only one or two dimensions of sensory variability, or it may be as large as 15–20 for products that vary on a larger number of sensory properties. DFC ratings on specific attributes should be collected on products that vary on a number of sensory properties.

Once the DFC ratings, and potentially the attribute descriptive information, are collected on the screened product set, two paths may be taken to establish the difference from control specifications. The screened set of samples can be submitted to a consumer test and the resulting acceptance ratings can be correlated with the DFC values obtained from the trained panel to establish the QC/sensory specifications. Alternatively, management can review the screened set of samples along with the DFC ratings obtained from the trained panel and set the QC/sensory specifications based on the range of difference they feel represents the product that the company intends to sell.

The approach used to test the samples with consumers is the same as was described in Section 17.2 and will not be repeated here. The resulting consumer acceptance ratings can be correlated with the DFC ratings of the trained panel obtained on the same samples. The QC/sensory specification limit is set to maintain the overall acceptance rating of the product at an acceptably high level (see Figure 17.6). In Figure 17.6, the company has chosen a 0.5-unit drop in acceptance as the action standard that represents a meaningful difference in overall acceptability compared to the control product. The critical DFC rating is obtained from the chart and, with additional input from management related to practicality and financial impact, is used to set the final DFC specification. Based on Figure 17.6, a DFC rating up to 4.9 would be acceptable. Management reviewed the consumer test results and the relationship between acceptance and the overall DFC ratings and decided to set the QC/sensory specification at an overall DFC value of 4.1. This was done in part based

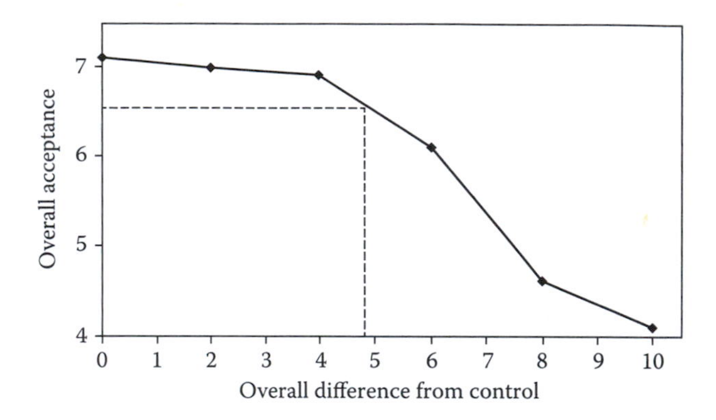

Figure 17.6 Overall difference-from-control vs. overall acceptance.

on the conservative nature of the company's management and because, after reviewing the products, management felt that the large majority of standard production would fall at or below an average DFC rating of 4.0.

Using an established action standard, such as a 0.5-unit drop, is recommended over the use of a statistically significant difference from the liking of the control product because the size of the difference that will be statistically significant depends on the number of respondents who participated in the consumer test. When extremely large numbers of respondents are involved, the size of the statistically significant difference may be trivially small in practical terms. Conversely, when only a small number of respondents are involved, the size of the statistically significant difference may be larger than the amount anyone associated with the product would be comfortable with. If possible, when planning the consumer test, the number of respondents should be set so that the size of the significant difference matches the size of the difference with which the organization is comfortable.

For products that exhibit variability on a variety of sensory properties, the same process of correlating the consumers' acceptance ratings with the attribute DFC ratings is used. Both the consumers' overall acceptance rating and the attribute acceptance ratings should be considered in the correlation analysis. It is recommended that somewhat greater weight be placed on the overall acceptance rating than on the attribute acceptance rating because consumers may react more strongly to differences from control on a specific attribute when their attention is drawn to it (by asking the attribute acceptance question) than they would if they were performing an overall assessment of acceptance. Regardless, the relationship between acceptance and difference from control is charted and specifications are developed for each key attribute (see Figure 17.7).

In Figure 17.7a, we observe that acceptability does not change as the size of the difference from control increases. No QC/sensory specification is needed for the flake size attribute. In Figure 17.7b, there is a clear relationship between acceptance and the amount of raisins. However, raisin counts are taken regularly in the QC lab, so there is no need to measure this attribute as part of the QC/sensory program. A specification is set for raisin firmness (Figure 17.7c) because there is a clear relationship between raisin firmness and acceptance, and no other method is being used to track this attribute.

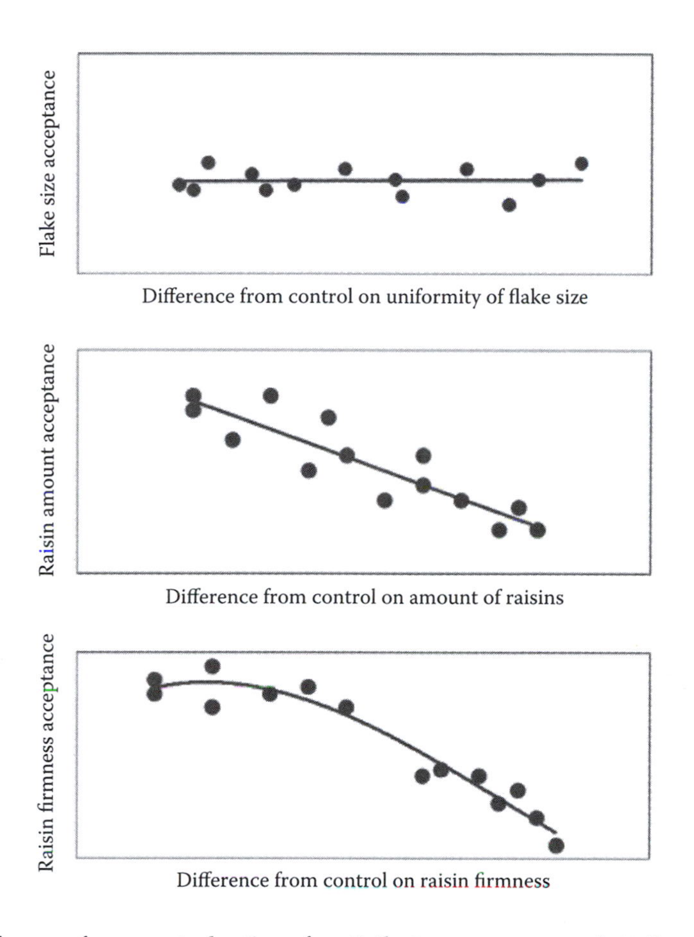

Figure 17.7 Difference-from-control ratings for attributes vs. consumers' attribute acceptance.

17.3.2 Implementing the In-Plant QC/Sensory Function

All of the considerations mentioned in Section 17.2.2 apply when implementing the in-plant difference-from-control QC/sensory function. In addition, it is critical that the panelists are intimately familiar with the DFC rating method and the sensory properties of the control product itself. The training should involve repeated exposures to the control product, both labeled and blind, and to a set of reference products that represent the entire range of the DFC scale that the panelists will encounter in practice. Regular feedback on the accuracy and reproducibility of each panelist's ratings is encouraged.

When appropriate, during the implementation phase it also is important to sensitize the panelists to the natural range of variation that may exist in the control product. Often, the control product is not a single point in the sensory space of the product but is instead a small region in the sensory space defined by natural variation in raw materials and processes. In practice, multiple control products may need to be included in the QC/sensory evaluations to account for this natural variation. Whether the control product is a highly reproducible product or one that consistently exhibits some amount of natural variation will influence the product sampling and data analyses associated with the in-plant difference-from-control QC/sensory program.

17.3.3 Product Sampling, Data Analysis, and Reporting

As with the attribute descriptive method, the samples evaluated by the difference-from-control QC/sensory panel should be drawn from production at the same time other QC samples are acquired. A common practice is to draw three samples per shift or per lot—early, middle, and late in the production lot.

For products that exhibit little batch-to-batch variation, a single control product can be included in the QC/sensory evaluations. For products that regularly exhibit perceptible batch-to-batch variation, multiple controls should be included in the evaluations. If the total number of samples is small enough for each panelist to evaluate the entire set, a complete-block serving protocol can be used. If, on the other hand, the total number of samples is too large to be evaluated by a panelist in a single session, a balanced incomplete block (BIB) serving protocol should be used (see Section 14.5.4).

The data analyses associated with the DFC program are described in Chapter 7. Methods associated with both the single control and multiple control sampling plans are presented. The panel's average DFC ratings for each sample are computed. If multiple control samples are included in the evaluations, the average of the control product is computed. If multiple production samples are included in the evaluations, the average and range of the production samples are computed. Two action standards are commonly applied to DFC ratings. The first is to statistically compare the average of the production samples to the average of the controls, using the appropriate ANOVA contrast calculation. If the production samples are significantly more different from the control than the control is from itself (i.e., the placebo effect), then that batch of production is deemed to be either out of spec or out of control, and remedial action may be required. The second approach is to use fixed cutoff points on the DFC scale without consideration of statistical significance. For example, on a 0 to 4 DFC scale, the following cutoff values might be chosen: An average rating between 0 and 1.5 would be considered acceptable production (i.e., green or go); averages between 1.5 and 2.5 would be considered marginal production (i.e., yellow or inspect further); and averages greater than 2.5 would be considered unacceptable production (i.e., red or no go). If consumer data were used to set specifications, they would be applied when defining these cutoff values. If the specifications were based on management input alone, the green-yellow-red cutoff values would be part of the management decision process when the program was initiated.

The panel coordinator should tabulate the average DFC ratings for the lot and compare the averages to the sensory specifications. Any lots with DFC values that fall outside of the QC/sensory specification are reported to management. As with the attribute descriptive method, if the organization has implemented an SPC program, the lot average and range can be plotted on control charts (similar to Figure 17.4) to monitor the changes in the DFC ratings from lot to lot over time. Standard SPC rules for identifying when assignable causes of variability are present can be applied to the QC/sensory data in the same way they would be applied to any other quantitative QC measure.

The reporting of QC/sensory results should be integrated into the standard QC reporting system so that the QC/sensory results can be considered along with all of the other QC measurements to determine the final disposition of a production lot.

17.4 IN–OUT METHOD

The in–out method is a variation on the "A"–"Not A" discrimination test (see Section 7.7) applied in a QC/sensory environment. In this method, trained panelists evaluate production samples and classify them as either "in" or "out" of the acceptable range of variation associated with the control or target product. The value used to determine the disposition of the lot is the percentage of the panel that classifies the production samples as being "in". The method is especially useful in detecting gross defects in raw materials and intermediate and finished products. The method applies best to products that exhibit variation in only a small number of sensory properties and ones for which the deviations from control are typically large and easily detectable. Simplicity is the primary advantage of the in–out method. It requires the least amount of panel training and, typically, the shortest amount of time needed to perform the QC/sensory evaluations themselves. The main disadvantage of the program is that it offers little or no information on the nature of the character of any product defects that are detected. The in–out method should be viewed as a decision-making tool rather than as a source of product information. Another disadvantage of the method is that because the raw data are in the form of counts (i.e., 0–1 data), a larger panel is required to obtain statistically sensitive results. In practice, large panels are not used with the in–out method. Instead, fixed action standards are applied to the percentage of "in" classifications, regardless of the statistical significance of the result. Lastly, the in–out panel requires careful monitoring and ongoing maintenance because the panelists' ratings are directly connected to the disposition of the product. Constant care must be exercised to ensure the panelists are assessing the products in an objective fashion and are not being biased by the pressure to meet product quotas or other, non-quality-related criteria.

17.4.1 Establishing Sensory Specifications

Unlike the attribute descriptive and DFC methods, consumer input is seldom used to set the specifications in the in–out method. Instead, specifications are set based on management input alone. Through intimate familiarity with the control product and an understanding of the most common sources of product defects, management identifies reference samples that represent both acceptable—"in"—and unacceptable—"out"—product characteristics.

Descriptive evaluations of an array of both "in" and "out" products can aid in the process by identifying key sensory attributes that affect quality and the intensity ratings associated with the boundaries between "in" and "out" products on each key attribute. This information is especially useful for maintaining a supply of reference standards used for panel training and ongoing QC/sensory operations.

During the specification setting process, management also needs to specify the action standards that will be applied when determining the disposition of the production samples evaluated in the QC/sensory program. Often a single go/no-go criterion is applied. The production lot or raw material is deemed acceptable whenever 50% or more of the panel classify the sample as "in". Alternatively, a green-yellow-red approach can be applied, where, for example less than 40% "in" is considered unacceptable (i.e., "red" or out of spec.); more than 60% "in" is considered acceptable (i.e., "green" or in spec.); and between 40% and 60% "in" is considered marginal ("yellow," requiring additional inspection).

17.4.2 Implementing the In-Plant QC/Sensory Function

As with the other QC/sensory methods discussed, the first two steps of the implementation process are to identify a QC/sensory coordinator, who will be responsible for the ongoing operation of the program, and to establish a QC/sensory panel that can reliably apply the in–out method to assess the quality of the production samples. Once a pool of potential panelists is identified, they should be screened for acuity and their ability to use the in–out assessment method, using the methods presented in Chapter 10.

The potential panelists who successfully complete the screening process are then exposed to the reference samples, identified during the specification setting process, that represent both "in" and "out" product. When available, the nature of the defect that makes the sample "out" can be presented and discussed (e.g., the presence of an off flavor in a raw material or a burnt note in a final product caused by overheating the product during production). During training, the panel should be exposed to all of the product characteristics that will lead to an "out" classification. Practice sessions in which both "in" and "out" samples are presented in a blind manner should be conducted to ensure that all of the panelists can reliably classify the reference samples correctly. Panel worksheets and score sheets similar to those used in the "A"–"Not A" test (see Section 7.7.4, Figure 7.15 and Figure 7.16) can be used for the in/out assessments performed by the QC/sensory panel. Appropriate remedial actions should be taken until all of the panelists can reliably classify the reference samples correctly.

17.4.3 Product Sampling, Data Analysis, and Reporting

As with the attribute descriptive and DFC methods, the samples that the in–out QC/sensory panel evaluates should be drawn from production at the same time other QC samples are acquired. A common practice is to draw three samples per shift or per lot—early, middle, and late in the production lot.

If the size of the in–out panel is large enough to warrant statistical analyses, the methods used for the "A"–"Not A" test (see Section 7.7.5) can be applied. A reasonable action standard for the statistical approach would be to accept a lot of production if significantly more than 40% (yellow acceptance) or 60% (green acceptance) of the panel rate the lot as being "in." As mentioned earlier, however, the size of a typical in–out panel is too small to yield sensitive statistical results. Instead, fixed action standards established during the specification setting process are used (e.g., the go/no-go or green-yellow-red action standard mentioned earlier).

Table 17.4 Results on Several Production Batches of Strawberry Yogurt Using the "In/Out" Method

	Percent "In"	
Batch	**Average**	**Range**
A3-0016	65.6	20
A3-0018	59.8	19
A3-0020	58.4	18
A3-0024	42.6*	23
A3-0026	44.8*	17
A3-0028	74.3	24

* Out of Specification (Percent "In" 50%)

The percentage of the panel that classifies each QC/sensory sample as "in" is computed. If multiple production samples are included in the evaluations, the average and range of the "in" percentages are computed across the multiple pulls for the lot being evaluated. The lot-average "in"% value is then compared to the action standard to determine the disposition of the lot.

The panel coordinator should tabulate the "in"% average and range for the lot and compare the averages the sensory specifications (see Table 17.4). Any lots that fall outside of the QC/sensory specification are reported to management. If the organization has implemented an SPC program, the lot average and range "in"% can be plotted on control charts (similar to Figure 17.4) to monitor the changes in the "in"% from lot to lot over time. Standard SPC rules for identifying when assignable causes of variability are present can be applied to the QC/sensory data in the same way they would be applied to any other quantitative QC measure.

The reporting of QC/sensory results should be integrated into the standard QC reporting system and considered along with all of the other QC measurements to determine the final disposition of a production lot.

REFERENCES

Amerine, M. A., E. B. Roessler, and F. Filipello (1959). Modern sensory methods of evaluating wines. *Hilgardia* 28: 477–567.

ASTM (1992). In J. Yantis (ed.), *MNL 14 The Role of Sensory Analysis in Quality Control*. West Conshohocken, PA: ASTM International.

Aust, L. B., M. C. Gacula, S. A. Beard, and R. W. Washam II (1985). Degree of difference test method in sensory evaluation of heterogeneous product types. *J Food Sci* 50: 511–3.

Carlton, D. K. (1985). Plant sensory evaluation within a multi-plant international organization. *Food Technol* 39(11): 130–3.

Chambers, E. IV (1990). Sensory analysis-dynamic research for today's products. *Food Technol* 44(1): 92–4.

Dawson, E. H. and B. L. Harris (1951). Sensory methods for measuring differences in food quality. *U.S. Depart Agri Agricult Inform Bull* 34: 1–134.

Dove, W. F. (1947). Food acceptability: Its determination and evaluation. *Food Technol* 1: 39–50.

Dziezak, J. D. (1987). Quality assurance through commercial labs and consultants. *Food Technol* 41(12): 110–27.

Hinreiner, E. H. (1956). Organoleptic evaluation by industry panels: The cutting bee. *Food Technol* 31(11): 62–7.

Hubbard, M. R. (1990). *Statistical Quality Control in the Food Industry.* New York: Van Nostrand Reinhold.

Mastrian, L.K. (1985). The sensory evaluation program within a small processing operation. *Food Technol* 39(11): 127–9.

Munoz, A. M., G. V. Civille, and B. T. Carr (1991). *Sensory Evaluation in Quality Control.* New York: Van Nostrand Reinhold.

Ough, C. S. and G. A. Baker (1961). Small panel sensory evaluations of wines by scoring. *Hildegardia* 30: 587–619.

Pangborn, R. M. and W. L. Durkley (1964). Laboratory procedures for evaluating the sensory properties of milk. *Dairy Sci* 26(2): 55–62.

Peryam, D. R. and R. Shapiro (1955). Perception, preference, judgment-clues to food quality. *Ind Qual Cont* 11: 1–6.

Platt, W. (1931). Scoring food products. *Food Ind* 3: 108–11.

Rutenbeck, S. K. (1985). Initiating an in-plant quality control/sensory evaluation program *Food Technol* 39(11): 124–6.

Sidel, J. L., H. Stone, and J. Bloomquist (1983). Industrial approaches to defining quality. In A. A. Williams and R. K. Adkins (eds.), *Sensory Quality in Foods and Beverages: Definitions, Measurement and Control* (pp. 48–57). Chichester, UK: Ellis Horwood.

Stouffer, J. C. (1985). Coordinating sensory evaluation in a multi-plant operation. *Food Technol* 39(11): 134–8.

18

Advanced Consumer Research Techniques

18.1 INTRODUCTION

The content covered in Chapter 13 is the "core" sensory consumer methodology for most sensory applications. Chapter 18 presents other approaches to understanding products and consumers and focuses on a deeper, more refined way to explore product and consumer understanding by linking innovation, creative understanding, and consumer data. The creative combination of tools from anthropology, ethnography, and creative problem solving insights can be used in combination with traditional sensory methods.

These additional approaches are meant to (1) create actionable output and (2) broaden the horizons of sensory professionals. A primary goal of sensory professionals is to add insight and value to product development and innovation. Lack of insightful and actionable data can slow this process. The tendency is to go back and apply many of the original tools that were previously discussed. To invigorate innovation, the tool box must be expanded. How and why should sensory methods be involved in innovation? Product developers and sensory professionals have a deep understanding of products. The research and development environment should be equipped with tools and approaches to link that technical knowledge with information about consumers. The core tool box can be augmented with approaches that enable better understanding of the consumer, the products, and their interactions.

Table 18.1 suggests approaches for innovation research.

The approaches found in Chapter 18 are based on some of the tools in preceding chapters. This chapter contains methods for seasoned research professionals who know and use these standard sensory methods. These new techniques are intended to be enhancements and advancements to the basic tools.

Table 18.1 New Approaches for Innovation Research

Situation	Suggested Approaches
New formats	Front-end innovation; sequence mapping; iconic experience; benefit perception; behavioral economics; in-context research
Concept development/ ideation	Front-end innovation; consumer cocreation; Kano; benefit perception; behavioral economics; in-context research
Product line expansion	Iconic experience; sequence mapping; consumer cocreation; category appraisal; Kano
New product category expansion	Sequence mapping; category appraisal; Kano; ad claims

18.2 FRONT END OF INNOVATION

18.2.1 Definition, Purpose, Outcome

Uncovering consumer needs often occurs in the beginning, at the front end of innovation. Typically, the research is conducted at the very early stage of a project, when planning is being carried out, initial market and technical feasibility is being assessed, and breakthrough ideas are being explored. Research at the front end of innovation is conducted before dollars are committed to detailed technical assessment, costly concept testing is executed, and/or significant manpower and out-of-pocket expenses are committed. These tools and techniques can be applied to understand the consumer early in the product development process *and* the subsequent stages of development.

Methods used are unique because they gather in-depth information on who the consumer really is, how and why they use products, and what they really like, dislike, and *need*. To capture this level of information, one must move beyond the standard, frequently used quantitative and qualitative approaches.

18.2.2 Applications

Research at the front end of innovation allows for the

- Exploration of consumers as purchasers of products with specific features or sensory properties identified
- Study of product functionality and ergonomics
- Determination of how a consumer is modifying a product or adapting usage to suit his/her needs
- Uncovering of attitudes, behaviors, and motivators within the culture
- Study of the consumers in their own environment through observational research to uncover unmet needs and desires

18.2.3 Tools and Techniques

There are many methods or techniques that can be used to uncover consumer's thoughts and ideas leading to new product ideas and beyond. The techniques that are used most often are qualitative in nature and occur in the field; however, quantitative approaches can also be effective. Consumers may be studied in context in their homes, on the street, in stores, or at point of purchase—when and where dollars are spent. Going to the field and observing consumers is often referred to as *ethnography* or *immersion*. When immersing oneself in the consumer's environment, information is gathered through observation and dialogue. Beyond the traditional techniques used to elicit information from consumers in focus groups or one-on-one interviews, information-gathering approaches that are used in support of the front-end innovation are often imagery-based and include, but are not limited to, compare-and-contrast methods, mind maps, word webs, and collages. Quantitative techniques that go beyond central location tests (CLTs) or home use tests (HUTs) to consider include online research and intrinsic/extrinsic studies. Online research provides early exploration into the design of concepts, attitudes, and behavioral research. Intrinsic/extrinsic research studies the essential aspects of a product along with the external motivators. See Section 13.7 for a detailed discussion of the use of Internet research at the front end of innovation.

As stated before, when studying consumers in context through observation, a deeper understanding is possible. This world approach helps to uncover actual behaviors. When conducting research of this type, there are two different paths that can be taken to capture the information. One approach has the researcher being a participant observer who watches without conversing with the consumer. The other approach has the researcher being an observational interviewer who actively interacts with the consumer, probing in-depth on areas of interest. Consider the degree of researcher and participant interaction in the following scenarios.

- Going into a home to observe a primary caregiver making lunch for their children, the observer studies what ingredients are used (bread, meat, condiments, and cheese), where the ingredients are stored (pantry and refrigerator), and how the sandwich is assembled (number of steps) and served. Multiple home visits would uncover differences in preferences as well as needs and behaviors.
- Watching and interviewing different women aged 18–21, 30–35, and 60–65 selecting and purchasing lipstick, gloss, or foundation would demonstrate different preferences based on age, lifestyle, and skin type. Further information related to the product use, what questions are asked, and what colors are selected is uncovered.
- Community narratives, also referred to as *storytelling*, are a qualitative research method in which a creative consumer group is encouraged to share experiences with products and to express their feelings. Community narrative techniques probe beyond surface consumer attitudes, behaviors, and feelings to allow researchers to learn at a deeper emotional level than that allowed by traditional qualitative methods. Consumers are encouraged to express their experiences and feelings, allowing motivations and unarticulated needs to be uncovered and new insights to emerge, often by building on other group members' ideas. Using literal and figurative exercises, the storytelling process provides focus on the researcher's initial questions while allowing spontaneous "verbal excursions" by group members. Understanding consumer responses on product sensory attributes is a major outcome.

18.2.4 Design of Front-End Innovation Research

The keys steps to working at the front end of innovation can be broken down into a framework called *I-SIGHT.* This is a dynamic framework that can be used throughout the product development process. Specific steps of this framework are

1. **I**nnovate: stimulate creativity and innovation through team building
2. **S**ynthesize current knowledge: summarize the known and the unknown, clarify facts and opinions
3. **I**dentify objectives: define opportunities and set objectives for the research
4. **G**enerate data through carefully designed research:
   a. Choose or create the right method to meet the project objective
   b. Take the path to conduct the research—the preparation phase
   c. Gather the information
5. **H**arvest ideas: uncover the truth underneath the data by organizing, sorting, and relating the information
6. **T**ake action: determine an action plan to move on to the next step of a project

The design of research at this stage of a project can take more time and thought than standard qualitative or quantitative research because it is specifically designed for the project or concept being studied.

18.2.5 Data Analysis and Mining

The analysis of front-end innovation data involves distilling the information into insights and possibilities or opportunities. Front-end innovation data is primarily qualitative and open to varying interpretations. Unlike quantitative data, the outcome is words, pictures, or stories—not quantitative with tables and statistical analysis. It is important to remember that front-end innovation research is exploratory and therefore allows the researcher to pursue several interpretations. The distillation of the data begins with putting aside personal biases and looking within and among all the collected data and asking questions about:

Commonalities: For instance, in a collection of individual collages, is there an underlying theme? Do one or two colors appear throughout the set? Do images, shapes, or objects repeat themselves? Are there commonalities in stories among subjects and do they use similar metaphors or words?

Missing information: What didn't people mention? What is avoided? What was uncomfortable for people to discuss? How are individuals compensating?

Sensory attributes: What attributes are mentioned or highlighted the most often? What attributes are never discussed? What product attributes frustrate people?

Interesting connections: How are the data connected? What sensory attributes are connected to key emotions? What connections are interesting? For example, every exclusive store in the mall has dark wood and gold tones; does that combination indicate exclusivity? What unusual relationships can you create by putting data together?

It is often useful to use a mapping technique (such as sequence mapping—see Section 18.3) to cluster collected data into manageable and thematic groups. Mapping approaches are primarily organization techniques. These techniques become very powerful and identify further insights when completed in a group setting. It is easier to look for insights within smaller sets of data, not to mention that the very act of organizing the data is an opportunity to answer the above questions.

18.2.5.1 Case Study: Understanding Consumer Perception of Crispy and Crunchy

18.2.5.1.1 Objectives

The objective of this experiment was to better understand consumer perception of crispy and crunchy as well as to assist in defining where crisp and crunch fall within Sensory Spectrum's fracturability scale.

18.2.5.1.2 Process

Sensory Spectrum's Community Narrative Panel (see Chapter 13, Section 13.4) was asked to define crispy and crunchy, outlining the differences between them. Products were shown, and panelists evaluated whether each was considered crispy, crunchy, or both. Each panelist placed the product within a map where they feel each fits. Once each product was evaluated and plotted, the panel discussed reasons for placement. Common characteristics between samples that were within the same quadrants were discussed.

18.2.5.1.3 Outcome

In this study, we learned that density, fracturability, noise level/pitch, and how labor intensive the product was determined crispiness or crunchiness for consumers. Persistence of crunch was a factor for consumers in selecting whether a product was crispy or crunchy—the longer the experience, the more it was considered crunchy as opposed to crispy. For the most part, the higher the fracturability, the crispier a product became for consumers (there are exceptions to this rule—i.e., ginger snaps).

18.3 SEQUENCE MAPPING

Throughout any day, consumers have thousands of touch points with a large variety of products. These moments in time range from the first visual interactions with a product on the store shelf through use to the sights and smells of disposal (Figure 18.1). At each point of interaction, a consumer will not only demonstrate a specific action or rubric but also register, if only within their own brain, the sensory characteristics of the product as well as the motivations and the emotions that are experienced. Sequence mapping is a method that allows the researcher to

1. Capture the individual consumer's actions (steps) along with their perceived sensory attributes, emotions, and motivations
2. Distill the collection of individual consumer information and create a collective map
3. Overlay descriptive analysis profiles across the stages of consumer-product interaction
4. Identify areas of opportunity for new products or product improvement

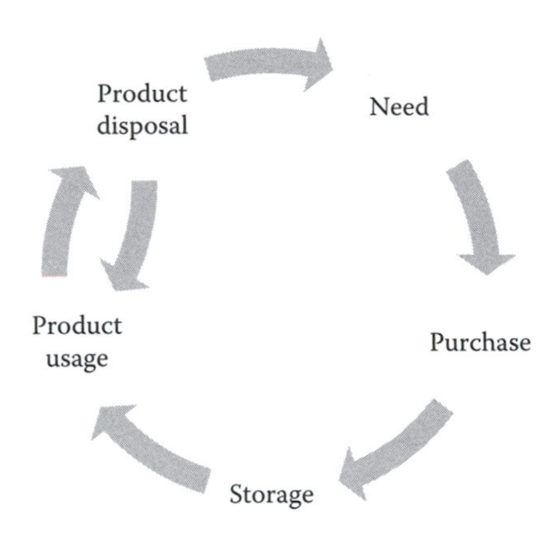

Figure 18.1 Sequence mapping captures the entire lifecycle of the product from detection of a need through repeated usage and product disposal.

A sequence map is best described as a complete and thorough events timeline, as heard and understood through the spoken and unspoken words of the consumer, designed to uncover hidden insights. It is the merging of thoughts, emotions, actions, and perceived sensory properties throughout the product lifecycle from early decision making through disposal. Example 18.1 and Map 18.1 depict an example of the research highlights from sequence mapping. The approach applies a series of qualitative techniques to individuals who meet a defined target group (e.g., acceptors, rejecters). The techniques include but are not limited to diaries, pictures, videos, home visits, point-of-purchase interviews, observational research, ethnography, in-context research, one-on-one interviews, focus groups, and so on. This data collection phase is typically conducted over 5–10 days.

The specific types of data collected for a sequence map include

1. Event
   a. High level category that names the event, e.g. purchase decision, product experience, overall satisfaction, etc.
   b. Purchase decision, product experience, overall satisfaction
2. Detailed event
   a. Detailed description of all possible actions within the event
   b. Allows multiple uses or segmentation, e.g. moms versus children, heavy users versus first time users, and so on
   c. Product interactions
3. Motivators
   a. Specific needs that motivate choices, steps
4. Emotions for events
   a. Specific emotions that describe how one feels during process
5. Sensory voice of the consumer
   a. Language used by the consumers to describe sensory properties
6. Descriptive attributes to describe the sensory experience

The data collected is qualitative in nature and may be generated within consumer shopping excursions, cohort group discussions, consumer qualitative groups, home visits, online research, consumer boards, and other social media outlets.

Example 18.1 Case Study: Creation of a Sequence Map

A leading manufacturer wants to create a product for women aged 24–50 years who have an on-the-go lifestyle. These individuals desire a sweet treat as part of a healthy lifestyle. Research is comprised of observational research in grocery, convenience, drug, and mass-merchandise stores, where point-of-purchase and one-on-one interviews are conducted among women who satisfy two segments: women who want indulgence and women who want healthy alternatives. The resultant map illustrates the motivations, emotions, and sensory attributes that are important to the consumer.

Map 18.1 demonstrates the first-level research highlights from the case study.

The sequence map results revealed that one product could be created to satisfy both niches: on-the-go indulgence and healthy living. To be successful, the product must have specific qualities, including

- A flavor that is clean, indulgent, and flavorful
- A creamy and/or crispy indulgent texture
- A clean aroma that is free from strong protein character or other nutrients such as soy, casein, vitamins, or minerals
- A high-quality milk chocolate
- No unpleasant aftertaste often associated with protein, soy, or vitamins

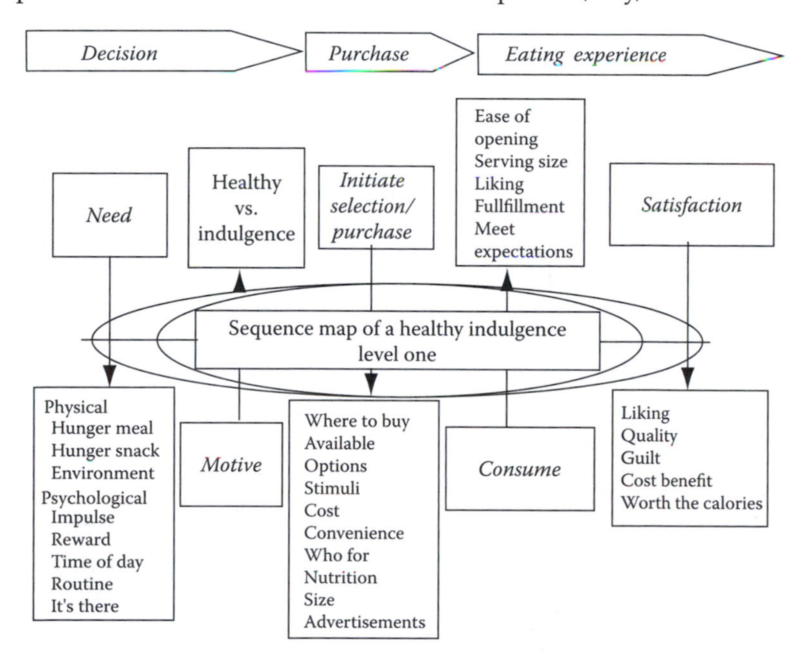

Map 18.1 Level one sequence map of research highlights for a healthy indulgence project.

- Smaller size, with a two- to three-ounce piece optimal
- Made with fresh ingredients and nutritionally balanced
- Added fiber is a plus
- Flavor options beyond chocolate:
    - Caramel nutty character, such as that delivered with salt and peanuts, toasted soy, or peanut butter; chocolate in combination with anything, including caramel, berries, peanut butter, or yogurt; berries, which in the right form are considered both indulgent and healthy
- Additional suggested forms:
    - Yogurt in a pudding tube for convenience
    - Layered products such as a ganache with a crispy center

Sequence mapping data can be used to support innovation. In the example above, identification of the emotions, motivations, and product attributes that deliver a healthy indulgence can help guide product development.

18.4 CAPTURING THE ICONIC EXPERIENCE

18.4.1 Definition and Purpose or Scope

The iconic experience is a qualitative approach developed to understand the deep emotional connection that consumers have to products that are special to them. It reveals the emotional language of the consumer and explores the sensory experience through a series of iterative qualitative research groups. The results uncover the product nuances that bring the brand "back home" to your consumer. Often, quantitative research offers solutions to many research questions, but it can fall short in aiding an understanding of the consumer response. The iconic experience approach taps into those responses and provides a full understanding of the sensory experience.

18.4.2 Applications

The iconic experience research method is appropriate when product developers are considering a significant change to or an extension of a product line. Additionally, if there has been a "dimming" of the signature experience—the experience that attracted consumers to the product in the past—this methodology will dive into these complex questions and provide solutions for both developers and marketing teams. The following are some possible project objectives where iconic experience may be an applicable solution.

- Uncovering the nostalgic original attributes, lost through cost reduction or systematic product quality erosion over time
- Defining and clarifying sensory properties that lead to higher consumer satisfaction
- Facilitating the creation of a winning product by uncovering optimal sensory properties
- Providing clear direction to product developers to create a product that satisfies consumer needs
- Uncovering consumer language to describe the products in the category

- Benchmarking prototypes against products currently in the marketplace by understanding strengths and weakness and applying what is learned to product design

18.4.2 Design of Research

The design of the iconic experience starts with commissioning the consumers to create or recreate the "authentic" version of the product. Through an iterative process, researchers break apart the language provided by the consumer, uncover sensory signals or cues to clarify consumer terms or attributes by using examples of products with the key attributes and gain a better understanding of the needs and desires of the consumer. This methodology affords the researcher the opportunity to listen more closely to the consumer.

Step 1	Objective setting and reference selection/review
Step 2	Understanding consumer experience
Step 3	Prototype design and review

Step 1: Objective Setting and Reference Selection/Review

Prior to gathering with consumers, a meeting is conducted to clarify objectives and action standards, define roles, and review test samples. Obstacles or limitations are also reviewed here by identifying any constraints such as raw materials, sourcing, plant and equipment issues, and so on. Discussion of areas of exploration and key questions are included at this time. Reference products are reviewed and selected based on anticipating attributes of interest to the consumers. Members of various teams should be present at this meeting: product development, marketing, research supplier, and moderator.

Step 2: Understanding Consumer Experience

Qualitative sessions, led by a trained moderator, are conducted with a target audience, which might be regular users and/or lapsed users. The purpose of this first round of groups is to pinpoint how consumers define the product. More specifically, the consumers are requested to evaluate references (selected in Step 1) and pinpoint the sensory properties that create the "iconic" experience that they recall. A series of exercises focusing on appearance, flavor, texture, skinfeel, packaging, and so on (dependent on the category) are utilized here to establish if the sensory properties of the references are appropriate.

Consumer feedback from this step of the research should include

- Cost of entry elements: Basic sensory properties or functionalities that are expected in a product.
- Key functions: Functional reasons consumers buy this product—image, perceived health benefits, appearance, flavor hit, satisfy craving, and so on. Together, the synergy of these elements is what will please more consumers, more of the time.
- Key drivers: Sensory properties that increase a product's overall acceptance.
- Added value/unexpected satisfiers: Elements that are not top of mind yet increase a product's overall interest and desirability.
- Undesirables or dissatisfiers: Attributes that lower a product's acceptance or pull it further from matching consumers' recollection.

Step 3: Prototype Design and Review
Qualitative sessions, led by a trained moderator, are conducted with the same consumers returning for a second round of testing. The purpose of this second round of groups is to review prototypes and further design the optimal product. Concepts may be presented here, with consumers being asked to respond to concept fit. Key sensory drivers that were defined in the first round are explored, and further development direction is provided.

18.4.3 Tools and Techniques

As in all qualitative research design, it is important that the moderator use the most appropriate and engaging tools and techniques. A working knowledge of sensory and descriptive information can also be leveraged, especially when probing with consumers. The following are some potential creative tools and techniques to utilize in iterative sessions.

- Prework assignments (consumers are asked to write a story, create a collage, or comment on a product they are currently familiar with)
- Framing the discussion, setting the stage (using ice breakers and warm-up exercises to establish the expectations of consumers, gain trust with the moderator, and build community)
- Establishing expectations, current needs (begin to unpack currently *unmet* needs)
- When evaluating each attribute, having consumers rank each sample on the same scale for proximity to their "ideal" or "best" version for that attribute
- Encouraging creativity by commissioning the consumers to build concepts or suggest product names

18.4.4 Data Analysis and Mining/Conclusions

Iconic experience methodology tells a more refined story of the optimal sensory properties that lead to higher consumer satisfaction. From this research, a hierarchy of features to deliver within the iconic product and/or brand are unveiled—with what is most important and most meaningful to the consumer becoming clearer. Product developers then have a sensory guide directing the creation of a product that satisfies these consumer needs while remaining within the range of sensory attributes consumers use to define the product as iconic or authentic.

As researchers, product developers, and marketers, our needs and desires are often not the same as those of our consumers. The outcome of the iconic experience methodology sheds light on multiple meanings held by consumers that do not represent the same definition used by product development.

18.5 CONSUMER COCREATION

Consumer cocreation is a highly experiential technique that integrates creative problem solving and sensory evaluation methodology within qualitative research. Consumers use physical stimuli to make prototypes or protocepts of the products they desire, working directly or indirectly with product developers. Cocreation with consumers allows the product developer to design a sensory experience that is highly influenced by what the target users would design for themselves.

Participants are screened for demographics, category use, articulation, and the ability to express their creativity in one or more ways. Participants are typically users of a category of products rather than a specific brand. Both established and innovative qualitative data-collection methods are used. While traditional focus groups allot only a few hours to probe the feelings and actions of representative consumer groups, this process has consumers work for 4–5 h as a team and in small group teams to create ideas and model products to provide direction and guidance to developers. In some cases, groups return to respond to the developer's interpretation of their model products and to refine the products further.

A key feature of cocreation is the direct connection between the company product developers and the consumers. The consumers meet the developers and understand that the session ideas and outcomes directly influence the development process and directly help real people bring one or more new products to market. Typically, the sponsoring company is not identified.

Figure 18.2 provides an overview of the process, and Figure 18.3 depicts an example of cocreation in action. After warming up the group, the session purpose and concept are shared and the developers or sponsor team are introduced. Consumers develop criteria for a successful product through facilitator-led divergent thinking tools. These are used along with the concept to judge ideas and products. Next, consumers assume the roles of chefs or developers. Using physical stimuli, small teams generate ideas and products to meet the criteria. An optional and beneficial step is to place a developer or client member

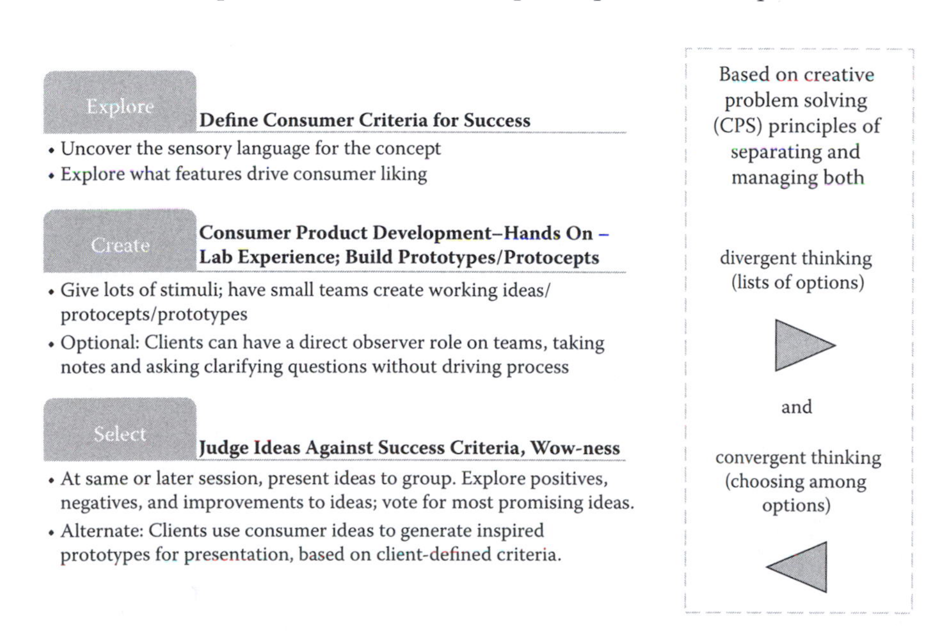

Figure 18.2 Schematic of the Spectrum cocreation process.

The consumer kitchen is prestaged with a variety of ingredients and their descriptions.

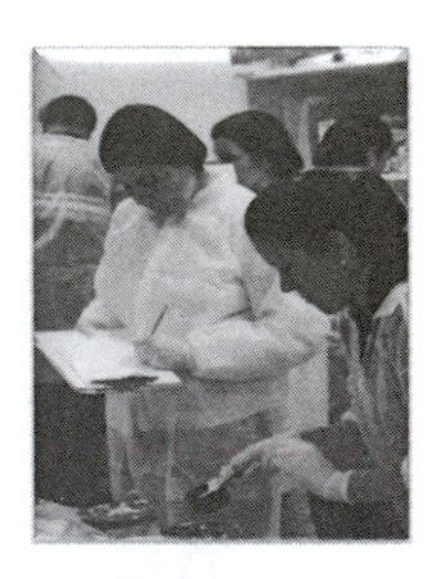

Product developers work side by side with consumers to record their recipes and probe on ingredient selection.

The consumers yield wonderful ideas for unique tasting cheese crackers that meet the concept and success criteria.

Figure 18.3 Consumer cocreation is used to develop new cheese cracker flavors.

on each team to scribe and ask questions during the creation phase. Because the goal of this phase is to strive for quantity and defer judgment, it is important that the client act as a scribe and observer only and that he/she does not provide ideas or limit the team. Chosen creations are seen and/or experienced by the other teams and the sponsor prior to being assessed against the previously developed success criteria. Positives, negatives, and ways to improve the idea are shared. Additionally, all ideas and products generated by the consumer teams are collected for later review by the sponsor. From here, the company can take the ideas and work them through its existing product development process or build workable prototypes inspired by the cocreation process and present them to the same participants for critique prior to quantitative consumer testing. Consumer cocreation is flexible and can be used at any point in the development process from product concept through prototype refinement or expansion.

In a Sensory Spectrum fielded case study presented at "A Sense of Diversity: Second European Conference on Sensory Consumer Science of Food and Beverages" (Civille et al., 2006), the consumer cocreation process was modeled from product concept through central location testing and analysis. Consumers were provided with a concept for a dessert pizza. Hands-on small group ideation followed thematic context activities and defining of success criteria. Creations were judged and the top four selected for a CLT. Differences in adults and teen reactions to the concept and products revealed refinement and positioning opportunities for age groups as well as the same winning product for both groups.

18.6 QUALITATIVE USE OF KANO METHODOLOGY

Kano is a quantitative technique for selecting worthwhile attribute features during the development of products or services. Created in the 1980s by Professor Noriaki Kano (Kano et al., 1984), the Kano Model of Customer Satisfaction divides product attributes into three types: Basic/threshold, performance, and excitement. Basic attributes are the "must-have" features that are expected—ones that consumers often take for granted and may

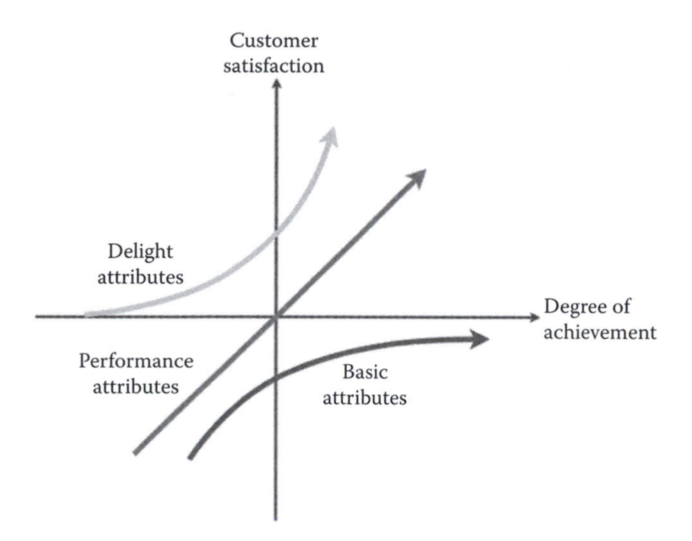

Figure 18.4 Kano model attribute execution and customer satisfaction for basic/threshold, performance, and excitement attributes. (From C. Holst. *UX and the Kano Model.* Baymard Institute) http://baymard.com/blog/kano-model. With permission; adapted from Kano, N., et al. (1984). *J Jpn Soc Qual Control* 14: 39–48. With permission.)

not even mention. They do not increase liking, but their absence or poor execution causes dissatisfaction. Performance attributes are value-added features that consumers tell you they want. They increase liking when the features deliver what the consumers expect. Excitement attributes are latent features that consumers cannot articulate or consciously imagine; absence of these features does not hurt satisfaction, and they generate delight and increase satisfaction when they are present. These are novel, unexpected features that provide a "wow" experience and serve to impress. Figure 18.4 of the Kano model demonstrates the relationship between customer satisfaction and how well the attribute is executed. The Kano model has many applications; here the focus is on product development.

To categorize features, information from surveys, customer complaints, competitive review, and other sources is selected to create a Kano customer satisfaction survey (CSS). The survey uses a series of paired "presence" and "absence" questions and a five-point hedonic scale. It is administered to a readable base of consumers in the target group. Collected data is evaluated to determine the customer requirements for the attributes studied.

Application of Kano is based on several principles: (a) Value attracts customers, (b) quality keeps customers and builds loyalty, (c) innovation is necessary to differentiate and compete in the market, (d) not all attributes hold the same value to customers, and (e) consumer needs evolve such that innovative attributes may become expected over time and new innovation is needed.

While the Kano model is executed as a questionnaire-driven technique, it often uses qualitative and quantitative data to identify the voice of the customer prior to creating the CSS. In contrast, qualitative application of the Kano model retains the concept of assigning product attributes as Basic/Must Have, Performance/Nice to Have, and Delighter/Remarkable while generating that output by continuing to work directly with the target user group in a qualitative setting. Such sessions allow for a more natural flow of questioning and avoid the awkward question format used in the CSS when asking about the absence of a feature (called the dysfunctional form of the question). It allows for immediate follow-up on the reasons consumers categorize as they do, often revealing more latent attribute features and new realizations either within the sessions or during postsession analysis. Sessions are more engaging when more hands on and less theoretical, so many types of stimuli are provided. These are the sensory link to encouraging and allowing consumers to better articulate what they want, do not want, would like to have, can live without, and identify as fundamental to the concept, as well as what does not exist but would improve or make them love it. The stimuli do not have to be optimized—far from it; sometimes showing something that is substandard is the best way to learn how something should be. Qualitative Kano provides guidance in selection of ingredients, processing techniques, packaging, and marketing. Qualitative Kano is most typically used prior to the development of prototypes, and session participants are often re-called to assess how well the prototypes developed meet the criteria established in the initial sessions and may be asked to provide guidance for optimizing the prototypes.

18.7 BENEFIT PERCEPTION BEYOND LIKING: FUNCTIONAL, EMOTIONAL, AND HEALTH AND WELLNESS BENEFITS

18.7.1 Definition and Purpose or Scope

For a long time, product development initiatives have relied mostly on consumer overall liking to strategically improve products. In traditional consumer testing, products are evaluated blind, and overall liking for the product is a key diagnostic. This is augmented with other product sensory diagnostics (e.g., flavor liking, texture liking, sweetness, creaminess, smoothness), which can be used to provide direction to product development and in turn generate higher sensory satisfaction.

In recent years, however, many product launches have failed in the marketplace even though hedonic performance was strong, bringing up many questions about whether overall liking was indeed a sufficient metric to predict product success. Many hypothesized that while overall liking was a big piece of the puzzle, understanding product performance against consumer expectations for brand image, product promises, and product sensory identity was also important to ensure product differentiation and avoid consumer cognitive dissonance.

18.7.2 Tools and Techniques

As consumers purchase products, they are not only seeking sensory satisfaction but also product experiences that address their specific needs. Depending on context and situation, they may be looking for products that deliver more strongly against perceived efficacy, performance or other functional benefits, anticipated emotional impact, or health-related added value. As a result, tools and techniques have been developed to specifically address those aspects of the product experience.

Hence the recent increased interest in incorporating instruments designed to quantitatively measure consumers' emotional reactions to products in sensory and consumer research. The state of research in this domain is constantly evolving, with techniques including

- Self-report through the use of adjective checklists, often adapted from academic and clinical frameworks to reflect the specific needs of commercial research and business applications
- Self-report through the use of physical stimuli such as the tactile (e.g., fabrics) or visual (e.g., images/characters), believed to be universally and unconsciously associated with various emotions
- Measurement of body responses to stimuli or concepts leveraging science and technology (e.g., skin conductance, eye tracking, vocal testing, etc.)

More tools and techniques designed to better understand emotional reactions to products beyond acceptance are expected to be developed in the future and incorporated into sensory testing.

18.7.3 Applications

The purpose of this research tool is to extend beyond consumer hedonics to reach a deeper understanding of why consumers purchase and repeat purchase products. It can be used to better understand other reasons consumers make purchase decisions and can even lead to a better understanding of sensory drivers for these holistic emotional, functional, or wellness perceived benefits. The following are some potential project objectives that could be addressed.

- Defining and clarifying emotional, functional, or wellness perceived benefits for a category
- Understanding specific attributes that deliver signals for perceived benefits
- Marketplace product understanding of targeted consumer perceived benefits
- Benchmarking prototypes against products currently in the marketplace by understanding strengths and weaknesses and applying what is learned to product design

18.7.4 Design of Research

The research may be designed to incorporate both qualitative and quantitative consumer testing techniques to get at the full scope of the question. Qualitative research explores the consumer perceived benefits of a product category, and quantitative research is used to both documents the product sensory attributes and to validates the interaction of perceived benefits and product attributes using consumers.

Step 1: Exploration of Benefits

Qualitative research with consumers may be used to explore and separate different perceived benefits from a product category. In this example, a community narrative panel (see Section 13.4) was used to discuss and separate perceived health and nutritional benefits from situational benefits in the yogurt category. The community narrative panel used creative qualitative techniques to generate perceived benefits when shown a range of different types of yogurt (Greek, light, low fat, European style, yogurt with extra probiotics). These benefits that were applicable to a general population were next incorporated into a consumer survey.

A survey was designed to assess which nutrition, sensory, or emotional attributes strengthen a perceived benefit for yogurt. The survey was completed using general population consumers that included a range of heavy yogurt users to non-yogurt users. Consumers (n = 586) were asked to sort a list of nutrition attributes (e.g., low fat, low sugar, high in calcium), a second list of sensory attributes (e.g., sweet, smooth texture, sour), and a third list of emotion attributes (e.g., is fun to eat, is comforting, is disgusting) into the importance of having that attribute in their yogurt for a stated perceived benefit (e.g., "it helps strengthen bones"). This exercise was repeated for all the different consumer perceived benefits generated from the community narrative panel.

Step 2: Assessment of Benefit, Sensory Attribute, and Emotional Relationships

The consumer quantitative survey data was analyzed using factor analysis (see Section 15.3) to group the different nutrition, sensory, and emotional attributes in relation to the perceived benefits. This analysis resulted in clusters showing which attributes were more correlated with the different consumer benefits. The prior community narrative panel results were used to decide whether an attribute evoked a negative or positive emotion.

Step 3: Sensory Competitive Benchmarking

Spectrum descriptive analysis (see Chapter 12) was used to profile a range of 15 different marketplace yogurts. The analysis captured the appearance, flavor, and texture of the yogurts. This data was analyzed using factor analysis to group the descriptive attributes with the yogurt samples. Next, the sensory signals and benefits from the consumer quantitative survey were added to this factor analysis map for an overlay understanding of which sensory descriptive attributes were correlated to a sensory, nutritional, or emotional attribute from consumers. Using both the descriptive attributes and the consumer attributes, three distinctly different yogurts were selected for the next phase of consumer research.

Step 4: Confirmation with Target Consumer

The community narrative panel was used as a final validation to check the prior analysis of merging the descriptive and consumer attributes and the resulting conclusions. Three different yogurts, that appeared in different quadrants on the factor analysis map, were selected from the merged descriptive and consumer attribute factor analysis and were presented to the community narrative panel for discussion. The panel went through an exercise that sorted the physical, nutritional, and emotional attributes into positive or negative buckets for the appearance, flavor, and texture attributes of each of the three yogurts. This was used to validate if the same correlations of descriptive to consumer attributes from survey research occurred while evaluating the actual yogurt sample.

In a next step, a larger quantitative consumer test could be fielded with actual product. The perceived benefits and products from the prior research steps would be useful to design both the questionnaire and the products to test in this larger scale study and allow quantitative confirmation for benefit perception in addition to perceived liking.

18.7.5 Conclusions

The research tool used as an example for exploring this topic shows an approach that utilizes a combination of qualitative, quantitative, expert descriptive panel, and consumer research to link specific product attributes to consumer perceived benefits. The data analysis tools used are similar to research comparing consumer liking to descriptive attributes (factor analysis to visualize the correlation and product groupings). The added research step that is needed is the qualitative exploration to generate the consumer perceived benefits. The consumer quantitative survey also helped with merging generated benefits with their importance in the yogurt category to further focus the research. This research can be applied to any benefit or product category to help with understanding of a benefit perception beyond liking. It also supports product development with specific sensory descriptive design direction that triggers a perceived benefit from the consumer. This case study was presented at the 2012 Pangborn conference.

18.8 BEHAVIORAL ECONOMICS

Behavioral economists strive to understand the motivations behind consumers' economic (i.e., purchase) decisions; whether the motivations are psychological, attitudinal, or cognitive.

Participants in economic experiments may respond to hypothetical choices in surveys (e.g., conjoint analysis) or may make nonhypothetical purchases in valuation tasks. In these methods, economists are often interested in determining consumers' willingness-to-pay values for value-added product(s). The nature of nonhypothetical, incentive-compatible valuation tasks parallels sensory evaluation, in that consumers physically experience the product, which makes these tasks potentially well suited to cross disciplines.

Incentive compatibility implies that the experiment is designed to elicit consumers' homegrown willingness-to-pay values and that there is no strategic advantage to under- or overbidding for the product. Experimental auctions are types of incentive-compatible valuation tasks and include the second price auction.

In the second price auction, consumers bid on the product of interest in small auction sessions (Vickrey, 1961). In each session, consumers submit their sealed and confidential willingness-to-pay values to the session moderator. The moderator reviews the values and selects the highest bidder as the winner of the session. The winner then pays the second highest price for the product. The cash exchange makes the task nonhypothetical, which eliminates hypothetical bias. Furthermore, paying the second highest price ensures that participants are not encouraged to under- or overbid. For example, Tom's true willingness-to-pay value for a juice fortified with Vitamin C is $5.00. If he bids $7.00 (an overbid), he may win the auction and have to pay $6.00 (the second highest bid in this scenario), which is more than his true willingness-to-pay value. If he bids $4.00 (an underbid), he may lose the auction even if the winner only pays $4.50. Compare this situation to the typical estate auction in which bidders compete with each other until only one remains (i.e., an English auction). In the English auction, bidders are incentivized to bid as low possible for as long as possible in hopes that the winning bid will be below their true willingness-to-pay value. In contrast to the English auction, the structure of the second price auction is conducive to unearthing homegrown willingness-to-pay values.

Valuation tasks are commonly applied to determine the premium consumers are willing to pay for new products and have been previously integrated into sensory evaluation experiments (Lange et al., 2002; Lawless et al., 2012; Xue et al., 2010). Recently, sensory evaluation and valuation tasks were combined to develop a hybridized form of penalty analysis, which calculates the penalty in dollars for a product not being "just-about-right" (Lawless, 2012). Using dollars in penalty analysis instead of overall liking is advantageous because dollars are less abstract and more readily understood among technical and business audiences. This type of analysis may provide a more comprehensive view of how consumers regard products.

Valuation tasks are not limited to the second price auction; they also include the nth price auction, the random nth price auction, and the Becker-Degroot-Marschak mechanism. The underlying principles of these methods resemble those of the second price auction, although the executions are slightly different. The various valuation tasks are summarized in other resources (Lusk and Shogren, 2008).

18.9 CATEGORY APPRAISALS, KEY DRIVERS STUDIES AND SENSORY SEGMENTATION

18.9.1 Definition and Purpose or Scope

Understanding the sensory features of products and how they relate to consumer perceptions is critical to developing successful new products, enhancing the appeal of existing products, and providing an edge in today's competitive marketplace. A category appraisal is a widely used process that allows just that. It integrates strategic product insights and consumer insights into one composite picture of the category (see Figure 15.14). Key drivers analysis applies external preference mapping to identify the sensory attributes that are most important to consumers and to develop predicted sensory profiles of the target products. For example, a preference map can be seen in Figure 18.5. The target products are the locations on the preference map that are predicted to be most well-liked by either the total respondent base or any of the demographic, attitudinal, or preference segments that are of interest to the researchers. The target products may fall at a place on the map where no actual products currently exist. Thus, through reverse engineering, the sensory profile of the virtual target product can be obtained, which is often called populating the "white space" on the map. Category appraisals are typically performed in established product categories to gain seminal understanding of both product characteristics and consumers' reactions to those characteristics. They are considered especially useful when one wants to become a key player in an existing category or when one wants to monitor shifts in the competitive landscape and the impacts of new product introductions in a dynamic category.

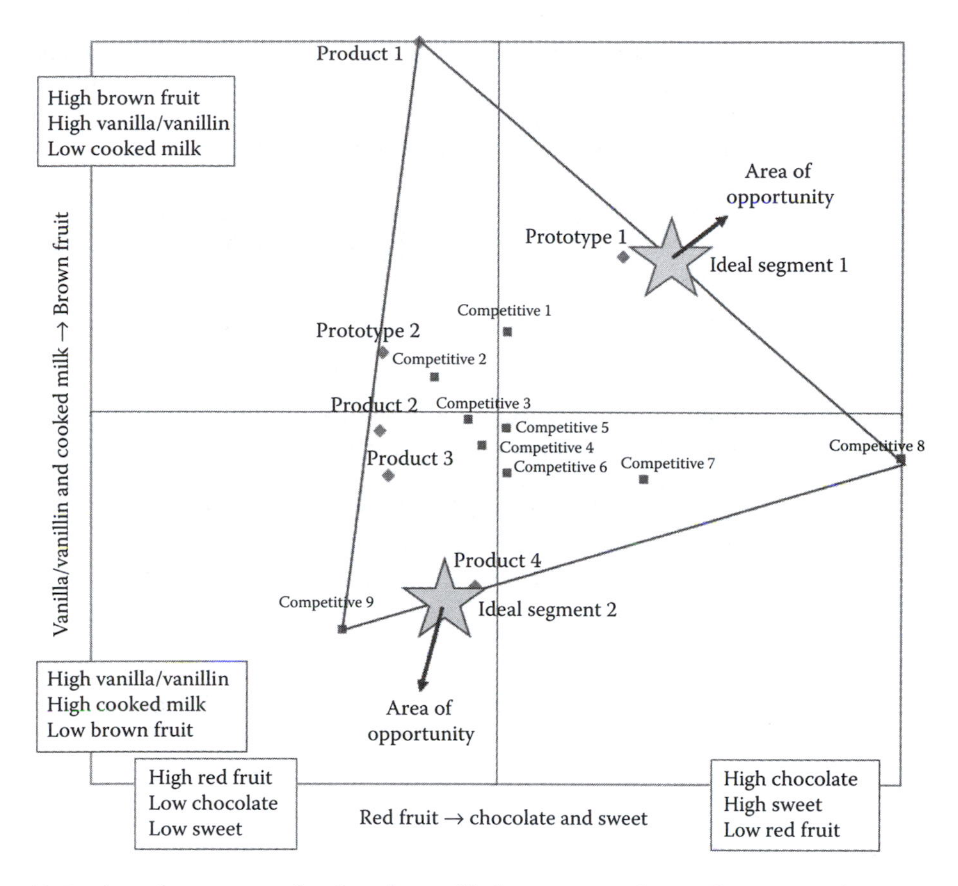

Figure 18.5 A preference map for chocolate milk that captures the product space of ready-to-drink chocolate beverages. The product space is defined by the samples chosen to appropriately represent the category. Prototype 1 is closest to the ideal product for Segment 1 and could be optimized by increasing the sweetness and chocolate impact. Segment 2 is best satisfied by Product 4, which could be less sweet and have lower chocolate impact to match the ideal.

18.9.2 Design and Benefits of the Research

A category appraisal is typically designed as a multistep approach, where each step leverages the findings of the previous phase to capture a more complete and complex picture of the category.

18.9.2.1 Phase I: Defining the Limits of the Category

This is the most critical phase of the research, wherein both the scope and objectives are defined. Key stakeholders are typically present to identify intended learning and expected outcomes. Examples of objectives may include

- Identifying key drivers of acceptance and benefit perception from consumer and descriptive data

- Understanding the competitive market and the place of a company's products and prototypes relative to its competitors' and the most-liked products within and across price points
- Determining consumer segments and what opportunities may exist to enhance acceptance
- Creating predictive models to give direction for formulating products that will be best-liked and/or perceived as delivering more of a specific benefit by consumers or consumer segments

More importantly, once clear objectives are defined, the scope of the research is discussed in terms of

- Products: The diversity of the product under investigation will determine the granularity of the findings. For example, if one defines the category as the "world of snacks" to include sweet and salty snacks, findings will be more general; whereas if one defines the category as the "world of chewy chocolate chip cookies" findings will be much more targeted. Very often, an initial product selection is made at this point to aid in the discussion and to ensure team alignment on what is considered within and outside the category.
- Market and consumers: As in any consumer initiative, respondent qualifications (geographic location, demographics, socioeconomic characteristics, usage, and attitudes toward the category, etc.) should be discussed at this point as well, as the consumer base will impact the insights that can be gained from the research.

18.9.2.2 Phase II: Documentation of Product Characteristics, Competitive Intelligence and Selection of Products for Consumer Testing

This phase is typically performed using descriptive analysis techniques and is designed according to the guidelines outlined in Chapter 11. The output provides

- Detailed documentation of the perceived qualitative (attributes) and quantitative (intensity) sensory characteristics of the products in the category. All sensory modalities relevant to the product category are typically included.
- Illustration of product differences.
- Information that can be related to consumer response and provide insights and assistance in decoding consumer information.

In addition to individual product profiles, perceptual maps can be generated using statistical techniques outlined in Chapter 15 (Section 15.3) in order to

- Visualize similarities and differences among products
- Highlight key sensory differentiators among products from various brands and market positions
- Highlight sensory redundancy and allow for the selection of products to be included in consumer testing that span the sensory space.

Many strategic product insights can be gathered from this phase. All products included in a category appraisal (typically between 8 and 30+ products) are fully documented descriptively, allowing for a good understanding of where differences and similarities across samples lie from a sensory standpoint. Inclusion of currently marketed products and prototypes is often recommended. Insights about sensory strategies and portfolio management can easily be uncovered looking at the data with several questions in mind:

- Do some brands/groups of products have a distinctive sensory signature?
- Do some brands/groups of products cover a wide area of the sensory space?
- Are there areas of the sensory space where no products currently exist?

18.9.2.3 Phase III: Determining Consumer Acceptance and Perception of the Products in the Category

This phase is designed to obtain consumer feedback on products and prototypes. It is typically a quantitative consumer test (central location test or home use test) but may be supplemented with qualitative research as well. Chapter 13 outlines the standard practices and guidelines for the design and execution of consumer research. Consumer response to the series of products is gathered and may include liking, expectations, perceived product properties, image, and benefits using consumer language.

A typical design for this phase may include

- $N = 150$ respondents who meet predetermined screening criteria; such a number is optimal to identify consumer segments in the population. The study might be conducted in multiple locations.
- $N = 12$–18 products optimally.
- The test is set up as balanced complete block design where each consumer tests each product to allow for consumer preference segmentation.
- Products are typically identified only by a three-digit code.

18.9.2.4 Phase IV: Identifying Key Drivers, Drivers of Benefit Perception, and Strategic Product Guidance

This phase relies on the preference mapping techniques outlined in Chapter 15 to identify critical descriptive features of the products that drive consumer acceptance. This includes consumer perception both overall and within each preference segment for all products in the product category. See Section 15.3 for a discussion of the statistical techniques used to determine the impact of product attributes on liking. Within the sensory space under investigation, target and opportunity areas are identified for product improvement and/or new product development, and the sensory specifications for increased product performance are determined for these area profiles.

Output from the research may include an ideal product profile that can easily be compared and contrasted with those of current products. By knowing how their current products compare to the target product and by understanding the importance that each sensory attribute has for acceptance, researchers can prioritize their product-improvement opportunities to maximize returns.

Identifying preference segments is an integral part of key drivers analysis. Preference segments are groups of respondents who, internally, have similar patterns of liking for the products but whose liking patterns differ from group to group. Identifying target products for each preference segment gives researchers a more realistic view of the possibilities that exist to satisfy the largest proportion of consumers, and they will also have a more realistic view of what they are giving up by selecting one set of product options over another. Specific recommendations for product improvement are given along with strategic recommendations for product optimization and prototype development within the context of the brand and its perceived image. Additional strategic guidance for new product development and line extension can also stem from this approach.

18.9.3 Conclusion

Overall, the approach provides consumer-focused, sensory-driven guidance for product development, tactical advice for product positioning, and priceless insights for new and strategic business opportunities.

18.10 AD CLAIMS

18.10.1 Introduction

Sensory evaluation, as a science, is often used to support advertising claims that have a sensory component. According to ASTM, an advertising claim is "a statement about a product that highlights its advantages, sensory or perceptual attributes or product changes or differences compared to other products to enhance its marketability" (ASTM E1958-12). In order to prove that the product in question has the qualities necessary to make the claim, a company must conduct tests to prove that its statements about the product are true. Without credible data from reliable tests, the company is liable to be held accountable by its competitors, the media (print, radio, TV or electronic), the National Advertising Division (NAD) of the Better Business Bureau, the National Advertising Review Board (NARB), or the courts (civil and federal). The risk of scrutiny requires careful planning and execution of tests to support ad claims.

18.10.2 Types of Claims

The basis of a claim is a clear statement of the assets or properties that the company wishes to tout, and which is agreed on by all the stakeholders in the company (research and development [R&D], marketing, legal, market research, etc.). After review and revision of the claims statement by this cross-functional team, the team can decide on the type of claim being made. The two primary types of claims are comparative (comparing two or more products) and noncomparative (a statement of the benefit of one product).

Within comparative and noncomparative claims, there are different claim statement types (see below). When a comparative claim is made, the claim can prove parity (that the products are similar or comparable) or superiority (that one sample is better than the other(s) in acceptance or some specific attribute).

- Noncomparative claims
  - Hedonic: "smells like spring", "makes your skin feel great", "tastes great"
  - Attribute/performance: "reduces odors", "naturally clean hair", "minty fresh breath", "fast acting"

- Comparative claims
  - Parity:
    - Hedonic: "feels as good as brand A"
    - Overall equality: "comparable to the national brand"
    - Attribute: "as fresh tasting as the leading brand"
    - Unsurpassed: "no one removes stains better"
  - Superiority:
    - Hedonic: "preferred to the leading brand", "new and improved"
    - Attribute: "longest protection of any deodorants", "crisper than the leading chip"

18.10.3 Types of Claims Testing

Before beginning the testing process, the team should assess if the claim is likely to be challenged. Factors influencing challenges to a claim are the advertising environment, the history of claims challenges between certain competitors, recent challenges and activities within the NAD, and the strength of the current claim. Once the cross-functional team has developed the claim statement and assessed the type of claim it can make, the team needs to determine the types of testing it needs to conduct to support the claim. One overarching factor is to make the claim test a *reasonable* test to adequately support the claim. This requires that the test represent all the aspects of the claim (including product, product usage, target consumer, and/or laboratory assessor) and utilize a test protocol that reflects the spirit of the claim in a rational and discerning way.

The testing options below are often called the legs of the stool—the more legs you have to support the claim, the more stable and substantial the claim (assuming all the testing is done correctly).

- *Material* claims are based on the physical, chemical, and/or geographical facts that underlie the substance of the product: "now with baking soda", "100% Florida Orange Juice"
- *Instrumental* claims are based on physical or chemical measurements: "thicker than Z's gravy", "reduces odor in the air"
- *Laboratory sensory test* claims are based on difference tests or descriptive tests conducted in a laboratory setting: "now with twice the beef flavor", "the most crisp chip"
- *Consumer (affective, preference, attribute, performance)* claims are based on tests with consumers of the product in question: "preferred to the leading brands", "leaves your skin aglow", "no product is rated better than ours"

18.10.4 Building the Case

Once the ad claim has been developed and the options for tests explored, the team needs to choose the appropriate test(s) and design the test protocols. Other sections in this book describe the test options for the laboratory and consumer test types (see Chapter 6) and the test controls for the subjects/assessors, samples, and facilities/site (see Chapter 3). Details for conducting these claims tests are also detailed in the ASTM document E1958-06. The parameters for selecting assessors (some consumer tests require hundreds of assessors from across the country), designing the test method, conducting the test (most preference tests for claims should have a "no preference" option), analyzing the data, drawing conclusions, and making recommendations are defined in the document. The entire case should be built on strong methods and protocols that are reasonable and documented. The objective from the outset is to build a case for the claim that will stand up to challenges later. Detailed documentation of a reasonable claim based on relevant testing and resulting information are critical elements in successful ad claims research.

18.10.5 Cautions and Things to Consider

A reasonable claim is based on choosing the wording of the claim to reflect the salient features of the product and the ability to prove that claim through proper testing. The product(s) chosen to represent those in the claim must be commercial products, within shelf-life range, collected across the selected geographic area, and used according to directions. The assessors and locations for the test(s) must represent the claim parameters, and the data analysis and interpretation need to follow ASTM's recommended guidelines for analysis and interpretation.

In order to reduce the chance of challenges, some practices to avoid include "fishing" for a claim by testing a wide array of attributes to find the one attribute that gives your product an advantage and basing the claim on that test; selecting your product from among the best samples and choosing your competitors from routine store pickups; assuming overall superiority from individual attribute superiority; using branded or concept aided tests, which are typical consumer research tests for marketing research but are not appropriate for claims.

Recent claims have embraced emotions, stated and unstated consumer needs, and clear scientific graphics to make the claim believable. These trends demonstrate a reduction in competitive claims and an attempt to reach the consumer on a different level. Ad claims may be as diverse and varied as the products they represent, but all claims should be grounded in the fundamental sensory testing principles described throughout this book.

ADDITIONAL RESOURCES

Front-End Innovation Additional References

Cooper, R. G. (2001). *Winning at New Products: Accelerating the Process from Idea to Launch* (3rd edn.). Cambridge, MA: Perseus Publishing.

Elmer, J. B. (1997). The fuzzy front end: Converting information streams to high potential product concepts. Presented at ASTM conference, San Diego, CA.

Hepner Brodie, C., and G. Burchill (1997). *Voices into Choices: Acting on the Voice of the Customer.* Madison, WI: Joiner Associates.

Koen, P. A., G. M. Ajamian, S. Boyco, A. Clamen, E. Fisher, S. Fountoulakis, A. Johnson, P. Puri, and R. Seibert (2002). Fuzzy front end: Effective methods, tools, and techniques. In *PDMA Tool Book for New Product Development*. New York: Wiley.

Mello, S. (2002). *Customer-Centric Product Definition: The Key to Great Product Development*. New York: Amacom.

Schrage, M. (2003). Daniel Kahneman: The thought leader interview. In *Strategy & Business* (pp. 1–36). New York: Booz, Allen & Hamilton.

Smith, P. G., and D. G. Reinertsen (1998). *Developing Products in Half the Time: New Rules, New Tools*. New York: Wiley.

Front-End Innovation Reading List

Qualitative Research, Information Gathering

Burchill, G., and C. Hepner Brodie (1997). *Voices into Choices: Acting on the Voice of the Customer*. Madison, WI: Joiner Associates.

Fontana, D. (1994). *The Secret Language of Dreams*. London: Duncan Baird.

Goleman, D. (1995). *Emotional Intelligence—Why It Can Matter More than IQ*. New York: Bantam Books.

Greenbaum, T. L. (1998). *The Practical Handbook and Guide to Focus Group Research*. Lexington, MA: D.C. Heath and Co.

Keirsey, D. (1998). *Please Understand Me II-Temperament, Character, Intelligence*. Del Mar, CA: Prometheus Nemesis.

Koppett, K. (2001). *Training to Imagine*. Sterling, VA: Stylus.

McQuarrie, E.F. (1998). *Customer Visits: Building a Better Market Focus* (2nd edn). Newbury Park, CA: Sage.

Mello, S. (2002). *Customer Centric Product Definition*. New York: Amacom.

Naparstek, B. (1997). *Your Sixth Sense—Activating Your Psychic Potential*. San Francisco, CA: HarperCollins.

Payne, S. (1957). *The Art of Asking Questions*. Princeton, NJ: Princeton University Press.

Popcorn, F. (1998). *Clicking: 17 Trends That Drive Your Business & Your Life*. New York: Harper Business.

Rogers, E. (1995). *Diffusion of Innovation*. New York: The Free Press.

Senge, P. (1990). *The Fifth Discipline—The Art and Practice of the Learning Organization*. New York: Doubleday Currency.

Stoller, P. (1989). *The Taste of Ethnographic Things*. Philadelphia: University of Pennsylvania Press.

Holistic Prototyping and Holistic Product Development

Cooper, R. G. (1993). *Winning at New Products*. Reading, MA: Addison-Wesley.

Cooper, R. G., and S. J. Edgett (1999). *Product Development for the Service Sector—Lessons from Market Leaders*. Cambridge: Perseus Books.

Qualitative Use of Kano Methodology Additional References

Mind Tools. 2015. Kano model analysis: Delivering products that delight. Accessed February 9, 2015. http://www.mindtools.com/pages/article/newCT_97.htm.

Stelick, A., K. Lopetcharat, and D. Paredes (2012). Kano satisfaction model. In *Product Innovation Toolbox*, Section 7.1. Oxford: Wiley-Blackwell.

Walder, D. (1993). Kano's model for understanding customer defined quality. *Center Qual Manag J* 39: 65–9.

Zultner, R. E., and G. H. Mazur (2006). The Kano Model: Recent developments. In *Presentation at the Eighteenth Symposium on Quality Function Deployment*, Austin, TX, December 2.

REFERENCES

ASTM E1958-12. 2012. Standard guide for sensory claim substantiation. West Conshohocken, PA: ASTM International. www.astm.org.

Civille, G. V., G. Gibbs, C. Dus, J. Heylmun, J. VanVisco, E. Wilson, B. Talbot, M. Rudolph, and A. Retiveau (2006). Consumer designed dessert pizza: Rapid development with community narrative consumer group. In *Presentation at a Sense of Diversity: Second European Conference on Sensory Consumer Science of Food and Beverages*, Hague, Netherlands, September 26–29.

Kano, N., N. Seraku, F. Takahashi, and S. Tsuji (1984). Attractive quality and must-be quality. *J Jpn Soc Qual Control* 14: 39–48.

Lange, C., C. Martin, C. Chabanet, P. Combris, and S. Issanchou (2002). Impact of the information provided to consumers on their willingness to pay for champagne: Comparison with hedonic scores. *Food Qual Prefer* 13: 597–608.

Lawless, L. J. R., R. M. Nayga, Jr., F. Akaichi, J. F. Meullenet, R. T. Threlfall, and L. R. Howard (2012). Willingness-to-pay for a nutraceutical-rich juice blend. *J Sens Stud* 27: 375–83.

Lawless, L. J. R. (2012). Consumer sensory preferences and willingness-to-pay for nutraceutical-rich fruit. Dissertation, University of Arkansas. Fayetteville: ProQuest/UMI.

Lusk, J., and J. F. Shogren (2008). *Experimental Auctions: Methods and Applications in Economic and Marketing Research*. New York: Cambridge University Press.

Vickrey, W. (1961). Counterspeculation, auctions, and competitive sealed tenders. *J Fin* 16: 8–37.

Xue, H., D. Mainville, W. You, and R. M. Nayga, Jr. (2010). Consumer preferences and willingness to pay for grass-fed beef: Empirical evidence from in-store experiments. *Food Qual Prefer* 21: 857–66.

19

Statistical Tables

Table 19.1 Random Orders of the Digits 1–9: Arranged in Groups of Three Columns

Instructions

(1) To generate a sequence of three-digit random numbers, enter the table at any location, for example, by closing the eyes and pointing. Without inspecting the numbers, decide whether to move up or down the column entered. Record as many numbers as needed. Discard any numbers that are unsuitable (out of range, came up before, etc.). The sequence of numbers obtained in this manner is in random order.

(2) To generate a sequence of two-digit random numbers, proceed as in (1), but first decide, for example, by coin toss, whether to use the first two or last two digits of each number taken from the table. Treat each three-digit number in the same manner, that is, discard the same digit from each. If a two-digit number comes up more than once, retain only the first.

(3) Random number tables are impractical for problems such as "place the numbers from 15 to 50 in random order." Instead, write each number on a card and draw the cards blindly from a bag or use a computerized random number generator such as PROC PLAN from SAS.®

862	245	458	396	522	498	298	665	635	665	113	917	365	332	896	314	688	468	663	712	585	351	847
223	398	183	765	138	369	163	743	593	252	581	355	542	691	537	222	746	636	478	368	949	797	295
756	954	266	174	496	133	759	488	854	187	228	824	881	549	759	169	122	919	946	293	874	289	452
544	537	522	459	984	585	946	127	711	549	445	793	734	855	121	885	595	152	237	574	611	145	784
681	829	614	547	869	742	822	554	448	813	976	688	959	714	912	646	873	397	159	155	136	463	363
199	113	941	933	375	651	414	891	129	938	862	572	698	128	363	478	214	841	314	437	792	874	926
918	481	797	621	743	827	377	916	966	429	657	246	423	277	685	533	937	223	582	946	323	626	519
335	662	875	282	617	274	635	379	287	791	334	139	117	963	448	957	451	585	821	829	267	512	638
477	776	339	818	251	916	581	232	372	374	799	461	276	486	274	791	369	774	795	681	458	938	171
653	489	538	216	446	849	914	337	993	459	325	614	771	244	429	874	557	119	122	417	882	714	769
749	824	721	967	287	556	628	843	725	731	553	253	183	653	988	431	788	426	875	838	457	927	475
522	967	259	532	618	624	396	562	134	563	932	441	834	787	231	958	232	537	439	956	531	345	352
475	172	986	859	925	932	282	924	842	642	797	565	399	896	596	282	441	784	258	684	625	662	291
894	333	612	728	869	487	741	259	476	127	286	736	257	168	847	316	969	692	786	549	949	559	526
116	218	464	191	132	218	573	786	258	296	471	372	618	935	353	747	123	863	644	161	793	196	847
381	641	393	375	354	193	165	615	587	384	119	187	965	572	112	695	615	941	361	375	376	871	633
968	755	847	643	773	765	439	478	611	978	868	898	546	319	775	169	896	275	513	222	114	233	184
742	421	226	286	522	618	471	218	397	745	461	477	478	535	957	674	132	228	442	225	444	171	151
859	878	392	311	659	772	935	447	834	117	658	161	754	654	176	883	855	195	637	751	586	948	513
964	593	137	574	288	994	582	961	746	336	983	782	611	988	833	265	969	584	564	683	197	214	326
177	636	674	897	167	157	856	524	662	598	145	926	362	777	415	931	313	317	195	137	959	536	985
228	755	915	955	946	233	647	653	425	674	719	543	549	826	669	429	576	773	756	392	632	725	879
591	214	851	669	394	349	299	192	179	264	332	294	896	299	782	397	791	659	921	569	811	683	762
636	167	789	438	413	565	118	889	253	452	577	859	125	141	241	746	444	841	313	446	225	362	248
415	982	543	743	835	826	364	776	988	923	224	615	283	462	328	512	228	466	278	874	373	499	437
383	349	468	122	771	481	723	335	511	889	896	338	937	313	594	158	687	932	889	918	768	857	694

Source: W. G. Cochran and G. M. Cox (1957). *Experimental Design*, New York: John Wiley & Sons.

Table 19.2 Standard Normal Distribution

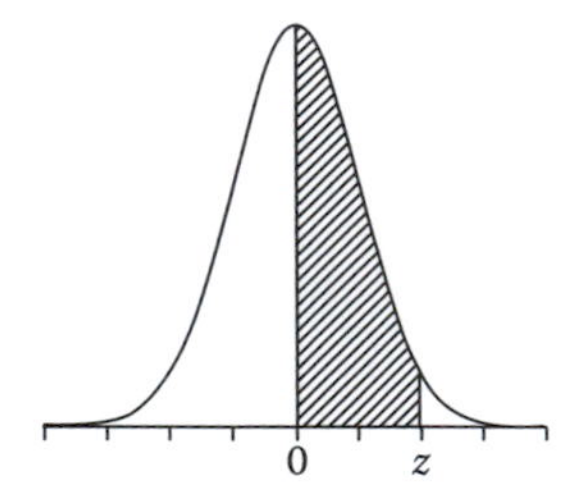

Instruction: See the Examples in Chapter 14.

Z	0.00	0.01	0.02	0.03	0.04	0.05	0.06	0.07	0.08	0.09
0.0	0.0000	0.0040	0.0080	0.0120	0.0160	0.0199	0.0239	0.0279	0.0319	0.0359
0.1	0.0398	0.0438	0.0478	0.0517	0.0557	0.0596	0.0636	0.0675	0.0714	0.0753
0.2	0.0793	0.0832	0.0871	0.0910	0.0948	0.0987	0.1026	0.1064	0.1103	0.1141
0.3	0.1179	0.1217	0.1255	0.1293	0.1331	0.1368	0.1406	0.1443	0.1480	0.1517
0.4	0.1554	0.1591	0.1628	0.1664	0.1700	0.1736	0.1772	0.1808	0.1844	0.1879
0.5	0.1915	0.1950	0.1985	0.2019	0.2054	0.2088	0.2123	0.2157	0.2190	0.2224
0.6	0.2257	0.2291	0.2324	0.2357	0.2389	0.2422	0.2454	0.2486	0.2517	0.2549
0.7	0.2580	0.2611	0.2642	0.2673	0.2704	0.2734	0.2764	0.2794	0.2823	0.2852
0.8	0.2881	0.2910	0.2939	0.2967	0.2995	0.3023	0.3051	0.3078	0.3106	0.3133
0.9	0.3159	0.3186	0.3212	0.3238	0.3264	0.3289	0.3315	0.3340	0.3365	0.3389
1.0	0.3413	0.3438	0.3461	0.3485	0.3508	0.3531	0.3554	0.3577	0.3599	0.3621
1.1	0.3643	0.3665	0.3686	0.3708	0.3729	0.3749	0.3770	0.3790	0.3810	0.3830
1.2	0.3849	0.3869	0.3888	0.3907	0.3925	0.3944	0.3962	0.3980	0.3997	0.4015
1.3	0.4032	0.4049	0.4066	0.4082	0.4099	0.4115	0.4131	0.4147	0.4162	0.4177
1.4	0.4192	0.4207	0.4222	0.4236	0.4251	0.4265	0.4279	0.4292	0.4306	0.4319
1.5	0.4332	0.4345	0.4357	0.4370	0.4382	0.4394	0.4406	0.4418	0.4429	0.4441
1.6	0.4452	0.4463	0.4474	0.4484	0.4495	0.4505	0.4515	0.4525	0.4535	0.4545
1.7	0.4554	0.4564	0.4573	0.4582	0.4591	0.4599	0.4608	0.4616	0.4625	0.4633
1.8	0.4641	0.4649	0.4656	0.4664	0.4671	0.4678	0.4686	0.4693	0.4699	0.4706
1.9	0.4713	0.4719	0.4726	0.4732	0.4738	0.4744	0.4750	0.4756	0.4761	0.4767
2.0	0.4772	0.4778	0.4783	0.4788	0.4793	0.4798	0.4803	0.4808	0.4812	0.4817
2.1	0.4821	0.4826	0.4830	0.4834	0.4838	0.4842	0.4846	0.4850	0.4854	0.4857
2.2	0.4861	0.4864	0.4868	0.4871	0.4875	0.4878	0.4881	0.4884	0.4887	0.4890
2.3	0.4893	0.4896	0.4898	0.4901	0.4904	0.4906	0.4909	0.4911	0.4913	0.4916
2.4	0.4918	0.4920	0.4922	0.4925	0.4927	0.4929	0.4931	0.4932	0.4934	0.4936
2.5	0.4938	0.4940	0.4941	0.4943	0.4945	0.4946	0.4948	0.4949	0.4951	0.4952
2.6	0.4953	0.4955	0.4956	0.4957	0.4959	0.4960	0.4961	0.4962	0.4963	0.4964
2.7	0.4965	0.4966	0.4967	0.4968	0.4969	0.4970	0.4971	0.4972	0.4973	0.4974
2.8	0.4974	0.4975	0.4976	0.4977	0.4977	0.4978	0.4979	0.4979	0.4980	0.4981
2.9	0.4981	0.4982	0.4982	0.4983	0.4984	0.4984	0.4985	0.4985	0.4986	0.4986
3.0	0.4987	0.4987	0.4987	0.4988	0.4988	0.4989	0.4989	0.4989	0.4990	0.4990

Table 19.3 Upper-α Probability Points of Student's t-distribution (Entries are $t_{\alpha:\nu}$)

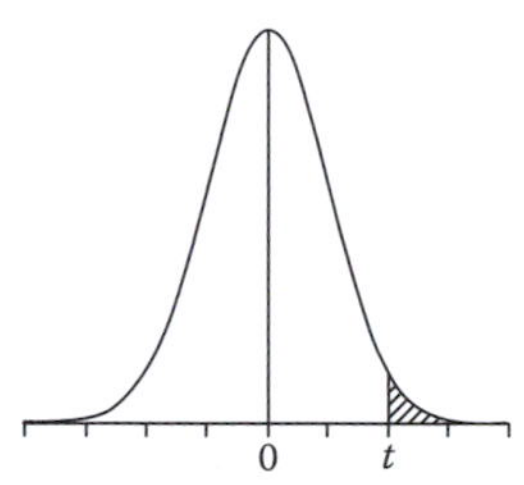

Instructions: (1) Enter the row of the table corresponding to the number of degrees of freedom (ν) for error.

(2) Pick the value of t in that row, from the column that corresponds to the predetermined α-level.

	α						
ν	**0.25**	**0.10**	**0.05**	**0.025**	**0.01**	**0.005**	**0.0005**
1	1.000	3.078	6.314	12.706	31.821	63.657	636.619
2	0.816	1.886	2.920	4.303	6.965	9.925	31.598
3	0.765	1.638	2.353	3.182	4.541	5.841	12.941
4	0.741	1.533	2.132	2.776	3.747	4.604	8.610
5	0.727	1.476	2.015	2.571	3.365	4.032	6.859
6	0.718	1.440	1.943	2.447	3.143	3.707	5.959
7	0.711	1.415	1.895	2.365	2.998	3.499	5.405
8	0.706	1.397	1.860	2.306	2.896	3.355	5.041
9	0.703	1.383	1.833	2.262	2.821	3.250	4.781
10	0.700	1.372	1.812	2.228	2.764	3.169	4.587
11	0.697	1.363	1.796	2.201	2.718	3.106	4.437
12	0.695	1.356	1.782	2.179	2.681	3.055	4.318
13	0.694	1.350	1.771	2.160	2.650	3.012	4.221
14	0.692	1.345	1.761	2.145	2.624	2.977	4.140
15	0.691	1.341	1.753	2.131	2.602	2.947	4.073
16	0.690	1.337	1.746	2.120	2.583	2.921	4.015
17	0.689	1.333	1.740	2.110	2.567	2.898	3.965
18	0.688	1.330	1.734	2.101	2.552	2.878	3.922
19	0.688	1.328	1.729	2.093	2.539	2.861	3.883
20	0.687	1.325	1.725	2.086	2.528	2.845	3.850
21	0.686	1.323	1.721	2.080	2.518	2.831	3.819
22	0.686	1.321	1.717	2.074	2.508	2.819	3.792
23	0.685	1.319	1.714	2.069	2.500	2.807	3.767
24	0.685	1.318	1.711	2.064	2.492	2.797	3.745
25	0.684	1.316	1.708	2.060	2.485	2.787	3.725
26	0.684	1.315	1.706	2.056	2.479	2.779	3.707
27	0.684	1.314	1.703	2.052	2.473	2.771	3.690
28	0.683	1.313	1.701	2.048	2.467	2.763	3.674
29	0.683	1.311	1.699	2.045	2.462	2.756	3.659
30	0.683	1.310	1.697	2.042	2.457	2.750	3.646
∞	0.674	1.282	1.645	1.960	2.326	2.576	3.291

Table 19.4 Percentage Points of the Studentized Range: Upper-α Critical Values for Tukey's HSD Multiple Comparison Procedure Instructions:

(1) Enter the section of the table that corresponds to the predetermined α-level.

(2) Enter the row that corresponds to the degrees of freedom for error from the ANOVA.

(3) Pick the value of q in that row from the column that corresponds to the number of treatments being compared.

ν	2	3	4	5	6	7	8	9	10	11	12	13	14	15	16	17	18	19	20
The entries are $q_{0.01}$ where $p(q < q_{0.01}) = 0.99$																			
1	90.03	135.0	164.3	185.6	202.2	215.8	227.2	237.0	245.6	253.2	260.0	266.2	271.8	277.0	281.8	286.3	290.4	294.3	290.0
2	14.04	19.02	22.29	24.72	26.63	28.29	29.53	30.68	31.69	32.59	33.40	34.13	34.81	35.43	36.00	36.53	37.03	37.50	37.95
3	8.26	10.62	12.17	13.33	14.24	15.00	15.64	16.20	16.69	17.13	17.53	17.89	18.22	18.52	18.81	19.07	19.32	19.55	19.77
4	6.51	8.12	9.17	9.96	10.58	11.10	11.55	11.93	12.27	12.57	12.84	13.09	13.32	13.53	13.73	13.91	14.08	14.24	14.40
5	5.70	6.98	7.80	8.42	8.91	9.32	9.67	9.97	10.24	10.48	10.70	10.89	11.08	11.24	11.40	11.55	11.68	11.81	11.93
6	5.24	6.33	7.03	7.56	7.97	8.32	8.61	8.87	9.10	9.30	9.48	9.65	9.81	9.95	10.08	10.21	10.32	10.43	10.54
7	4.95	5.92	6.54	7.01	7.37	7.68	7.94	8.17	8.37	8.55	8.71	8.86	9.00	9.12	9.24	9.35	9.46	9.55	9.65
8	4.75	5.64	6.20	6.62	6.96	7.24	7.47	7.68	7.86	8.03	8.18	8.31	8.44	8.55	8.66	8.76	8.85	8.94	9.03
9	4.60	5.43	5.96	6.35	6.66	6.91	7.13	7.33	7.49	7.65	7.78	7.91	8.03	8.13	8.23	8.33	8.41	8.49	8.57
10	4.48	5.27	5.77	6.14	6.43	6.67	6.87	7.05	7.21	7.36	7.49	7.60	7.71	7.81	7.91	7.99	8.08	8.15	8.23
11	4.39	5.15	5.62	5.97	6.25	6.48	6.67	6.84	6.99	7.13	7.25	7.36	7.46	7.56	7.65	7.73	7.81	7.88	7.95
12	4.32	5.05	5.50	5.84	6.10	6.32	6.51	6.67	6.81	6.94	7.06	7.17	7.26	7.36	7.44	7.52	7.59	7.66	7.73
13	4.26	4.96	5.40	5.72	5.98	6.19	6.37	6.52	6.67	6.79	6.90	7.01	7.10	7.19	7.27	7.35	7.42	7.48	7.55
14	4.21	4.89	5.32	5.63	5.88	6.08	6.26	6.41	6.54	6.66	6.77	6.87	6.96	7.05	7.13	7.20	7.27	7.33	7.39
15	4.17	4.84	5.25	5.56	5.80	5.99	6.16	5.31	6.44	6.55	6.66	6.76	6.84	6.93	7.00	7.07	7.14	7.20	7.26
16	4.13	4.79	5.19	5.49	5.72	5.92	6.08	6.22	6.35	6.46	6.56	6.66	6.74	6.82	6.90	6.97	7.03	7.09	7.15
17	4.10	4.74	5.14	5.43	5.66	5.85	6.01	6.15	6.27	6.38	6.48	6.57	6.66	6.73	6.81	6.87	6.94	7.00	7.05
18	4.07	4.70	5.09	5.38	5.60	5.79	5.94	6.08	6.20	6.31	6.41	6.50	6.58	6.65	6.73	6.79	6.85	6.91	6.97
19	4.05	4.67	5.05	5.33	5.55	5.73	5.89	6.02	6.14	6.25	6.34	6.43	6.51	6.58	6.65	6.72	6.78	6.84	6.89
20	4.02	4.64	5.02	5.29	5.51	5.69	5.84	5.97	6.09	6.19	6.28	6.37	6.45	6.52	6.59	6.65	6.71	6.77	6.82
24	3.96	4.55	4.91	5.17	5.37	5.54	5.69	5.81	5.92	6.02	6.11	6.19	6.26	6.33	6.39	6.45	6.51	6.56	6.61
30	3.89	4.45	4.80	5.05	5.24	5.40	5.54	5.65	5.76	5.85	5.93	6.01	6.08	6.14	6.20	6.26	6.31	6.36	6.41
40	3.82	4.37	4.70	4.93	5.11	5.26	5.39	5.50	5.60	5.69	5.76	5.83	5.90	5.96	6.02	6.07	6.12	6.16	6.21
60	3.76	4.28	4.59	4.82	4.99	5.13	5.25	5.36	5.45	5.53	5.60	5.67	5.73	5.78	5.84	5.89	5.93	5.97	6.01
120	3.70	4.20	4.50	4.71	4.87	5.01	5.12	5.21	5.30	5.37	5.44	5.50	5.56	5.61	5.66	5.71	5.75	5.79	5.83
∞	3.64	4.12	4.40	4.60	4.76	4.88	4.99	5.08	5.16	5.23	5.29	5.35	5.40	5.45	4.49	5.54	5.57	5.61	5.65

The entries are $q_{0.05}$ where $p(q < q_{0.05}) = 0.95$

1	17.97	26.98	32.82	37.08	40.41	43.12	45.40	47.36	49.07	50.59	51.96	53.20	54.33	55.36	56.32	57.22	58.04	58.83	59.56
2	6.08	8.33	9.80	10.88	11.74	12.44	13.03	13.54	13.99	14.39	14.75	15.08	15.38	15.65	15.91	16.14	16.37	16.57	16.77
3	4.50	5.91	6.82	7.50	8.04	8.48	8.85	9.18	9.46	9.72	9.95	10.15	10.35	10.53	10.69	10.84	10.98	11.11	11.24
4	3.93	5.04	5.76	6.29	6.71	7.05	7.35	7.60	7.83	8.03	8.21	8.37	8.82	8.66	8.79	8.91	9.03	9.13	9.23
5	3.64	4.60	5.22	5.67	6.03	6.33	6.58	5.80	6.99	7.17	7.32	7.47	7.60	7.72	7.83	7.93	8.03	8.12	8.21
6	3.46	4.34	4.90	6.30	5.63	5.90	6.12	6.32	4.49	6.65	6.79	6.92	7.03	7.14	7.24	7.34	7.43	7.51	7.59
7	3.34	4.16	4.68	5.06	5.36	5.61	5.82	6.00	6.16	6.30	6.43	6.55	6.66	6.76	6.85	6.94	7.02	7.10	7.17
8	3.26	4.04	4.53	4.89	5.17	5.40	5.60	5.77	5.92	6.05	6.18	6.29	6.39	6.48	6.57	6.65	6.73	6.80	6.87
9	3.20	3.95	4.41	4.76	5.02	5.24	5.43	5.59	5.74	5.87	5.98	6.09	6.19	6.28	6.36	6.44	6.51	6.58	6.64
10	3.15	3.88	4.33	4.65	4.91	5.12	5.30	5.46	5.60	5.72	5.83	5.93	6.03	6.11	6.19	6.27	6.34	6.40	6.47
11	3.11	3.82	4.26	4.57	4.82	5.03	5.20	5.35	5.49	5.61	5.71	5.81	5.90	5.98	6.06	6.13	6.20	6.27	6.33
12	3.08	3.77	4.20	4.51	4.75	4.95	5.12	5.27	5.39	5.51	5.61	5.71	5.80	5.88	5.95	6.02	6.09	6.15	6.21
13	3.06	3.73	4.15	4.45	4.69	4.88	5.05	5.19	5.32	5.43	5.53	5.63	5.71	5.79	5.86	5.93	5.99	6.05	6.11
14	3.03	3.70	4.11	4.41	4.64	4.83	4.99	5.13	5.25	5.36	5.46	5.55	5.64	5.71	5.79	5.85	5.91	5.97	6.03
15	3.01	3.67	4.08	4.37	4.59	4.78	4.94	5.08	5.20	5.31	5.40	5.49	5.57	5.65	5.72	5.78	5.85	5.90	5.96
16	3.00	3.65	4.05	4.33	4.56	4.74	4.90	5.03	5.15	5.26	5.35	5.44	5.52	5.59	5.66	5.73	5.79	5.84	5.90
17	2.98	3.63	4.02	4.30	4.52	4.70	4.86	4.99	5.11	5.21	5.31	5.39	5.47	5.54	5.61	5.67	5.73	5.79	5.84
18	2.97	3.61	4.00	4.28	4.49	4.67	4.82	4.96	5.07	5.17	5.27	5.35	5.43	5.50	5.57	5.63	5.69	5.74	5.79
19	2.96	3.59	3.98	4.25	4.47	4.65	4.79	4.92	5.04	5.14	5.23	5.31	5.39	5.46	5.53	5.59	5.65	5.70	5.75
20	2.95	3.58	3.96	4.23	4.45	4.62	4.77	4.90	5.01	5.11	5.20	5.28	5.36	5.43	5.49	5.55	5.61	5.66	5.71
24	2.92	3.53	3.90	4.17	4.37	4.54	4.68	4.81	4.92	5.01	5.10	5.18	5.25	5.32	5.38	5.44	5.49	5.55	5.59
30	2.89	3.49	3.85	4.10	4.30	4.46	4.60	4.72	4.82	4.92	5.00	5.08	5.15	5.21	5.27	5.33	5.38	5.43	5.47
40	2.86	3.44	3.79	4.04	4.23	4.39	4.52	4.63	4.73	4.82	4.90	4.98	5.04	4.11	5.16	5.22	5.27	5.31	5.36
60	2.83	3.40	3.74	3.98	4.16	4.31	4.44	4.55	4.65	4.73	4.81	4.88	4.94	5.00	5.06	5.11	5.15	5.20	5.24
120	2.80	3.36	3.68	3.92	4.10	4.24	4.36	4.47	4.56	4.64	4.71	4.78	4.84	4.90	4.95	5.00	5.04	5.09	5.13
∞	2.77	3.31	3.63	3.86	4.03	4.17	4.29	4.39	4.47	4.55	4.62	4.68	4.74	4.80	4.85	4.89	4.93	4.97	5.01

(Continued)

Table 19.4 (Continued) Percentage Points of the Studentized Range: Upper-α Critical Values for Tukey's HSD Multiple Comparison Procedure Instructions:

(1) Enter the section of the table that corresponds to the predetermined α-level.

(2) Enter the row that corresponds to the degrees of freedom for error from the ANOVA.

(3) Pick the value of q in that row from the column that corresponds to the number of treatments being compared.

ν	2	3	4	5	6	7	8	9	10	11	12	13	14	15	16	17	18	19	20
The entries are $q_{0.10}$ where $p(q < q_{0.10}) = 0.90$																			
1	8.93	13.44	16.36	18.49	20.15	21.51	22.64	23.62	24.48	25.24	25.92	26.54	27.10	27.62	28.10	28.54	28.96	29.35	29.71
2	4.13	5.73	6.77	7.54	8.14	8.63	9.05	9.41	9.72	10.01	10.26	10.49	10.70	10.89	11.07	11.24	11.39	11.54	11.68
3	3.33	4.47	5.20	5.74	6.16	6.51	6.81	7.06	7.29	7.49	7.67	7.83	7.98	8.12	8.25	8.27	8.48	8.58	8.68
4	3.01	3.98	4.59	4.03	5.39	5.68	5.93	6.14	6.33	6.49	6.65	6.78	6.91	7.02	7.13	7.23	7.33	7.41	7.50
5	2.85	3.72	4.26	4.66	4.98	5.24	5.46	5.65	5.82	5.97	6.10	6.22	6.34	6.44	6.54	6.63	6.71	6.79	6.86
6	2.75	3.56	4.07	4.44	4.73	4.97	5.17	5.34	5.50	5.64	5.76	5.87	5.98	6.07	6.16	5.25	6.32	6.40	6.47
7	2.68	3.45	3.93	4.28	4.55	4.78	4.97	5.14	5.28	5.41	5.53	5.64	5.74	5.83	5.91	5.99	6.06	6.13	6.19
8	2.63	3.37	3.83	4.17	4.43	4.65	4.83	4.99	5.13	5.25	5.36	5.46	5.56	5.64	5.72	5.80	5.87	5.93	6.00
9	2.59	3.32	3.76	4.08	4.34	4.54	4.72	4.87	5.01	5.13	5.23	5.33	5.42	5.51	5.58	5.66	5.72	5.79	5.85
10	2.56	3.27	3.70	4.02	4.26	4.47	4.64	4.78	4.91	5.03	5.13	5.23	5.32	5.40	5.47	5.54	5.61	5.67	5.73
11	2.54	3.23	3.66	3.96	4.20	4.40	4.57	4.71	4.84	4.95	5.05	5.15	5.23	5.31	5.38	5.45	5.51	5.57	5.63
12	2.52	3.20	3.62	3.92	4.16	4.35	4.51	4.65	4.78	4.89	4.99	5.08	5.16	5.24	5.31	5.37	5.44	5.49	5.55
13	2.50	3.18	3.59	3.88	4.12	4.20	4.46	4.60	4.72	4.83	4.93	5.02	5.10	5.28	5.25	5.31	5.37	5.43	5.48
14	2.49	3.16	3.56	3.85	4.08	4.27	4.42	4.56	4.68	4.79	4.88	4.97	5.05	5.12	5.19	5.26	5.32	5.37	5.43
15	2.48	3.14	3.54	3.83	4.05	4.23	4.39	4.52	4.64	4.75	4.84	4.93	5.01	5.08	5.15	5.21	5.27	5.32	5.38
16	2.47	3.12	3.52	3.80	4.03	4.21	4.36	4.49	4.61	4.71	4.81	4.89	4.97	5.04	5.11	5.17	5.23	5.28	5.33
17	2.46	3.11	3.50	3.78	4.00	4.18	4.33	4.46	4.58	4.68	4.77	4.86	4.93	5.01	5.07	5.13	5.19	5.24	5.30
18	2.45	3.10	3.49	3.77	3.98	4.16	4.31	4.44	4.55	4.65	4.75	4.83	4.90	4.98	5.04	5.10	5.16	5.21	5.26
19	2.45	3.09	3.47	3.75	3.97	4.14	4.29	4.42	4.53	4.63	4.72	4.80	4.88	4.95	5.01	5.07	5.13	5.18	5.23
20	2.44	3.08	3.46	3.74	3.95	4.12	4.27	4.40	4.51	4.61	4.70	4.78	4.85	4.92	4.99	5.05	5.10	5.16	5.20
24	2.42	3.05	3.42	3.69	3.90	4.07	4.21	4.34	4.44	4.54	4.63	4.71	4.78	4.85	4.91	4.97	5.02	5.07	5.12
30	2.40	3.02	3.39	3.65	3.85	4.02	4.16	4.28	4.38	4.47	4.56	4.64	4.71	4.77	4.83	4.89	4.94	4.99	5.03
40	2.38	2.99	3.35	3.60	3.80	3.96	4.10	4.21	4.32	4.41	4.49	4.56	4.63	4.69	4.75	4.81	4.86	4.90	4.95
60	2.36	2.96	3.31	3.56	3.75	3.91	4.04	4.16	4.25	4.34	4.42	4.49	4.56	4.62	4.67	4.73	4.78	4.82	4.86
120	2.34	2.93	3.28	3.52	3.71	3.86	3.99	4.10	4.19	4.28	4.35	4.42	4.48	4.54	4.60	4.65	4.69	4.74	4.78
∞	2.33	2.90	3.24	3.48	3.66	3.81	3.93	4.04	4.13	4.21	4.28	4.35	4.41	4.47	4.52	4.57	4.61	4.65	4.69

Table 19.5 Upper-α Probability Points of χ^2-distribution (Entries are $\chi^2_{\alpha:\nu}$)

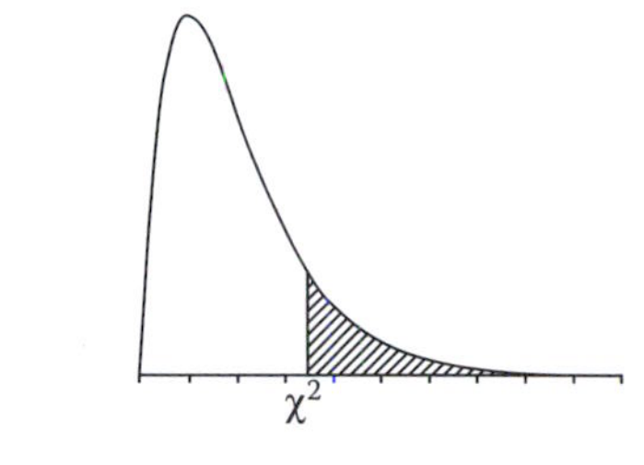

Instructions: (1) Enter the row of the table corresponding to the number of degrees of freedom (ν) for χ^2.
(2) Pick the value of χ^2 in that row, from the column that corresponds to the predetermined α-level.

	α												
ν	0.995	0.990	0.975	0.950	0.900	0.750	0.500	0.250	0.100	0.050	0.025	0.010	0.005
1	0.0000393	0.000157	0.000982	0.00393	0.0158	0.102	0.455	1.32	2.71	3.84	5.02	6.63	7.88
2	0.0100	0.0201	0.0506	0.103	0.211	0.575	1.39	2.77	4.61	5.99	7.38	9.21	10.6
3	0.0717	0.115	0.216	0.352	0.584	1.21	2.37	4.11	6.25	7.81	9.35	11.3	12.8
4	0.207	0.297	0.484	0.711	1.06	1.92	3.36	5.39	7.78	9.49	11.1	13.3	14.9
5	0.412	0.554	0.831	1.15	1.61	2.67	4.35	6.63	9.24	11.1	12.8	15.1	16.7
6	0.676	0.872	1.24	1.64	2.20	3.45	5.35	7.84	10.6	12.6	14.4	16.8	18.5
7	0.989	1.24	1.69	2.17	2.83	4.25	6.35	9.04	12.0	14.1	16.0	18.5	20.3
8	1.34	1.65	2.18	2.73	3.49	5.07	7.34	10.2	13.4	15.5	17.5	20.1	22.0
9	1.73	2.09	2.70	3.33	4.17	5.90	8.34	11.4	14.7	16.9	19.0	21.7	23.6
10	2.16	2.56	3.25	3.94	4.87	6.74	9.34	12.5	16.0	18.3	20.5	23.2	25.2
11	2.60	3.05	3.82	4.57	5.58	7.58	10.3	13.7	17.3	19.7	21.9	24.7	26.8
12	3.07	3.57	4.40	5.23	6.30	8.44	11.3	14.8	18.5	21.0	23.3	26.2	28.3
13	3.57	4.11	5.01	5.89	7.04	9.30	12.3	16.0	19.8	22.4	24.7	27.7	29.8
14	4.07	4.66	5.63	6.57	7.79	10.2	13.3	17.1	21.1	23.7	26.1	29.1	31.3
15	4.60	5.23	6.26	7.26	8.55	11.0	4.3	18.2	22.3	25.0	27.5	30.6	32.8
16	5.14	5.81	6.91	7.96	9.31	11.9	15.3	19.4	23.5	26.3	28.8	32.0	34.3
17	5.70	6.41	7.56	8.67	10.1	12.8	16.3	20.5	24.8	27.6	30.2	33.4	35.7
18	6.26	7.01	8.23	9.39	10.9	13.7	17.3	21.6	26.0	28.9	31.5	34.8	37.2
19	6.84	7.63	8.91	10.1	11.7	14.6	18.3	22.7	27.2	30.1	32.9	36.2	38.6

(*Continued*)

Table 19.5 (Continued) Upper-α Probability Points of χ^2-distribution (Entries are $\chi^2_{\alpha:\nu}$)

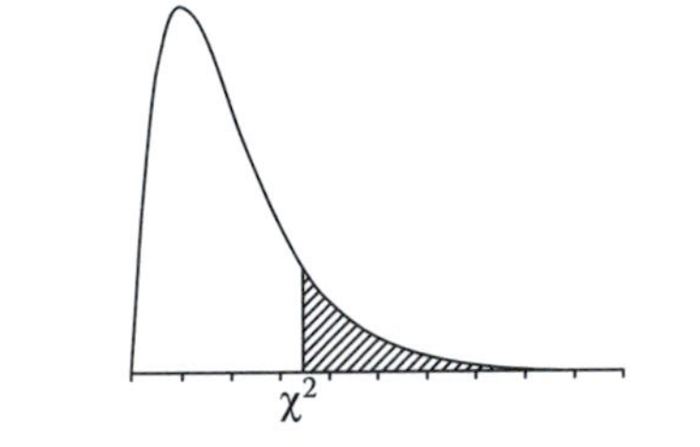

Instructions: (1) Enter the row of the table corresponding to the number of degrees of freedom (ν) for χ^2.
(2) Pick the value of χ^2 in that row, from the column that corresponds to the predetermined α-level.

	α												
ν	0.995	0.990	0.975	0.950	0.900	0.750	0.500	0.250	0.100	0.050	0.025	0.010	0.005
20	7.43	8.26	9.59	10.9	12.4	15.5	19.3	23.8	28.4	31.4	34.2	37.6	40.0
21	8.03	8.90	10.3	11.6	13.2	16.3	20.3	24.9	29.6	32.7	35.5	38.9	41.4
22	8.64	9.54	11.0	12.3	14.0	17.2	21.3	26.9	30.8	33.9	36.8	40.3	42.8
23	9.26	10.2	11.7	13.1	14.8	18.1	22.3	27.1	32.0	35.2	38.1	41.6	44.2
24	9.89	10.9	12.4	13.8	15.7	19.0	23.3	28.2	33.2	36.4	39.4	43.0	45.6
25	10.5	11.5	13.1	14.6	16.5	19.9	24.3	29.3	34.4	37.7	40.6	44.3	46.9
26	11.2	12.2	13.8	15.4	17.3	20.8	25.3	30.4	35.6	38.9	41.9	45.6	48.3
27	11.8	12.9	14.6	16.2	18.1	21.7	26.3	31.5	36.7	40.1	43.2	47.0	49.6
28	12.5	13.6	15.3	16.9	18.9	22.7	27.3	32.6	37.9	41.3	44.5	48.3	51.0
29	13.1	14.3	16.0	17.7	19.8	23.6	28.3	33.7	39.1	42.6	45.7	49.6	52.3
30	13.8	15.0	16.8	18.5	20.6	24.5	29.3	34.8	40.3	43.8	47.0	50.9	53.7

Table 19.6 Upper-α Probability Points of F-distribution (Entries are $F_{\alpha:v1,v2}$)

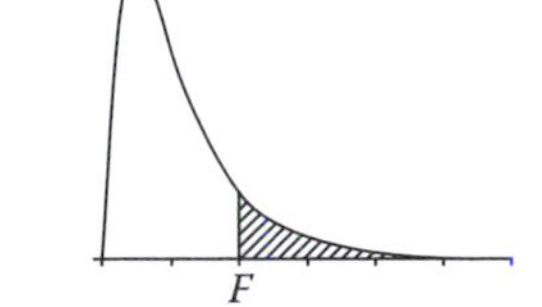

Instructions:
(1) Enter the section of the table corresponding to the predetermined for α-level.
(2) Enter the row that corresponds to the denominator degrees of freedom (υ_2)
(3) Pick the value of F in that row, from the column that corresponds to the numerator degrees of freedom (υ_1).

	ν_1																		
ν_2	1	2	3	4	5	6	7	8	9	10	12	15	20	24	30	40	60	120	∞
$\alpha = 0.10$																			
1	39.86	49.50	53.59	55.83	57.24	58.20	58.91	59.44	59.86	60.19	60.71	61.22	61.74	62.00	62.26	62.53	62.79	63.06	63.33
2	8.53	9.00	9.16	9.24	9.29	9.33	9.35	9.37	9.38	9.39	9.41	9.42	9.44	9.45	9.46	9.47	9.47	9.48	9.49
3	5.54	5.46	5.39	5.34	5.31	5.28	5.27	5.25	5.24	5.23	5.22	5.20	5.18	5.18	5.17	5.16	5.15	5.14	5.13
4	4.54	4.32	4.19	4.11	4.05	4.01	3.98	3.95	3.94	3.92	3.90	3.87	3.84	3.83	3.82	3.80	3.79	3.78	3.76
5	4.06	3.78	3.62	3.52	3.45	3.40	3.37	3.34	3.32	3.30	3.27	3.24	3.21	3.19	3.17	3.16	3.14	3.12	3.10
6	3.78	3.46	3.29	3.18	3.11	3.05	3.01	2.98	2.96	2.94	2.90	2.87	2.84	2.82	2.80	2.78	2.76	2.74	2.72
7	3.59	3.26	3.07	2.96	2.88	2.83	2.78	2.75	2.72	2.70	2.67	2.63	2.59	2.58	2.56	2.54	2.51	2.49	2.47
8	3.46	3.11	2.92	2.81	2.73	2.67	2.62	2.59	2.56	2.54	2.50	2.46	2.42	2.40	2.38	2.36	2.34	2.32	2.29
9	3.36	3.01	2.81	2.69	2.61	2.55	2.51	2.47	2.44	2.42	2.38	2.34	2.30	2.28	2.25	2.23	2.21	2.18	2.16
10	3.29	2.92	2.73	2.61	2.52	2.46	2.41	2.38	2.35	2.32	2.28	2.24	2.20	2.18	2.16	2.13	2.11	2.08	2.06
11	3.23	2.86	2.66	2.54	2.45	2.39	2.34	2.30	2.27	2.25	2.21	2.17	2.12	2.10	2.08	2.05	2.03	2.00	1.97
12	3.18	2.81	2.61	2.48	2.39	2.33	2.28	2.24	2.21	2.19	2.15	2.10	2.06	2.04	2.01	1.99	1.96	1.93	1.90
13	3.14	2.76	2.56	2.43	2.35	2.28	2.23	2.20	2.16	2.14	2.10	2.05	2.01	1.98	1.96	1.93	1.90	1.88	1.85
14	3.10	2.73	2.52	2.39	2.31	2.24	2.19	2.15	2.12	2.10	2.05	2.01	1.96	1.94	1.91	1.89	1.86	1.83	1.80
15	3.07	2.70	2.49	2.36	2.27	2.21	2.16	2.12	2.09	2.06	2.02	1.97	1.92	1.90	1.87	1.85	1.82	1.79	1.76
16	3.05	2.67	2.46	2.33	2.24	2.18	2.13	2.09	2.06	2.03	1.99	1.94	1.89	1.87	1.84	1.81	1.78	1.75	1.72
17	3.03	2.64	2.44	2.31	2.22	2.15	2.10	2.06	2.03	2.00	1.96	1.91	1.86	1.84	1.81	1.78	1.75	1.72	1.69
18	3.01	2.62	2.42	2.29	2.20	2.13	2.08	2.04	2.00	1.98	1.93	1.89	1.84	1.81	1.78	1.75	1.72	1.69	1.66

(Continued)

Table 19.6 (Continued) Upper-α Probability Points of *F*-distribution (Entries are $F_{\alpha:v1, v2}$)

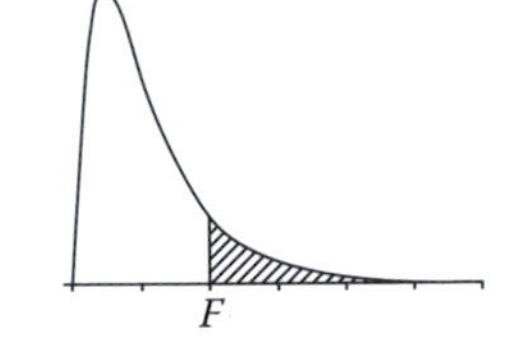

Instructions: (1) Enter the section of the table corresponding to the predetermined for α-level.
(2) Enter the row that corresponds to the denominator degrees of freedom (υ_2)
(3) Pick the value of *F* in that row, from the column that corresponds to the numerator degrees of freedom (υ_1).

ν_2	ν_1: 1	2	3	4	5	6	7	8	9	10	12	15	20	24	30	40	60	120	∞
α = 0.10																			
19	2.99	2.61	2.40	2.27	2.18	2.11	2.06	2.02	1.98	1.96	1.91	1.86	1.81	1.79	1.76	1.73	1.70	1.67	1.63
20	2.97	2.59	2.38	2.25	2.16	2.09	2.04	2.00	1.96	1.94	1.89	1.84	1.79	1.77	1.74	1.71	1.68	1.64	1.61
21	2.96	2.57	2.36	2.23	2.14	2.08	2.02	1.98	1.95	1.92	1.87	1.83	1.78	1.75	1.72	1.69	1.66	1.62	1.59
22	2.95	2.56	2.35	2.22	2.13	2.06	2.01	1.97	1.93	1.90	1.86	1.81	1.76	1.73	1.70	1.67	1.64	1.60	1.57
23	2.94	2.55	2.34	2.21	2.11	2.05	1.99	1.95	1.92	1.89	1.84	1.80	1.74	1.72	1.69	1.66	1.62	1.59	1.55
24	2.93	2.54	2.33	2.19	2.10	2.04	1.98	1.94	1.91	1.88	1.83	1.78	1.73	1.70	1.67	1.64	1.61	1.57	1.53
25	2.92	2.53	2.32	2.18	2.09	2.02	1.97	1.93	1.89	1.87	1.82	1.77	1.72	1.69	1.66	1.63	1.59	1.56	1.52
26	2.91	2.52	2.31	2.17	2.08	2.01	1.96	1.92	1.88	1.86	1.81	1.76	1.71	1.68	1.65	1.61	1.58	1.54	1.50
27	2.90	2.51	2.30	2.17	2.07	2.00	1.95	1.91	1.87	1.85	1.80	1.75	1.70	1.67	1.64	1.60	1.57	1.53	1.49
28	2.89	2.50	2.29	2.16	2.06	2.00	1.94	1.90	1.87	1.84	1.79	1.74	1.69	1.66	1.63	1.59	1.56	1.52	1.48
29	2.89	2.50	2.28	2.15	2.06	1.99	1.93	1.89	1.86	1.83	1.78	1.73	1.68	1.65	1.62	1.58	1.55	1.51	1.47
30	2.88	2.49	2.28	2.14	2.05	1.98	1.93	1.88	1.85	1.82	1.77	1.72	1.67	1.64	1.61	1.57	1.54	1.50	1.46
40	2.84	2.44	2.23	2.09	2.00	1.93	1.87	1.83	1.79	1.76	1.71	1.66	1.61	1.57	1.54	1.51	1.47	1.42	1.38
60	2.79	2.39	2.18	2.04	1.95	1.87	1.82	1.77	1.74	1.71	1.66	1.60	1.54	1.51	1.48	1.44	1.40	1.35	1.29
120	2.75	2.35	2.13	1.99	1.90	1.82	1.77	1.72	1.68	1.65	1.60	1.55	1.48	1.45	1.41	1.37	1.32	1.26	1.19
∞	2.71	2.30	2.08	1.94	1.85	1.77	1.72	1.67	1.63	1.60	1.55	1.49	1.42	1.38	1.34	1.30	1.24	1.17	1.00
α = 0.05																			
1	161.4	199.5	215.7	224.6	230.2	234.0	236.8	238.9	240.5	241.9	243.9	245.9	248.0	249.1	250.1	251.1	252.2	253.3	254.3
2	18.51	19.00	19.16	19.25	19.30	19.33	19.35	19.37	19.38	19.40	19.41	19.43	19.45	19.45	19.46	19.47	19.48	19.49	19.50

3	10.13	9.55	9.28	9.12	9.01	8.94	8.89	8.85	8.81	8.79	8.74	8.70	8.66	8.64	8.62	8.59	8.57	8.55	8.53
4	7.71	6.94	6.59	6.39	6.26	6.16	6.09	6.04	6.00	5.96	5.91	5.86	5.80	5.77	5.75	5.72	5.69	5.66	5.63
5	6.61	5.79	5.41	5.19	5.05	4.95	4.88	4.82	4.77	4.74	4.68	4.62	4.56	4.53	4.50	4.46	4.43	4.40	4.36
6	5.99	5.14	4.76	4.53	4.39	4.28	4.21	4.15	4.10	4.06	4.00	3.94	3.87	3.84	3.81	3.77	3.74	3.70	3.67
7	5.59	4.74	4.35	4.12	3.97	3.87	3.79	3.73	3.68	3.64	3.57	3.51	3.44	3.41	3.38	3.34	3.30	3.27	3.23
8	5.32	4.46	4.07	3.84	3.69	3.58	3.50	3.44	3.39	3.35	3.28	3.22	3.15	3.12	3.08	3.04	3.01	2.97	2.93
9	5.12	4.26	3.86	3.63	3.48	3.37	3.29	3.23	3.18	3.14	3.07	3.01	2.94	2.90	2.86	2.83	2.79	2.75	2.71
10	4.96	4.10	3.71	3.48	3.33	3.22	3.14	3.07	3.02	2.98	2.91	2.85	2.77	2.74	2.70	2.66	2.62	2.58	2.54
11	4.84	3.98	3.59	3.36	3.20	3.09	3.01	2.95	2.90	2.85	2.79	2.72	2.65	2.61	2.57	2.53	2.49	2.45	2.40
12	4.75	3.89	3.49	3.26	3.11	3.00	2.91	2.85	2.80	2.75	2.69	2.62	2.54	2.51	2.47	2.43	2.38	2.34	2.30
13	4.67	3.81	3.41	3.18	3.03	2.92	2.83	2.77	2.71	2.67	2.60	2.53	2.46	2.42	2.38	2.34	2.30	2.25	2.21
14	4.60	3.74	3.34	3.11	2.96	2.85	2.76	2.70	2.65	2.60	2.53	2.46	2.39	2.35	2.31	2.27	2.22	2.18	2.13
15	4.54	3.68	3.29	3.06	2.90	2.79	2.71	2.64	2.59	2.54	2.48	2.40	2.33	2.29	2.25	2.20	2.16	2.11	2.07
16	4.49	3.63	3.24	3.01	2.85	2.74	2.66	2.59	2.54	2.49	2.42	2.35	2.28	2.24	2.19	2.15	2.11	2.06	2.01
17	4.45	3.59	3.20	2.96	2.81	2.70	2.61	2.55	2.49	2.45	2.38	2.31	2.23	2.19	2.15	2.10	2.06	2.01	1.96
18	4.41	3.55	3.16	2.93	2.77	2.66	2.58	2.51	2.46	2.41	2.34	2.27	2.19	2.15	2.11	2.06	2.02	1.97	1.92
19	4.38	3.52	3.13	2.90	2.74	2.63	2.54	2.48	2.42	2.38	2.31	2.23	2.16	2.11	2.07	2.03	1.98	1.93	1.88
20	4.35	3.49	3.10	2.87	2.71	2.60	2.51	2.45	2.39	2.35	2.28	2.20	2.12	2.08	2.04	1.99	1.95	1.90	1.84
21	4.32	3.47	3.07	2.84	2.68	2.57	2.49	2.42	2.37	2.32	2.25	2.18	2.10	2.05	2.01	1.96	1.92	1.87	1.81
22	4.30	3.44	3.05	2.82	2.66	2.55	2.46	2.40	2.34	2.30	2.23	2.15	2.07	2.03	1.98	1.94	1.89	1.84	1.78
23	4.28	3.42	3.03	2.80	2.64	2.53	2.44	2.37	2.32	2.27	2.20	2.13	2.05	2.01	1.96	1.91	1.86	1.81	1.76
24	4.26	3.40	3.01	2.78	2.62	2.51	2.42	2.36	2.30	2.25	2.18	2.11	2.03	1.98	1.94	1.89	1.84	1.79	1.73
25	4.24	3.39	2.99	2.76	2.60	2.49	2.40	2.34	2.28	2.24	2.16	2.09	2.01	1.96	1.92	1.87	1.82	1.77	1.71
26	4.23	3.37	2.98	2.74	2.59	2.47	2.39	2.32	2.27	2.22	2.15	2.07	1.99	1.95	1.90	1.85	1.80	1.75	1.69
27	4.21	3.35	2.96	2.73	2.57	2.46	2.37	2.31	2.25	2.20	2.13	2.06	1.97	1.93	1.88	1.84	1.79	1.73	1.67
28	4.20	3.34	2.95	2.71	2.56	2.45	2.36	2.29	2.24	2.19	2.12	2.04	1.96	1.91	1.87	1.82	1.77	1.71	1.65
30	4.17	3.32	2.92	2.69	2.53	2.42	2.33	2.27	2.21	2.16	2.09	2.01	1.93	1.89	1.84	1.79	1.74	1.68	1.62
40	4.08	3.23	2.84	2.61	2.45	2.34	2.25	2.18	2.12	2.08	2.00	1.92	1.84	1.79	1.74	1.69	1.64	1.58	1.51
60	4.00	3.15	2.76	2.53	2.37	2.25	2.17	2.10	2.04	1.99	1.92	1.84	1.75	1.70	1.65	1.59	1.53	1.47	1.39
120	3.92	3.07	2.68	2.45	2.29	2.17	2.09	2.02	1.96	1.91	1.83	1.75	1.66	1.61	1.55	1.50	1.43	1.35	1.25
∞	3.84	3.00	2.60	2.37	2.21	2.10	2.01	1.94	1.88	1.83	1.75	1.67	1.57	1.52	1.46	1.39	1.32	1.22	1.00

(Continued)

Table 19.6 (Continued) Upper-α Probability Points of F-distribution (Entries are $F_{\alpha:v1,v2}$)

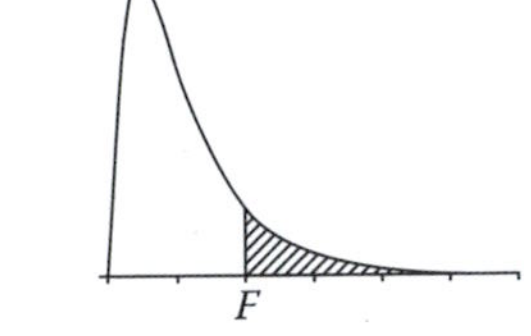

Instructions: (1) Enter the section of the table corresponding to the predetermined for α-level.
(2) Enter the row that corresponds to the denominator degrees of freedom (υ_2)
(3) Pick the value of F in that row, from the column that corresponds to the numerator degrees of freedom (υ_1).

ν_2	ν_1: 1	2	3	4	5	6	7	8	9	10	12	15	20	24	30	40	60	120	∞
α = 0.01																			
1	4052	4999.50	5403	5625	5764	5859	5928	5982	6022	6056	6106	6157	6209	6235	6261	6287	6313	6339	6366
2	98.50	99.00	99.17	99.25	99.30	99.33	99.36	99.37	99.39	99.40	99.42	99.43	99.45	99.46	99.47	99.47	99.48	99.49	99.50
3	34.12	30.82	29.46	28.71	28.24	27.91	27.67	27.49	27.35	27.23	27.05	26.87	26.69	26.60	26.50	26.41	26.32	26.22	26.13
4	21.20	18.00	16.69	15.98	15.52	15.21	14.98	14.80	14.66	14.55	14.37	14.20	14.02	13.93	13.84	13.75	13.65	13.56	13.46
5	16.26	13.27	12.06	11.39	10.97	10.67	10.46	10.29	10.16	10.05	9.89	9.72	9.55	9.47	9.38	9.29	9.20	9.11	9.02
6	13.75	10.92	9.78	9.15	8.75	8.47	8.26	8.10	7.98	7.87	7.72	7.56	7.40	7.31	7.23	7.14	7.06	6.97	6.88
7	12.25	9.55	8.45	7.85	7.46	7.19	6.99	6.84	6.72	6.62	6.47	6.31	6.16	6.07	5.99	5.91	5.82	5.74	5.65
8	11.26	8.65	7.59	7.01	6.63	6.37	6.18	6.03	5.91	5.81	5.67	5.52	5.36	5.28	5.20	5.12	4.03	4.95	4.86
9	10.56	8.02	6.99	6.42	6.06	5.80	5.61	5.47	5.35	5.26	5.11	4.96	4.81	4.73	4.65	4.57	4.48	4.40	4.31
10	10.04	7.56	6.55	5.99	5.64	5.39	5.20	5.06	4.94	4.85	4.71	4.56	4.41	4.33	4.25	4.17	4.08	4.00	3.91
11	9.65	7.21	6.22	5.67	5.32	5.07	4.89	4.74	4.63	4.54	4.40	4.25	4.10	4.02	3.94	3.86	3.78	3.69	3.60

12	9.33	6.93	5.95	5.41	5.06	4.82	4.64	4.50	4.39	4.30	4.16	4.01	3.86	3.78	3.70	3.62	3.54	3.45	3.36
13	9.07	6.70	5.74	5.21	4.86	4.62	4.44	4.30	4.19	4.10	3.96	3.82	3.66	3.59	3.51	3.43	3.34	3.25	3.17
14	8.86	6.51	5.56	5.04	4.69	4.46	4.28	4.14	4.03	3.94	3.80	3.66	3.51	3.43	3.35	3.27	3.18	3.09	3.00
15	8.68	6.36	5.42	4.89	4.56	4.32	4.14	4.00	3.89	3.80	3.67	3.52	3.37	3.29	3.21	3.13	3.05	2.96	2.87
16	8.53	6.23	5.29	4.77	4.44	4.20	4.03	3.89	3.78	3.69	3.55	3.41	3.26	3.18	3.10	3.02	2.93	2.84	2.75
17	8.40	6.11	5.18	4.67	4.34	4.10	3.93	3.79	3.68	3.59	3.46	3.31	3.16	3.08	3.00	2.92	2.83	2.75	2.65
18	8.29	6.01	5.09	4.58	4.25	4.01	3.84	3.71	3.60	3.51	3.37	3.23	3.08	3.00	2.92	2.84	2.75	2.66	2.57
19	8.18	5.93	5.01	4.50	4.17	3.94	3.77	3.63	3.52	3.43	3.30	3.15	3.00	2.92	2.84	2.76	2.67	2.58	2.49
20	8.10	5.85	4.94	4.43	4.10	3.87	3.70	3.56	3.46	3.37	3.23	3.09	2.94	2.86	2.78	2.69	2.61	2.52	2.42
21	8.02	5.78	4.87	4.37	4.04	3.81	3.64	3.51	3.40	3.31	3.17	3.03	2.88	2.80	2.72	2.64	2.55	2.46	2.36
22	7.95	5.72	4.82	4.31	3.99	3.76	6.59	3.45	3.35	3.26	3.12	2.98	2.83	2.75	2.67	2.58	2.50	2.40	2.31
23	7.88	5.66	4.76	4.26	3.94	3.71	3.54	3.41	3.30	3.21	3.07	2.93	2.78	2.70	2.62	2.54	2.45	2.35	2.26
24	7.82	5.61	4.72	4.22	3.90	3.67	3.50	3.36	3.26	3.17	3.03	2.89	2.74	2.66	2.58	2.49	2.40	2.31	2.21
25	7.77	5.57	4.68	4.18	3.85	3.63	3.46	3.32	3.22	3.13	2.99	2.85	2.70	2.62	2.54	2.45	2.36	2.27	2.17
26	7.72	5.53	4.64	4.14	3.82	3.59	3.42	3.29	3.18	3.09	2.96	2.81	2.66	2.58	2.50	2.42	2.33	2.23	2.13
27	7.68	5.49	4.60	4.11	3.78	3.56	3.39	3.26	3.15	2.06	2.93	2.78	2.63	2.55	2.47	2.38	2.29	2.20	2.10
28	7.64	5.45	4.57	4.07	3.75	3.53	3.36	3.23	3.12	3.03	2.90	2.75	2.60	2.52	2.44	2.35	2.26	2.17	2.06
29	7.60	5.42	4.54	4.04	3.73	3.50	3.33	3.20	3.09	3.00	2.87	2.73	2.57	2.49	2.41	2.33	2.23	2.14	2.03
30	7.56	5.39	4.51	4.02	3.70	3.47	3.30	3.17	3.07	2.98	2.84	2.70	2.55	2.47	2.39	2.30	2.21	2.11	2.01
40	7.31	5.18	4.31	3.83	3.51	3.29	3.12	2.99	2.89	2.80	2.66	2.52	2.37	2.29	2.20	2.11	2.02	1.92	1.80
60	7.08	4.98	4.13	3.65	3.34	3.12	2.95	2.82	2.72	2.63	2.50	2.35	2.20	2.12	2.03	1.94	1.84	1.73	1.60
120	6.85	4.79	3.95	3.48	3.17	2.96	2.79	2.66	2.56	2.47	2.34	2.19	2.03	1.95	1.86	1.76	1.66	1.53	1.38
∞	6.63	4.61	3.78	3.32	3.02	2.80	2.64	2.51	2.41	2.32	2.18	2.04	1.88	1.79	1.70	1.59	1.47	1.32	1.00

Table 19.7 Minimum Number of Assessments in a Triangle Test (Entries are n_{α,β,p_d}.)
Entries are the sample sizes (n) required in a Triangle test to deliver sensitivity defined by the values chosen for α, β, and p_d. Enter the table in the section corresponding to the chosen value of p_d and the row corresponding to the chosen value of α. Read the required sample size, n, from the column corresponding to the chosen value of β.

	β							
α	0.50	0.40	0.30	0.20	0.10	0.05	0.01	0.001
$p_d = 50\%$								
0.40	3	3	3	6	8	9	15	26
0.30	3	3	3	7	8	11	19	30
0.20	4	6	7	7	12	16	25	36
0.10	7	8	8	12	15	20	30	43
0.05	7	9	11	16	20	23	35	48
0.01	13	15	19	25	30	35	47	62
0.001	22	26	30	36	43	48	62	81
$p_d = 40\%$								
0.40	3	3	6	6	9	15	26	41
0.30	3	3	7	8	11	19	30	47
0.20	6	7	7	12	17	25	36	55
0.10	8	10	15	17	25	30	46	67
0.05	11	15	16	23	30	40	57	79
0.01	21	26	30	35	47	56	76	102
0.001	36	39	48	55	68	76	102	130
$p_d = 30\%$								
0.40	3	6	6	9	15	26	44	73
0.30	3	8	8	16	22	30	53	84
0.20	7	12	17	20	28	39	64	97
0.10	15	15	20	30	43	54	81	119
0.05	16	23	30	40	53	66	98	136
0.01	33	40	52	62	82	97	131	181
0.001	61	69	81	93	120	138	181	233
$p_d = 20\%$								
0.40	6	9	12	18	35	50	94	153
0.30	8	11	19	30	47	67	116	183
0.20	12	20	28	39	64	86	140	212
0.10	25	33	46	62	89	119	178	260
0.05	40	48	66	87	117	147	213	305
0.01	72	92	110	136	176	211	292	397
0.001	130	148	176	207	257	302	396	513

Table 19.7 (Continued) Minimum Number of Assessments in a Triangle Test (Entries are n_{α,β,p_d}.)
Entries are the sample sizes (n) required in a Triangle test to deliver sensitivity defined by the values chosen for α, β, and p_d. Enter the table in the section corresponding to the chosen value of p_d and the row corresponding to the chosen value of α. Read the required sample size, n, from the column corresponding to the chosen value of β.

	β							
α	**0.50**	**0.40**	**0.30**	**0.20**	**0.10**	**0.05**	**0.01**	**0.001**
$p_d = 10\%$								
0.40	9	18	38	70	132	197	360	598
0.30	19	36	64	102	180	256	430	690
0.20	39	64	103	149	238	325	439	819
0.10	89	125	175	240	348	457	683	1011
0.05	144	191	249	325	447	572	828	1178
0.01	284	350	425	525	680	824	1132	1539
0.001	494	579	681	803	996	1165	1530	1992

Table 19.8 Critical Number of Correct Response in a Triangle Test (Entries are $x_{\alpha,n}$)
Entries are the minimum number of correct responses required for significance at the stated α-level (i.e., column) for the corresponding number of respondents, n (i.e., row). Reject the assumption of "no difference" if the number of correct responses is greater than or equal to the tabled value.

	α								α						
n	0.40	0.30	0.20	0.10	0.05	0.01	0.001	n	0.40	0.30	0.20	0.10	0.05	0.01	0.001
								31	12	13	14	15	16	18	20
								32	12	13	14	15	16	18	20
3	2	2	3	3	3	—	—	33	13	13	14	15	17	18	21
4	3	3	3	4	4	—	—	34	13	14	15	16	17	19	21
5	3	3	4	4	4	5	—	35	13	14	15	16	17	19	22
6	3	4	4	5	5	6	—	36	14	14	15	17	18	20	22
7	4	4	4	5	5	6	7	42	16	17	18	19	20	22	25
8	4	4	5	5	6	7	8	48	18	19	20	21	22	25	27
9	4	5	5	6	6	7	8	54	20	21	22	23	25	27	30
10	5	5	6	6	7	8	9	60	22	23	24	26	27	30	33
11	5	5	6	7	7	8	10	66	24	25	26	28	29	32	35
12	5	6	6	7	8	9	10	72	26	27	28	30	32	34	38
13	6	6	7	8	8	9	11	78	28	29	30	32	34	37	40
14	6	7	7	8	9	10	11	84	30	31	33	35	36	39	43
15	6	7	8	8	9	10	12	90	32	33	35	37	38	42	45
16	7	7	8	9	9	11	12	96	34	35	37	39	41	44	48
17	7	8	8	9	10	11	13	102	36	37	39	41	43	46	50
18	7	8	9	10	10	12	13	108	38	40	41	43	45	49	53
19	8	8	9	10	11	12	14	114	40	42	43	45	47	51	55
20	8	9	9	10	11	13	14	120	42	44	45	48	50	53	57
21	8	9	10	11	12	13	15	126	44	46	47	50	52	56	60
22	9	9	10	11	12	14	15	132	46	48	50	52	54	58	62
23	9	10	11	12	12	14	16	138	48	50	52	54	56	60	64
24	10	10	11	12	13	15	16	144	50	52	54	56	58	62	67
25	10	11	11	12	13	15	17	150	52	54	56	58	61	65	69
26	10	11	12	13	14	15	17	156	54	56	58	61	63	67	72
27	11	11	12	13	14	16	18	162	56	58	60	63	65	69	74
28	11	12	12	14	15	16	18	168	58	60	62	65	67	71	76
29	11	12	13	14	15	17	19	174	61	62	64	67	69	74	79
30	12	12	13	14	15	17	19	180	63	64	66	69	71	76	81

Note: For values of n not in the table, compute $z = (k-1(1/3)n)/\sqrt{(2/9)n}$, where k is the number of correct responses. Compare the value of z to the α-critical value of standard normal variable, that is, the values in the last row of Tables 19.3 ($z_\alpha = t_{\alpha,\infty}$).

Table 19.9 Minimum Number of Assessments in a Duo–Trio or One-Sided Directional Difference Test (Entries are n_{α,β,p_d})
Entries are the sample sizes (n) required in duo–trio or one-sided directional difference test to deliver the sensitivity defined by the values chosen for α, β, and p_d. Enter the table in the section corresponding to the chosen value of p_d for duo–trio test or p_{max} for a directional difference test and the row corresponding to the chosen value of α. Read the required sample size, n, from the column corresponding to the chosen value of β.

	β							
α	**0.50**	**0.40**	**0.30**	**0.20**	**0.10**	**0.05**	**0.01**	**0.001**
$p_d = 50\%$ $p_{max} = 75\%$								
0.40	2	4	4	6	10	14	27	41
0.30	2	5	7	9	13	20	30	47
0.20	5	5	10	12	19	26	39	58
0.10	9	9	14	19	26	33	48	70
0.05	13	16	18	23	33	42	58	82
0.01	22	27	33	40	50	59	80	107
0.001	38	43	51	61	71	83	107	140
$p_d = 40\%$ $p_{max} = 70\%$								
0.40	4	4	6	8	14	25	41	70
0.30	5	7	9	13	22	28	49	78
0.20	5	10	12	19	30	39	60	94
0.10	14	19	21	28	39	53	79	113
0.05	18	23	30	37	53	67	93	132
0.01	35	42	52	64	80	96	130	174
0.001	61	71	81	95	117	135	176	228
$p_d = 30\%$ $p_{max} = 65\%$								
0.40	4	6	8	14	29	41	76	120
0.30	7	9	13	24	39	53	88	144
0.20	10	17	21	32	49	68	110	166
0.10	21	28	37	53	72	96	145	208
0.05	30	42	53	69	93	119	173	243
0.01	64	78	89	112	143	174	235	319
0.001	107	126	144	172	210	246	318	412

(Continued)

Table 19.9 (Continued) Minimum Number of Assessments in a Duo–Trio or One-Sided Directional Difference Test (Entries are n_{α,β,p_d})
Entries are the sample sizes (n) required in duo–trio or one-sided directional difference test to deliver the sensitivity defined by the values chosen for α, β, and p_d. Enter the table in the section corresponding to the chosen value of p_d for duo–trio test or p_{max} for a directional difference test and the row corresponding to the chosen value of α. Read the required sample size, n, from the column corresponding to the chosen value of β.

		β							
α		**0.50**	**0.40**	**0.30**	**0.20**	**0.10**	**0.05**	**0.01**	**0.001**
	$p_d = 20\%$ $p_{max} = 60\%$								
0.40		6	10	23	35	59	94	171	282
0.30		11	22	30	49	84	119	205	327
0.20		21	32	49	77	112	158	253	384
0.10		46	66	85	115	168	214	322	471
0.05		71	93	119	158	213	268	392	554
0.01		141	167	207	252	325	391	535	726
0.001		241	281	327	386	479	556	731	944
	$p_d = 10\%$ $p_{max} = 55\%$								
0.40		10	35	61	124	237	362	672	1124
0.30		30	72	117	199	333	479	810	1302
0.20		81	129	193	294	451	618	1006	1555
0.10		170	239	337	461	658	861	1310	1905
0.05		281	369	475	620	866	1092	1583	2237
0.01		550	665	820	1007	1301	1582	2170	2927
0.001		961	1125	1309	1551	1908	2248	2937	3812

Table 19.10 Critical Number of Correct Responses in Duo–Trio and One-Sided Directional Difference Test (Entries are $x_{\alpha,n}$)

Entries are the minimum number of correct responses required for significance at the stated α-level (i.e., column) for the corresponding number of respondents, *n* (i.e., row). Reject the assumption of "no difference" if the number of correct responses is greater than or equal to the tabled value.

	α								α						
n	0.40	0.30	0.20	0.10	0.05	0.01	0.001	*n*	0.40	0.30	0.20	0.10	0.05	0.01	0.001
								31	17	18	19	20	21	23	25
2	2	2	—	—	—	—	—	32	18	18	19	21	22	24	26
3	3	3	3	—	—	—	—	33	18	19	20	21	22	24	26
4	3	4	4	4	—	—	—	34	19	20	20	22	23	25	27
5	4	4	4	5	5	—	—	35	19	20	21	22	23	25	27
6	4	5	5	6	6	—	—	36	20	21	22	23	24	26	28
7	5	5	6	6	7	7	—	40	22	23	24	25	26	28	31
8	5	6	6	7	7	8	—	44	24	25	26	27	28	31	33
9	6	6	7	7	8	9	—	48	26	27	28	29	31	33	36
10	6	7	7	8	9	10	10	52	28	29	30	32	33	35	38
11	7	7	8	9	9	10	11	56	30	31	32	34	35	38	40
12	7	8	8	9	10	11	12	60	32	33	34	36	37	40	43
13	8	8	9	10	10	12	13	64	34	35	36	38	40	42	45
14	8	9	10	10	11	12	13	68	36	37	38	40	42	45	48
15	9	10	10	11	12	13	14	72	38	39	41	42	44	47	50
16	10	10	11	12	12	14	15	76	40	41	43	45	46	49	52
17	10	11	11	12	13	14	16	80	42	43	45	47	48	51	55
18	11	11	12	13	13	15	16	84	44	45	47	49	51	54	57
19	11	12	12	13	14	15	17	88	46	47	49	51	53	56	59
20	12	12	13	14	15	16	18	92	48	50	51	53	55	58	62
21	12	13	13	14	15	17	18	96	50	52	53	55	57	60	64
22	13	13	14	15	16	17	19	100	52	54	55	57	59	63	66
23	13	14	15	16	16	18	20	104	54	56	57	60	61	65	69
24	14	14	15	16	17	19	20	108	56	58	59	62	64	67	71
25	14	15	16	17	18	19	21	112	58	60	61	64	66	69	73
26	15	15	16	17	18	20	22	116	60	62	64	66	68	71	76
27	15	16	17	18	19	20	22	122	63	65	67	69	71	75	79
28	16	16	17	18	19	21	23	128	66	68	70	72	74	78	82
29	16	17	18	19	20	22	24	134	69	71	73	75	78	81	86
30	17	17	18	20	20	22	24	140	72	74	76	79	81	85	89

Note: For values of n not in the table, compute $z = (k - 0.5n)/\sqrt{0.25n}$ where k is the number of correct responses. Compare the value of z to the α-critical value of a standard normal variable, that is, the values in the last row of Table 19.3 ($z_\alpha = t_{\alpha,\infty}$).

Table 19.11 Minimum Number of Assessments in a Two-Sided Directional Difference Test (Entries are $n_{\alpha,\beta,p_{max}}$)
Entries are the sample sizes (n) required in two-sided directional difference test to deliver the sensitivity defined by the values chosen for α, β, and p_{max}. Enter the table in the section corresponding to the chosen value of p_{max} and the row corresponding to the chosen value of α. Read the required sample size, n, from the column corresponding to the chosen value of β.

	β							
α	0.50	0.40	0.30	0.20	0.10	0.05	0.01	0.001
$p_{max} = 75\%$								
0.40	5	5	10	12	19	26	39	58
0.30	6	8	11	16	22	29	42	64
0.20	9	9	14	19	26	33	48	70
0.10	13	16	18	23	33	42	58	82
0.05	17	20	25	30	42	49	67	92
0.01	26	34	39	44	57	66	87	117
0.001	42	50	58	66	78	90	117	149
$p_{max} = 70\%$								
0.40	5	10	12	19	30	39	60	94
0.30	8	13	18	22	33	44	68	102
0.20	14	19	21	28	39	53	79	113
0.10	18	23	30	37	53	67	93	132
0.05	25	35	40	49	65	79	110	149
0.01	44	49	59	73	92	108	144	191
0.001	68	78	90	102	126	147	188	240
$p_{max} = 65\%$								
0.40	10	17	21	32	49	68	110	166
0.30	13	20	29	42	59	81	125	188
0.20	21	28	37	53	72	96	145	208
0.10	30	42	53	69	93	119	173	243
0.05	44	56	67	90	114	145	199	176
0.01	73	92	108	131	164	195	261	345
0.001	121	140	161	188	229	267	342	440
$p_{max} = 60\%$								
0.40	21	32	49	77	112	158	253	384
0.30	31	44	66	89	133	179	283	425
0.20	46	66	85	115	168	214	322	471
0.10	71	93	119	158	213	268	392	554
0.05	101	125	158	199	263	327	455	635
0.01	171	204	241	291	373	446	596	796
0.001	276	318	364	425	520	604	781	1010

Table 19.11 (Continued) Minimum Number of Assessments in a Two-Sided Directional Difference Test (Entries are $n_{\alpha,\beta,p_{max}}$)
Entries are the sample sizes (n) required in two-sided directional difference test to deliver the sensitivity defined by the values chosen for α, β, and p_{max}. Enter the table in the section corresponding to the chosen value of p_{max} and the row corresponding to the chosen value of α. Read the required sample size, n, from the column corresponding to the chosen value of β.

	β							
α	**0.50**	**0.40**	**0.30**	**0.20**	**0.10**	**0.05**	**0.01**	**0.001**
p_{max} = 55%								
0.40	81	129	193	294	451	618	1006	1555
0.30	110	173	254	359	550	721	1130	1702
0.20	170	239	337	461	658	861	1310	1905
0.10	281	369	475	620	866	1092	1583	2237
0.05	390	497	620	786	1055	1302	1833	2544
0.01	670	802	963	1167	1493	1782	2408	3203
0.001	1090	1260	1461	1707	2094	2440	3152	4063

Table 19.12 Critical Number of Correct Responses in a Two-Sided Directional Difference Test (Entries are $x_{\alpha,n}$)
Entries are the minimum number of correct responses required for significance at the stated α-level (i.e., column) for the corresponding number of respondents, *n* (i.e., row). Reject the assumption of "no difference" if the number of correct responses is greater than or equal to the tabled value.

	α								α						
n	0.40	0.30	0.20	0.10	0.05	0.01	0.001	*n*	0.40	0.30	0.20	0.10	0.05	0.01	0.001
								31	19	19	20	21	22	24	25
2	—	—	—	—	—	—	—	32	19	20	21	22	23	24	26
3	3	3	—	—	—	—	—	33	20	20	21	22	23	25	27
4	4	4	4	—	—	—	—	34	20	21	22	23	24	25	27
5	4	5	5	5	—	—	—	35	21	22	22	23	24	26	28
6	5	5	6	6	6	—	—	36	22	22	23	24	25	27	29
7	6	6	6	7	7	—	—	40	24	24	25	26	27	29	31
8	6	6	7	7	8	8	—	44	26	26	27	28	29	31	34
9	7	7	7	8	8	9	—	48	28	29	29	31	32	34	36
10	7	8	8	9	9	10	—	52	30	31	32	33	34	36	39
11	8	8	9	9	10	11	11	56	32	33	34	35	36	39	41
12	8	9	9	10	10	11	12	60	34	35	36	37	39	41	44
13	9	9	10	10	11	12	13	64	36	37	38	40	41	43	46
14	10	10	10	11	12	13	14	68	38	39	40	42	43	46	48
15	10	11	11	12	12	13	14	72	41	41	42	44	45	48	51
16	11	11	12	12	13	14	15	76	43	44	45	46	48	50	53
17	11	12	12	13	13	15	16	80	45	46	47	48	50	52	56
18	12	12	13	13	14	15	17	84	47	48	49	51	52	55	58
19	12	13	13	14	15	16	17	88	49	50	51	53	54	57	60
20	13	13	14	15	15	17	18	92	51	52	53	55	56	59	63
21	13	14	14	15	16	17	19	96	53	54	55	57	59	62	65
22	14	14	15	16	17	18	19	100	55	56	57	59	61	64	67
23	15	15	16	16	17	19	20	104	57	58	60	61	63	66	70
24	15	16	16	17	18	19	21	108	59	60	62	64	65	68	72
25	16	16	17	18	18	20	21	112	61	62	64	66	67	71	74
26	16	17	17	18	19	20	22	116	64	65	66	68	70	73	77
27	17	17	18	19	20	21	23	122	67	68	69	71	73	76	80
28	17	18	18	19	20	22	23	128	70	71	72	74	76	80	83
29	18	18	19	20	21	22	24	134	73	74	75	78	79	83	87
30	18	19	20	20	21	23	25	140	76	77	79	81	83	86	90

Note: For values of *n* not in the table, compute $z = (k - 0.5n)/\sqrt{0.25n}$, where *k* is the number of correct responses. Compare the value of *z* to the α/2-critical value of a standard normal variable, that is, the values in the last row of Table 19.3 ($z_{\alpha/2} = t_{\alpha/2,\infty}$).

Table 19.13 Minimum Number of Assessments in a Two-out-of-Five Test (Entries are n_{α,β,p_d})

Entries are the sample sizes (n) required in a two-out-of-five test to deliver sensitivity defined by the values chosen for α, β, and p_d. Enter the table in the section corresponding to the chosen value of p_d and the row corresponding to the chosen value of α. Read the required sample size, n, from the column corresponding to the chosen value of β.

		β							
α		0.50	0.40	0.30	0.20	0.10	0.05	0.01	0.001
	$p_d = 50\%$								
0.40		3	4	4	5	6	7	9	13
0.30		3	4	4	5	6	7	9	16
0.20		3	4	4	5	6	7	12	18
0.10		3	4	4	5	8	9	15	18
0.05		3	6	6	7	8	12	17	24
0.01		5	7	8	9	13	14	22	29
0.001		9	9	12	13	17	21	27	36
	$p_d = 40\%$								
0.40		4	4	5	6	7	9	12	20
0.30		4	4	5	6	7	9	15	23
0.20		4	4	5	6	7	12	15	23
0.10		4	4	5	9	10	15	18	30
0.05		6	7	7	11	13	18	24	33
0.01		8	9	12	14	18	23	30	42
0.001		12	13	17	21	26	31	41	54
	$p_d = 30\%$								
0.40		5	5	6	8	9	11	20	30
0.30		5	5	6	8	9	15	24	35
0.20		5	5	6	8	13	15	28	39
0.10		5	5	9	11	17	22	32	47
0.05		7	8	12	14	20	26	39	54
0.01		13	14	18	23	30	36	49	69
0.001		21	22	27	32	42	49	66	87
	$p_d = 20\%$								
0.40		6	7	8	10	13	21	38	59
0.30		6	7	8	10	18	26	43	69
0.20		6	7	8	15	22	30	53	79
0.10		10	11	17	23	31	40	62	94
0.05		13	19	24	27	40	53	76	108
0.01		24	30	36	43	57	70	99	136
0.001		38	48	55	67	81	99	129	172

(*Continued*)

Table 19.13 (Continued) Minimum Number of Assessments in a Two-out-of-Five Test (Entries are n_{α,β,p_d})
Entries are the sample sizes (n) required in a two-out-of-five test to deliver sensitivity defined by the values chosen for α, β, and p_d. Enter the table in the section corresponding to the chosen value of p_d and the row corresponding to the chosen value of α. Read the required sample size, n, from the column corresponding to the chosen value of β.

	β							
α	**0.50**	**0.40**	**0.30**	**0.20**	**0.10**	**0.05**	**0.01**	**0.001**
$p_d = 10\%$								
0.40	9	11	13	22	40	60	108	184
0.30	9	16	19	34	54	80	128	212
0.20	14	22	31	47	73	99	161	245
0.10	25	38	54	70	103	130	206	297
0.05	41	55	70	94	127	167	244	249
0.01	77	98	121	145	192	233	330	449
0.001	135	158	187	224	278	332	438	572

Table 19.14 Critical Number of Correct Responses in Two-out-of-Five Test (Entries are $x_{\alpha,n}$) Entries are the minimum number of correct responses required for significance at the stated α-level (i.e., column) for the corresponding number of respondents, n (i.e., row). Reject the assumption of "no difference" if the number of correct responses is greater than or equal to the tabled value.

	α								α						
n	0.40	0.30	0.20	0.10	0.05	0.01	0.001	n	0.40	0.30	0.20	0.10	0.05	0.01	0.001
								31	4	5	5	6	7	8	10
								32	4	5	6	6	7	9	10
3	1	1	2	2	2	3	3	33	5	5	6	7	7	9	11
4	1	2	2	2	3	3	4	34	5	5	6	7	7	9	11
5	2	2	2	2	3	3	4	35	5	5	6	7	8	9	11
6	2	2	2	3	3	4	5	36	5	5	6	7	8	9	11
7	2	2	2	3	3	4	5	37	5	6	6	7	8	9	11
8	2	2	2	3	3	4	5	38	5	6	6	7	8	10	11
9	2	2	3	3	4	4	5	39	5	6	6	7	8	10	12
10	2	2	3	3	4	5	6	40	5	6	7	7	8	10	12
11	2	3	3	3	4	5	6	41	5	6	7	8	8	10	12
12	2	3	3	4	4	5	6	42	6	6	7	8	9	10	12
13	2	3	3	4	4	5	6	43	6	6	7	8	9	10	12
14	3	3	3	4	4	5	7	44	6	6	7	8	9	11	12
15	3	3	3	4	5	6	7	45	6	6	7	8	9	11	13
16	3	3	4	4	5	6	7	46	6	7	7	8	9	11	13
17	3	3	4	4	5	6	7	47	6	7	7	8	9	11	13
18	3	3	4	4	5	6	8	48	6	7	8	9	9	11	13
19	3	3	4	5	5	6	8	49	6	7	8	9	10	11	13
20	3	4	4	5	5	7	8	50	6	7	8	9	10	11	14
21	3	4	4	5	6	7	8	51	7	7	8	9	10	12	14
22	3	4	4	5	6	7	8	52	7	7	8	9	10	12	14
23	4	4	4	5	6	7	9	53	7	7	8	9	10	12	14
24	4	4	5	5	6	7	9	54	7	7	8	9	10	12	14
25	4	4	5	5	6	7	9	55	7	8	8	9	10	12	14
26	4	4	5	6	6	8	9	56	7	8	8	10	10	12	14
27	4	4	5	6	6	8	9	57	7	8	9	10	11	12	15
28	4	5	5	6	7	8	10	58	7	8	9	10	11	13	15
29	4	5	5	6	7	8	10	59	7	8	9	10	11	13	15
30	4	5	5	6	7	8	10	60	7	8	9	10	11	13	15

Note: For values of n not in the table compute $z = (k - 0.1n)/\sqrt{0.09n}$, where k is the number of correct responses. Compare the value of z to the α-critical value of standard normal variable, that is, the values in the last row of Tables 19.3 ($z_\alpha = t_{\alpha,\infty}$).

Table 19.15 Critical Number of Correct Responses in a Specified Tetrad Test (Entries are $x_{\alpha,n}$) Entries are the minimum number of correct responses required for significance at the stated α-level (i.e., column) for the corresponding number of respondents, *n* (i.e., row). Reject the assumption of "no difference" if the number of correct responses is greater than or equal to the tabled value.

	α								α						
n	0.40	0.30	0.20	0.10	0.05	0.01	0.001	*n*	0.40	0.30	0.20	0.10	0.05	0.01	0.001
								31	7	7	8	9	10	11	13
								32	7	7	8	9	10	12	14
3	2	2	2	2	3	3	-	33	7	8	8	9	10	12	14
4	2	2	2	3	3	4	4	34	7	8	8	10	10	12	14
5	2	2	2	3	3	4	5	35	7	8	9	10	11	12	14
6	2	2	3	3	4	4	5	36	7	8	9	10	11	13	15
7	2	3	3	3	4	5	6	42	9	9	10	11	12	14	16
8	2	3	3	4	4	5	6	48	10	10	11	12	13	15	18
9	3	3	3	4	4	5	7	54	11	11	12	14	15	17	19
10	3	3	4	4	5	6	7	60	12	12	13	15	16	18	21
11	3	3	4	4	5	6	7	66	13	14	15	16	17	19	22
12	3	4	4	5	5	6	8	72	14	15	16	17	18	21	24
13	3	4	4	5	6	7	8	78	15	16	17	18	20	22	25
14	4	4	4	5	6	7	8	84	16	17	18	19	21	23	26
15	4	4	5	5	6	7	9	90	17	18	19	21	22	25	28
16	4	4	5	6	6	8	9	96	18	19	20	22	23	26	29
17	4	5	5	6	7	8	9	102	19	20	21	23	24	27	30
18	4	5	5	6	7	8	10	108	20	21	22	24	26	28	32
19	4	5	5	6	7	8	10	114	21	22	23	25	27	30	33
20	5	5	6	7	7	9	10	120	22	23	24	26	28	31	34
21	5	5	6	7	7	9	10	126	23	24	25	27	29	32	36
22	5	5	6	7	8	9	11	132	24	25	27	29	30	33	37
23	5	6	6	7	8	9	11	138	25	26	28	30	31	35	38
24	5	6	6	7	8	10	11	144	26	27	29	31	33	36	40
25	6	6	7	8	8	10	12	150	27	28	30	32	34	37	41
26	6	6	7	8	9	10	12	156	28	29	31	33	35	38	42
27	6	6	7	8	9	10	12	162	29	30	32	34	36	39	44
28	6	7	7	8	9	11	12	168	30	31	33	35	37	41	45
29	6	7	7	8	9	11	13	174	31	32	34	36	38	42	46
30	6	7	8	9	10	11	13	180	32	34	35	37	39	43	47

Note: For values of *n* not in the table, compute $z = (k - n/6)/\sqrt{5n/36}$, where *k* is the number of correct responses. Compare the value of *z* to the a-critical value of standard normal variable, that is, the values in the last row of Tables 19.3 ($z_\alpha = t_{\alpha,\infty}$).

20

Practical Sensory Problems

SCENARIO 1

In an effort to diversify its product line from just grain-based snacks, a major consumer packaged goods (CPG) corporation acquired a yogurt company and plans an extensive line of yogurt-based foods and beverages. The research and development (R&D) and marketing divisions are charged to understand the yogurt category and contribute to the strategy and success in the development of new products and their campaigns in the market place.

In what ways can the sensory department work with R&D and marketing to achieve success for the new brands to be developed? Which sensory method or approach would be appropriate? State the project objective and the test objective.

The project objective is to understand the yogurt category; the test objectives are to define the yogurt characteristics and map the landscape. Additionally, the sensory group should plan to contribute to the company's understanding of how consumers "feel" about different yogurts and what experiences they provide to the consumer (health, convenience, satisfaction of hunger, etc.).

To understand the yogurt market, it is necessary to collect all or most of the products in the category. Start with one version (plain yogurt or yogurt with fruit) and shop in the local stores and on the Internet to learn the different brands and their product offerings. Buy a wide representative sample and prepare to taste them all. If the company has a descriptive panel, engage several members to work with R&D to first screen the chosen array and develop a preliminary lexicon. If your panel is inexperienced, the panelists cannot be expected to generate the lexicon without some input from the literature and the American Society for Testing and Materials (ASTM) lexicon (Lawless and Civille, 2013; ASTM, 2011). Without a descriptive panel, the researchers will have difficulty capturing the appearance, aroma, flavor, and texture characteristics of the yogurts. This piece is critical to defining the target space in which the company wants to compete.

Describe the testing methodology. Address any special considerations based on the details of the scenario.

Choose the category to study (plain yogurt) and collect the products. The controls involve choosing brands, ages, and types of yogurt that represent the competitive set. Keep samples refrigerated and serve in coded cups for all sessions, including the screening sessions with R&D, the panel development of the lexicon (which may take three to five sessions), and the final evaluation of all brands and types using the lexicon (which should take several sessions).

Outline the test design. Where will the test be conducted? Who are the assessors, and how many will there be? How will the samples be handled and presented to the assessors?

A total of 10 trained descriptive panelists participate in the test. The test is conducted in a sensory descriptive panel room with controls such as conditioned air and fragrance-free facilities. Product controls include coded cups for samples held in refrigerated conditions and within shelf life. Care is needed in preparing and serving yogurt since the gel breaks with scooping, and samples need to be served immediately and not held refrigerated after scooping.

What sort of data will be collected, and how will it be analyzed? What are the results of the test, and what can you conclude from the statistical analysis (if providing data or results)? Interpret the results of the statistical analysis in the context of the test objective and the project objective. What are your recommendations for the company?

The final lexicon (and no lexicon is ever final) should be completed with definitions and references for all the attributes and with procedures for evaluation of the appearance and texture terms. Descriptive panelists will rate each attribute on a 0–15 scale, and individual/consensus data is collected (see Chapters 11–12).

The descriptive data is analyzed with analysis of variance (ANOVA) and mapped using principal component analysis (PCA; see Sections 15.2.1.2 and 15.3.1) or partial least-squares (PLS) regression (see Section 15.2.2.3). Flavor and aroma can be mapped together; texture and appearance can be mapped together. The maps are used to understand the sensory yogurt space: where competitors fit on that map, where the white space exists or not, and where there are opportunities for the company to succeed.

At this point, a consumer test is recommended to understand the consumer responses and possible segments. This test can include questions on attitudes and usage, emotions, perceived benefits, and sensory attributes associated with the products. See Section 18.9 for more details on category appraisals and Section 15.3 for a discussion of the statistical analysis involved. Combining both sensory descriptive data and consumer liking data enables the discovery of sensory segments and key drivers in the product category.

How might this scenario be addressed differently?

The company may choose to let marketing drive the initiative and study the consumer marketing segments; in no way is this a substitute for the aforementioned R&D research, but these approaches can be combined to better understand the interface of product and consumer.

What could future testing involve?

Collaboration with marketing to determine the company winner(s), how it can best be marketed, and how to advertise using sensory signals that drive consumer purchase behavior.

SCENARIO 2

A supplier of flavors and ingredients needs to swap out a finished formula coolant for a customer's product. The finished product in question is mouthwash. The current coolant provides a unique cooling/burning sensation in the mouth while swishing as well as a lingering cooling experience that lasts up to 20 min. As the sensory scientist on this project, you need to supply data that prove there is no discernable difference between the current and new coolant ingredient in the final flavored mouthwash formula.

Which sensory method or approach would be appropriate? State the project objective and the test objective.

Project Objective: To determine if the new coolant ingredient can be substituted for the existing coolant ingredient in a finished mouthwash formula.

Test Objective: To determine if two mouthwash samples made with two different coolant ingredients can be distinguished by flavor in the mouth and residual flavor.

Solution: The standard discrimination tests will not work for this scenario. The lingering cooling of the mouthwash prevents side-by-side discrimination testing as an option. It also prevents using a reference training sample that is used for some discrimination testing. You decide to run a simple difference test and set up the protocol so only one sample will be evaluated at a time.

Describe the testing methodology. Are there any special considerations based on the details of the scenario?

A simple difference test determines whether there is a sensory difference between two products. Due to the lingering effects of the cooling product, multiple samples should not be presented simultaneously. A triangle test or duo–trio test are inappropriate.

Samples: Mouthwash A—current formula coolant; mouthwash B—new formula coolant.

Protocol: Panelists will come to the sensory laboratory and sample one mouthwash at a time. They will be instructed to return and sample the second mouthwash 4 hours later. Times will be scheduled for both evaluations.

Outline the test design, including the test controls and product controls. Where will the test be conducted? Who are the assessors, and how many will there be? How will the samples be handled and presented to the assessors?

Test design: A total of 100 untrained consumer participants will complete the test. Fifty of the participants will receive matched pairs (A/A or B/B) and 50 of the participants will receive unmatched pairs (A/B and B/A).

Sample presentation design will be as follows: 25 pairs of A/A; 25 pairs of B/B; 25 pairs of A/B; and 25 pairs of B/A. Presentation order will be balanced.

Special instructions on panelist worksheet:

1. You will be evaluating two mouthwash samples at two different session times.
2. To evaluate the mouthwash sample, place the entire volume in your mouth and swish around for 30 s. Pay special attention to the intensity and balance of mint flavor, cooling, and burning this sample has while swishing in your mouth. Expectorate in provided cup. Start your provided timer as soon as you expectorate the sample.
3. Carry the timer with you and note down on the worksheet when you feel the cooling has dissipated from the mouth.
4. Come back to your 2nd scheduled evaluation time and repeat the process with the 2nd mouthwash sample.
5. Determine if the samples are the same/identical or different.
6. Mark your response below.

What sort of data will be collected, and how will it be analyzed? What are the results of the test, and what can you conclude from the statistical analysis? Interpret the results of the statistical analysis in the context of the test objective and the project objective. What are your recommendations for the company?

Record whether assessors said matched and unmatched pairs were the same or different. Use chi-squared analysis and set an alpha level of 0.05. The raw data is presented in Table 20.1.

Chi Square Tabulation Table

		Subjects Received		
		Matched pair AA or BB	**Unmatched pair AB or BA**	**Total**
Subjects said:	Same	23	22	45
	Different	27	28	55
	Total	50	50	100

The tabulated chi-squared = 0.04 in this example. This is lower than T = 3.84, the critical chi-squared value from Table 19.5 for alpha = 0.05 and df = 1. See Section 7.6 and Example 7.10 or Example 14.6 for worked examples of the analysis.

Therefore, the mouthwash samples are not be significantly different.

The business recommendation would be to substitute the new coolant ingredient for the existing coolant in the tested mouthwash formula.

How might this scenario be addressed differently? What could future testing involve?

Another approach would be to train a small panel on intensity of cooling. This would need to be done in a session prior to the sample testing. A small group of N = 30 participants who have been prescreened for ability to detect and rank order cooling intensities could be trained to rate cooling intensity in the two mouthwash samples. The training session would consist of showing the panelists different intensity levels of cooling across an anchored intensity scale. The scale would be demonstrated as intensities of external

Table 20.1 Raw Data

Participant	Sample Pair Evaluated	Response	Correct
1	AA	Same	Yes
2	BB	Different	No
3	AB	Same	No
4	BA	Same	No
5	AA	Different	No
6	BB	Same	Yes
7	AB	Same	No
8	BA	Different	Yes
9	AA	Different	No
10	BB	Same	Yes
11	AB	Different	Yes
12	BA	Same	No
13	AA	Same	Yes
14	BB	Same	Yes
15	AB	Different	Yes
16	BA	Same	No
17	AA	Same	Yes
18	BB	Same	Yes
19	AB	Different	Yes
20	BA	Same	No
21	AA	Same	Yes
22	BB	Different	No
23	AB	Same	No
24	BA	Same	No
25	AA	Different	No
26	BB	Same	Yes
27	AB	Different	Yes
28	BA	Different	Yes
29	AA	Same	Yes
30	BB	Different	No
31	AB	Same	No
32	BA	Same	No
33	AA	Different	No
34	BB	Different	No
35	AB	Same	No
36	BA	Same	No
37	AA	Same	Yes
38	BB	Different	No
39	AB	Different	Yes

(Continued)

Table 20.1 (Continued) Raw Data

Participant	Sample Pair Evaluated	Response	Correct
40	BA	Same	No
41	AA	Different	No
42	BB	Different	No
43	AB	Same	No
44	BA	Same	No
45	AA	Different	Yes
46	BB	Different	No
47	AB	Different	Yes
48	BA	Same	No
49	AA	Different	No
50	BB	Same	Yes
51	AB	Different	Yes
52	BA	Different	Yes
53	AA	Different	No
54	BB	Same	Yes
55	AB	Same	No
56	BA	Same	No
57	AA	Same	Yes
58	BB	Different	No
59	AB	Different	Yes
60	BA	Same	No
61	AA	Different	Yes
62	BB	Different	No
63	AB	Different	Yes
64	BA	Different	Yes
65	AA	Different	No
66	BB	Same	Yes
67	AB	Same	No
68	BA	Same	No
69	AA	Different	No
70	BB	Different	No
71	AB	Different	Yes
72	BA	Different	Yes
73	AA	Same	Yes
74	BB	Same	Yes
75	AB	Different	Yes
76	BA	Different	Yes
77	AA	Different	No
78	BB	Same	Yes
79	AB	Different	No

Table 20.1 (Continued) Raw Data

Participant	Sample Pair Evaluated	Response	Correct
80	BA	Different	Yes
81	AA	Different	No
82	BB	Different	No
83	AB	Different	No
84	BA	Same	No
85	AA	Different	No
86	BB	Different	No
87	AB	Different	Yes
88	BA	Different	Yes
89	AA	Same	Yes
90	BB	Same	Yes
91	AB	Different	Yes
92	BA	Different	Yes
93	AA	Same	Yes
94	BB	Same	Yes
95	AB	Different	Yes
96	BA	Different	Yes
97	AA	Different	No
98	BB	Same	Yes
99	AB	Different	No
100	BA	Different	Yes

references as well as for cooling. For example, the panel could be trained to recognize Sweet 5, Sweet 10, Cooling 5, and Cooling 10 (see Chapter 12).

For the sample evaluation on a separate day/session, the panelists would be asked to anchor their intensities using the Sweet 5 and Sweet 10 references. They would then proceed to rate the cooling in the mouth and after expectoration of a mouthwash sample. The two mouthwash samples would still be evaluated on separate sessions at least 4 h apart. The panelists would be shown the sweet intensity references at both mouthwash sample evaluations.

The advantage of this method over the simple difference test is that if the samples are different, you will have information as to direction of difference and whether the cooling difference occurs in the mouth or in the aftertaste or both.

The version takes additional training time (approximately 1 training session), but requires a smaller N. This approach would be especially beneficial if several coolant evaluations might be needed for the business in the future.

SCENARIO 3

Company XYZ, a fragrance house, has recently received multiple briefs from potential clients to develop fragrance submissions for the bodywash category. Potential clients' success criteria include alignment with concepts that may emphasize both emotional and functional benefits along with sensory appeal. Based on this situation, Company XYZ wishes to gain seminal information about the category and understand how different fragrance accords may impart different functional and emotional benefits to the base they are included in.

In what ways can the sensory department work with R&D and marketing to achieve success for the newly developed brands?

The research is intended to be both exploratory and seminal with the intent to not only provide guidance to R&D but also to be leveraged by marketing and sales. As such, information that is relevant to those functions can be gathered, and the project initiation phase should strive to ensure that all key stakeholders' interests are represented.

Which sensory method or approach would be appropriate? State the project objective and the test objective.

Company XYZ wishes to demonstrate which underlying fragrance accords and individual fragrance notes drive perception of benefits in bodywash applications.

Project Objective: In order to do so, the company has developed a variety of fragrance accords that could be incorporated in a bodywash base. A descriptive panel has documented the characteristics of the fragrance accords to supplement information from the perfumers.

Test Objective: Company XYZ now wishes to relate the fragrance accords to a series of perceived functional benefits and emotional connections within the context of the bodywash category. In order to address that objective, a multistep approach is decided on:

- Step 1: Methodology development to ensure that proper protocols are in place for obtaining reliable information prior to implementation of a large quantitative test
- Step 2: Quantitative consumer testing is then executed to assess the benefits (functional and emotional) associated with various fragrance accords
- Step 3: Analysis and reporting to mine the data collected and identify those characteristics that may drive consumer perception of benefits

Describe the testing methodology. Address any special considerations based on the details of the scenario.

Step 1: Methodology Development

While initial discussions with the project team may help decide some of the methods to be used (e.g., sniff test vs. in-use test, central location vs. home use, number of participants, and screening criteria for participants), a methodology development step is often necessary if questions around study design still exist. The primary objective of the methodology development phase is to ensure that proper protocols and questions are in place to ensure reliable information prior to conducting a large quantitative test with consumers. Questions around test design and methods may include

- Questions surrounding consumer language—how do consumers talk about products in the category? What are the expected benefits?
- Questions surrounding questionnaire designs—what scale type and length is most appropriate for capturing consumer information?
- General questionnaire design—should overall liking and strength be asked concurrently or separately to benefit questions, at the beginning of the questionnaire or at the end?
- Questions surrounding the test design—how many samples can be evaluated within a session?

An example of frequently used methods for this objective includes qualitative approaches (such as focus groups) and quantitative pretesting of the developed questionnaire.

In this case study, two to three focus groups can easily be conducted with 8–12 target consumers in each group. The groups are designed to gather information about consumer language and expectations of benefits in the category. Through those qualitative sessions, consumers are asked to identify their expectations for the category as well as react to various stimuli (fragrance accords, sniff test) and to highlight their expected benefits, thereby generating language around emotional, functional, and sensory benefits and developing hypotheses linking scent properties to such benefits.

The language highlighted in the sessions can then be used to design a questionnaire for a larger consumer quantitative test. Sample questions may include

Assessment of functional benefits: This bodywash will… clean my skin, leave my skin smooth, leave my skin soft, be gentle on my skin, leave my skin moisturized, leave my skin refreshed, be harsh on my skin, nourish my skin, repair my skin, protect my skin. Negative product characteristics such as "irritate my skin" may also be included.

Emotional: Using this bodywash will make me feel… soothed, relaxed, energized, sensual, romantic, feminine, masculine, natural, content, happy, fresh, sophisticated, active/energetic, calm, refreshed…

If questions remain about scale types and length and about the structure of the questionnaire, the questionnaire can then be pretested. An example of design for such a pretest would include

- N = 60 consumers, representative of the target population
- Seven fragrance accords, all presented in a nondescript squeeze bottle identified by a three-digit code
- Four versions of the questionnaire, where benefits are measured using
  - 5-point intensity scale (not at all → extremely)
  - 7-point intensity scale (not at all → extremely)
  - 5-point Likert scale (disagree strongly → agree strongly).
  - 7-point Likert scale (disagree strongly → agree strongly).
- Overall liking and fragrance strength are also measured in each of the four versions of the questionnaire

Each consumer is asked to attend two 60 min sessions, during which they are exposed to two of the four questionnaires (balanced incomplete block design). In each session, consumers are asked to complete the questionnaire for each of the seven fragrances. At the end of each session, a short exit questionnaire is administered where consumers report on the complexity of the task, fatigue, and optimal number of evaluations per 60 min session. The information is then mined to identify optimal question types and scales to capture consumer perception of functional and emotional benefits.

Outline the test design. Where will the test be conducted? Who are the assessors, and how many will there be? How will the samples be handled and presented to the assessors?

Step 2: Quantitative Consumer Testing

The primary objective of this step is to assess the benefits (functional and emotional) associated with various fragrance accords.

- Method: In this case, a sniff test is selected (no in-use component).
- Stimuli: 18 fragrance accords evaluated out of 4-oz squeezable plastic bottles with flip-caps (material was tested to ensure that no contamination with plastic notes occurred within 48 h of sample preparation).
- Respondents: N = 250 respondents, mix of gender, age, ethnicity, 10 geographic locations within the United States, current users of fragranced bodywash.
- Test design: Each respondent evaluates all 18 fragrances over 3 days (one 60-min session per day, 6 samples per session) following a balanced complete block design.
- Questionnaire includes assessment of emotional benefits first (5-point intensity scale), followed by functional benefit (5-point intensity scale), followed by overall liking (7-point intensity scale) and overall fragrance strength (7-point intensity scale). Within the benefit statements, order of benefits is rotated.

What sort of data will be collected, and how will it be analyzed? What are the results of the test, and what can you conclude from the statistical analysis? Interpret the results of the statistical analysis in the context of the test objective and the project objective. What are your recommendations for the company?

Step 3: Analysis and reporting

Once data is collected, data is summarized using univariate techniques such as analysis of variance (ANOVA). This allows for a benefit profile to be provided for each of the fragrance accords included in the test as well as highlighting similarities and differences among fragrances.

The data can then be represented to highlight some key insights. An example of graphical output may include Figure 20.1.

Additionally, multivariate statistics such as factor analysis can also be applied to the data with the intent of generating benefit maps that summarize consumer benefit perceptions. Examples of such maps are presented in Figures 20.2 and 20.3.

Similar output can be generated for consumer subgroups of interest. Due to the relatively large sample size, age, gender, geographic location, and usage groups can be analyzed separately. Further information can also be overlaid on top of the maps, such as liking, fragrance strength, and fragrance characteristics identified by the descriptive

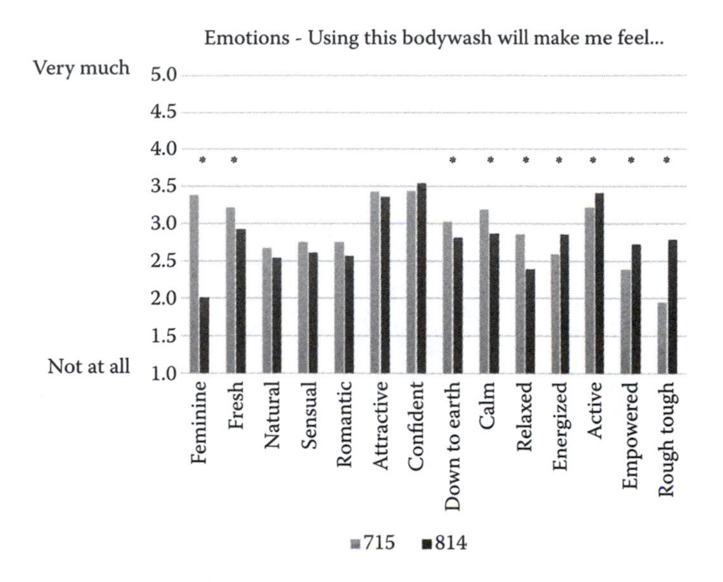

Figure 20.1 Example of emotional profile for two fragrances (sample 715 and sample 814). *Note*: * indicates attributes for which a significant difference is noted among the two fragrances at the 95% confidence level.

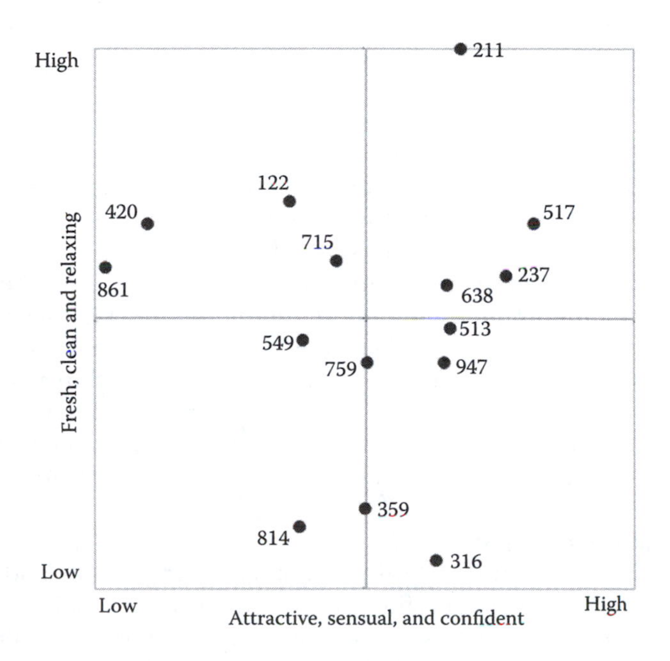

Figure 20.2 Example of functional and emotional benefit map. *Note*: In an actual report, three-digit codes would be replaced by actual fragrance accord to provide directions for product development.

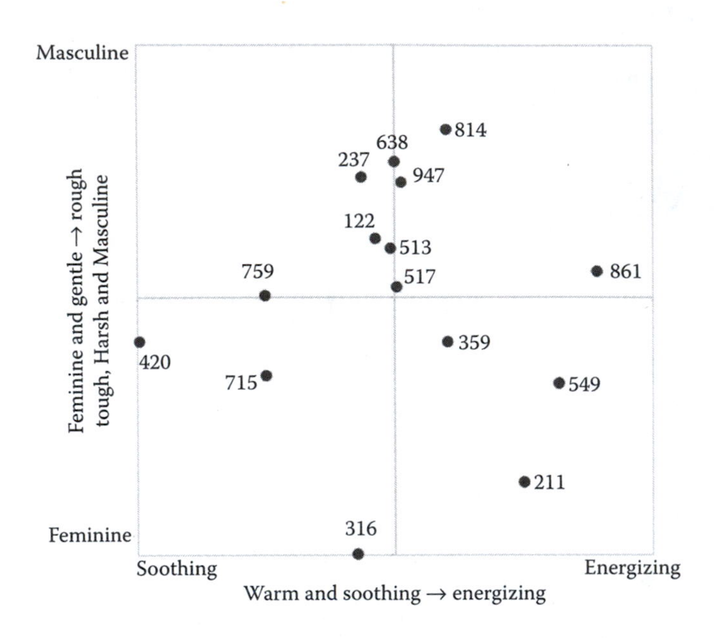

Figure 20.3 Example of function and emotional benefit map. Note: In an actual report, three-digit codes would be replaced by actual fragrance accord to provide directions for product development.

panel, to provide further direction for product development or guidance when a customer asks for a fragrance that delivers on specific benefits. Additionally, the information can also be leveraged in many other ways as an educational tool internally or as a marketing/ sales tool during customers' visits.

How might this scenario be addressed differently?

In a scenario like this one, many factors can affect the design of the study. It is very important to have an initial project meeting with key players to identify needs and anticipated learnings and outcomes from the study. Depending on internal knowledge, the methodology development phase may or may not be necessary or may be approached differently (e.g., mining Internet reviews or social media data for language rather than conducting focus groups, or leveraging published literature for optimal testing methodology). Also, in this case, it was decided that the fragrances be evaluated in a sniff test condition at a central location facility following a complete block design, but a monadic test in the home, with fragrances incorporated in a bodywash base, may have provided additional insights. Budget and time constraints certainly need to be weighed up as well.

What could future testing involve?

The study provides fundamental information about fragrances and their associated benefits. It is unknown whether the findings would be translatable to categories other than bodywash, so similar initiatives can be conducted with other personal or home care applications. Furthermore, it may be interesting to see if and how benefit maps may shift as different fragrance trends emerge in the market over time.

SCENARIO 4

Product developers at a natural foods company have been given the task of developing prototypes for a new product category: a shelf-stable grated Italian cheese. As the product category is not new to the consumer, just new to the company, product developers need to understand the sensory attributes that consumers expect to be in a high quality, all natural grated Italian cheese. In addition, "authenticity" in flavor is a value for the company, so ultimately the product launched needs to be perceived by the consumers as having authentic flavor. R&D team have developed three prototypes representing different flavor types. They are looking for further sensory design guidance from consumers for the flavor direction and have asked the sensory scientists on their team to help them obtain that guidance.

In what ways can the sensory department work with R&D and marketing to achieve success for the new brands to be developed?

Before designing the research, the sensory scientist facilitates an information gathering (data dump) session. Within this session, information is shared on the scope of the category space, market leaders, past consumer research learning, and sensory panel data, especially for suspected key attributes.

Which sensory method or approach would be appropriate? State the project objective and the test objective.

The project objective is to develop the launch product for the shelf-stable grated Italian cheese category that is regarded by consumers to be high quality and natural and to have authentic "Italian cheese" flavor.

The sensory scientist recommends sensory based qualitative research with frequent users of the category. The test objectives are twofold: (1) to understand how consumers perceive and describe the sensory experiences associated with high quality, authentic, Italian cheese flavor; and (2) to determine the flavor direction to proceed and choose one to two prototypes for further development for the eventual marketing based tests.

Outline the test design. Where will the test be conducted? Who are the assessors, and how many will there be? How will the samples be handled and presented to the assessors?

Test Design:

Number of qualitative sessions: 6
Location of qualitative sessions: 3 sessions in region 1; 3 sessions in region 2
Length of each session: 2.5 hours
Number of consumers per session: 8–10
Consumer demographics:

- Women, 25–54
- HH Income $50K+ ($30K + for single income)
- Households of 2 or more
- Mix of women with no children, with a child 6–12, and with a child 13+

- Primary grocery shopper, 50% or more purchases at traditional grocery
- Heavy shelf-stable grated cheese users (purchase 8oz canisters at least 4× per year or 16 oz canisters 2× per year)
- Use grated cheese more than 1× per week
- Creative cooks—at least 2 per group
- Articulate; pass security screen; some post-high-school education

Samples:

Sample ID	Code
Prototype 1: 100% Source 1	536
Prototype 2: 100 % Source 2	829
Prototype 3: 50/50 Blend Source 1 and 2	428
Market Leader 1	315
Market Leader 2	741
Private Label (negative control)	253

Sample preparation and presentation:

- Serve 2 tbsp. of cheese in coded 3 oz. soufflé cup with lid. Prepare pasta per package directions and add olive oil. Keep pasta warm. For each sample, provide
    - 2 tbsp. grated cheese in coded cup
    - ½ cup Pasta in Styrofoam bowl with lid
    - Fork, sample spoon, and napkin for each sample

Moderator Guide

Guide for Grated Italian Cheese Groups
Arrive 15 minutes prior–Re-screen to start on time

I. Introduction **(15 minutes)**

Hello. Thank you for agreeing to participate in this study. For the next 2.5 h we will be tasting and discussing grated Italian cheese. We want to learn as much as we can about what you expect from Italian cheese and how you talk about what it looks like and tastes like, and what the texture is like. We want to know what are the must haves, what is nice to have, and what you are willing to give up/compromise on. You will have the opportunity to taste and look at a number of different Italian cheeses as the session progresses.

Some brief ground rules—Please turn off your cell phones so that we do not get interrupted. Restrooms are located in the hallway; feel free if you need to get up and stretch your legs. We have individuals behind this mirror that are very interested in what you have to say. For research purposes, we are recording this information using the camera and microphones you see above you. [Confirm permission.] In order for all of your opinions to be heard, it is important that we speak one at a time and be respectful of others. We are not affiliated with any of the products you will be looking at and trying as we are hired as a third party. You can feel free to speak honestly; in fact, that's what I really need for today to be successful.

As we go forward, others in our group may have opinions that differ from your own. That's OK—remember each of you represents thousands of people who couldn't be here. It's OK to disagree—it's not OK to fight (insert laughter). Many of the products seen here may not be on the market. It is very important that you remember the confidential nature of this research and the form you signed.

- Let us begin by learning more about you. Please tell us your name, who does the majority of cooking in your household, if you have children living at home (how many and what their ages are), and the last time you had Italian cheese and how was it used/served.
- Possible probe questions: What are the other ways you use Italian cheese? (usage occasions including topper, ingredient in recipes [type], olive oil dip.) Do you use different Italian cheeses for different occasions [grated, shredded, refrigerated, shelf stable, block]?

I. Introduction of D.R.I.V.E. (20 Minutes)

In order for us to understand what you want in grated Italian cheese, we'd like to take you through a brief exercise.

Explain use of D.R.I.V.E. by Thinkx—use photos with DRIVE process to elicit creative responses

During this exercise, one moderator is probing and one is capturing feedback on large post-its on the wall.

What does Italian cheese do?

What does it have to look like, smell like, taste like, feel like in mouth and on food, look like in food?
What should it dispense like?

What are the restrictions on Italian cheese (what shouldn't it do)?

What should it not look like, smell like, taste like, feel like in mouth and on food, look like in food?
Are there flavors that it should definitely not have?
Are there texture/feels that it should definitely not have?
Are there visual looks that it should definitely not have?
What should it not dispense like?
What does it have to avoid?

What are consumers willing to invest?

What are you prepared to invest in time, effort to find, determining product quality, etc.?
If you can't find your Italian cheese in the store you are in, what are you prepared to invest to get it?
What are some brands that you are willing to invest in when it comes to pasta and Italian cheese? What other brands do you invest in?

What are the consumers' values regarding Italian cheese?

Let's talk about values in terms of quality. In making food choices, where is quality important to you? Where is it not or less important?
What defines a high quality Italian cheese?
What is your current brand of Italian cheese and how would you describe its quality?
How important is it for you to have Italian cheese in your house?

What are the essentials of a Italian cheese?

What are the nonnegotiable/must haves for your Italian cheese?
How will you know you want to buy this?
What would make you want to keep this in the house?
What would make you recommend this to friends?
What would make you buy this if it was not on sale or at a higher price point?
Look over the list from the other points and see what should be included in essentials.

II. MOMENT TO MOMENT—Warm Up (5 minutes)

Demonstration—explain moment to moment exercise by giving them 2–3 tablespoons of chocolate sprinkles in a soufflé cup. Review appearance, flavor and texture. Explain that we would like to get their feedback by looking at and tasting a variety of Italian cheeses that have different sensory properties. Discuss how we will be evaluating the cheeses: cold and hot for appearance, and hot for flavor and texture on pasta.

III. DEEP DIVE PRODUCT TASTING—FLAVOR and TEXTURE Focus with Appearance (Hot)—focus on look on pasta/melt, brief mention of AROMA (65 minutes)

For the approximately the next hour, you will have the opportunity to taste a number of different grated Italian cheeses. We will see a mix of currently marketed Italian cheese as well as some new products.

Each respondent is served a coded sample of Italian cheese (serve Market Leader 1 first).

Sample ID	Code	Order
Prototype 1: 100% Source 1	536	2
Prototype 2: 100% Source 2	829	4
Prototype 3: 50/50 Blend Source 1 and 2	428	5
Market Leader 1	315	1
Market Leader 2	741	3
Private Label (negative control)	253	6

Each respondent will be provided with:

2 tbsp of Italian cheese in a lidded, coded cup
A small bowl of cooked pasta (oil on pasta) (1/2 cup pasta, lidded)
Plastic teaspoon and fork (containers placed on table—self serve)

Respondents are instructed to try the cheese using the following procedures, writing notes on the worksheets provided as they go along:

- Smell what's in the cup and describe what you smell

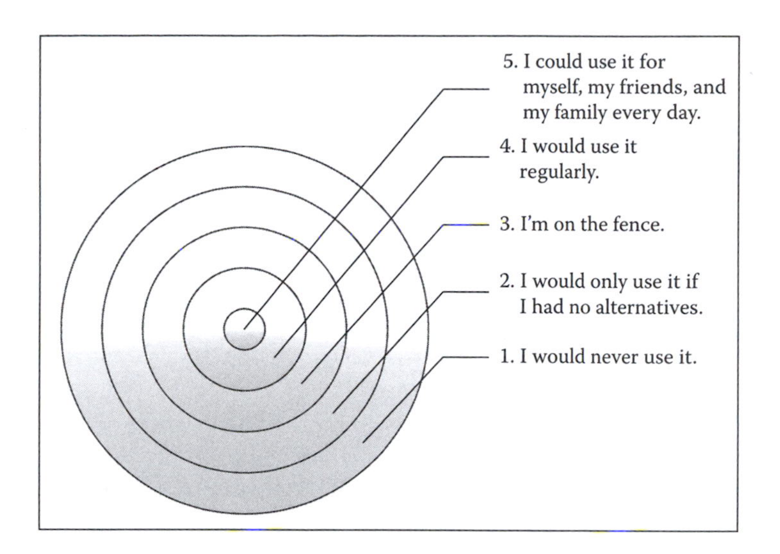

Figure 20.4 Panelist target worksheet.

- Put your desired amount of cheese on the pasta, up to 2 spoonfuls
- Smell the bowl of cheese and pasta and describe what you smell
- Taste the cheese on the pasta and describe what you taste and feel

[Explain for first sample only, then continue to us throughout.] Please look at the worksheet we are handing out and follow along as we explain the process. As you taste the samples, make notes on the bottom of your worksheet. After you have tasted the sample and made your notes, go to the target on the easel, and using the colored dot stickers we provide, **please indicate or mark how this Italian cheese compares to your "best grated Italian cheese".** Remember that we need your honest assessment. Don't worry if your opinion is different than someone else's— it's your opinion so you can't be wrong.

[Respondent will receive a worksheet based on the information below in Figure 20.4—look will be different.]

Rating Scale	Reaction after Tasting
	(After eating on pasta)
1	I would never use it
2	I would only use it only if I had no alternatives
3	I'm on the fence
4	I would buy and use it regularly
5	I could use it for myself, my friends, and my family every day

[Respondents will mark target after writing notes but before discussion. WRITE—VOTE—DISCUSS]

[Key probes are in bold.]

- After respondents finish writing notes, probe on responses and record additional sensory attributes on large wall post-it notes. Ask 2–3 to provide responses and then ask for anything else to add, missing attributes. Probe for depth of sensory properties.

- Probe on aroma attributes for cheese and for cheese on pasta—see flavor attributes below for examples.
- Probe on flavor attributes such as cheesy, buttery, sweet, bitter, salt, sour, strong, mild, nutty, fresh, greasy/oily flavor, toasted cheese, milky, savory, sharpness, bite, fruity, off flavors (consumers drive language used).
- **Probe on "what does bite/sharp mean"**
- Probe on overall perception of cheese character.
- Probe on perception of quality of the cheese.
- Probe on texture attributes such as softness/hardness, chewiness, graininess, rate of melt, dry versus oily feel, powdery, rubbery (consumers drive language used).
- **If it comes up, probe on fluffy texture versus opposite—what does it indicate about flavor and texture and quality?**
- Briefly probe on appearance on pasta—include color, particle size, look on pasta, melt.
- Capture language on wall or easel. Revisit—make sure we have captured everything.
- Probes may include: Would you use this cheese instead of the one you usually use? How would you use this cheese?
- Probe on scores. What are your reasons for your score? What would make this a 5?
- **Would you replace the grated cheese you are currently using (i.e., Kraft) with this one?**
- **Who in your family would enjoy this cheese? Does anything about it change who would eat it?**

For respondents: Review your notes and at the bottom of your worksheet rate this sample on how it compares in quality to your regular grated Italian cheese (Use a 5 point anchored JAR scale with center point of About the Same Quality.

- Discuss and probe on why.
- As appropriate, probe on how a higher quality Italian cheese would fit in their life.

Tell respondents that the first sample they tasted is going to serve as a comparison point or reference or benchmark for the additional samples. For all additional samples, we will be following the same process of smelling, looking, tasting, and writing, and as appropriate also thinking about how the samples compare to this first one that you tasted.

Repeat process with samples 2–6.

[If time is available—after completion of this section and before appearance—allow brief stretch/bio break. Otherwise, let people leave briefly as needed.]

What sort of data will be collected, and how will it be analyzed? What are the results of the test and what can you conclude from the analysis? Interpret the results in the context of the test objective and the project objective. What are your recommendations for the company?

Results Summary

The summary of results is generated through qualitative data (see Tables 20.2 through 20.5) and careful interpretation of the combined responses from all focus groups.

Overall

Prototype 3 is the prototype to pursue for next level of development and testing.

Texture in the mouth was the surprise, critical element during the sessions. While consumers had relatively little unaided description of desired texture other than "good" and "fresh", the diversity of the sample set provided the needed contrast for consumers to expand their responses to more clearly define desirable characteristics. *They want a product that is creamy, melts in food, and says cheesy, with no residual particles.* (Market Leader 2 delivers the creamy, melty texture)

Seen blind, many consumers recognize Market Leader 1 outright or as a typical grated Italian cheese. The target exercise reveals that consumers do recognize its deficiencies and that better products can exist.

Texture in Mouth

- Texture is *critical* to product liking, although most consumers did not have specific texture attributes top of mind in the D.R.I.V.E. exercise.
- Consumers are interested in a grated Italian cheese that melts into warm food, providing a signal of creamy. The market leaders best deliver on this signal.
- Prototypes 1, 2, and 3 were described as grainy, gritty, or sandy with low melt on pasta and leaving particles in the mouth.

Appearance

- Consumers were more likely to reject samples that were very yellow or seemed darkened. They saw such products as old, dry, or stale.
- Clumps are an accepted part of grated Italian cheese, but too many clumps or harder clumps indicate products that are dry or old.
- Powdery is an accepted signal for some consumers (like market leaders); some visual texture is OK but product should not look grainy. Powdery signals "canned" cheese, whereas "fluffy" signals freshly ground and fresh.

Flavor

- Consumers are unclear what real Italian cheese tastes like and vary greatly in their perception of the "sharpness" of each sample.
- Consumers want an identifiable Italian cheese flavor that should include saltiness, some sharpness, and cheese flavor.
- Most consumers find Market Leader 1 (blinded) to be similar in quality to what they usually use, and most are familiar with the flavor and recognize it as the type of grated Parmesan cheese they usually use.
- Market Leader 2 was almost universally liked by the consumers. This product delivers high saltiness with no objectionable flavors; liking may have been driven by product texture in mouth and saltiness in the simple pasta delivery system.
- Prototype 1 had a flavor that was universally disliked.
- Prototype 2 had a low intensity flavor seen as bland by the consumers.

Table 20.2 From DRIVE: Consumer Essentials for Grated Italian Cheese

• Made of Parmesan	• Creamy texture on hot food	• Versatile for all uses
• Mild Parmesan taste	• Creamy, smooth texture	• Affordable
• Fresh look	• Good texture	• Easy to use shaker
• Fresh taste	• Fluffiness	• Heritage
• Freshness	• Moist look and feel	• Less preservatives
• Smells good	• Not too many clumps	• Shelf life
• Tastes good	• Texture to match food	
• Right amount of salt	• Softer texture	

Table 20.3 The Sensory Consumer Attributes—What Grated Parmesan Must Deliver

Appearance	Flavor	Texture in Mouth
Light, creamy color	Mild Parmesan flavor	Creamy texture
Few clumps	Slight tang or sharpness	Melt in hot food
Clumps that break apart easily	Clear cheese flavor	Most texture disappears into food—not distinct
Slightly moist look	Moderate flavor strength	Retains some texture in food—creaminess or feel of soft chewy pieces
Fluffy appearance signaling freshly grated	Aroma and flavor similar in characteristics and strength	Some larger lumps that yield to pressure is ok
Lack of powdery, sawdust look	Salty flavor	Slightly coarse grind that is fluffy, not powdery
Slightly coarse grind	Salt-cheese flavor balance	*Not grainy, sandy, or gritty*
No dry pieces	Slight sweetness	*No rubbery or chewy particles*
No dark edges or dried look	Some rich flavor	*No particle residue in mouth after swallowing*
No grainy or pebbly look	*No strong acid*	
Some visibility when mixed in hot food	*No strong buttery note*	
	No chemical note	

- Prototype 3 offered a flavor profile many consumers liked.

Tables 20.2 through 20.5 summarize the results.

How might this scenario be addressed differently?
Social media data mining could be coupled with descriptive analysis so patterns of likes and dislikes can be identified.

What could future testing involve?
Use of the very simple delivery system of oiled pasta was positive overall, providing for a good understanding of flavor. However, the use of oiled pasta was atypical for most consumers and left them wondering how the cheeses would perform in a more typical food like a red sauce or alfredo/white sauce. Set 2 cheeses should be tested in a more advanced model system, either internally or as part of additional qualitative or quantitative consumer research.

Table 20.4 How Do Samples Hit the Target? (N~47)

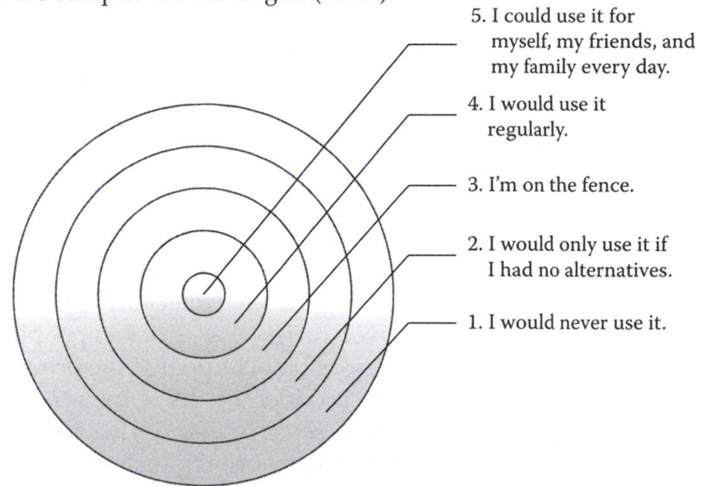

		Rating									
Code	**Samples**	**1**	**1.5**	**2**	**2.5**	**3**	**3.5**	**4**	**4.5**	**5**	**Top 2 Ratings**
315	Market leader 1	1	2	7	4	11	6	15	1	0	16
253	Private label	12	3	14	1	12	0	5	0	0	5
536	Prototype 1	8	0	13	0	9	3	13	0	1	14
829	Prototype 2	15	0	11	7	7	1	6	0	0	6
428	Prototype 3	2	0	3	0	12	0	15	0	7	22
434	Market leader 2	1	0	3	0	4	5	21	1	12	34

Table 20.5 How Do Samples Compare in Quality to Consumers' Regular Grated Italian Cheese? (N~47)

Code	**Samples**	**Much Lower**	**Somewhat Lower**	**About the Same**	**Somewhat Higher**	**Much Higher**	**Top 2 Ratings**
315	Market leader 1	5	15	22	5	0	5
253	Private label	19	16	9	2	0	2
536	Prototype 1	10	13	12	11	0	11
829	Prototype 2	18	16	6	6	0	6
428	Prototype 3	2	8	10	14	3	17
434	Market leader 2	1	6	4	23	12	39

SCENARIO 5

A manufacturer of frozen pot pies wants to determine if their current product design is "on target" in terms of maximizing consumer acceptance. Further, if it is determined that the current product needs to be modified, the manufacturer wants to understand the changes that need to be made and the size of the improvement in consumer acceptance that can be expected.

In what ways can the sensory department contribute to gaining this critical product information? State the project objective and the test objective.

To understand the pot pie market it is necessary to collect all or most of the products in the category. Select the largest selling flavor variety (e.g., chicken pot pies) and collect as diverse an array of commercial products as possible. The goal of the product selection step is to maximize the sensory variability in the product set, not to focus on category leaders. Submit all of the products for evaluation by a sensory descriptive panel, either the company's internal panel, if one exists, or a qualified, external panel. Obtain a comprehensive descriptive profile of each product in the study.

In parallel with the sensory descriptive evaluations, the products also should be submitted for consumer testing. Overall liking is the primary response of interest, but a small number of attribute hedonics and attribute intensity questions also can be collected. The attribute intensities provide a sanity check to confirm that the difference reported by consumers on familiar and widely understood attributes (e.g., sweetness, saltiness, sauce clarity) matches the differences among the products reported by the sensory descriptive panel.

The sensory profiles are then merged with the consumer information using preference mapping. (See Section 15.3.2). The preference map will illustrate the differences in the sensory properties of the products, will locate the point on the map that is predicted to be most well-liked by consumers, and will produce a predicted sensory profile for this target product.

Project Objective: To determine if the company's current chicken pot pie product needs to be modified to improve its consumer acceptance and, if so, what changes need to be made.

Test Objectives:

1. Map the sensory space of the chicken pot pie market with descriptive analysis.
2. Measure the acceptability of all of the products in the study.
3. Combine the sensory and consumer information to understand the sensory attributes that drive liking, develop the sensory profile of the target product and its predicted liking rating, and compare the sensory profile of the current product to that of the target product to define the opportunities that exist to improve the current product. The liking rating of the current product can be compared to the predicted liking rating of the target product to determine if the potential increase in liking is sufficient to warrant the redesigning of the current product.

Describe the testing methodology, including the test controls and product controls. Address any special considerations based on the details of the scenario.

Choose the category to study (chicken pot pies) and collect the products. The controls involve choosing brands that represent as diverse a set of sensory differences as exists in the marketplace. For both the sensory evaluations and the consumer test, prepare the products by strictly adhering to the manufacturers' preparation instructions. Using the same preparation methods in both tests is critical to ensure that the products have the same sensory properties in the tests. Because of the large number of products involved (typically 10–15), controls need to be in place to ensure that sensory fatigue is minimized.

If possible, each respondent in the consumer test should evaluate all of the test products. This may require a multiday consumer test. Doing so provides the opportunity to identify preference segments among the consumers, which may reveal the need to produce multiple products to satisfy all consumers (or to make an informed business decision about which segment to pursue and which to give up).

Outline the test design. Where will the test be conducted? Who are the assessors, and how many will there be? How will the samples be handled and presented to the assessors?

The sensory descriptive tests are conducted in a facility that has adequate kitchen equipment to bake all of the pot pies in a reliable and consistent manner. The facility also must have a separate panel room with controls such as clean air. Product controls include coded plates or bowls for serving the pot pies and a fixed regimen for timing how long the pies have been out of the oven before they must be evaluated.

Similar controls must be in place for the consumer test. Respondents should be chosen to participate in the consumer test based on the company's marketing strategy. Because of the large number of respondents and thus the large number of pies that need to be prepared, the kitchen facility at the consumer testing site must have an adequate number of similar ovens to ensure that all of the products are prepared and served according to instructions.

What sort of data will be collected, and how will it be analyzed? What are the results of the test and what can you conclude from the statistical analysis? Interpret the results of the statistical analysis in the context of the test objective and the project objective. What are your recommendations for the company?

The descriptive data are analyzed with ANOVA. The average or consensus profiles of the products are used to generate a perceptual map of the product space using principal component analysis (PCA; see Sections 15.2.1.2 and 15.3.1). All sensory modalities can be mapped together. The maps are used to understand the sensory space of chicken pot pies: what attributes account for the major sources of sensory variability in the chicken pot pie category, where the current product is on the map, and where the competitors are.

The consumer data also are analyzed with ANOVA to determine which products differ significantly in acceptability. If each consumer evaluates all of the test products, the consumer liking ratings can be submitted to a cluster analysis to determine if preference segments exist (See Section 15.3.2.2). If so, average liking ratings for all of the products should be computed for each segment.

The factor scores of the products from the perceptual map are used as the independent variables in a regression analysis. The dependent variables are the average liking ratings of the products for the total respondent base and all subgroups of respondents the company is interested in studying. Typically the analysis begins with a second-order polynomial regression model, and a variable-selection procedure (e.g., backward elimination) is used to ensure that only significant predictors remain in the final model (See Section 15.2.2.1.2). The resulting models are used to locate the point on the perceptual map that is predicted to maximize consumer liking. This point is labeled the target product. The predicted liking score of the target product is noted and its sensory profile is estimated using reverse engineering (Section 15.3.2.4).

Recommendations are formulated by first determining if the difference between the liking rating of the current product and that of the target product is big enough to warrant reformulating the current product. If so, the recommended changes should be prioritized to focus on the sensory attributes of the current product that (1) matter to consumers and (2) are currently "off target".

How might this scenario be addressed differently?

The approach just described is a form of external preference mapping. The company may choose to perform internal preference mapping or PLS regression (Section 15.3.1) to accomplish its goal.

What could future testing involve?

Descriptive evaluations of prototypes should be conducted to determine if the reformulated prototypes have moved closer to the target profile. When a promising prototype is developed, it should be submitted to a confirmatory consumer test to verify that a higher level of acceptability has been achieved.

REFERENCES

ASTM Stock #DS72 (2011). Lexicon for sensory evaluation: Aroma, flavor, texture, and appearance. ASTM International, West Conshohocken, PA, 2011, www.astm.org

Lawless, L. J. R. and G. V. Civille (2013). Developing lexicons: A review. *J Sen Stud* 28: 270–81.

ADDITIONAL QUALITATIVE REFERENCES

Bystedt, J., S. Lynn, and D. Potts (2003). *Moderating to the Max.* Ithaca: Paramount Market.

Hurson, T. (2007). *Think Better: An Innovator's Guide to Productive Thinking.* New York: McGraw-Hill.

INDEX

B

C

D

P

S

T

U

V

W